Portland Public Library

A History of Lumbering in Maine 1861-1960

David C. Smith

University of Maine Studies No. 93
University of Maine Press, Orono, Maine, 1972

Copyright by Maine Studies, University of
Maine at Orono in 1972

Printed by the University of Maine at Orono
Printing Office on University Text, manufactured by
S. D. Warren Paper Co., Cumberland Mills, Maine

THIS BOOK IS FOR SYLVIA

"E'en cold Caucasean rocks with trees are spread,
And wear green forests on their hilly head.
Though bending from the blast of eastern storms,
Though shent their leaves, and scattered are their arms,
Yet heaven their various plants for use designs—
For houses, cedars—and, for shipping, pines."

Virgil, GEORGICS, II
(Dryden's translation)

FOREWORD

Nearly every person would confess to a desire to write a book, but it is a very few who have had the opportunity to present that book as their doctoral dissertation as this work originally was. The author was fortunate enough to have spent his boyhood in the Maine woods, time which was full of enjoyment and little worry about the woods, how they came to be, or about their meaning. Later, when I came to academic life, I wanted to study those woods, learn about their relationship to the Maine economy, and to the social structure of the area where life had been so pleasant in earlier times.

Historians of Maine have not written much of the Maine woods. The story of conservation movements, the practice of forestry, lumbering in general, the passing of the old sawmills, and overall the utilization of the greatest resource of the state was nowhere presented except in the vaguest of generalities. More important for some was the fact that while lumbering in all of its forms dominated the Maine economic and social structure, great lumber firms and land holders seemed to dominate the industry. A few said they also dominated the legislature, the government, and, in fact, all forms of human behavior. The cynical among these critics felt no other way was possible. That story was not told either, and it certainly was part of the history of the Maine woods.

A work which would attempt to answer some of these questions, and which would indicate some of the complex relationships involved was needed. This book then is an attempt to make real boyhood dreams, and to bring alive the grunt of the axeman, the thrust of the cant dog, the scream of the sawmill, and the smell of the pulp and paper mills.

Maine people are noticeably reticent about their private lives, and for that reason, the documentary record is more difficult to locate than in some other places. Even today the Coburn papers are not open, and the Pingree papers are still unavailable. Both collections of papers may modify this story somewhat, but only in the sense that they span both the pine and spruce eras, and in the latter case come into the pulp wood period as well. It appears that the Pingree papers may merit book-length study, and such a book would be widely useful. It is my feeling though that study of these, or any other papers which may be freed for scholars through the agency of this work, will not change the story as presented here. More work needs to be done on land ownership, and especially on the transfer of lands after the heyday of the pine period in the late 1840's. Work also is needed on industries more peripheral to the business, such as the manufacture of sawmills and other machinery. The Hinckley-Egery papers may be useful in this area, and it is my intention to investigate this possibility presently. Finally, it is to be hoped that the papers of the Great Northern Paper Company and other large firms may be more widely opened for historical investigation, and equally that the papers of important Maine individuals such as Percival Baxter be preserved. The tragedies which happened to the early papers of Llewellyn Powers, Austin Cary, and others should not be allowed to reoccur. Such establishments as the Lumberman's Museum in Patten, Maine, or the projected Lumberman's Museum at the University of Maine, could form a nucleus for archival work. Such work might be undertaken in conjunction with the universities of the Maritime Provinces, or of New Hampshire. Scholarly work is needed in Maine history as much as anywhere. It is hoped that this work will aid in stimulating such efforts, and it is with that thought in mind that it is presented.

ACKNOWLEDGMENTS

As with all writers my debt to others is unrepayable. Such listing as follows is only to acknowledge the existence of the debt, not to meet its obligations. To the members of my graduate committee at Cornell University, Professors Paul Gates, Edward Fox, Walter LeFeber, and the late Clinton Rossiter goes the intellectual thanks of the student, and in addition, to Professor Gates, the thanks which are due for the hours of drudgery spent in dragging this work into some semblance of the English language. Professors LaFeber, David B. Davis, and John G. B. Hutchins have read this work and I have profited immensely from their comments. Others who have read parts or all of this manuscript, and who have offered advice on its contents, include Professors Richard Schadt, Oneonta State Teachers College; James Crouthamel, Hobart and William Smith Colleges; Robert Thomson, Edward Schriver, and Edward D. Ives, University of Maine; and Gwilym R. Roberts, Farmington State Teachers College. Leslie Decker of the University of Oregon and Joseph Miller of the Yale Forestry School are others who have given me much time and knowledge. My intellectual debt to all these is very great indeed.

The historian leans always on those who are in charge of, or who otherwise control manuscript and archival materials. My dependence is very great. The Great Northern Paper Company, especially John T. Maines, Vice President—Woodlands, and the late William Hilton, gave me freely of their time and facilitated use of some of the firm's holdings in answer to my specific questions. The S. D. Warren Company, Westbrook, Maine, opened their archives completely to me, and Alfred Wilson, then editor of their journal, *Warren's Standard*, gave me time, knowledge, and entertainment in my pursuit of the role his company played in the early days of the pulp and paper industry.

Mona Wakefield White, Veazie, Maine, insured the salvaging of the Wakefield papers, and made possible their retention in my hands. Her niece, Georgia Stacy, of Burlington, Maine, was instrumental in the saving of these papers. The late Mrs. Bradley Sawyer, Bangor Historical Society, provided me access to relevant papers, as did the City Clerk in Bangor, and County Clerk of Penobscot County. Although perplexed as to the purpose of search they cheerfully obtained the requisite papers for me and provided me with space to work. Olive Smythe guided my work in the papers of the Bangor Public Library, as did Lilian Tschamler in the Forest Commissioner's Office.

Answers to questions were always forthcoming on my frequent trips to Augusta. The Forest Commissioner, Austin Wilkins, turned his office space over to me for work on more than one occasion.

Elizabeth Ring of the Maine Historical Society was also helpful. Edith L. Hary, Law Librarian of the Maine State Library, unthreaded some knotty legal problems at a late stage of this work. The late Earle Clifford of the South Paris Public Library Board of Trustees provided me with work space when that library was closed, and later the staff of that library insured that my work proceeded with dispatch. The staffs of the Maine Historical Society were always helpful, and cheerfully answered my somewhat esoteric questions. Muriel Hodge, research librarian at Hobart and William Smith Colleges, is one of the best in the country at her business, and my obligations to her are so numerous as to defy description. The Records and Manuscripts Division of the National Archives was, as always, more than helpful. The staff at the Fogler Library, University of Maine, simply make research a pleasure. David Tolman, Arline Thomson, and the Staff of the University of Maine Press are real professionals.

Mr. K. W. Matheson, Vice-President Fraser Companies, Ltd., graciously provided me with office space, and allowed me to read and use the unpublished history of his company. The staff of *Paper Trade Journal* and especially my friend, John C. W. Evans, turned their library over to my use for a period of intensive study. The late Ballard F. Keith of Bangor also gave me library space, and further brought to his office the relevant documents in his custody. There are many others besides this partial list, all of whom contributed more than routinely to his work.

Discussion and conversation about the work was important throughout its progress. Caleb Scribner, Patten, Maine; I. H. Bragg, Patten, Maine; Harold Noble, Topsfield, Maine; William Hilton, Bangor, Maine; and Louis Freedman, Old Town, Maine, all spent time with me yarning about their days in the woods, and mine, cheerfully submitting to detailed questioning, and offering corrections both orally and by letter as to my interpretation of men or events. It is an indication of the feebleness of time that only "Hallie" will see this book. I owe much to a discussion in a Bangor park with an anonymous passer-by of the artistic and physical merits of Tefft's statue, "The River-drivers". My typists, Rosemary Currie, and Susan Shaw Rocha helped make life bearable. Parts of chapters 2 and 3 were given as the Annual Address to the Maine Historical Society in June, 1965. Parts of chapters 9 and 10 have appeared, in different form, in *Forest History* and in my *History of Paper Making in the United States*.

To the people who taught me at one time or another the use of the axe, cant dog, spud, saw, and pulp-hook also are given thanks, for in the real sense of the words, without them this work would not have been possible. To the John J. Nissen Baking Company, and the men of Local 334, Bakers and Confectionery Workers of America, Brewer, Maine, go heartfelt gratitude and friendship. The long days and nights in the bakery come to fruition with this work.

Of course, to my parents and to my wife's parents, my debt is, as always, both economic and personal, as well as social. For my wife, Sylvia, and our children, Clayton and Katherine, the book was written. All contributed to this work, but only I am at fault. The building of a debt does not transfer to the shoulders of the lender any responsibility. For failure of fact, for inability to interpret, for lack of imagination—to all these I plead guilty, and accept all responsibility.

Geneva, New York
Bangor, Orono,
Topsfield, Maine

CONTENTS

Foreword	VII
Acknowledgments	IX

Chapter		PAGE
I.	The Maine Woods	1
II.	Life in the Woods	11
III.	Logging the Maine Forests, 1860-1890	37
IV.	Driving the Maine Rivers, 1860-1890	63
V.	The Manufacture of Lumber, 1860-1890	107
VI.	Shipping the Product	137
VII.	Lumbering and Land Sales, 1869-1900	173
VIII.	The Lumbering Economy, 1860-1890	211
IX.	Pulp and Paper Come to The Northeast, 1865-1890	233
X.	The Big Companies Come, 1890-1905	247
XI.	The Companies Gain Control of the Rivers, Part I	263
XII.	The Companies Gain Control of the Rivers, Part II	301
XIII.	Conservation and Forestry Come to the Maine Woods, 1865-1930, Part One	333
XIV.	Conservation and Forestry Come to the Maine Woods, 1865-1930, Part Two	357
XV.	The Maine Woods in the Twentieth Century	383
XVI.	Factors in the Modern Maine Woods Industry, 1930-1960	419

Appendix I.	The Log Drive on the Kennebec, 1866-1912	431
Appendix II.	Penobscot River Navigation Dates, 1818-1904	432
Appendix III.	Ice-Free Dates—St. John River at Fredericton Boom, 1875-1915	436
Appendix IV.	Ice-Free Dates—Moosehead Lake, 1860-1900	438
Bibliography		441
Index		463

LIST OF ILLUSTRATIONS

Bangor Lumber Mill—Possibly Hathorn Mill Near High Head	Bangor Public Library	Frontispiece
Logging Camp in Autumn	Larson Collection	following page 144
A Horse Hovel	Larson Collection	
Woodscook at Work	NAFOH*	
Lunching Out—Mid-morning	NAFOH	
Interior of Camp	NAFOH	
Necessary Camp Furniture	NAFOH	
Loggers Scrimshaw Spruce Gum Books	NAFOH	
Blacksmith at Work Author's Grandfather —	David Smith	
Hauling With Ox-Team	NAFOH	following page 160
Handloading Birch Bolts on Logging Sleds	Larson Collection	
Lombard and Crew, Grindstone, mid 1920's	NAFOH	
Tractor Accident	NAFOH	
Tractor With 25 Sleds Little St. John Pond	Great Northern Paper Company	
Brook Drive, Carding the Ledges	Larson Collection	
High Water Slewgundy Heater Mattawamkeag River	NAFOH	
Gulf Hagas	Great Northern Paper Company	

*Northeast Archives of Folklore and Oral History

Headworks in Operation	NAFOH	following page 288
Grand Falls, N. B., Log Jam Early Print, 1875	Webster Collection	
Ripogenus Flume	Great Northern Paper Company	
Towboat A. B. Smith Chesuncook Lake	Great Northern Paper Company	
Wing Jams on the St. Croix River	NAFOH	
Rafting on the River St. Johns River, 1914	NAFOH	
The Booms at Fredericton Early Print, around 1880	Webster Collection	
The First Leary Raft An Artist's Conception	Webster Collection	
Argyle Boom	NAFOH	following page 304
Schooner Loading Lumber Ritchie's Wharf	Larson Collection	
Logs in Mill Pond and Shingled Up for Winter, St. John's River, 1914	NAFOH	
Hell's Half Acre Bangor, 1884	James Mundy	
Suppliers to the Woodsmen	James Vickery	
Opening of the European and North American Railway	*Canadian Illustrated News*	
New Uses for Wood Pulp Paper December 1885	*Century Magazine*	
The Houses Lumber Built Broadway in Bangor	James Vickery	

XVI

1

THE MAINE WOODS

The "northeast corner" of the United States is concerned primarily with one raw material and its manufacture. For that reason the central fact about Maine[1] throughout its history is that Maine is lumber and lumber is Maine. The life of every person who lives in this area is intimately tied to the forest and the forest frequently supplies him with his livelihood. In good times the society is fat, in poor times lean. When one travels through the countryside the forest is his constant companion. It has always been so and bids fair to remain that way. Many have been the efforts to change this and none have had any real effect. Therefore the human geography of the northeast corner is understood only by the impact of the forest, and the story of Maine is really told primarily in these terms. The burden of the present work, which deals with the lumbering industry of the region, will be to document this idea, among others, as it discusses first the lumbering industry, then the shift to pulpwood operations and the manufacture of paper, and finally the efforts of the inhabitants of the area to preserve and maintain their central way of life, the forest.

Every early visitor to the area was cognizant of the woods. Captain John Smith,[2] Captain George Weymouth,[3] Timothy Dwight,[4] Alexander Baring,[5] Talleyrand,[6]—all were in Maine and all paid their respects to the pervading influence of the forest. It was the central factor for these visitors, all of whom were interested in the possibility of settlement, and the potential of agricultural life. Early attempts at settlement failed or succeeded only as the pioneers were willing to deal with this elemental fact, the forest. Thoreau perhaps caught the feeling best, when, on the occasion of his first visit to the Maine woods, he noted in his journal,[7] "Each one's world is but a clearing in the forest, so much open and inclosed ground."

Political boundaries are not often drawn to fit either economic or geographic units. Thus it is with Maine. The story of Maine lumbering does not just include the political area of Maine, for the lumbermen did not, and could not recognize these boundaries as worthwhile. On the west the area bounded in this study must include the Androscoggin river, which drains not only a part of Maine but also the land to the east of the White Mountain watershed in New Hampshire.

As one moves west to east from the Androscoggin, coming closer to the North woods, the lumberman, and later the historian, finds his course laid for him in the river openings into the forest: first the Saco, then the Presumscott, the Androscoggin in Maine, the Kennebec, the Penobscot, the Union, the Narraguagus, the Machias, the St. Croix, and finally the St. John. This last river rises in Maine, flows as the boundary between Maine and New Brunswick—again on the watershed between the Atlantic and the Saint Lawrence—and finally comes to sea in New Brunswick proper. The area included in this study of the northeast corner is not then only Maine but also parts of New Hampshire and New Brunswick, as well as a small portion of Quebec.[8] This region is a contiguous economic and geographical unit. Its economy is the forest, and its geography is the upland plateau of New England.

The area[9] between the White Mountains and the Laurentian watersheds is a broad plateau extending east and north from the White Mountains, sloping gradually to the sea and to the rivers, and finally setting off the Maritimes from the rest of Canada. The plateau is broken with many mountains, particularly in the western part of the area. Horsebacks (*eskers* or *kames*), residual deposits of gravel left by the receding glaciers, undulate roughly west to east across the area. Other distinct evidence of glacial action is the 2,500 or so lakes and ponds in the area, and the nearly 5,500 rivers and streams flowing through it. These streams are rapid, having cut narrow valleys, and in the spring when the ice leaves and the snow melts, they were used for transport of the forest products to the mill towns situated throughout the area. In fact, the area is so well watered that only one township, so-called "Dry Town", could not have its logs driven out on its own water.[10] The climate is that of the typical maritime variety on the east side of a great land mass. The mean annual temperature is about 42 degrees Fahrenheit with cool summers, and long, often severe, winters. Precipitation amounts annually to about 45 inches with approximately a fourth of it coming in the form of snow.[11] Westveld and his associates[12] found three zones of forest vegetation in the northeast corner. The first zone, that of spruce-fir-northern hardwoods, runs from the Berkshires in western Massachusetts, along the highlands through the Green and White

Mountain ranges into northeastern Maine and is found from sea level to an elevation of 4,500 feet. The soils are strongly podzolized. The hardwoods in zone one include beech, yellow birch and sugar maple, as well as paper birch, aspen, and red maple in higher elevations and in swamps and bogs. The soft woods (Maine natives use the term "black growth") are pure spruce and fir on the high slopes, as old-field associates or in swamps. White pine is an associate in old fields. In the swamps black spruce is the key species with tamarack (hackmatack—juniper) and northern white cedar often present. There is relatively little red spruce. Hemlock is present at lower elevations and to the east is predominant. Abandoned cleared land grows up to red and white spruce quite quickly. Balsam fir is present as well, but it deteriorates rather rapidly.[13]

Zone two, northern hardwoods-hemlock-white pine, which is really a continuation of the hardwoods section of zone one (what in Maine are usually referred to as "hardwood ridges"), runs through most of western Massachusetts to south and central Vermont and New Hampshire into southwestern Maine in a belt to the south of zone one. The soils are less podzolized. The tree cover is predominantly beech, sugar and red maple, and yellow birch, with the soft woods such as hemlock, white pine and an occasional spruce and fir. Old field growth is predominantly white pine with hemlock and an infrequent spruce. Fire stands fill quickly with aspen and paper birch. The best development of this zone is on lower hills and ridges to the north.[14]

Zone three, transition hardwoods-white pine-hemlock, covers some of southwestern Maine and most of southern New Hampshire in the region covered by this work. This was the area called the white pine region as it was nearly all white pine original growth. The white pine comes back onto the land after agricultural abandonment, and it is frequently a successor to cut over hardwood land, especially on the sandier soils. The soils of the region are predominantly brown podzols.[15]

Practically all the area was once forest, and although Wilkins found that in 1930 only about 78 percent of the state's area was then forest-covered, this was after 300 years of forest "mining". In terms of the ratio of forest area to total land area Maine led the union in 1930, with New Hampshire a close second. Of a total land area of 19,132,000 acres, 14,988,289 were in forest land.[16] In New Hampshire 4,400,000 of its 5,779,840 acres were forested, which amounted to 76 percent of that state's total area.

As quickly as land once used for agricultural purposes was abandoned, which happened in rapid and continuous fashion after 1840, the forest reclaimed it.[17] In Maine in 1930, 63.45 percent of the forest land was wild land and 36 percent was in farm wood land or wood lots. More than 97 percent of the forest land was privately owned, and 32 people or firms owned about 85 percent of the forest land in the state.[18] Thus the story of lumbering in this area is not only one of constant usage, but of constant usage by private individuals. Maine followed a policy of rapid disposal of her public lands, although the practice was interspersed with various attempts at disposal not for forestry purposes but for agricultural settlement.[19]

The amount of wild lands had increased since 1860 as farming was abandoned under the dual attack of the rich western farm lands being opened by the national government after 1860, and the "siren call" of the growing eastern cities after the Civil War. In 1860 there were 8,473,861 acres of wild lands in the state, all located in eight counties, Aroostook, Franklin, Hancock, Oxford, Penobscot, Piscataquis, Somerset, and Washington.[20] This study will concern itself primarily from the lumbering standpoint with these eight counties, although it will spill over into the other areas, especially in the time when poplar was the predominant species used in the manufacture of pulp and paper.

In 1860 about 3,200,000 acres of this wild land was still owned by the state of Maine;[21] nearly all of it was disposed of by 1890, and much of the early story in this period is concerned with the twin problems of making a highly competitive business pay while at the same time obtaining the land available for as little as possible from the state. Maine tried to attract settlers in this period, but the competition from cheap western public lands was too great, and many who found lumbering itself unprofitable were able to recoup their losses by disposing of their lands to the great paper companies which came increasingly to dominate the Maine lumbering scene after 1890.

Throughout the period since the Civil War, in one form or another whether it was logging, land ownership, sales, speculation, the pulp paper revolution, sawmilling, purveying to the logger or sale of the manufactured product, the larger percentage of the population of the area was as concerned with lumbering as it had been when Maine broke away from Massachusetts and became independent in 1820.

The Maine woods

Until the Revolution, settlement in Maine was sporadic and confined to the coast. Indian raids, and the difficulty of maintaining life in the forest land, precluded much beyond token settlement.[22] By 1784, with the war over, immigrants began to penetrate the Maine wilderness to find homes. A land office was set up by Massachusetts to dispose of land to the settlers, and land warrants were issued to veterans of the Revolution for Maine lands. These early efforts were capped by the sale in 1792 of a million acres of land on the Kennebec River, and another million on the Penobscot, to William Bingham of Philadelphia. Other lands were purchased by General Henry Knox.[23] This policy of selling land in large chunks created dissension among the settlers, and incipient warfare raged on Knox's lands. The land agent for Bingham was able to keep the area under Federalist domination until 1805 and even to prevent the erection of a separate state of Democratic persuasion. After 1805, though, Maine became increasingly a Democratic area, and finally through a compromise devised by William King, Maine's first governor, and the Bingham heirs, Maine was admitted to the union in 1820.[24]

Many towns were set up in this period, nearly all of them along the coast or at the head of navigation, and a flourishing coastal trade in fish, timber, masts, and wheat was set up. A fairly extensive trade with the West Indies in these products was begun.

The first census in 1790 showed a population of 96,540 within the present area of Maine. Most of this population was concentrated in the present day counties of York and Cumberland, with a few scattered settlements and houses inland on the river openings. Most of these people were concerned with getting a living from the wilderness, by sawing a little lumber stolen from the proprietary lands,[25] and raising some wheat. In the more settled areas they were also developing the coastal and shipping trades and doing the necessary lumbering not only for their own use and for the ships even then building along the coast, but also for sale in the Boston, New York, Philadelphia, and West Indian markets.

The population had risen to 298,335 in 1820, and by 1840 it was just over half a million (501,793). In 1820 Maine had only four towns over 3,000 inhabitants with Portland, the largest, having 8,581. Eleven other towns were close to 2,500. More than half of the people lived west of the Kennebec and south of the Sandy River. Most of the rest of the state was a forested wilderness. Total export from the state amounted to about seven million dollars, nearly all of which was a product of the forest.[26]

There were 726 sawmills and 524 grist mills scattered through the inhabited areas. There were also 248 tanneries which utilized the hemlock bark from the forest and their products were valued at about $382,000. Farming was the principal occupation with lumbering second, although the two were often interwoven with the farmer working in the woods in winter, his slack season.[27]

Lumbering was done in a rudimentary fashion. About the only tree cut was the pine, and it was felled with the axe. Teams of oxen dragged the huge logs to the streams, and in the spring they were driven to the mill on the freshet. The mills, nearly all of them driven by water power, sawed the logs into boards with up and down rip saws, which took a big kerf and were extremely slow in moving through the logs. Relatively little of the state's forest wealth was touched, even by the time of the Civil War. The pine was cut heavily, but not much else, and the primary tree in the forest cover was the spruce. About the only changes were the development of steam sawmills in the larger centers, with slightly faster and sharper saws, and the gradual movement of the loggers further into the wilderness. Still, lumbering in 1860 was practically the same as it had been in 1820.[28]

In the years before the Civil War the output generally increased each year, although there were bad years as in 1833 when a bad freshet downeast created large losses. Good years were 1837, 1841 ("at an unprecedented rate"), 1843, 1847, 1849, ("best year for lumbering ever"), and 1850, ("flourishing").[29] Of course newer areas were being opened up as railroads were built inland, the Androscoggin was dammed, and settlers moved north.[30] Not all years were good, however. Some began to call for using such woods as hemlock and some small amounts began to come to sawmills downstream.[31] Increasingly, many in the state could think of the woods as one poet did just prior to the Great Rebellion.[32]

> Up, Brothers, join our march tonight!
> The crinkling snow is sparkling bright;
> The ringing echoes far prolong
> The chorus of our wild road song;
> And the startled deer from his covert springs,
> And off to his mountain fastness hies
> Where, silver white, Katahdin lies,
> Aglow with the full moon's rays!
>
> One gentle thought to those we leave!
> They'll miss us sore, come fall of eve,
> For maiden dreams, from scenes more gay,
> To forest camps shall often stray.

The Maine woods

> And we—we will close with the wintry blast,
> As it whistles our forest dwelling past,
> A Song to tell to the rushing storm
> That still the Logger's heart is warm
> And true to the far away!

The general situation also did not change much in Maine until the Civil War. Farming remained the major occupation. The textile boom fostered by the War of 1812 involved Maine capital and Maine workmen along the Androscoggin and Saco Rivers, but most citizens were farmers, fishermen, or lumbermen and occasionally all three. The timberlands in the state provided a ripe location for speculation for some, culminating in the great speculation of 1835. The states of Maine and Massachusetts attempted to attract settlers to the public lands, but the unsettled nature of the boundary before 1842 (with its resulting border incidents), the policy of favoring the big capitalists on timberlands, and the constant competition of "Ohio Fever" tended to thwart the desires of individual farmers, and incidentally of those in the state who regarded this area as a potential Garden of America.[33]

By 1860 it had become clear to many, if not most, that the predominant feature of northeast corner geography, the forest, was destined to remain. Most of the state remained unsettled, despite all efforts to change this; the fortunes of the state rose and fell as did the lumber market, and even as it was in 1604 most of Maine was a trackless forested frontier. How best to master this frontier and not ruin it, to utilize it and still maintain it, to have one's cake and cut it too, this was the central problem of life in the region and it is a problem that never has been settled to everyone's satisfaction.

NOTES

[1] By Maine I mean the area covered in this study, from the waters of the Androscoggin on the west to the waters of the St. John on the east. For a detailed description see below.

[2] Captain John Smith, *The Generall Historie of Virginia, New England, and the Summer Iles* . . . , London, 1624, especially Chapter Six.

[3] For Weymouth, see *Collections* of the Maine Historical Society, First Series, VI, 291-318; V, 307-338. For the early explorations and visits see Henry S. Burrage, *Beginnings of Colonial Maine*, Portland, 1914.

[4] His rather splenetic comments appear in *Travels in New England and New York*, New Haven, 1821, II, 233-6.

[5] Baring's description is in Allis, editor, *William Bingham's Maine Lands*, Baring to Hope and Company, May 26, December 3, 1796, I, 651-6, II, 777,

784-8. See also Benjamin Lincoln to William Bingham, February 26, 1793, reprinted in *Ibid.*, I, 176-87, especially 182-7.

[6] Talleyrand's journal is reprinted in American Historical Association, *Annual Report*, Washington, 1941.

[7] *The Journal of Henry David Thoreau*, I, 47, (May 5, 1838). For other comments by Thoreau not in his *Maine Woods*, see *Ibid.*, V, IX, passim. In the *Maine Woods*, the relevant passage is at 107-110.

[8] The best early defense of including this area in one unit is in Jeremy Belknap, *The History of New Hampshire*, Dover, 1812 (in my edition), III, 62.

[9] The best description appears in *Geographie Universelle*, publiee sous la direction de P. Vidal De La Blache et L. Gallois, Tome XIII, *Amérique Septentrionale*, Part One, *Généralités—Canada*, Part Two, *Etats-Unis*, by Henri Baullig, Librairie Armand Colin, 1935, 1936; especially Part One, Chapters 1-6, 12, 16; Part Two, Chapters 18, 32. The best English introduction is *Maine, A Guide Down-East*, Federal Writers Project, Boston, 1937.

[10] Dry Town is T12R10, WELS, in Aroostook County. Pulpwood could could be driven here with the aid of splash dams. In the future, the abbreviation, WELS, is understood. It means West from the east line of the state, and is a survey designation. Others which are used are TS (Titcomb Survey); IP (Indian Purchase); OIP (Old Indian Purchase); NWP (North of the Waldo Patent); ED (Eastern Division Bingham's Penobscot Purchase); MD (Middle Division Bingham's Penobscot Purchase); ND (Northern Division Bingham's Penobscot Purchase); SD (Southern Division Bingham's Penobscot Purchase); NBKP (North of Bingham's Kennebec Purchase); WBKP (West of Bingham's Kennebec Purchase); NBPP (North of Bingham's Penobscot Purchase); BKP WKR (Bingham's Kennebec Purchase, West of the Kennebec River); BKP EKR (Bingham's Kennebec Purchase, East of the Kennebec River); EPR (East of the Penobscot River). The state was surveyed in various sections and at various times, and wild land is still designated by the original surveys. When townships have another name, such as Chain of Ponds Town (T2R6 WBKP) I shall endeavor to give both, because there are duplicates in the lumbering soubriquets. On Dry Town see Austin Wilkins, *The Forests of Maine—Their Extent, Character, Ownership and Products*, Bulletin #8, Maine Forest Service (1932), and Wilkins, (Forest Commissioner) to author, August 21, 1962.

[11] C. F. Brooks, "New England Snowfall," *Geographic Review*, III, 1917, 222-240, and Charles B. Fobes, "Snowfall in Maine," *Geographical Review*, Vol. 32, No. 2. [April, 1942].

[12] Marinus Westveld, et al, "Natural Forest Vegetation Zones of New England," *Journal of Forestry*, Vol. 54, no. 5 (May, 1956), 332-8, and map reproduced there "Natural Forest Vegetation Zones of New England," compiled by Committee on Silviculture, Society of American Foresters, New England Section, 1955.

[13] *Ibid.*, 333, 335.

[14] *Ibid.*, 333, 335.

[15] *Ibid.*, 335. Zone four is Central hardwoods-hemlock-white pine, zone five is Central hardwoods-hemlock, and zone six is pitch pine-oak. None of these zones is located in the area of this work.

[16] Wilkins, *op. cit.*, 20. The land area of the state is classified in the following table adopted from *Ibid.*

The Maine woods

Land Area Classification — Maine — 1930

Type	Acreage	Percentage
Forest	14,968,289	78.3
Farm	2,160,491	11.3
Waste	1,738,000	9.1
Urban, etc.	246,020	1.3

Not a county in the state had less than 43% of its land as forest. These percentages were: Androscoggin, 43.5% Aroostook, 84%; Cumberland, 56%; Franklin, 82%; Hancock, 85%; Kennebec, 43%; Knox, 60%; Lincoln, 59%; Oxford, 79%; Penobscot, 78%; Piscataquis, 90%; Sagadahoc, 52%; Somerset, 83%; Waldo, 48%; Washington, 88%; and York, 67%. Figures in the 1970's are even higher.

[17] See, on this point, P. L. Butterick, "Forest Growth on Abandoned Agricultural Land," *Scientific Monthly*, V, (July, 1917); Roland M. Harper, "Changes in Forest Area in New England in Three Centuries," *Journal of Forestry*, XVI, (April, 1918), and Ben Ames Williams, "The Return of the Forest," *Literary Digest*, XCIV, (September 10, 1927).

[18] Wilkins, *op. cit.*, 30.

[19] See my "Maine and its Public Domain: Land Disposal on the Northeastern Frontier" in David Ellis, ed., *The Frontier in American Development*, Cornell University Press, Ithaca, 1969, 113-137.

[20] Annual Message of the Governor, *Public Documents*, 1860, I.

[21] *Ibid.*, this included about 1,000,000 acres that Maine had bargained already, to be conveyed upon payment; 240,000 acres upon which stumpage permits had been sold; 1,500,000 acres which the state still owned outright; and 450,000 acres which had been sold by Massachusetts to be conveyed upon payment. By 1868, 1,186,485 acres had either been sold or set aside for special purposes, leaving about one million acres, of which 700,000 went to the European and North American Railway, of which more below.

[22] The standard history of Maine is Joseph W. Williamson, *A History of Maine*, 2 volumes, 1821, second edition 1828, and of New Hampshire, Jeremy Belknap, *A History of New Hampshire*, 3 volumes, 1795, second edition 1812.

[23] See Frederick W. Allis, editor, *William Bingham's Maine Lands*, two volumes, 1956.

[24] Ronald F. Banks, *Maine Becomes a State*, Wesleyan University Press, 1970.

[25] For David Cobb's difficulties with lumber thieves and his comments on the settlers see *Bingham's Maine Lands*, I, *passim*.

[26] Statistics from *Historical Statistics of the United States*, Washington, 1957, and Moses Greenleaf, *A Survey of the State of Maine*, etc., Portland, 1829, 274-83.

[27] Clarence Albert Day, *A History of Maine Agriculture 1604-1860*, University of Maine Studies, Second Series, no. 68, especially Chapter 16. Bangor *Whig and Courier*, April 18, 1844. Sometimes there were so many farmers unemployment was bad. See *Whig and Courier*, April 24, 1843.

[28] The best contemporary account of the logging is in John S. Springer, *Forest Life and Forest Trees*, New York, 1851. A history of Maine logging

from 1820 to 1861 has been written by Richard S. Wood, *A History of Lumbering in Maine, 1820-1861*, Orono, 1935. A recent useful account is Robert Dana Stanley, "The Rise of the Penobscot Lumber Industry to 1860," M.A. Thesis, University of Maine, August, 1963. Good contemporary accounts include Bangor *Whig and Courier*, August 14, 1843; January 5, 1844 (A speech by Nicholas G. Norcross to the Bangor Mechanics Association); July 14, 1845; Bangor *Democrat*, January 16, 1844; Bangor *Journal*, February 29, 1844, February 16, 1854.

[29] *Maine Farmer*, April 22, 1833, citing *Hancock Advertiser*; December 19, 1837 (citing annual production of over 400,000,000 feet in state; Franklin *Register*, January 2, 1841, citing Skowhegan *Sentinel*, November 20, 1841 citing Piscataquis *Herald*, December 18, 1841, citing Skowhegan *Clarion* "lumbering mania"; December 14, 1843, letter from "S", no. 7; Farmington *Chronicle*, August 19, 1847; April 19, 1849; October 17, 1850, citing Calais *Advertiser*.

[30] Oxford *Democrat*, November 18, 1853, discussing work on Umbagog Lake and the river. Oxford *Democrat*, October 27, 1854; January 12, September 21, December 21, 28, 1855. Lumbering was down by more than a half; the wharves at Bangor and Calais were full, and the next year was slim due to the overstock.

[31] *Drew's Rural Intelligencer*, May 12, 1855, letter from A.R.

[32] *Drew's Rural Intelligencer*, April 5, 1856, poem by Godfrey Greylick and named "The Logger's Song."

[33] See my "Maine and Its Public Domain," *op. cit.*, and my "Toward A Theory of Maine History—Maine's Resources and the State," in Arthur Johnson, ed., *Explorations in Maine History, Miscellaneous Papers*, Orono, 1970, 44-64.

2

LIFE IN THE WOODS

> "Come along to the shady vales of 'Suncook,'
> And swamp those logs for me."
>
> —OLD WOODS SONG

The period from 1860 to 1890 in the lumber business was one of transition. New areas were being opened for lumbering, areas more difficult of access than those cut in the palmy pine days. The competition from city industries and western lands meant that lumbermen had to provide better conditions of work. Later in the period a growing social consciousness also tended to advance working conditions. At the same time the operator often found himself in severe competition with the pineries of Minnesota and Wisconsin, and the economic squeeze was always present. There was barely a year during this time in which the newspapers did not complain of the prospects for the loggers. Pulpwood operations would solve this problem for many in the woods, but not without much difficulty, and for most not until after 1890. Severe economic problems and ever multiplying social changes dominate the period.

Most casual observers regard Maine lumbering as something which took place before the Civil War. The best history of Maine lumbering ends in 1861.[1] The most well-known commentaries are of the pre-Civil War era,[2] and even the standard popular history of the "lumberjack" says that although the Bangor industry reached its height in 1862 its importance to the national scene was much diminished by the time of the Civil War.[3] However, it is statistical fact that more lumber was cut in Maine after 1861 than before and the biggest years in the Maine woods were in the last decade of the nineteenth and the first decade of the twentieth century. Although wide pine boards might not have been the product, spruce and later pulpwood kept the Maine economy afloat. In fact, at this writing, lumbering is still more important than fishing or tourism. The best evidence we have for this is contained in the following table which lists the lumber surveyed at Bangor from very early in the industry's history until a time when most lumber cut went to the pulp grinder rather than the sawmill. The most interesting fact shown by the table, in addition to the obvious one of industry growth, is how spruce came to dominate completely the Maine woods.

Lumber Surveyed at Bangor, 1832-1905[4]

Year	Pine	Spruce	Hemlock, etc.	Total
		(Feet Board Measure)		
1832				37,556,093
1833				44,000,845
1834				30,756,558
1835				67,431,699
1836				50,841,756
1837				61,976,832
1838				74,020,409
1839				90,767,789
1840				70,717,421
1841				82,338,639
1842				112,341,566
1843				120,137,126
1844				116,788,121
1845				154,834,849
1846				140,064,864
1847				191,136,272
1848				213,051,235
1849				160,418,808
1850				203,754,201
1851	143,586,200	47,567,682	10,851,948	202,005,830
1852	124,399,736	63,859,929	11,129,757	199,389,422
1853	92,484,711	78,087,096	12,370,477	182,942,284
1854	93,446,799	53,564,196	12,580,342	159,591,337
1855	123,026,157	78,337,283	10,305,753	211,669,193
1856	102,411,667	66,526,983	11,322,580	180,262,230
1857	75,816,045	56,735,284	12,557,680	145,109,009
1858	69,453,844	62,045,696	16,166,907	147,666,447
1859	84,704,700	78,066,187	15,275,553	178,046,440
1860	98,401,676	88,285,040	14,662,811	201,349,527
1861	48,238,957	72,928,910	9,874,824	131,042,691
1862	61,725,787	90,865,804	7,471,392	160,062,983
1863	63,544,438	110,304,467	16,823,364	190,672,269
1864	54,846,506	106,774,936	12,814,830	174,436,272
1865	48,296,222	107,505,867	14,078,934	169,881,023
1866	63,575,411	154,971,243	19,000,952	237,547,606
1867	51,207,174	139,445,478	15,820,706	206,483,358
1868	50,309,399	152,931,455	17,553,912	220,794,766
1869	40,980,911	133,756,757	16,103,240	190,840,908
1870	30,030,000	149,103,192	22,881,000	202,014,192
1871	42,383,000	163,121,675	21,987,000	227,491,675

Life in the woods 13

Year	Pine	Spruce	Hemlock, etc.	Total
1872	46,150,000	176,933,649	23,370,000	246,553,649
1873	32,586,848	129,277,908	17,337,192	179,202,348
1874	24,178,309	135,226,015	17,382,608	176,786,932
1875	22,335,849	116,664,487	15,662,793	154,663,129
1876	19,615,572	82,087,987	13,417,632	115,121,191
1877	14,704,152	85,480,149	17,683,444	117,867,745
1878	19,479,497	81,358,056	21,302,775	122,140,328
1879	17,959,415	91,907,627	12,695,220	122,562,262
1880	17,668,651	91,573,149	14,208,737	123,450,537
1881	33,732,101	104,704,537	15,912,159	154,348,797
1882	33,408,035	122,548,230	16,154,829	172,111,094
1883	26,522,485	115,348,484	19,392,223	161,263,192
1884	24,718,767	84,425,303	16,169,276	125,313,346
1885	30,480,937	94,446,522	17,867,104	142,794,563
1886	28,603,785	100,905,443	17,055,420	146,564,466
1887	29,103,725	102,746,234	17,792,578	149,647,537
1888	30,942,687	114,348,153	19,473,695	164,764,535
1889	27,685,394	121,659,086	20,665,903	170,210,383
1890	28,255,256	129,541,485	21,310,006	179,106,727
1891	23,114,771	118,205,741	23,664,844	164,985,356
1892	26,896,302	105,044,377	28,453,079	160,393,758
1893	22,425,974	81,400,612	25,447,931	129,274,517
1894	25,369,893	116,969,664	18,934,467	161,274,024
1895	27,189,050	91,488,448	25,513,996	144,191,494
1896	23,229,739	90,449,002	24,270,204	137,948,945
1897	25,935,354	118,007,612	25,817,117	169,760,083
1898	22,501,025	95,167,159	26,656,559	144,324,743
1899	23,246,498	133,234,823	25,001,268	181,482,569
1900	22,543,902	102,465,989	17,689,361	142,699,252
1901	22,723,741	81,971,594	16,259,562	120,954,897
1902	21,755,767	87,793,649	15,218,230	124,767,546
1903	31,778,278	101,985,724	22,745,196	156,509,198
1904	28,303,861	106,602,644	28,778,965	163,685,470
1905				185,000,000

Maine might better have been called the "Spruce Tree State."

In the period from 1860 to 1890 the methods of logging changed but little from those described by Springer at mid-century. More horses than oxen were used at the end of the period, and the saw was replacing the ax in the woods. The food was getting better, and the camps were not quite the hovels that had housed man and cattle in Stephenson's day. Still, not much had changed. The axeman on Bingham's lands in 1820 would have been able to make his way in 1890.

The operator still looked for a good "logging chance." On the Kennebec and Androscoggin he probably worked for the landowner.

The Coburns, Bradstreets, Coe and Pingree, and the Berlin Mills Company owned most of the land they operated, and they hired crews to cut the lumber they thought they could saw and sell at a profit. On the Penobscot in the north country and to a great extent downeast,[5] the operator bought his stumpage from the great landowners, Pingree again, the Hersey heirs, and such individuals as Llewelyn Powers from Houlton. In either case the first step was to explore the area. One man, or two, went out, usually in June or July, perhaps later, with tea, pork, matches, and the ever-present axe. "They just walked in, looked it over; if there was enough, they cut it!! There were a heluva lot of bulls in those days if a man didn't know what he was doing."[6] As time went on, and the economics of lumbering began to pinch the middleman, cruising became more precise, but before 1890 it was a rare cruiser who really knew how much lumber there was, how much reproduction was coming along, or the other items always present in today's cruising. It was much more important to determine where the drivable streams were (and how much work they needed) to map out the necessary logging roads and plan the best locations for the various camps in the tract. Frequently it was these things that really determined whether the location was a good "chance" or not.[7]

Once the explorer had made his report, including his estimate of the merchantable timber, to either the landowner, or the prospective operator, and the decision was taken to send in a crew to work, several steps followed. Where streams needed splash dams built, or abutments to ward off logs on curves, men had to be sent in to construct them. Often men would be in the woods as early as June to start the work. These small dams built of logs and rocks would facilitate the brook drives. In later years as the crews began to operate further in the bush and the streams grew smaller, dynamite might be used to widen a stream bed in order to open up the timber. Such was the case on Wassataquoik Stream and on the Pleasant River, above the Gulf, in the eighties.

The same crews that were facilitating the driving usually had the task of swamping out roads for the winter's logs. It was customary to cut and yard the logs until the time for hauling. Of course, camps had to be constructed. Small poles, perhaps four to six inches in diameter, were cut. These poles, usually spruce or cedar, were laid on the ground and fitted and notched log cabin style. Frequently one side was higher than the other, but seldom much over seven feet high, giving a pitch to the roof and obviating the necessity of a roof tree.

Life in the woods

The roof was usually of cedar splints (splits) about two inches thick, and four to five feet long. The logs were chinked with moss. As the winter wore on snow would cover much if not most of the structure and give excellent insulation. Underneath the snow and over the splints heavy fir boughs were laid. This formed an air space which made the insulation better and insured that melting snow on the roof did not freeze at night, thus backing up and creating leaks in the roof. The best logging chances would allow a camp to be used for as long as three years and this meant making it snug and tight.

In the sixties most camps had a fire in the center of the building with a poor chimney leading through the roof. By the nineties many camps had stoves, one for cooking and one for heat, at either end of the camp. These sheet iron or cast iron monsters, almost always manufactured by Wood-Bishop Company of Bangor, threw a tremendous amount of heat in an enclosed space. Many camps were constructed so that there were two doors. Inside the first door was a three or four foot space, called the dingle, where meat was hung, and other food kept. By the nineties most camps had some beef, and a side of beef hung here was under refrigeration in the Maine woods. The molasses barrel, the pork barrel, extra flour, and so on were also kept in the dingle. Through the second door to the left was the wood dingle. Here the great logs for the open fire, or the stove wood in better days, were stored. The camp was usually kept very warm to help combat the extreme cold of the woods. To the right of the door was usually the cookroom where the cook held forth with his minions, the cookees. Sometimes, the cook's quarters were at the far end of the room. In either case shelves were constructed for the tin plates, dippers, mugs and cutlery. In the early days the cooks needed fewer implements. Two big kettles for the beans, and a flouring board for the biscuit were enough. Flour and the other ingredients were kept in a barrel from which the cook worked. In the early days sourdough bread was the rule; later most cooks used saleratus and cream of tartar. Baking soda did not come until late. In a good camp there might be molasses cake, or even gingerbread or molasses cookies. By the nineties such delicacies[8] were the rule, not the exception. All were made on the same baking board, and it was not unusual to find pieces of biscuit in the molasses cake, or cake in the biscuit. The cooks ruled the camps, and most insisted on absolute silence during the meals. The food varied little. Beans, pork, and biscuit were usual. In the days of the big fire the beanhole beans were cooked in the camp itself, in later days in a hole outside. One of the cookee's duties was to keep the fire going. In some camps by the nineties items like preserved fruits and codfish were often served. Probably almost as much salt fish was used in the camps as salt pork or salt beef, especially as much of the crew personnel was made up of men from New Brunswick or Prince Edward Island, where salt fish was a diet staple. Northern camps served a fair amount of "Canada beef" or moose or deer venison. It doubtless came as a welcome change to the ordinary diet.

The camp scene was much the same everywhere. A floor of poles, sometimes hewed level, at other times covered with the boards of packing cases holding the food, stretched to the end wall except for the area of the fire. On one side would be the rough table, constructed again of cedar splints, with a full length bench, often on both sides. On the other was the bunk. It was the general rule throughout this period to have a common bunk, which was filled with fresh boughs on Sundays. The men also shared a common coverlet. Two long pieces of blanket material, twenty-five or thirty feet long, were sewed together and stuffed with straw or occasionally cotton batting. In front of the bunk was found the deacon seat. This bench, made of logs about fifteen inches through, split, shaved smooth, and mounted on legs, was the social center of the camp. Here, in the evening, the men would sit, talk, smoke, and discuss the day's work. The only other item of mention in the camp would be a fir tree with the limbs cut off near the trunk and stuck near the fire to hang mittens, stockings, and caps to dry. In larger camps and later in the century bunks and deacon seats were located on both sides of the camp.

Outside the camp and a few feet away was constructed, in the same fashion as the men's camp, a camp or hovel for the animals. This too had its dingle for the hay, and oats if there were horses. Depending on the location some might be true hovels, that is, with one side open to weather. They were called hovels however they were constructed.

Once the camps were constructed, and the plans made for the winter's operations, a crew was hired. They usually traveled overland beyond the railroad or river steamer, arriving in the woods in late October or early November. Further logging roads had to be swamped, and the supplies toted in. Most supplies were toted after the snow came, but if it was an area near towns or villages some might be brought by buckboard before first snow. Once the snow came the winter's work began in earnest.

Life in the woods 17

It was usual for the cook to rouse the men a half hour or an hour before sunrise. After a quick and hearty meal of salt pork and beans with biscuits, the choppers went to the woods, usually arriving with the sun. The trees were felled, limbed, and if necessary, cut into log lengths. In earlier times the logs were then dragged to the landings. At this time the general rule was to yard out the logs, fifty or sixty logs to a yard, and when the amount contracted for was felled, all men in the crew participated in the hauling to the stream or lake from which they would be driven. By the end of the century, most choppers used the double-bitted ax. One side was ground thin for chopping, the other had a thicker bevel, and was used for limbing or "knotting." This was to prevent breakage in frozen fir or hemlock knots. The standard handles were 32 inches long, with a notch at two feet in order to measure the logs.[9] Each crew had about six men to a yard: two choppers, one knotter, one swamper, one teamster, and one yard roller. The choppers usually faced each other on the same tree and worked in the same cut. The knotter limbed out the trees, while the swamper kept down the underbrush so men could work easily. The teamster and yard roller hauled the logs and piled them in the yards for easier handling.

If the men were lucky and had a good cook there might have been doughnuts or cookies to fill one's pockets before leaving in the morning. If not, the next meal came when the cookee appeared at the chopping location with more beans, cold biscuit, and hot tea (if the location wasn't too far away). The food usually arrived around eleven o'clock. The loggers might have built a fire to warm the food upon arrival (even in thirty below temperatures the body was warm doing this work). After this meal the work continued until the end of the daylight hours, when the woodsmen would make their way back to the camp, usually arriving just at dark (four-to-five p.m. in the Maine winter). After another huge meal of the same food there might be time to smoke a pipe, swap a few lies, or wash out a pair of socks before bedtime, which usually came before nine o'clock. Some woodsmen knitted socks or mittens of their own in the evening; still others played cards—the ubiquitous lumbering games: cribbage, sixty-three, red dog, or pitch. Often there was some singing in the lumber camps, especially if a Larry Gorman was present in the crew.[10]

In March all hands turned to and moved the logs from the yard to the landings, either at the stream, or on the lake, to wait for the ice to go out. Frequently the loggers were finished and came out of the woods to enjoy themselves before the drives started. In northern Maine it would be late April or early May before the ice would leave and allow the drive to come along.

Sunday was a day of diversion. One could sleep late, for after rising, only boughing the bunks, washing and mending clothes, and greasing boots took up the time. Some went hunting, did a bit of trapping, or perhaps sought out spruce gum.[11] Carving containers for the gum whiled away the time as well. Occasionally we hear of other diversions. In some areas there apparently was a ceremony for green hands similar to "crossing the line." The neophyte was "shaved" with an iron hoop or broom, the lather being soap suds or even molasses.[12] If the camp had a practical joker poor unfortunates might find the runners of their sleds frozen in, sleigh bells muffled, harnesses unhooked, or nearly anything else to "jill-poke" a rival's outfit.[13] Some woods bosses would ramrod their teamsters by putting up a score each Saturday night with the amount hauled each week after the man's name, with a prize to the winner.[14] The men themselves sometimes wagered among themselves as to the amount cut, or hauled, with the losers "paying the bitters" in Bangor in the spring.[15] The work was hard, and there was little else to do. The crews, especially in the later period, were rather poor specimens, many of them. In fact, it was the fashion to speak of your crew as three crews, one going to Bangor, one coming from Bangor, and one working. The jovial lumberman can, and should be, confined to novels.[16]

Life was difficult in the woods, and there was little time for play. The loggers life consisted of work, sleep, and "Beans and brown bread for breakfast, brown bread and beans for dinner, and a mixture of the two for supper," [17] and little else. The only real changes which occurred in this period were that the horse began to replace the ox for hauling, and the distance of the haul began to increase. Frequently the haul would be three or four miles, not one.[18] One heard more and more of the one or one and a half turn road, rather than the three or four turn road of earlier times. The old bob or bunk sled began to go, and by the end of the period one observer describes them as "a thing of the past." This was due to the replacement of oxen with horses.[19] By the end of the period beef had begun to replace pork as a food staple also. Beef was cheaper with the new methods of refrigeration, and fresh western plate beef occasionally appeared on the logger's menu.[20] The work itself remained as difficult, and perhaps even became more so as the logging chances

Life in the woods 19

were found farther and farther from the best drivable streams, and it took more logs to make a thousand.

The men themselves changed as well. The logging crews, which for years had been made up of Bangor rivermen and farmers, now came to be more and more Provincemen, that is, natives of New Brunswick, Nova Scotia, or Prince Edward Island. As a song of the day[21] had it:

> Oh, the boys of the Island
> They feel discontent.
> The times they are hard,
> And they can't make a cent;
> So, says Rory to Angus,
> "Here we're doing no good.
> Let's go over to Bangor,
> And work in the woods."
>
> Now the boys of the Island
> Will work cheap, you bet,
> Fifteen dollars a month
> Is the wages they get.
> See their socks and their mittens
> All knitted three-ply,
> You can tell by their duds
> That they come from P.I.

The Canadians were in the Maine woods early. The *Commercial* reported that most of the bark peelers at Winn in the summer of 1873 were Frenchmen from Madawaska and Edmundston. They were described as a hardy lot and good workers at low wages.[22] That was the big complaint, they depressed the wages and deprived the Maine logger of his livelihood. The complaints intensified and increased in the 'eighties,' and are but an example of the economic squeeze on the industry. Editorials, bitter, scornful, pungent, and even acid, fill the newspapers. Provincemen, and their horses, came to be the symbol of discontent among those trying to cope with the changing economic situation. As a correspondent put it in his anonymous letter from Calais in 1882:[23]

> There is much comment on that species of protection that allows hundreds of horses to be brought here openly from New Brunswick, bonded at small cost, and allowed to work through the lumbering season and return home in the spring. Something over a hundred span were bonded here the past week, and will be engaged during the winter hauling bark for F. Shaw and Bros. These horses come from St. John and Miramachi, and are owned by small farmers who have no use for them during the winter months, did not our custom laws kindly invite

them to 'Come o'er to Ameriky' where they work for just sufficient to keep themselves and teams until the opening of farming operations in the spring.

The teams came, and what was worse, food and provisions[24] came as well.

> Rumors say that large quantities of grain are smuggled across the borders by our neighbors across the line for use by provincial lumber concerns that are operating on this side. Our lumber is of no benefit to us; it is cut and hauled and driven down the streams by province parties who bring their lumber and go back home again. If the province teams were not allowed to come over here and work our farmers could get employment for their teams. Our position in this regard is about the same as the Chinese in California. Where is our protection for the laboring man that we have heard so much about from Republican orators during the last campaign?

Any item was seized upon as evidence of the wickedness of this practice. The *Argus* described a visit to one of these logging camps while drawing a moral for its readers.[25]

> I visited the sleeping quarters, an unfinished attic room, with 25 or 30 so-called beds, for the comfort and rest of from 50 to 60 boarders. There was no ventilation and—well, it was evident that this room had been and still was occupied. Our visit here was brief. We were not anxious to remain. Yet since these lumbermen entered the camp last autumn there had not been a case of sickness. They have no holidays—except Sunday—no books, no papers, nor do they want them. Sunday they spend as they desire. It is doubtful if the manner in which it is spent would have been approved by the Puritan fathers, but I doubt if any fifty men in New England are happier than these fifty. If 'happiness' is indeed 'one's end and aim' then that 'end and aim' may be found in Lowe and Burbank's grant, Coos County, New Hampshire. The lumber manufacturers have been demanding that the tariff on lumber should not be removed. They demand protection for American labor, but the American labor in this lumber camp is composed almost entirely by Canucks, who have come down from Canada for the winter. It is a question whether we can afford to have our forests destroyed for the sake of protecting such labor.

The editorialist went on to say that the same situation was true in Maine. Still, "the enterprising Bluenose" came with his teams and men, and, according to Maine newspapers, good Maine woodsmen had their choice of "sucking their thumbs or going west."[26] The hiring of the Provincemen was a regular business. Bangor boarding houses recruited them, housed and fed them, and then contracted them out to Penobscot operators. Many of the men went to Gloucester

Life in the woods 21

and worked on the fishing schooners, or shipped out of Bangor in the lumber trade during the summer.[27] They were disliked by the Mainers, but the Provincemen continued to come. Some advocated the prohibition of the export of round logs,[28] in the hope of stimulating sawmills in the northern regions. Others brought forth the old idea of purchasing the Maritimes in order to end these problems,[29] annexing all of Canada,[30] signing another reciprocity treaty,[31] or even asking help from the state and federal government to aid the depressed regions to build up pulp, paper, saw and shingle mills.[32] Even the Indians who had been a source of supply along the Penobscot waters, found themselves forced to turn to agriculture in larger numbers because to the Provincemen's competition.[33]

The great northeast area was an economic unit, if it was not a political one, and in these changing times, economics ruled politics, as always. If the Provincemen worked for lower wages, they would be hired. The pinch of lumbering prices worked the same havoc on one side of the line as the other, and the more competitive labor won out.

Wages ruled low throughout most of the period. In 1860, David Libbey had received $2 a day for his work, but he was an above-average worker.[34] Twenty to thirty dollars a month and board would have been a closer figure in the period up to 1872. With the coming of the depression of the seventies wages dropped sharply. In 1875 jobs were reported scarce and wages at eight to sixteen dollars a month.[35] At times men went into the woods for as low as five dollars a month;[36] it was either this or the souphouse, which did a thriving business in 1877.[37] It was not until the eighties that wages returned to the pre-depression levels. Wages were reported in the early eighties as ranging from eighteen to thirty dollars or even thirty-five dollars a month.[38] They did not remain at this height long. Twelve to twenty dollars was the usual quotation throughout the rest of this time, and at the end of it, wages were thirteen to seventeen dollars on the Penobscot, with plenty of men, and few jobs.[39] Bangor opened another souphouse.[40]

Another difficult problem for the operator was the securing of supplies for his men and teams. It took a very large amount of supplies to get out logs. One observer thought that for every million feet of logs obtained, one thousand bushels of oats, nine barrels of pork and beef, ten tons of hay, and forty barrels of flour were consumed. These figures are for an operation using two pair of horses. Using these figures, the cost of supplies amounted to about four dollars per thousand.[41] These supplies did not include rum for the men. It disappeared from the woods menu in this period.[42]

The larger operators ran farms in or near the woods for their crews, but the smaller operators had to depend on hauling the goods upriver. On the St. John the fall saw a procession of flatboats and scows going up the river laden with hay, supplies, and horses. These flat-bottomed boats, which did not draw much over eight or ten inches of water, could carry twelve and one half tons of weight, or one hundred twenty-five 200 pound barrels. Above railroad connections these boats, dragged by heavy horses, 1,300 to 1,500 pounds, walking tandem on the towpaths, went their way up as far as the Baker Branch. When the autumn freeze came early conditions for both men and beasts were frightful.[43] On the Penobscot the small operator usually purchased his goods in Old Town,[44] where they had been brought by train. Until 1868 when the European and North American railway bought the line out, these goods went from Old Town to Passadumkeag, Winn, and even Mattawamkeag, by steamer.[45] After 1868 they went as far as they could by train. From then on it was into the woods via the buckboard, or after the snow came, the sleigh.[46] The life of the "toter" was none too pleasant, and he too found himself in an economic squeeze from time to time.[47] It took a good deal of toting to get the goods in. On the operations of Quinn and Mitchell at Wytopitlock in 1890 the men consumed a barrel of flour every two and one half days. In each week they used 400 pounds of beef, four bushels of beans, one hundred pounds of lard, and seven pounds of tea. There were sixty men on the operation, getting out 800,000 feet of spruce, and 7,000 cords of pulpwood for the mill at Enfield.[48]

The large operators usually got some of their supplies from big farms that they maintained in the woods. Here, in addition to the crops, it was possible to summer horses and oxen as well. The Coburn's big farm on the Moose River produced 700 bushels of oats, 125 bushels of beans, 800 bushels of buckwheat, 300 bushels of turnips, 750 bushels of potatoes, and 250 tons of hay in 1875, for instance.[49] Some smaller operators also used their home farms to raise their supplies. B. F. Osgood of Prentiss, operating on the Mattawamkeag River with forty men, summered twenty-four horses on his farm at Prentiss. In addition, he and his eleven sons raised from 1,900 to 2,700 bushels of oats each year, and cut about 100 tons of hay.[50] This is a good example of the combination lumberman-farmer so prevalent in the Maine woods.

Life in the woods 23

More important and larger was the famous Chamberlain Farm at Chamberlain Lake. The Coe-Pingree interests had cleared the land and constructed the buildings after the Telos Canal had opened that area up to Penobscot operators in 1846. The buildings were giant affairs, all made of eight inch squared timbers and shingled. The main house was one story until 1901. In 1888 the seven men employed raised 700 bushels of oats, and 75 tons of hay, as well as summering the stock throughout the 600 acres of cleared land.[51]

Other big farms were the Grant Farm, with its 200 acres, although it was primarily used after our period,[52] Trout Brook Farm in Patten with one hundred acres cleared,[53] Silver Lake Farm near Katahdin Iron Works, and the Todd Farm at Tomah Stream, on the St. Croix.[54] E. W. Loveland, of Bangor, ran a farm for himself just outside Sebois. For the northern waters there were supplies available at the Depot Farm,[55] and the settlers in the back country lived off the lumbermen. Sixteen families at Oxbow Plantation in 1884 raised 501 bushels of wheat, 1,000 bushels of buckwheat, 6,560 bushels of oats, and 205 bushels of peas, nearly all of which was destined for the Aroostook logger.[56]

Conditions of work improved little in this time. It was impossible to change some of the distinguishing characteristics of Maine lumbering. For the man building dams, or camps, or later the bark peelers, there was the constant torture of the flies. Hemlock bark, boiled down to make essence for the tanneries had to be peeled from the logs when the sap was running, from perhaps the first of May to the end of July. Later when pulpwood operations began, poplar had to be peeled in order to drive it, and by the twentieth century, much of the softwood used for pulp was being peeled in the woods as well. Peeling time coincided with fly time. The men were tortured with bites and stings of the black fly, mosquito, moose fly, and the worst offender of all, the midge. The only palliative was to shave the head high above the ears, and put on layers of tar, mixed with bear grease or lard. This did not solve the problem; it did, however, make work at this season possible.[57]

In the winter the main problem was the weather. The Maine woods get extremely cold. In 1884 temperatures more than 40 degrees below zero were recorded for a week.[58] Mixed with the cold weather frequently were severe storms, high winds, and deep snow, as in the year 1885.[59] Perhaps the worst year was 1887. There were terrible storms, with the deepest snows in years. Lumbermen did not get their expected cuts. Horses had to be killed in the woods as supplies could not be got to them; some lumbermen hauled their horses out on sledges to save them, and a newspaper proposed a type of snowshoe for them.[60] One newspaper remarked:[61]

> Don't you forget it, young man, and be sure and make a record of it; or, three score years from now, when you tell the story to a generation unborn, they will not believe it, that on the tenth day of March in the year of our Lord 1887, the ground throughout eastern Maine was covered with snow to the depth of seven feet.

Perhaps the worst winter in the woods in the nineteenth century was in 1887 with its combination of deep snow, high winds, and low temperatures. Snow to equal that depth did not again occur until 1963.

There were other natural, or climatic disasters. In 1869 there was the famous Saxby Gale, so-called because it was predicted by Lieutenant N. S. Saxby, of the British navy. It apparently was a hurricane or tornado, with a tidal wave. Tremendous blowdowns resulted all over Washington County and into New Brunswick.[62] There were further great gales in 1872 and 1883. The 1872 gale in northern Maine, was centered in T15R6, and the 1883 hurricane blew down a swath of trees from the Kennebec to the West Branch of the Penobscot. Above Ambejejus Lake it leveled trees so that the observer could see fourteen lakes and ponds not normally visible.[63] The 1884-1885 cut on the Penobscot was primarily blown down trees as the owners attempted to salvage what they could from the wreckage.[64] In addition to the weather, the bud-worm attacked the spruce and a saw fly the hackmatack intermittently during this period.

There were other dangers as well. Smallpox was prevalent from time to time, and once the State Board of Health recommended mass vaccination of all lumbermen to help control the disease.[65] Fires in logging camps were fairly common.[66]

> A lumbering camp on Ripogenus Stream in T4R11 owned by H. K. Robinson of Brewer was destroyed by fire Monday forenoon, the 12th. Nothing was saved—camping tools, bedding, clothes of the crew, and some money all being destroyed.

This fire even served as a vehicle for a lyricist.[67]

Of course, for the unfortunate lumberman who happened to get sick, or more often hurt himself with an ax or a falling tree, there was the long overland trip by sled to Greenville, or another nearby point, if the rude woods remedies did not prevail.

Life in the woods 25

Although most operators were concerned only with the physical well-being of the men in the woods, some interested themselves in their moral welfare. This interest was to be one of the major changes in the twentieth century. It came early though. In 1872 Reverend James Cameron of Greenville supplied Bibles, testaments, tracts, and other periodicals to some thirty logging camps in his area, and asked for donations from people who were interested.[68] The Reverend Charles Whittier, brother of the famous poet, served as a missionary visitor to logging camps in Washington County for years,[69] and at the end of this period the State Convention of W. C. T. U. discussed the moral condition of the lumbermen, and agreed to employ a man to collect and send reading matter to the camps, as well as doing "evangelical work." [70]

Nearly everyone in Maine was interested in news of the logging camps, and newspapers gratified that interest by constantly printing news of records set by men chopping, of teams hauling, or of monster trees which had met the woodsman's ax. These were the records of the day, records to be broken year after year. A few of these reports will suffice to show the impact of the industry on the region. For instance, Abram Sibley of Lowell was given the title of "King Bee" for chopping in 1872. In twenty-two days he cut 642 logs of pine, spruce and juniper, on T20, all this while he was assisting to load.[71]

Some of the loads hauled were enormous. Many loads of five or six logs scaled 2,500 or even 3,000 feet. The largest noted in the period was a load of spruce handled by one team of horses at Van Buren which scaled 4,357 feet.[72] Oxen also hauled big loads. At the W. J. Reed operation on Chamberlain Lake in 1889 one four ox team hauled three loads in one day, forty-seven spruce sticks totaling 9,232 feet, and another team of four oxen hauled three loads, forty-two sticks which scaled out 8,910 feet.[73] When the logs were moved from the yards to the landings a day's work might be extra large. Charles Chase, of Woostock, New Brunswick, working with one pair of horses and ten men, moved 264 logs of spruce, scaling 30,254 feet, on to the ice of the St. Croix River.[74] Over the entire season the work done was often prodigious. Eight horses on T8R18, working for the northern Maine firm of Cunliffe and Stevens, hauled, in ninety-six days, 8,870 spruce trees and 1,502 pine sticks, measuring in all 1,346,800 feet. One pair of the horses was twenty-five years old.[75] In 1879 Nichols, Smith and Sons, hauling from Perkins Stream Town, worked eighty-four days. With one four ox team, and three two horse teams they moved 2,300,000 feet onto landings. Forty men had cut and yarded this lumber from two to three and one half miles from the stream. The oxen could move 4,700 feet at a load; the horses from 2,200 to 2,700 feet.[76]

In addition to the stories about giant loads, or master choppers, there was always the report of another huge tree, each one larger than the last. The tree which made two to three thousand board feet was not unusual. Nearly every winter saw the report of at least one pine or spruce which scaled this large.[77] Some were even larger. For instance, the men for the Lewiston Stream Mill cutting at Livermore in 1872 got a master pine. This tree measured five feet and eight inches across the butt. The first log was seventeen feet long and four feet through. Here the tree crotched. The four trees which branched off were 110, 110, 109, and 84 feet long, to the eight inch diameter mark. The entire tree scaled out at 6,532 feet.[78]

The next year another nearly as large was discovered. A newspaper hailed its discovery with the following squib:[79]

> Warren Smith's team on Twp 22 lately hauled on to the Sebago Lake a tree which at full measurement, scaled 6,125 feet. The butt log, 19 feet long, measured 1,900 feet. Who Comes Next??

Other big trees continued to fall during this period, although none as large as these two monsters of the forest. One in the city of Bangor itself, at Six-Mile Falls, gave a log 68 feet long. This one had little taper: it was twenty-eight inches in diameter at the butt, fifteen inches at the center, and thirteen at the top. It was naturally saved for a mast.[80] Others were recorded at Sanford,[81] Milford,[82] T11R14,[83] Butler's Mills,[84] and Columbia Falls.[85] As the Argus said in reviewing the big trees discovered one winter:[86]

A FEW OF 'EM STILL LEFT

> A white spruce log scaling 1,968 feet, and a pine log 60 feet long, 30 inches in diameter, and scaling 4,500 feet, were cut recently on E. S. Coe's land on T11R14 and hauled out. C. W. Ryerson of Norway recently cut a pine tree that scaled 2,784 feet. The first out of 12 feet in length scaled 815 feet and the first three cuts of 40 feet scaled 2,093. And, on T36 (Washington County) a pine tree was cut for Holway, Sullivan and Co. of Machias from which were taken 17 logs, 294 feet total length, and 2,950 feet of sound timber.

During the entire period from 1860 to 1890 men went into the woods hoping to find such master trees, cut them, and haul them out. These

Life in the woods 27

were great days in the Maine Woods. A revolution was coming; in fact, by the end of this period it had already started. Still, most lumbermen were concerned with their great trees, and with the economic squeeze.

NOTES

[1] Richard G. Wood, *A History of Lumbering in Maine 1820-1861*, Orono, 1935.

[2] John S. Springer, *Forest Life and Forest Trees, Comprising Winter Camp-Life Among the Loggers, and Wild-Wood Adventures With Descriptions of Lumbering Operations on the Various Rivers of Maine and New Brunswick*, New York, 1851; Henry D. Thoreau, *The Maine Woods*, Boston, 1868, and Isaac Stephenson, *Recollections of a Long Life*, 1829-1915, Chicago, (privately printed), 1915.

[3] Stewart H. Holbrook, *The American Lumberjack* (Holy Old Mackinaw) New York, 1938, 1962. Holbrook is as responsible as anyone for the term lumberjack. It was never used in the Maine woods. The man was always a "logger", "lumberman", or "chopper". Holman Day also used the term in his novels about the Maine woods, but the usage still sticks in the throat of a Mainer, and one seldom hears it, except to show one's ignorance, or to impress the casual summer tourist. For a contemporary view, New York *Evening Post*, May 22, 1865; December 13, 1866.

[4] Table compiled from Reports of the Surveyor General, Penobscot County, and James Defenbaugh, *History of Lumbering*, Chicago, 1907. Surveyors were first spoken of in the Maine statutes in 1829. See *Public Laws, 1829*, chapter CCCCXLVII, 1218-9. At first these officials were optional, later they became mandatory where lumber was shipped. Their duty was to insure quality control. Much of the pine surveyed was two or three years old, and dried to be used for box boards, after 1880. These figures do not include any lumber used within Penobscot County, nor do they reflect lumber used for pulp purposes. They are not the current years cut, as some have supposed, but rather are only the amount of sawed lumber put up for sale in a given year. July, August, and September were the biggest months, and although lumber was surveyed all year round, the great bulk of it was handled when the river was open for navigation. The monthly figures for 1872, the greatest year, follow.

Lumber Surveyed, 1872, Penobscot County

January 1 to June 1	39,830,617	
June	38,257,975	The river closed
July	33,233,722	to navigation
August	36,233,287	December 7, 1872
September	34,719,052	
October	40,161,380	
November	25,017,036	Figures for other
December	5,010,580	areas in the state,
Total	246,453,649	see the appendix.

[5] *The Industrial Journal*, October 26, 1888. Allan Eric, "Life on the St. Croix." This was the pseudonym of their regular correspondent from Boston. The article is a description of a cruising expedition in Washington County, which included "Eric." It was not done very scientifically even at this date. The cruisers climbed trees and viewed the land, and traversed the areas of interest. They were much more interested in setting their logging roads, landing areas, and camp sites than anything else.

[6] Interview, Harold Noble, Topsfield, Maine, July, 1962; Caleb Scribner, Patten, Maine, August, 1963.

[7] There was ordinarily no redress for a bad cruising job. The exception was on the state lands which were to be sold primarily for timber. For instance, the survey on NW½ 5R15 and 6R15 was made at a distance, and the report of lumber was inaccurate. Upon appeal a genial legislature made up the difference by refunding some monies paid in to the purchaser. Senate *Document* 11, 25th Legislature, Senate, 1845.

[8] One recipe for molasses cake read, "Take a gallon of molasses, one quart of lard, flour enough to thicken, a handful (two spoonsful) of spices, and three spoonsful of soda. Mix well and bake." *The Industrial Journal*, April 3, 1896. The spices were mixed in Bangor by the Three Crow Spice Company and sold in five pound tins to the lumbering concerns.

[9] See especially George Smith, "King Axe", *St. Croix Observer*, Vol. 38, no. 7, January-February, 1963.

[10] Larry Gorman, perhaps the most famous Maine logger, was a poet and lyricist of some ability. A recent biography discusses this side of the Maine woods. Edward D. Ives, *Larry Gorman: The Man Who Made The Songs*, Bloomington, 1964.

[11] *The Industrial Journal*, October 2, 1891, Allan Eric, "Diversions of the Lumber Camp."

[12] This ceremony is described in *The Loggers*, Boston, 1870, 17 and Atkinson, *Grand Lake Stream Plantation*, 14-5.

[13] *The Northern*, Vol. VI, no. 1, (April, 1926). Hugh Seavey, "The Bucking Board". This is an excellent article describing camp life in the late 'eighties. Jill-poke is a Maine expression meaning to mess anything. A jill-poke is one who is usually mixed up most of the time. The ceremony of shaving the green hands must have been confined to Washington County, as I have only found it described from that area.

[14] Seavey, *op. cit.*, and interview Harold Noble, July, 1962.

[15] "To pay the bitters" was another Maine expression of that time. It meant to buy the whiskey. The standard invitation was, "Do you smile?" See F. H. Eckstorm, *The Penobscot Man, Bangor*, 1903, 1926.

[16] The best, and most easily available description of a logging camp, this one on the Androscoggin River, Bemis Stream, near Upper Richardson Lake, is in F. L. Barker, *Lakes and Forests as I Have Known Them*, 54-9. Other good contemporary accounts are *Drew's Rural Intelligencer*, March 31, 1855; Phillips *Phonograph*, December 9, 16, 23, 30, 1887; January 13, 20, February 10, 24, March 9, 1888, a continuing series of letters on the Rangeley area camps. As evidence for the singing statement the one really great logging song is "The Jam at Gerry's Rock." I have never seen reference to this song in the contemporary literature of the time. The one song which does appear, and through-

Life in the woods

out the period, in the popular literature is "We'll Roam the Wild Woods Over." The first and last verses follow. For Gerry's Rock and the other songs see F. H. Eckstorm, *Minstrelsy of Maine*, Boston, 1927.

The Loggers Boast

Come all ye sons of freedom through-
out the State of Maine,
Come all ye gallant lumbermen, and
listen to my strain;
On the banks of the Penobscot, where
the rapid waters flow,
O! we'll range the wild woods over,
and a-lumberin' will go.
And a-lumberin' we'll go,
so a-lumberin' will go
O! we'll range the wild woods over,
While a-lumbering we go.

When the white frost gilds the valleys,
the cold congeals the flood;
When many men have naught to do to
earn their families bread;
When the swollen streams are frozen,
and the hills are clad with snow,
O! we'll range the wild woods over, etc.

And when upon the long-hid soil the
white Pines disappear,
We will cut the other forest trees,
and sow whereon we clear;
Our grain shall wave o'er valleys rich,
our herds bedot the hills,
When our feet no more are hurried on
to tend the driving mills;
Then no more a-lumbering go,
so no more a-lumbering go,
O! we'll tell our wild adventures over,
And no more a-lumbering go!

[17] Portland *Eastern Argus*, December 8, 1884. Comment from a lumberman in Athens, Somerset County. Not all agreed as one cited such amenities as beef, pies, and doughnuts. Farmington *Chronicle*, April 4, 1875 (Dead River camps).
[18] Bangor *Daily Commercial*, January 11, 1872. "Sol Rounds" discussing changes on the East Branch, Penobscot River in the last decade.
[19] *Ibid.*, and Portland *Eastern Argus*, November 28, 1887.
[20] Bangor *Daily Commercial*, February 6, 1886, *The Industrial Journal*, February 3, 1888, editorial on the consumption of beef in the logging camps.
[21] Roland F. Gray, *Songs of the Maine Lumberjack*, University of Pennsylvania Press, 1926, 52.

[22] Bangor *Daily Commercial*, June 4, 1873.

[23] Portland *Eastern Argus*, January 1, 1883, letter dated Calais, December 30, 1882. On January 13, 1883 the *Argus* reported five hundred horses entering at Houlton. The *Commercial*, February 8, 1888 said that these teams drove the pay for teams down from $45 a month to $30 in the Winn area, another tannery town.

[24] *Ibid.*, February 17, 1886. The same article attacked the Swedes for the same reason. The Presque Isle *North Star*, February 20, 1883 covered the same ground as they said that the Provincemen established prices and drove out the small local operators. The money in the area in circulation was all New Brunswick money. "The entire lumber business of Aroostook . . . pays tribute to St. John. . . ." Also Phillips *Phonograph*, February 11, 1887; January 20, 1888 on provincemen, horses, and supplies.

[25] Portland *Daily Argus*, March 1, 1883. It ought to be remarked that newspapers were using this to belabor their Republican opponents, and in the *Argus'* case to help foster conservation arguments, of which more later.

[26] These comments come from Caribou in a letter from "R" printed in the Bangor *Daily Commercial*, October 28, 1884. Also see *The Industrial Journal*, January 16, 1885, on the Prince Edward Islanders.

[27] Bangor *Daily Commercial*, December 5, 1882. One house brought in 484 "buckwheat eaters" that fall. Also *ibid.*, February 2, 1886, an interview with the proprietor of a Washington Street hotel who had sent 286 men in from his hotel, 250 of them from P.E.I. On this same point see *ibid.*, April 7, 10, 1886 and Portland *Eastern Argus*, December 27, 1886. This pattern of work was established by Yankees, see Franklin *Reporter*, November 16, 1843 for an early account.

[28] Bangor *Daily Commercial*, October 22, 1887, their Caribou correspondent.

[29] *Ibid.*, November 1, 1887, in an editorial. There was more comment on November 19, 1887, and on this day the *Argus* reprinted the views of Portland and St. John businessmen, most of whom were favorably impressed by the idea.

[30] Bangor *Daily Commercial*, November 15, 1887, editorial. On the previous day they had disposed of the idea expressed by their correspondent in note 28.

[31] Same editorial in *Commercial* which discussed reciprocity. They thought reciprocity would hasten annexation.

[32] *Ibid.*, November 25, 1887. "R" from Caribou again, apparently safer in this than in his original idea.

[33] *Report of the Indian Agent*, 1877 (Augusta, 1878) *Public Documents*, State of Maine, 1878.

[34] F. H. Eckstorm, *David Libbey, Penobscot Lumberman and River Driver*, Vol. IV of *True American Types*, Boston, 1904, 47-8.

[35] Bangor *Daily Commercial*, November 27, December 22, 1875. Presque Isle *Sunrise*, December 27, 1867; January 3, 1868, quotes $15-$20 month. Farmington *Chronicle* December 16, 1875.

[36] This was due to the extreme depression and the fact that the men were hired in January and February, Bangor *Daily Commercial*, January 6, 1879.

Life in the woods 31

[37] See *ibid.*, January 13, 16, 19, 22, 29, March 2, May 15, 16, 1877. There was also a souphouse in Portland, run by the WCTU, with the soup at 5¢ a pint. See *ibid.*, January 22, 1877. The souphouse in Bangor fed an average of 500 per day all spring. Meetings were held and resolutions were passed, but no work was to be had.

[38] Portland *Eastern Argus*, January 7, 1879. Wages this winter were ten to fifteen dollars a month. Bangor *Daily Commercial*, December 16, 1882, January 15, 1883, and Portland *Eastern Argus*, November 15, 1883 for the higher figures.

[39] Bangor *Daily Commercial*, December 27, 1883; November 22, 1889; October 19, 1894, and November 2, 1894.

[40] *Ibid.*, January 22, 1894. This one was supported in part by private charity.

[41] Portland *Eastern Argus*, April 19, 1876, quoting the *Northwestern Lumberman*, April 8, 1876.

[42] For the coming of temperance in the St. John woods see Hannay, *New Brunswick, op. cit.*, II, 174-5. In an interview a man who has known the woods throughout the twentieth century said, "No rum in my time. My father told of loading provisions, though, with the first item a sixty gallon barrel of rum." Interview, Harold Noble, Topsfield, Maine, July, 1962. Temperance was attempted earlier. Bangor *Whig and Courier*, March 6, 1834 for an abortive discussion.

[43] Lucius L. Hubbard, *Woods and Lakes of Maine*, Boston, 1883 (Second Edition), 181.

[44] In the Bangor Public Library, shelved with the papers of Joseph Porter, is an account book of Rideout and Brothers, grocers of Old Town. Page after page is an account of goods shipped upriver and nearly all these goods are consigned to lumbermen on the Passadumkeag, Mattawamkeag, or Penobscot Rivers, during the period 1865-1875.

[45] For the early history of these steamboats see David Norton, *Sketches of Old Town*, Bangor, 1878, 41-3. We have fair records for the year 1864 for one operator, Joseph Porter, of Burlington, who operated on the Passadumkeag. The steamers *Aroostook*, and *W. M. Ray* hauled his goods from Milford to Passadumkeag, after they had come from Bangor on the railroad. His consignments consisted of:

August 25	1 bundle
September 5	1 hogshead
	12 barrels
	5 bags
	2 shafts
	4 wheels
September 10	29 barrels
September 16	1 barrel
September 19	20 bags
September 20	1 stove
	1 roll zinc
	3 joints
	1 elbow pipe

September 21	1 table
	1 bundle chains
	2 boxes
September 22	17 barrels
September 29	1 hogshead
September ??	13 barrels
October 6	1 keg
	4 boxes tea
October 12	10 bags
	8 bundles of fish
October 20	1 bag
	1 bundle of fish

These were all on one waybill dated October 20, 1864. The total of goods came to 118,000 pounds, and the freightage was $18.86. In addition Porter sent up another barrel on October 6 which cost him 26¢, and 80 bags of corn, weighing 864 pounds on October 18, which cost him $1.20. This cost was low, as in 1859, he paid $15.80 to transport 2 boxes, 17 barrels, 2 firkins, and 1 chest of tea, which weighed 4,860 pounds, from Old Town to Winn. See Waybill dated November 18, 1859. These are all in Bangor Historical Society.

[46] Where possible men went in via steamboat, and later across the ice of Moosehead Lake. The Mount Kineo House made a good place to break the trip either for rest or meals. When the men were coming out sometimes as many as 75 were there for dinner, with their horses. At other times the traffic was steady at 9 or 10 each day. The steamers were involved fairly early. See *Maine Farmer*, November 11, 1836. For the Mount Kineo House I used the "Mt. Kineo House Guest Ledgers, 1882, 1883," and "Hotel Register", September, 1885 to August, 1887." These were loaned to me by Duke McKeil.

[47] Sometimes the toter was hard-pressed by unscrupulous individuals who attempted a corner on supplies, as with pork on the St. John in 1880. Bangor *Daily Commercial*, October 9, 1880. The prevailing rates for hauling by 1891 were $6 a ton from Mattawamkeag to Patten, and the toters lost money, and threatened a strike, which did not materialize. The *Commercial* advised them, ". . . just learn to labor and to wait." January 22, 1891. Toters brought the mail which was a boon. *Ibid.*, January 11, 1872. Farm areas near the Rangeley Lakes survived on goods, and toting. See Phillips *Phonograph*, February 26, 1886 (on Kingfield where 46 oxen and 54 horses were hauling lumber out and supplies in); October 15, 1886, February 11, 1887. Later the big firms had some large toting contracts. For instance, F. A. Gilbert signed a contract for the Great Northern Paper Company with Charles Gaudier of St. George, Quebec on September 18, 1902 to unload, house, and tote all logging supplies from St. Francis (Beauceville) to T6R18 for $10 a ton and $3 a month rent for his storehouse. Quoted in *Pittston Farm Weekly*, April 16, 1964.

[48] Bangor *Daily Commercial*, December 26, 1890.

[49] Portland *Eastern Argus*, November 15, 1875; Bangor *Daily Commercial*, November 9, 1875.

[50] *Ibid.*, December 24, 1884, January 2, 1886; Portland *Eastern Argus*, January 22, 1886.

Life in the woods

[51] Hubbard, *op. cit.*, 76, and Hubbard's *Guide to Moosehead Lake and Northern Maine*, 3rd edition, A. Williams and Co., Boston, 1882, 107. "At the great Cross-lake, brings one to Chamberlain Farm, where the most necessary articles of camp fare are usually to be had—at a good round price." Also see Bangor *Daily Commercial*, October 25, 1888. There is an excellent and detailed description by A. G. Hempstead, "A Visit to Chamberlain Farm," *The Northern*, Vol. VII, no. 8, November, 1927, 7, 14-5. Thoreau also was there, see *The Maine Woods*, 303, 327-8, Riverside edition, Boston, 1896.

[52] *The Industrial Journal*, November 19, 1897 has a description.

[53] A good description is in Bangor *Daily Commercial*, September 4, 1875.

[54] The Silver Lake Farm, or Hotel as it was known, opened up after J. W. Palmer came in on the Pleasant River. They had 200 acres of intervale land, with 150 acres in grass, and 50 in vegetables. The farm had three barns, and the land was all ditched and drained. The proprietors put 200 cords of stable manure on their fields in 1887, produced by 50 sheep, 10 horses, and 8 cows. Fifteen hands produced 1,800 bushels of oats, 1,000 bushels of potatoes, and 200 tons of hay. In 1891 they produced 250 tons of hay, 750 bushels of oats, and 600 bushels of potatoes. Bangor *Daily Commercial*, February 15, 1887, and the *Industrial Journal*, November 13, 1891. The Todd farm was really mammouth. It was not infrequent that 300 head of oxen summered there, and it took twenty-five men to do the haying. It was maintained as a depot camp for their operations, with a big slaughter house, toting camps, and central wangan. William Todd, *Todd's of the St. Croix Valley*, Mount Carmel, Connecticut (privately printed), 1924, 18-9, and interview, Harold Noble, Topsfield, Maine, July 1962. Another large farm was the Pittston Farm. Log buildings were erected before 1879. At one time, near the end of its farm usage, the stables could house 128 horses, and 4 cows. The barns held 550 tons of hay. Close to a 100 men could sleep there. The basement held 80 cords of firewood, and a water tower 9,000 gallons of water. As late as 1924 this farm produced 5,255 bushels of potatoes, and 1,116 bushels of oats. *Pittston Farm Weekly*, June 20, 27, August 8, 1963. Sometimes large meadows were hayed, and were called "farms". See, for instance, the account in "Account of an Excursion to Mount Katahdin, Maine," by J. W. Bailey in *Silliman's Journal*, reprinted in *Maine Farmer*, June 6, 13, 1837. In 1890 some crews cut 100 tons of wild hay on T10 R9. Cited in *Pittston Farm Weekly*, November 2, 1965.

[55] Loveland raised 3,000 bushels of potatoes, 2,500 bushels of grain, and summered over 30 head of cattle and 30 horses in 1885. Bangor *Daily Commercial*, September 24, 1885. The Depot Farm was at the lower end of Long Lake on the road from Seven Islands on the main river St. John. Hubbard who described it in his *Moosehead Lake*, 123, liked it better, or the prices at any rate, than he did the Chamberlain Farm.

[56] Bangor *Daily Commercial*, December 30, 1884. Some homesteaders on the northern Maine lands depended on the loggers for their survival.

[57] See for particularly detailed comments Springer, *op. cit.*, 56, and J. W. Bailey, *The River St. John*, Boston, 1894, 163-4. The word midge is pronounced mindge or minge in Maine. The Indians called them "no-see-ums."

[58] Bangor *Daily Commercial*, January 26, 1884. The coldest was 46 below at Mattawamkeag, but even Old Town registered 37 below.

[59] Portland *Eastern Argus*, March 19, 1885. The average snowfall at Pittston Farm in the twentieth century was 105 inches, although several years totalled 170 inches.

[60] Bangor *Daily Commercial*, March 10, 11, 1887. *The Industrial Journal*, April 1, 1887. The *Journal* said such snowshoes had been used with success in northern California. I know of no such use in Maine or New Hampshire. Another account of horse slaughter because of the snow is in Wilton *Record*, April 14, 1887.

[61] Bangor *Daily Commercial*, March 11, 1887.

[62] The best descriptions of the effect of this tornado are in Samuel Lane Boardman, M.S., *The Naturalist of the St. Croix*, Bangor, 1903, 56-8, especially the letter printed there from Boardman to Spencer F. Baird of the Smithsonian Institute, October 14, 1869, and in Lewis B. Fisher, *The Story of a Downeast Plantation*, Chicago, 1914, 155-7. This was the end of lumbering in Charlotte, Twp 3, M.D., according to Fisher, 157.

[63] Bangor *Daily Commercial*, March 8, 1884, an excellent description.

[64] *Ibid.*, July 15, 1883 (on the 1872 gale), and March 18, 1885 on the salvage of the 1883 timber.

[65] Letter A. D. Young to all lumbermen, September 29, 1885, printed in *ibid.*, October 1, 1885. An earlier smallpox epidemic occurred in 1847-8 near Moosehead and in Bangor. Farmington *Chronicle*, February 3, 1848.

[66] Portland *Eastern Argus*, January 23, 1874. Earlier some had been terribly destructive of loggers camps and supplies. Farmington *Chronicle*, June 21, 28, 1845 on Cold Stream Town.

[67] "The Burning of Henry K. Robinson's Camp in 1873". See *Minstrelsy of Maine*, 48-51.

[68] Bangor *Daily Commercial*, March 29, 1872. Another early venture, although little came of it was in 1835. Penobscot *Freeman*, February 16, 1835. W. R. Cross to ? Chandler, March 19, 1874 in my possession, describes vividly a missionary trip to the loggers.

[69] Minnie Atkinson, *Hinckley Township or Grand Lake Stream Plantation*, Newburyport, Massachusetts, 1920, 15.

[70] Bangor *Daily Commercial*, September 30, 1892.

[71] Bangor *Daily Commercial*, March 19, 1872. For general remarks on the rivalry of different choppers and teams see Allan Eric, "Winter Life Among the Loggers in the Wild Woods of Northern Maine," *The Industrial Journal*, August 24, 1888.

[72] Bangor *Daily Commercial*, May 13, 1874. For other large loads see *ibid.*, February 19, 20, 1878; April 9, 1879, and February 26, 1886.

[73] *Ibid.*, March 5, 1889. These logs ran 2 and 3 to the M.

[74] *Ibid.*, December 26, 1877.

[75] Portland *Eastern Argus*, April 23, 1872.

[76] Bangor *Daily Commercial*, April 19, 1879. The oxen girt 7 feet and two inches. One team of horses weighed 2,700 pounds, the other 2,200. A famous early team, also girting 7 feet each, lived twenty years and during most of that time hauled masts, spars, and ship timber from Chesterville to the Kennebec. *Drew's Rural Intelligencer*, December 8, 1855.

Life in the woods 35

[77] See e.g., Portland *Eastern Argus*, January 20, 1872; Bangor *Daily Commercial*, January 19, February 3, 1872; February 6, 1873. Farmington *Chronicle*, May 13, 1875.

[78] *Ibid.*, April 2, 1872. The largest one noted in the state's history by the author was one cut at Liberty in early 1834. This pine had a butt log 7 feet in diameter, and it was 330 feet high. The tree scaled 10,619 board feet. *Christian Intelligencer and Eastern Inquirer*, January 31, 1834.

[79] Portland *Eastern Argus*, February 27, 1873.

[80] *Ibid.*, December 10, 1874.

[81] *Ibid.*, February 2, 1881.

[82] *Ibid.*, February 22, 1881.

[83] *The Industrial Journal*, March 17, 1882. They were still getting them here two years later. One spruce 44 feet long and 30 inches at the top scaled 2,080 feet. Bangor *Daily Commercial*, February 23, 1884.

[84] *Ibid.*, January 1, 1889.

[85] *Ibid.*, March 2, 1893. This tree, cut by A. T. Worcester on T24 M.D., made 13 logs, and scaled 3,590 feet. It was a pine.

[86] Portland *Eastern Argus*, February 29, 1884. Even today it is possible to find along the banks of the smaller streams remains of giant four and five foot diameter butt logs which were just too large to drive. Even though these logs look today like giant humps of moss, the knots have not rotted so it is possible to see how huge they must have been. I have seen pine in the Maine woods which would go five feet across the stump, and spruce which would go four, within the past decade.

3

LOGGING THE MAINE FORESTS, 1860-1890

Since the middle of the 1830's large-scale lumbering had been the chief enterprise in Maine. The profits had been large; the future had nearly always been bright. Few would have thought that anything else could be true. It was thus unexpected when a Maine newspaper, located in the heart of the business, encouraged its readers to "have two strings to your bow." The editorialist, peering into the future, went on to urge diversification,[1] as he said:

> Bangor is too much dependent upon one interest for her prosperity. Should our lumber business, from some unseen cause, meet with disaster in any one year, our prosperity would be most affected.

This prophetic judgment was ignored. Why should such a profitable enterprise be curtailed by the relocation of funds? Profits were there to be had simply for the taking. Since the beginning of the Civil War Maine lumbermen had been most prosperous. The yearly cut of the state was in the neighborhood of a billion board feet of lumber, and its value was always close to $20 million.[2] With the minor embarrassment of a shortage of river drivers in 1861 and 1862 the prosperity continued unabated until 1873. Thereafter lumbermen found themselves in increasing competition with those who had gone west before them[3] and with the increasing magnetic pull of the western country. Partially because of the economic impact of this great depression, and partially because of scientific developments, the end of the depression found a revolution taking place in the Maine woods. This was the revolution created by the shift to pulp and paper manufacture rather than ordinary logging operations. This chapter will review the history of these thirty transition years in order to ascertain the general state of the business just prior to that revolution. An economic analysis will follow in later chapters.

Until the depression, money was to be made, at least when the weather was favorable. For instance, as late as the end of January in 1870 it appeared to be a bad year. One newspaper was extremely worried.[4]

> It is estimated that a foot of snow in New England just now would be worth $30,000,000, or $2,500,000 an inch. Millions of feet of logs are lying in the forests . . . , awaiting the fall of enough snow to enable them to be hauled to market. The oxen and teams . . . are eating their heads off in the stables; teamsters are idle, and the owners of the lumber are troubled to get the money they need to meet their obligations with, because their prosperity is not saleable where it lies.

It snowed heavily three days later, and more fell in February. By the eighth of February the snow lay eighteen inches deep in Oxford County and the season was saved.[5] The lumber all came to the mills.

This was a boom year, and so were the next two. However, 1872 was the last of the boom years. When the loggers came out of the woods the state cut was estimated at close to 900 million feet.[6] These logs did not sell well and state lumbermen in desperation issued a call for a convention to meet in Bangor in 1873. The call for unity, signed by ninety-five different firms in Maine and New Brunswick, urged discussion of measures to assist the trade.[7]

When the convention met, nearly every lumberman of consequence in the northeast was in attendance. A standing committee, representative of every sawmill city, was appointed, and charged with the duty of setting a standard scale and survey and maintaining a standard price. None of their attempts were successful. The depression of the seventies had caught up with the industry. Lumbermen had many logs on hand as well as much lumber. The table, next page, shows the state of the industry on the eve of the depression.[8]

After 1872 it is probable that the total state cut did not rise above 700 million feet, with the possible exceptions of the relatively good years 1881-1883, until the pulpwood revolution in the nineties. After this time, until World War I, and the opening of the Panama Canal, the state cut was 900 million or a billion board feet a year. The biggest year was 1909, when the state cut was officially estimated at 1,111,565,000 board feet.[9]

Weather was always a factor. Thus, in the era of smaller cuts, some winters were better than others. Eighteen seventy-four and 1875 were reasonably good years, although 1876 was poor.[10] In 1877 many lumbermen did not operate in the woods, "as they can purchase what logs they want at a cheap rate."[11] The business was still estimated to be the largest in the state though, even greater than cotton textiles, and worth then about $12,000,000 a year.[12] The seasons continued unfavorable in 1877 and 1878, except in the St. John area.[13] Prices ruled low, provisions were high, and the snow did

Logging the forests

Lumber Industry — Maine and New Hampshire — Spring of 1873

River	Old Logs 1871-2	New Cut 1872-3	Total Logs on Hand	1872 Cut	Sawed Lumber on Hand Spruce	Pine	Hem.	Total
			(Millions of board feet)					
Kennebec	35	65	100	150	8.6	4.7		13.3
Penobscot	60	115	175	225	4	6	4	14
Calais	16	65	81	100	5.5	3.3	7	15.8
Dennysville	1	9	10	22	2	.1	.7	2.8
Machias	8	23.3	31.3	33	2.5	2.5	.9	5.9
Ellsworth	12	28.5	40.5	60		.3	2.2	2.5
Cherryfield	5.2	14	19.2	23	.7	2.5		3.2
Androscoggin	5	41	46	55	1	.4		1.4
Portland		89*	89*	114*	9	1	.3	10.3
St. John in Maine			75**	100**				
TOTALS	142.2	449.8	667	882	33.3	20.8	15.1	69.2

*Portland figures are lumber brought by rail from Quebec and/or New Hampshire.
**St. John figures are an estimate based on Portland *Eastern Argus,* July 10, 1874.

not fall in the expected quantities. By 1879 the depression had reached its lowest point. Provisions were lower than for forty years past, and lumbermen went to the woods for $5 a month. The cut was larger than for some years as a result throughout 1879 and 1880.[14] The next two years were marked by drought. Partially open winters, with poor sledding, drove log prices up, and sales were brisk.[15] The demand was greater than the loggers could supply.[16] There followed bad years of 1883 and 1884, marked by serious damage by the budworm, and heavy loss from down timber.[17] Half a billion feet was a good cut by the end of the decade.[18] Lumbering outlooks continued to be described as "rather slim" or "very poor". Some went lumbering to "save their outfits".[19]

This was at the end of the period, however, and the pulp revolution had, in fact, taken place. Between 800 and 900 million were being felled and driven each year in the early nineties. The 1893 cut included 573 million board feet of spruce, 107 million pine, 100 million hemlock, 61 million cedar, and 7 million hardwood for a total of 849 million.[20] Samuel Boardman, who was one of the foremost lumbermen, and a practicing naturalist and conservationist, estimated at about this same time that the average annual cut on Maine rivers would run just over 850,000,000.[21] His figures, which are the best guide now available, are reproduced in the following table.

Average Maine Lumber Cut, 1890-1900

River Basin	Cut in Millions	Remarks
Saco and Piscataqua	30,000,000	25,000,000 pine
Great/Little Ossipee	5,000,000	
Mousam	10,000,000	
Androscoggin	200,000,000	125,000,000 pulpwood
Kennebec	140,000,000	40,000,000 pulpwood
Penobscot	200,000,000	50,000,000 pulpwood
Union	30,000,000	
Narraguagus	36,000,000	
St. Croix	20,000,000	
Machias	15,000,000	
Upper St. John	90,000,000	
Mednuxneag	10,000,000	
Aroostook	36,000,000	
Inland waters	40,000,000	all transported by rail
Totals	862,000,000	250,000,000 pulp (estimate)

The river basins in the southern half of Maine were the first to feel the pulp paper impact. The predominant tree here had always been pine, and the area had been heavily cut. The second growth pine and spruce was useful for the early days of pulp making when the machines were smaller and larger trees had to be split before being ground. The Presumpscot River, which had not had a timber cut since about 1852, found itself being used as a driving stream for poplar coming down to S. D. Warren in Westbrook. The Androscoggin also began to be used for pulpwood drives, both poplar and spruce. However, most of the wood used by these mills came overland by railroad.

Even though the pulpwood was coming, much long log cutting, driving, and sawing continued. There were always small logging operations on the Piscataqua, Mousam, and Saco, but after the price fall in 1872, only those on the Saco amounted to very much. The last

Logging the forests

really big cut was apparently in 1877 when 35 million feet was cut on the tributaries and driven into Saco and Biddeford, about 10 million of it to one firm.[22] The size of the drive astonished one observer.[23]

> One would naturally think that in a few years everything suitable for lumber along the Saco and its tributaries would be cut and used, but it is said that the increase about equals the yearly consumption. In the town of Fryeburg alone, the estimated amount of lumber to be cut is 50 million feet. Much more than this is growing for future use.

This area, around Hiram, Fryeburg, and Bartlett, New Hampshire, continued to produce 3 or 4 million feet each year for the Saco mills,[24] mostly spruce and hemlock.[25] Some of the cut was hardwood, in and around such towns as Denmark[26] and Steep Falls.

Most operations were small and varied. The following table, drawn from the operations of a firm in Steep Falls in 1883, is typical of work done in southern Maine at the end of our period.[27]

Typical Small Firm — Mixed Operations

Item	Destination	Remarks
500 cords stove lumber	Moody's Mills, Steep Falls	
300 cords pulpwood	S. D. Warren	by railroad
250 cords stovewood	Portland	by railroad
650 cords stovewood	Steep Falls parties	
250 cords peg wood	Bartlett, N. H.	by team
200,000 feet long lumber	Saco	by spring drive

Total — 2,350 cords or approximately 1,175,000 board feet

The Androscoggin River was a different matter. Here the trees were larger, and the cut remained high. Many operations were large. H. Winslow and Company of Berlin Falls, one of the predecessors of the Berlin Mills Company, had such an operation in 1861. The firm put in five camps and used 102 oxen, 12 horses, and 187 men on the Magalloway. Twenty teams hauled supplies. An observer noted the emphasis on spruce. "Most of the lumber is spruce for which it seems there is an increasing demand." [28]

Other big operators were the Lewiston Steam Mill Company and later, the Berlin Mills Company. In the years just after the Civil War these firms took out between 7 and 10 million each year.[29] The 1869 cut on the Androscoggin is summarized below.[30]

Androscoggin River Operations — 1869

Employer	Location of Operation	Amount
Lewiston Steam Mill Company	Swift River	11,500,000
Sturdevant and Flint	Magalloway	3,000,000
H. Winslow and Company	Umbagog Lake	2,500,000
Toothaker and Coe	Mooselookmeguntic	7,000,000
W. W. Briggs	Errol, N. H.	2,500,000
Berlin Mills Company	Double Diamond	8,000,000
Miscellaneous small firms		1,000,000
Total		35,500,000 board feet

In the last big years before both the price fall and the pulpwood transformation the cuts were 53,000,000 and 47,000,000. Of this, about 20 million was sawed in Lewiston; the rest in Berlin Falls.[31]

By the time of the shift lumbermen were beginning to move into more difficult terrain. Some of the lumber took two years to drive. New roads were swamped out (cleared for logging purposes) near Mooselookmeguntic Lake to open this virgin area for the ax.[32] Still, there was little else to be done for work in this area. As one sarcastic observer put it, "Logging is the only visible means of support here winters, and pays pretty well for those who go into it with both eyes open and their eye teeth cut." [33]

In the eighties more and more of the cuts, which began to increase in size, went to pulp. The biggest operators, the Berlin Mills Company and the Lewiston firms, still cut long logs, but they too handled much pulp. By 1890 these firms, cutting between 25 and 30 million feet each, annually, were putting half at least of their cut into pulp stock.[34] A large amount of timber remained, but it would go nearly all into pulp stock. In 1895 the land owners on the river estimated that there remained 3,600,000,000 feet in Maine and 900,000,000 feet in New Hampshire.[35]

Small scale lumbering did continue all over the area during the same time. Much of it was not driven, but went on the railroad to its destination. Perhaps as typical as any was an operation at Waterford in 1874, oak to one mill for staves, pine and hemlock to a second for

Logging the forests 43

boards, birch to a third for clothespins, and poplar to a nearby pulp mill.[36] Ship timbers were cut,[37] and firewood was hauled.[38] In Portland itself at the very end of this period one man got 200 M of good pine.[39] A typical railroad operation occurred at Bethel near the end of the period. A firm contracted to cut 3,000,000 feet of spruce, and 1,000 cords of birch and poplar. Their crews totaled 75 men, 50 oxen, and 34 horses. Supplies came in over the Grand Trunk Railroad and the lumber went out over it. The operation was successful, and the lumber had all left the woods by the end of January.[40]

Operations on the Kennebec watershed were usually larger than on the Androscoggin. Sixty-five million feet was a small year and 100 million was an average year in this time. Nearly all of it up to the early nineties was long lumber; little pulpwood was cut until the very end of the period.

The big firm in the early years was that of Abner and Philander Coburn. The firm had had its start in the 1830's, but other interests were claiming the attention of the Coburn family by the Civil War. They owned about 400,000 acres of land on the Kennebec and during the Civil War years were still getting about 24,000,000 feet a year on their own lands. A typical operation for them in this time was that of 1866. The firm used 228 oxen, and 152 horses as well as 860 men to get their 25,000,000 feet.[41]

The Coburns soon sold their land to the textile firm of Sprague and Sons of Rhode Island and Augusta, Maine. The Coburns took a mortgage on the lands at the sale, and when Sprague failed in 1873, thus helping to bring on the depression of the 1870's, the Coburns were forced to foreclose. After the foreclosure the parent firm, The Coburn Land Company, issued orders to their men to come out of the woods.[42] This act triggered the hard years in the Maine woods as many operators tended to follow the Coburn example. Thereafter the Coburns sold stumpage and did little operating themselves. The owners of the firm died and their wills were contested, and the contest hung on for a long time in Maine courts. Eventually much of this land fell into the hands of the Great Northern Paper Company and Hollingsworth and Whitney (now Scott Paper Company).

The table which follows gives some indication of the Kennebec cut in the years after the depression. These figures are not the Kennebec drive for which figures are given later, but the estimated cut by the operators themselves.[43]

Kennebec Operations — 1875-1894

Year	Amount	Largest Operation	Remarks
1875	120,000,000		
1876	50,000,000		
1877	not known	25,000,000	European deal
1878	78,000,000	12,500,000	European deal
1879	65,000,000		
1880	79,000,000	8,000,000	
1881	not known		
1882	not known		
1883	125,000,000		
1884	135,000,000		
1885	90,000,000	9,000,000	virgin pine
1886	125,000,000	20,000,000	pulpwood
1887	105,000,000		
1888	100,000,000	18,000,000	
1889	102,000,000		
1890	not known		
1891	124,000,000		
1892	111,500,000	7,800,000	
1893	111,000,000		35 million of
1894	110,000,000		this was pulp

As prices fell so did wages. Lumbermen went across the border and hired French Canadians to harvest the crop.[44] One newspaper made a gloomy survey of the river prospects:[45]

> Lumber rules very low, the demand for lumber is limited, the foreign demand for deals having fallen off greatly, but the supply of logs in the boom is not large, so there must necessarily be enough cutting done to keep the mills along as they cannot stop without disastrous sacrifices. Labor, hay, grain, flour, pork, and all sorts of supplies that go in to make up a lumbering operation are plenty and cheap, and lumber can be got in at a very low figure.

Some firms were operating at such a scale in the north country during these depression times that they surveyed and swamped out a tote road from Moose River to Lake Megantic in Quebec. The Lake Megantic farmers, apparently also hard-hit by the depression, assumed half the cost of the survey and the cost of the road. The lumbermen estimated that it would save them from $12 to $15 a ton in transport cost over the older route from Skowhegan to The Forks.[46] Supplies were always a problem in this north country. The Bangor and Piscataquis Railroad helped somewhat; during October of 1880 it hauled 450 tons of lumbermen's supplies for the Moosehead region.[47]

Logging the forests

By the middle eighties the cut on the Kennebec was beginning to reflect the pulpwood revolution, and the hard times of the seventies were nothing but a harsh memory. This situation was not true in the area which Maine natives call "downeast". That area, the watersheds of the Union, Narraguagus, Machias, and St. Croix Rivers, felt the depression most of all. The land is among the poorest in the state, and it was heavily cut before the Civil War. Generally speaking, Washington County, which covers most of the area, has a heavy growth of hemlock, and it was quite dependent on tanneries.

As Davis has said,[48] the lumber industry on the St. Croix reached its peak between the panic of 1857 and the crash of 1873. In 1866 there were 500 teams and 2,000 men in the woods,[49] and in 1868 Alexander "Boss" Gibson, the great New Brunswick lumberman, sent 1,500 men and 300 teams up the river from Calais. Although he took $90,000 worth of supplies with him, he still intended to purchase all of his hay and oats from local farmers, which must have been a boon.[50]

Even in 1870 it was estimated that 3,000 men and 1,000 horses were in the woods in Washington County,[51] In 1872, the last good year, 35,000,000 feet came down the river to Machias, and 55,000,000 to Ellsworth.[52] In 1873 the cut was halved,[53] and by 1876 all the rivers in the area had much reduced cuts, some said the smallest amount in thirty or perhaps fifty years.[54] Lumbering operations were described as "generally unsuccessful".[55]

The worst year of the century for this section of Maine was 1879. Wages were as low as 75¢ a day for bark haulers, and men worked as choppers for $9 a month. A correspondent reported these facts with this comment:[56]

> Much suffering is reported among the poorer classes. Some families have been without shoes or stockings all winter, being unable to even provide food. It is the hardest winter ever experienced in this vicinity.

This marked the nadir of the depression, however. In 1880 things began to look up for the lumberman, the Washington County cut was estimated at 129 million feet and 35 million came down the Union River to Ellsworth.[57]

The cuts never were to be very big again, with perhaps 100,000,000 feet being the average over the entire county during the next decade. The figures are scattered, and the operations were usually small, each firm taking out a million or so feet.[58]

Machias River Cut — 1883-1892

Year	Amount	Remarks
1883	29,000,000	440 men, 235 animals
1884	15,000,000	
1885	23,000,000	7 million from T36, M.D.
1886	not known	
1887	not known	
1888	13,500,000	
1889	not known	
1890	18,000,000	362 men, 174 animals
1891	not known	
1892	15,500,000	259 men, 127 animals

The St. Croix was better off than the Machias, Narraguagus, and Union Rivers, but here the lumbering was all controlled by a few large firms, from both the Maine and New Brunswick sides of the river.[59] The river produced 62,000,000 feet in 1883, and 52,000,000 in 1884. Between a fifth and a fourth came from New Brunswick lands.[60] The cuts remained about the same, and by 1893 had stabilized at about 30,000,000 feet each year.[61] The pulpwood revolution had not come, and the logs had stopped. A chronic depression had set in "down-east".

When one speaks of Maine lumbering the normal reaction is to think of Bangor and especially the Penobscot River. Although the river itself is not as long as the St. John, nor does it spill as much water into the Atlantic, with its tributaries the Piscataquis, Mattawamkeag, Passadumkeag, East Branch, and West Branch to name only the more important, it does make up a very large drainage system.[62]

As this river system drains an area which is heavily wooded, with some pine, but mostly spruce, plus the usual intermixture of hemlock, hackmatack (juniper-larch-tamarack), and such hardwoods as beech and birch, it has always been a premium area for lumbering. Because of these great resources, and the relative ease of transport, the ten year depression after 1873 had generally less effect on Penobscot waters than on any river system in the Maine woods. Even so the cut was diminished during this time. From 1860 to 1872, with the exception of the first two years of the war, the cut ran big, seldom much under 200,000,000 feet a year, and in the two best years of the century 1866 and 1872, it neared the 250,000,000 foot level. The lowest total was in 1861 but this was a result of losing part of the drive from lack of men, as the war tolled woodsmen away to military life.

Logging the forests

After 1872 the cut was never quite as high, although 200 million feet was again reached. The depression froze out many of the smaller operators, and the heavy toll taken of the woods before also served to limit the cut. The trees were smaller and newer and more difficult areas were opened to the ax. The amounts of capital necessary for these new operations militated against huge cuts often, and it was not until labor saving devices could be introduced into the woods in the twentieth century that the giant operation returned.

It is not until the coming of the depression that extensive information is available for the Penobscot, as lumbering before was a way of life and needed little comment. When the economic pressure began to tell on the operators then there is more comment as one method or another was attempted to alleviate the situation. Lumber was always king, and nearly every town was completely dependent on the business. As a reporter remarked late in the period,[63] about one such Penobscot town:

> Patten is the center of extensive lumbering operations. The lumber cut in that vicinity will be very much above the average and probably never before have the woods thereabouts been the scene of so much business as now.

He elaborated his point with the following evidence:

> It is estimated that 350 horses and 4,050 men are engaged in lumbering in that vicinity, while 100 horses and 40 men additional are employed in hauling supplies . . . . About 500 tons of hay and 40,000 bushels of oats will be required to feed the horses and 110 tons of provisions will be necessary to supply the men. To this 50 tons will have to be added for spring driving purposes, besides boats, cant dogs, etc., making upwards of 1,400 tons in weight to be moved through Patten before the snow leaves the ground. It will require nearly $125,000 to pay for labor, $90,000 for hay, grain, and hauling and at least $25,000 for provisions for the men, making a total of $240,000. The distribution of such a large sum of money should result beneficially to all classes.

His last laconic line described accurately the way it was when times were good. This story is typical of and can be duplicated for all the other small towns in the region. It is not hard to understand why everyone was worried when the lumber cuts began to diminish.

Operators on the Penobscot were usually people cutting on stumpage contracts. Most contracts were between one million and five million feet. These operators lived usually in Bangor, Brewer, Old Town, and Orono, and nearly as can be determined, did their business on borrowed money, often, but not always from the landowner. When one of these money lender-land owners, Major J. L. Smith, of Orono, ran as a Greenback candidate for Governor in 1879, his opponents attacked him for being a "shylock" in his money lending. His debtors, who included nearly every Penobscot operator of importance, signed a communication to the newspapers listing their total borrowings and attesting to his good faith, honesty and integrity.[64]

Operations like those hinted at in the newspaper discussion of the money-lending side of the business meant much to the towns of these operators. When the depression occurred these towns suffered, and they welcomed any evidence of resumption of business. In describing such resumption in 1877 one correspondent remarked:[65]

> People of Orono are rejoicing at revival in the lumber business. Several of our foremost lumbermen have remained inactive during the past three or four winters, mainly because the products of an operation—conducted in the most economical manner—would be insufficient to meet the expense incurred by such investment, but now, as the scarcity of logs will necessarily increase the demand another year, our former passive operators are well-pleased with the outlook and are already sending men and teams up the Penobscot with a vengeance.

This revival was confined primarily to the West Branch, and although the totals did not reach earlier figures, they were an improvement. The West Branch totals appear in tabular form in the chapter on river driving. The story of that tributary properly belongs in a latter part of the story, after 1890. With steamboats on the lakes, and tramways and sluices in the woods to move the logs the cut could skyrocket to the old-time goals of 55 and 60 million a year on this branch. Much, if not most of this went for pulp, though, and it was, in many respects, modern lumbering.

The East Branch of the Penobscot was a more difficult place to lumber, and to drive, and the cuts were less. Twelve million feet was a good total during the peak of the depression, and it was not until the mid-eighties that 25 million was reached.

By 1887 though the East Branch cut had risen to nearly 40 million feet, although the largest operation was only two million, as M. L. Jordan cleaned up the remnants of the Tracey and Love cuttings. The last years of the period are outlined for this branch in the following table,[66] next page.

On the Mattawamkeag the story is much the same. The cut, if anything, was a bit higher than the East Branch, but it varied greatly, and it was not until the nineties that it grew to be quite large—30,250,-

Logging the forests

East Branch — Penobscot River Operations — 1884-1889

Year	Amount	Operations	Men	Horses	Cattle
1884	18,700,000				
1885	30,000,000				
1886	not known				
1887	37,500,000	28		340	12
1888	not known				
1889	12,500,000		325	108	18

000 feet in 1892 and 22,500,000 feet in 1894.[67] Mills at Kingman and Island Falls took much of this, so it is difficult to estimate the total cut. Figures of the logs driven over the Mattawamkeag Lake Dam, which are some indication of the cut, at least in the northernmost regions, are available.[68] They show a slow and steady decline, with 269,911,692 feet being driven from 1865 to 1913.

Of all the branches of the Penobscot River the one on which the most detailed information is available is the Passadumkeag, one of the smallest in size and drainage area. It is the nearest to Bangor and thus its lumber was cut early, driven to the boom first, and generally reflected the vicissitudes of the industry quite well. For that reason, and perhaps because the newspaper correspondents were better, we have quite a detailed record and thus are able to follow the economic ups-and-downs with some precision. The table reproduces the information available.[69]

Passadumkeag Operations — 1877-1887

Year	Amount	Operations	Men	Horses	Cattle
1877	4,500,000	32	191	154*	
1878	7,850,000	32	185	151*	
1879	4,000,000				
1880	7,000,000	30	211	77	52
1881	8,000,000	36	311	140	42
1882	7,500,000	34	250	100	51
1883	5,500,000	25	273	109	48
1884	10,000,000				
1885	4,900,000				
1886	4,000,000	31	208	103	22
1887	9,085,000	41		126	43
Approximate Total	72,335,000	*Total animals			

This is the traditional story of the Maine woods in miniature. The area was heavily lumbered, but there was little else to do. Today it still shows the extremely heavily cutting. Of much more interest than this rather dismal picture of heavy cutting and small profits are the attempts of the Penobscot lumbermen to open up new areas not before cut for their timber riches.

With the coming of lower prices, townships with heavy stands of virgin timber, most of which had not been cut because of the difficulty of access, or of the lack of good driveable streams, came into prominence. F. A. Reed and Company of Springfield took 45 horses and 18 men near Sandy Stream at the foot of Mount Katahdin in the early seventies. They lumbered several years in the area after expending $3,000 on widening and deepening the brook for driving purposes. The growth was mostly spruce and the two million feet or so that they took each year commanded premium prices.[70] By the late seventies other firms were building dams and widening the Upper Sourdnahunk above Katahdin to get pine which had never seen the axe.[71] This was truly "opening up the swamp". Still, these operations, important as they were in this time of weak prices, did not compare with two which captured the thoughts of many woodsmen in the 1880's.

The first of these was the operation conducted by the Palmer family of Bangor[72] above Gulf Hagas on the West Branch of the Pleasant River. The "Gulf" was a famous place. Precipitous banks from ninety to one hundred and fifty feet high with a five foot shelf two-thirds of the way down looked over boiling rapids some 1,000 feet long. The stream flowed straight, then sharply to the right, and as abruptly back to the left in the course of its travels between the cliffs. A bridge spanned the water for the use of river drivers who were let down with ropes to break jams which occurred on the rocks.[73]

The lumber above the gulf was excellent, but it had never been cut because of the difficulty of getting men, teams, and provisions into the area, and the logs out. Joab Palmer, who had been an East Branch Penobscot and Mattawamkeag operator, decided to open up this section. In 1853 a dam had been built there forty feet high and fifty rods long and in 1853 and 1854 about 350,000 pine logs each year were driven down, but this had barely touched the outskirts. A new dam company was formed, the Pleasant River Dam Company, with Palmer as president, and the other investors, C. H. Bennett, Daniel Hinckley, John Morison, and J. P. Welker, as directors. About $15,000 capital was raised. Palmer and his crew went up in June of 1879 to construct new dams, and to blast out the worst of the obstructions, and straighten out the worst curve. He estimated that their work would open up 100,000,000 feet of virgin timber.[74]

Logging the forests 51

Six weeks of work the first year widened the river from eight feet to twenty-seven feet at the worst place, and the initial crew of sixty men went immediately to felling as soon as the dam was built. That year they operated with two camps, one of seventy men and one of forty-two. Supplies were toted from Katahdin Iron Works. They got about seven million feet.[75]

The next year more work was done on the river, and with 100 horses and 275 men Palmer cut 10 million feet, over half of it pine. By this time he had five large camps and two blacksmith shops. He figured his expenses at about $400 a day, as his yearly supplies amounted to 500 barrels of flour, 200 of pork, 100 of salt beef, 400 bushels of beans, 2,500 bushels of corn, and 230 tons of hay. Oats, and other goods, were in proportion.[76]

Others followed Palmer into the Pleasant River section once it was opened up. Besides the Palmer camps there were eight other operations, with 275 more men, and 50 more horses. These other camps were cutting twelve and one half million feet and nearly half of it was pine.[77]

In 1882 the operation began to attract more attention. Palmer operated four camps, with 160 men and 80 horses. The slopes were so steep that four-horse teams were used, and in many places the crews had to resort to snubbing out the logs. Their supplies were still hauled from the Iron Works by big contract toters. With a month still to go they had hauled 270 barrels of flour, 135 barrels of pork, 300 bushels of beans, 2,000 gallons of molasses, 300 bushels of potatoes, 3 tons of fresh beef, 110 tons of hay, and 5,500 bushels of corn and oats.[78]

By this time the Palmers had blasted out a good channel in the river, constructed a series of roll dams, and extended the one big dam at the head with its gate. It was thought such a hard chance that Palmer paid only 75¢ per M for his stumpage. Other stumpage, of poorer lumber, was $2.50 and $3.00 per M. He never minced any words in conversation, and his comments on this operation, and the resulting drive are worth preserving.[79]

> We have done a fair winter's work. We have cut eight million of spruce, and one half million of pine. We also had one million left up here last year and we shall drive two million feet for Murphy, making nearly twelve million feet. We shall use about two hundred and fifty men driving this spring. I have about fifty going up in this train. My head man stays with them and he will keep them quiet. At Brownville we shall have carts ready to haul them to the woods.

By the time they get to Brownville they won't care what they ride in. We shall have to load some of them as we would barrels of pork. The power to discharge men isn't the secret of governing them. You can't do anything with them unless you get their good will. They must know that you will have just what you wish done. We pay our head man $1,200 a year, and if he is at home a week or two he doesn't lose any time. He knows how to dispose of his men to the best advantage. Yes, they will feel pretty bad when they are getting over this spree. We give them Jamaica ginger. I sent up ten dozen bottles the other day. I have sent up in all as many as thirty dozen bottles. The season is a little later this year. We commenced to drive April 13th last year and this is my first crew. We were forty-three days getting to Old Town last year, ...

In 1883 Palmer took out eight million feet of spruce, 1,400,000 feet of poplar for the Penobscot Chemical Fibre Corporation, and 200,000 feet of cedar,[80] and in 1884, his son, Walter, who operated three camps of 140 men and sixty horses, cut six million feet.[81] By 1885 the Palmers had skimmed the cream of the Pleasant River and had gone back to the East Branch, at Rockabema. Other operators had cut nine million feet on the Pleasant River in 1885. The largest had 2,400,000, and the rest were small.[82] The new dam probably did not produce the 100 million feet as foretold, but it undoubtedly came close to 80 million. Good profits were still to be had in the Maine Woods, but it took capital and men willing to take risks, both of which were often in short supply. But the Palmers were such men.

Another area which lay fallow until men and capital intervened was that watered by the Wassataquoik Stream, about twenty-five miles from the East Branch of the Penobscot River in extremely difficult terrain. Lumbermen had long known of this forest wealth, but it was not until F. H. Todd and Sons, of St. Stephen, N. B., and Calais Maine, purchased the land that anything was done. In 1881 sixty men went in in June to blast rocks, build dams, and generally make the area open to woods operations.[83]

This initial work was of little account and the area was not yet opened. In 1883, however, money was expended and the work began to progress. About $2,000 was expended by the Todds that year, and $10,000 was earmarked for the next. One hundred and twenty men and seventy horses went in to begin the operations. Supplies were a problem. Hay and grain they were able to purchase from farmers, but not the rest, and eventually the Todds bought their goods in Portland, shipped them to Calais by schooner, then to Mattawamkeag via the New Brunswick and Maine Central railroads, and finally in to the

Logging the forests

operations with teams. Todd and Sons contracted with Tracey, Murchie and Love to do the cutting and hauling, seven million feet the first winter. It was estimated that the township involved, T5R10, held 100 million feet of softwood.[84]

In 1884, with the new expenditures, the Wassataquoik began to boom. Tracey and Love (Murchie had dropped out), took contracts from the Todds to handle sixteen million feet that year, and five to six million feet a year for the next five years. The outlay, by this time, was over $90,000.[85]

In 1886 Tracey and Love had three hundred men and one hundred and twenty horses in for the ten million feet they cut.[86] It was too much for the one firm, however. Some of the contract was filled by Jordan and Ballard of Old Town who cut two million in 1887. S. F. Tracey, the only survivor of the original three, cut four million feet that year.[87] Profits were probably not as high on the "Sattacook", as they had been on the Penobscot. The down timber was not as valuable, and the great distance to drive pushed the cost of the lumber too high. Still, this operation, along with Palmer's, mark the end of the old regime on Penobscot waters. Mechanized logging and hauling would prove to be the final answer to the inexorable economics of the northern Maine woods, but it would be two decades before that answer would be available.

The farther north one pushed the more difficult the economic squeeze came. As a result, the cut on the Aroostook, Mednuxnekeag, and above all the St. John, were of interest, if not of as much importance as those we have been discussing. The north country was wild country, and even after "the Aroostook War" was settled, little or no lumbering took place in these waters. It was not until the pine began to disappear, after 1850, on the main New Brunswick rivers that the large operators turned their attention to Maine spruce. The St. John is an extremely long river—Springer called it the "Mississippi of the East"—and it waters an immense amount of land. Starting near Penobscot waters just north of Moosehead Lake, it winds its way north, picking up tributaries as it goes, and then east and south, until it finally reaches the sea at St. John, New Brunswick, some four hundred and fifty miles away. Water was at a premium in these far north areas and the exploitation of the St. John was retarded by the digging in 1846 of the Telos Canal, which diverted some of its waters to the Penobscot.

It took men of great vision to utilize these lumber districts given all the difficulties set in their way. Once, however, the famous New

Brunswicker "The Main John" Glazier had safely taken a drive over the Grand Falls of the St. John, lumbering took an upward swing.[88] During the years of the American Civil War lumbering in this area languished. The northern Maine woods became the haven for "scalleywags" escaping the draft, and with England in an economic depression because of the lack of American cotton, cuts were much less. After the Civil War, though, the cutting of the northern Maine woods moved rapidly, with the cut on St. John-Aroostook-Fish River waters sometimes totalling perhaps 100 million feet a year. Probably, however, the cut averaged nearer 80,000,000. When the panic came in 1873 it at first did not bother the New Brunswickers who came up the St. John to lumber. They advertised in that first bleak winter in Maine newspapers for workmen to come to their city, work in their sawmills and busy themselves sawing Maine spruce into deal for the English.[89]

The Canadians, however, faced severe competition from the Baltic, a competition which began to be serious at about this same time. As a result, by 1874-75 the American depression, which by this time was world-wide or nearly so, began to tell on America's Canadian friends and rivals.[90] Still operations were large; Cunliffe and Stevens cut twenty million feet on the St. John in 1873, and the Aroostook waters cut in 1874 was estimated at seventy-five million feet, worth two million dollars.[91] The next year the depression began to tell. As a correspondent from Fort Kent said:[92]

> The lumber business looks dull and unpromising. Only one party has made any move towards cutting lumber, Cunliffe and Stevens. The rivers have been very dry this season, and so have many of the men who have lumbered on them, and my judgment is that they had better keep dry for the next winter, live on what they have, and not throw it away in lumbering. Many small parties spoil the whole thing by the way they manage the cutting of spruce and making short lumber. Many go on the long drives down the river, the lumber does not sell, and they have to come home and have nothing to live on during the long summer.

This was the period when the State of Maine was attempting to compete with Washington in giving away land to settlers and opening Aroostook County to agriculture, and the dual attack caused the lumbering industry to continue to linger. It was not until 1877[93] that things began to look up. In that year D. S. King and Sons of St. John had 10 camps operating out of Caribou on the Aroostook River, with three hundred and ten men and one hundred and twenty-two horses.[94] Cun-

Logging the forests 55

liffe and Stevens put sixty men and thirty-five teams on the St. John as well.[95] The snow was just right for hauling and lumbering operations were described as "unusually active." [96]

By the 1880's the north country was back near the 100,000,000 foot mark each year. A distinguishing characteristic of the North woods was the size of the operation. Ten, fifteen, even twenty million feet per operator was not unusual on the Aroostook, Mednuxnekeag, and St. John.[97] The biggest operators consistently were Hayford and Stetson of St. John and Bangor. In 1882-3, for instance, they operated on six townships and cut ten million feet of spruce as well as five million feet of cedar. Three hundred men, eighty-six horses and twenty-six oxen did the work.[98] Every man was a New Brunswicker, and their supplies all crossed the line as well. The cut in this time was located primarily in the Fish and Aroostook River areas. Those who went further up stream ran the risk of losing their drive in low water, and often bankruptcy resulted.[99] The far upper St. John would await the coming of upriver mills and later pulp mills for their lumbering. This change was not far away.

Throughout this period the cut remained about the same— twenty-five to thirty million feet on the Aroostook annually, fifteen to twenty million feet on the Mednuxnekeag, two to three million feet on the Bridgewater stream, and fifty to sixty million on the St. John, with about a third of this from the Fish River. These are figures of logs cut in Maine and driven to New Brunswick sawmills. The total cut in northern areas of Maine was ordinarily about 100,000,000 feet, counting the small local mills, and on the St. John all told, on both sides of the line, probably between 160 and 170,000,000 feet. The total for two years when customs records were available bear this estimate out.[100]

Maine Lumber Driven To New Brunswick Mills

Year	Amount
1891	105,280,551 feet
1892	97,694,811 feet

By this time log lumbering was nearing its end. As one moved north the evidence of coming change was perhaps less compelling, but nevertheless an air of inevitability hung over the woods.

NOTES

[1] Bangor *Daily Commercial*, February 2, 1872. Not an entirely new sentiment, however, see Bangor *Whig & Courier*, July 8, 1847; Bangor *Democrat*, February 17, June 2, 16, 1851.

[2] Oxford *Democrat*, March 2, 1866. Their estimate for the 1865-1866 cut was 1,150,000,000 board feet, "a normal year."

[3] Bangor *Daily Commercial*, February 2, 1872.

[4] Portland *Eastern Argus*, January 26, 1870.

[5] *Ibid.*, February 8, 9, 1870.

[6] Bangor *Daily Commercial*, April 13, 1872, quoting the Lewiston *Journal*. *Kennebec Journal*, January 4, 1872 said the Machias was an exception. Also see for the good year Farmington *Chronicle*, March 28, 1872.

[7] Bangor *Daily Commercial*, February 7, 8, 1873. Those signing the call, and the text of the proclamation appear both days. For the poor year 1873, Farmington *Chronicle*, January 13, (on Coburns); 16 (on Androscoggin cut); August 3, September 18, 1873 (on Downeast cut).

[8] Bangor *Daily Commercial*, February 20, 21, 22, 1873 for the meetings. The table derives basically from a report issued by the convention.

[9] Henry Bake Steer, *Lumber Production in the United States, 1790-1946*, Washington, 1948, 11. The Census Bureau estimated the cut, and in the decennial years it was: 1879—580 million; 1889—658 million; 1899—785 million. From 1900 to 1910 the cut was always between 800 million and 1,111 million.

[10] Portland *Eastern Argus*, March 5, 1874; March 10, November 16, 1875; Bangor *Daily Commercial*, August 22, 1874; and *Argus*, April 5, 25, 1876. Farmington *Chronicle*, January 1, 22, 1874; April 1, 1875.

[11] *Argus*, September 6, 1876.

[12] Centennial Address by Ex-Governor Chamberlain to Joint Session, Maine Legislature, *House Journal*, 1877, 212 ff. His statement is at 291.

[13] Portland *Eastern Argus*, March 16, August 2, 1877. *The Kennebec Journal*, December 25, 1877 made the following estimates of the cut.

Lumber Cut of 1878

River Basin	1878 Projected Cut	Remarks
Kennebec	50,000,000	down 10 million
Penobscot	100,000,000	
St. Croix	65,000,000	
St. John	200,000,000	nearly twice 1877
Saco	18,000,000	
Narraguagus	13,000,000	up one half
Union	30,000,000	down one fourth
Androscoggin	28,000,000	
Sandy River	1,000,000	first since 1872
Totals	505,000,000	

Logging the forests 57

Also see *Argus*, February 16, 1878 which describes horses as coming out of the woods. There was not enough snow.

[14] Bangor *Daily Commercial*, January 6, 1879. The $5 men were hired in Calais. Also, October 9, 1880. *Katahdin Kalendar*, March 27, 1880 (heavy operations).

[15] Portland *Eastern Argus*, December 31, 1881; Bangor *Daily Commercial*, December 31, 1881; January 28, 1882. The Penobscot cut was down from 150 million to 100 million, diminishing from 40 to 30 million on the Mattawamkeag, on the West Branch from 50 to 35 million and from 60 to 35 million on the East Branch, Passadumkeag, and Piscataquis tributaries.

[16] Bangor *Daily Commercial*, January 28, 1882.

[17] Portland *Eastern Argus*, January 12, 1883 on the drought; Bangor *Daily Commercial*, August 15, 1883 on the budworm; *Argus*, November 15, 1883, March 19, 1885, and Bangor *Commercial*, March 8, 1884 and March 18, 1885 on the hurricane and the blowdown timber. After some estimates on the damage the *Argus*, April 14, 1884 estimated the winter's cut at 268,000,000: St. Croix—20 million, Penobscot—65 million, and the Aroostook and St. John—183 million. This estimate would have been close to 150,000,000 under the norm.

[18] Portland *Eastern Argus*, November 28, 1887.

[19] Bangor *Daily Commercial*, October 5, 15, 17, November 8, 10, 1891.

[20] Robert Treat Whitehouse, "The Commercial History of Maine," in *The New England States*, Boston, 1897, 1221. The poplar for pulp and birch for bolts would probably be in addition. One might assign a tentative figure of 5 and 15 million here. Others agreed, for instance, St. John *Daily Sun*, March 29, 1899.

[21] Samuel L. Boardman, "The Lumber Industry of Maine," in *ibid.*, 1230-41.

[22] Bangor *Daily Commercial*, June 21, 1877 (the date the main drive arrived in Biddeford Pool). The previous year had been a poor one, though. See Portland *Eastern Argus*, February 17, 1876.

[23] Bangor *Daily Commercial*, June 21, 1877.

[24] Bangor *Daily Commercial*, February 13, March 12, 1879; Portland *Eastern Argus*, December 23, 1873, December 30, 1880. Oxford *Democrat*, April 28, 1879.

[25] Portland *Eastern Argus*, January 3, 1881. Occasionally they found some fair pine, as in this operation at Fryeburg, 200,000 with only 5 sticks to the M.

[26] Bangor *Daily Commercial*, February 7, 1882, quoting the Ossipee *News*. Much of the hardwood went for clothespins, caskets, and tool handles.

[27] Portland *Eastern Argus*, April 25, 1883. Pine operations in this period of a million feet were front page news, if the cut was for long lumber. For instance, *ibid.*, January 10, 1887, for one in Biddeford itself.

[28] Oxford *Democrat*, January 18, 1861, letter from John Wilson to editor. The camps were at T5, R2, 3, 4 WBKP, the Dartmouth College Grant and Wentworth's Location, in New Hampshire. They found a good deal of hackmatack.

[29] Oxford *Democrat*, June 12, 1863, May 20, 1864, April 27, 1866, December 18, 1869, January 28, February 11, 1870.

[30] Oxford *Democrat*, June 18, 1869.

[31] Oxford *Democrat*, January 16, 1872, January 14, 1873; Bangor *Daily Commercial*, January 23, 1872; Portland *Eastern Argus*, February 6, 1872, January 13, 1873. One firm, operating at Success Township, New Hampshire, showed how life in the woods was changing in these times. This firm, Hamlin, Twitchell, and Bickford, with 16 men, 8 oxen and 4 horses, cut about 1,500,000 feet each year. They kept a cow in the woods for milk, and served beefsteak, potatoes, and pie as well as the ubiquitous pork and beans. The boss allowed no profanity or card playing, and there was Bible reading and prayer each night before bedtime. Oxford *Democrat*, March 4, April 1, 1870, February 24, 1871.

[32] Oxford *Democrat*, March 24, 1874; Portland *Eastern Argus*, October 20, 1876; April 6, December 7, 1877; February 4, 5, December 25, 1878. Lewiston *Journal*, October 19, 1876; Bangor *Daily Commercial*, December 26, 30, 1878; February 15, 1879 quoting the Lewiston *Journal*.

[33] Oxford *Democrat*, March 30, 1875; letter from "S" describing conditions at Milan, New Hampshire.

[34] Bangor *Daily Commercial*, March 15, 20, November 7, 1879. December 12, 1880 (this is an excellent description of the Lewiston loggers leaving for the woods which was colorful, ceremonious, and liquid.); January 11, 1881, November 22, 1887, March 6, 1888, October 10, 1890; Portland *Eastern Argus*, November 6, 19, 1879; January 21, June 30, 1887; *The Industrial Journal*, April 26, 1889.

[35] J. A. Pike to Austin Cary, December 9, 1895, reprinted in Cary, "Spruce on the Androscoggin," *Forest Commissioner's Report*, III, 1894, 149. In 1887 there were 10 operations, employing 218 men, 100 horses, and 14 oxen in the Rangeley area. The next year 25 operations used 462 men, 170 horses, and 80 oxen, Phillips *Phonograph*, February 4, 1887; February 24, 1888.

[36] Oxford *Democrat*, February 24, 1874. They used 16 men, 18 oxen, and 4 horses.

[37] *Ibid.*, February 29, 1876. They got some spars 22 inches in diameter and 80 feet long. One pine was 165 feet high, and it was still 18 inches through at 105 feet.

[38] Portland *Eastern Argus*, January 1, 1881, commenting on firewood cut at Brunswick.

[39] Bangor *Daily Commercial*, March 9, 1889. The pine was cut at Cape Elizabeth.

[40] *The Industrial Journal*, February 2, 1884. Their supplies included 120 tons of hay, 1,000 bushels of corn, 2,000 bushels of oats, 75 barrels of flour, 20 barrels of pork, 25 barrels of beef, and 50 bushels of beans. Labor costs of the operation were estimated at about $6,000.

[41] Portland *Eastern Argus*, December 6, 1866.

[42] Oxford *Democrat*, November 25, 1873. They printed a letter, Coburn Land Company to Wethern and Parsons, one of their contract toters. The firm was hauling from Skowhegan to T3R4 BKP WKR. The firm is interesting because they used mules for hauling as well as horses and oxen. The letter said, "Suspend further operations. The Coburn Land Co. have suspended operations entirely for the present." They had cut less than before though, Farmington *Chronicle*, January 13, 1873.

[43] Portland *Eastern Argus*, March 11, April 19, 1876; August 10, November 10, 1876; November 12, December 28, 1878; November 7, 8, 17, 1879;

Logging the forests 59

November 13, 1880; March 6, 1883; March 16, 1885; October 29, November 30, 1886; November 17, 1887; Bangor *Daily Commercial* August 15, 1874; March 5, April 12, May 2, November 14, 1879; January 20, 1883; March 16, 27, November 19, 25, 1885; March 26, April 17, December 20, 1886; March 7, 1887; March 23, December 18, 1889; February 16, 1892; December 8, 1893; February 20, 1894; *The Industrial Journal*, March 1, 1895; Somerset *Reporter*, November, 1878, November 1879; Franklin *Journal*, January 17, 1883; Bath *Independent*, December 18, 1886; Gardiner *Reporter*, March 5, 1887.

[44] Bangor *Daily Commercial*, August 15, 1874; Portland *Eastern Argus*, March 11, April 19, 1876.

[45] Somerset *Reporter*, quoted in Bangor *Daily Commercial*, November 4, 1878. This was the worst winter in the state during the period 1860-95.

[46] Portland *Eastern Argus*, November 8, 1879.

[47] *Ibid.*, November 13, 1880. Some new land was opened by these tote roads and the railroad transportation of supplies. In 1884-1885 a strip 12 miles long and 7 miles wide on the South Branch of the Dead River furnished 18 million feet of virgin pine to 5 operators. Bangor *Daily Commercial*, November 25, 1885. Some operations were still fairly large as when 9 camps housed 298 men, 160 horses, and 80 oxen to take out 11 million feet in Stratton, Phillips *Phonograph* March 26, 1886. Most were smaller, 300 cords of birch, 500,000 of pine for instance, as in Carthage the same year. Wilton *Record*, December 9, 1886.

[48] Harold H. Davis, *An International Community on the St. Croix*, Orono, Maine, 1953, 205.

[49] Kennebec *Journal*, May 25, 1866 quoted in *ibid.*, 205.

[50] Oxford *Democrat*, January 3, 1868.

[51] Eastport *Sentinel*, September 14, 1870, quoted in Davis, *op. cit.*, 213.

[52] Bangor *Daily Commercial*, January 23, 1872, on Machias; June 3, 1872 on the Union River, quoting the Ellsworth *American*. The Ellsworth cut had been 35,000,000 in 1871. Kennebec *Journal*, January 4, 1872.

[53] Portland *Eastern Argus*, March 17, 1873, quoting the *American*. Farmington *Chronicle*, September 18, 1873.

[54] Portland *Eastern Argus*, December 2, 1875; March 11, 1876 quoting the Calais *Times* (on the St. Croix), and December 7, 1876, quoting the Machias *Union*, on the Machias, and the comment about "smallest amounts."

[55] Portland *Eastern Argus*, January 29, 1877, quoting the Ellsworth *American*. They quoted them in much the same vein the next year, see *Argus*, March 19, 1878.

[56] Bangor *Daily Commercial*, February 15, 1879; Portland *Eastern Argus*, March 15, 1879 the quotation from "Cherryfield Items." This was the year when a few green hands went to the woods for as low as $5 a month downeast. See *Commercial*, January 6, 1879 for this report.

[57] Portland *Eastern Argus*, October 28, 1879, December 3, 1879 quoting the Calais *Advertiser*; Bangor *Daily Commercial*, March 9, 1880 and April 9, 1881 quoting the Machias *Union*.

[58] Bangor *Daily Commercial*, January 25, 31, 1883; October 29, 1884 quoting the Machias *Republican*, April 1, May 16, 1885; April 9, 1888; December 12, 1889; December 17, 1891; March 17, 1892; Portland *Eastern Argus*, January 24, March 29, April 4, 1883; April 3, 1885.

[59] There is a description of some of these firms in Davis, *op. cit.*, 269-274.
[60] *Ibid.*, Appendix C-17, 321.
[61] *Ibid.*, Appendices C-18, C-19, 321-2.
[62] The best description of the Penobscot, with detailed maps and photographs, is *Water Resources of the Penobscot River Basin*, H. K. Barrows, and C. C. Babb, for the United States Geological Survey, (Water Supply Papers #279), Washington, 1912.
[63] *The Industrial Journal*, January 21, 1887.
[64] Bangor *Daily Commercial*, September 3, 1879.
[65] *Ibid.*, October 15, 1877.
[66] Bangor *Daily Commercial*, December 3, 1884; January 20, April 4, 1885; February 15, 1887; January 26, 1889. Portland *Eastern Argus*, February 17, 1887. In 1892, 17,500,000 feet was felled by 340 men. March 1, 1892.
[67] Bangor *Daily Commercial*, March 21, 1892; February 19, 1894. The largest operations were in the 3-4 million foot class, and most of it went for pulpwood.
[68] These records are in the custody of Ballard F. Keith, of Bangor, the last clerk of the concern, and I am indebted to him for their use. The biggest years are 1867, 1870, 1872, and 1878. The smallest were 1895, 1897, 1901, 1906, and 1912. It is unfortunate that so few of these records are extant. The story could be more accurate if these records were available.
[69] Bangor *Daily Commercial*, December 27, 1877; March 13, 1879; January 26, 1880; January 19, 1881; February 14, 1882; January 18, 1883; December 3, 1884; January 20, April 4, 1885; January 5, 1886; February 15, 1887. In 1893 the cut was just over 7 million feet from 23 woods operations. The next year though only two and one half million feet came out from ten camps. March 17, 1893; February 20, 1894.
[70] Bangor *Daily Commercial*, March 28, 1874.
[71] *Ibid.*, June 21, 1879.
[72] Old Joab Palmer, the founder of the family, is perhaps as famous for his lumbering as he is for the fact that on his tombstone, still to be seen in a Bangor cemetery, is carved his log mark.
[73] The best contemporary description is in *Hubbard's Guide to Moosehead Lake and Northern Maine*, by Lucius L. Hubbard, 3rd edition, A. Williams and Company, Boston, 1882, 151-154. Hubbard was there in 1880. There is a picture of Billings Falls in the "gulf" at 154. The steepest bank of all was named "Hammond Street Pitch", obviously by homesick Bangorians.
[74] Palmer had cut 2 million feet on the Mattawamkeag in 1879. See Bangor *Daily Commercial*, April 16, 1879. This was mostly spruce. On the new company and its plans see ibid., April 29, 1879.
[75] *Ibid.*, August 13, 1879 on the dam building and February 9, 1880, on the first winter's operation.
[76] On the second year see *ibid.*, January 25, 1881, February 21, 1881. They cut 12 million feet on the College Grants this season.
[77] *Ibid.*, February 17, 1881. The other operations were Con Murphy, E. B. Quimby, Robinson and McCusick, N. S. Drummond, Ira Weymouth, Cunningham and Stevens, Howard, and E. L. Chase. See also Portland *Eastern Argus*, February 25, 1881.
[78] The Palmers were good feeders. Their menu featured baked beans,

Logging the forests 61

fresh roast beef and pork, corned beef, boiled salt codfish, salt mackerel, soups, stews, biscuit, doughnuts, gingerbread, pie, and tea. *The Industrial Journal*, March 10, 1882.

[79] The Bangor *Daily Commercial*, April 25, June 8, 1882. The interview took place in Bangor as the Bangor and Piscataquis train prepared to haul Palmer's crew back to drive the winter's cut. The drive came in in 37 days.

[80] Portland *Eastern Argus*, May 10, 1883 and *The Industrial Journal*, May 18, 1883.

[81] Bangor *Daily Commercial*, January 11, 1884.

[82] On the Rockabema cutting, see *ibid.*, January 30, 1885. On the Gulf, see ibid., May 22, 1885. Three million feet of this drive stopped in Dover and Foxcroft to be sawed. In 1890 Walter Palmer, with 125 men and 30 horses put 3 million feet in the Trout Brook, on the East Branch, still a difficult area to work. Bangor *Daily Commercial*, April 5, 1890.

[83] Portland *Eastern Argus*, September 14, 1881.

[84] Portland *Eastern Argus*, December 19, 1883 and *The Industrial Journal*, December 7, 1883.

[85] Bangor *Daily Commercial*, September 27, 1884. Tracey built a house and moved his family into the woods, along with a schoolteacher for the children. *Ibid.*, December 18, 1884. The 1884 spring drive had not come down because of low prices, and it, with the new cut, meant that the 1885 drive was over 17,000,000 feet. Nearly all of it came in, and observers estimated that 100,000,000 feet remained on the township. *Ibid.*, January 12, May 30, 1885.

[86] *Ibid.*, December 23, 1886.

[87] On Jordan and Ballard see the comments by "G. W. B." in *ibid.*, January 15, 1887. On S. F. Tracey see *ibid.*, February 15, 1887 in the detailed summary of the work on the East Branch. Tracey had 50 horses and Ballard 25 for their cut.

[88] For general remarks on this history see Fred H. Phillips, "New Brunswick Lumber Kings," Saint John *Telegraph Journal*, February 13, 1937, and "The 'Main John' Glazier," *ibid.*, February 27, 1937. He was called "the main John" to distinguish him as the premier Glazier. I am indebted to Mr. Phillips for copies of his articles.

[89] Bangor *Daily Commercial*, May 16, 1874. The first "crop" for many settlers was hand-rived cedar shingles. Presque Isle *Sunrise*, November 8, 15, 1867; March 27, 1868.

[90] On this see Clapham, *op. cit.*, II, *passim*, and Portland *Eastern Argus*, March 19, 1873.

[91] Bangor *Daily Commercial*, May 31, 1873 and Portland *Eastern Argus*, July 10, 1874.

[92] Bangor *Daily Commercial*, October 17, 1874.

[93] *Ibid.*, September 18, 1877 letter from "J".

[94] *Ibid.*, December 31, 1877.

[95] *Ibid.*, October 18, 1877.

[96] Portland *Eastern Argus*, October 27, 1877; February 20, April 5, May 9, 1878. The comment is from the May 9th issue. In addition to the other operators Johnson and Phair of Presque Isle had eight million feet on the Aroostook, as did Daniel Dudley. Moore of Fredericton, N. B. had 12 million cut in the Presque Isle area, as well as other small drives on the St. John.

[97] Bangor *Daily Commercial*, October 22, January 21, December 12, 1880; March 9, October 21, 1881; February 9, December 20, 1882; February 27, 28, April 4, 23, 1883; April 18, 1891 are all accounts of operations. The budworm gave trouble in 1881 in T16 and 17R7. *Ibid.*, March 9, 1881.

[98] *Ibid.*, December 20, 1882.

[99] Portland *Eastern Argus*, June 16, 1887, for a discussion of the failure of Eaton and Dickey, a Frenchville firm, who experienced this difficulty. This failure had a profound, shocking effect on the North Country.

[100] Bangor *Daily Commercial*, December 10, 1891 for the customs figures, and December 31, 1892; March 17, 1892 for further discussion of these figures, and other similar estimates. Three years later the cut was much less though. St. John *Daily Sun*, February 11, 1895.

4

DRIVING MAINE RIVERS, 1860-1890

WHEN THE DRIVE COMES DOWN[1]

Come all ye gallant shanty boys and listen while
    I sing!
We've worked six months in cruel frosts but now
    we'll have our fling.
The ice is black and rotten and the rollways is piled high.
So boost upon your peavey sticks while I do tell you why.

For it's break the rollways out, me boys, and let the
    big sticks slide!
And file your calks and grease your boots and start upon
    the drive!
A hundred miles of water is the nearest way to town:
    So tie into the tail of her and keep her hustling down!

When the drive comes down, when the drive comes down,
    Oh, it's then we've paid our money and it's then we
    own the town!
All the gutters run with whiskey when the shanty boys so
    frisky
Set their boot calks on the sidewalks when the drive
    is down!

But break the rollways out, me lads, and let the big
    sticks slide!
For one man killed within the woods, ten's drownded
    in the drive!
So make your peace before we take the nearest way to
    town,
While the lads that be in heaven watch the drive go down!

Between 1860 and 1890, logs were nearly always floated or driven down the streams from where they were cut to the boom or holding ground where they could be sorted for delivery to the owner. By the end of the period newer methods were being advocated which would avoid the expense of the drive: moving the logs by railroad,[2] and, more importantly, moving the sawmills and paper mills to the logs. The railroads in particular made a strong attempt to prove a case as to economy in the first decade of the twentieth century, but moving the mills to the logs was cheaper, and this method finally won out in the battle of transportation. When this occurred not everyone was happy over the mills moving upriver, and for twenty years a titanic struggle was waged to control the rivers—a struggle which was won in every case by the up-river mills, although not without considerable expense and much ill-feeling. This ill-feeling has not entirely left the Maine woods.

The two most important items in any river drive were first, control and conservation of the water, and second, a crew of men who would do the actual work. The first depended on the dams, which held back the water on nearly every driveable stream for the purpose of insuring a steady supply of water once the ice had melted. Such dams were usually constructed by persons interested in the lumbering in that area. On the brooks, or at the outlets to small ponds, dams were small affairs, ordinarily constructed of logs, perhaps in the form of crib work, and filled with stone, with a gate in the center whereby the water could be released as it was needed. On larger lakes dams were much more elaborate and were often constructed as investments by capitalists. Such a dam was built at the outlet to Mattawamkeag Lake by the Mattawamkeag Lake Dam Company. This dam, originally built in 1860, paid the holders of its fifty shares twenty-six dividends amounting to $432 for each share between 1862 and 1908, when it came into the hands of the Penobscot Chemical Fibre Corporation.[3]

Where a dam was large enough a sluice gate or way was built in it and tolls were charged as the logs passed through. The tolls on the Mattawamkeag Lake Dam were 1¢ per M for each $100 expended in construction of the dam. Smaller dams were called roll dams as on a high pitch of water, logs would float up and over the obstruction without the necessity of sluicing. When a lumberman was planning to operate in a new area he often found it necessary to build small splash dams for his own purpose on the brook drive, and sometimes it was possible to combine the necessary abutments on the curves of his brook with this sort of dam. A splash dam is usually just a few logs put in place to hold back a small amount of water. Once the logs had gone by this place the dam would be taken out to push the logs even more rapidly on to a larger stream. On the smaller driveable streams near old cuttings even today it is possible to see the remains of abutments, splash dams, and in places even larger dams which are no longer needed to control the waters.

The second important item on the drive was the men who moved the logs. Although river driving today is almost universally pictured as a highly romantic occupation, it is difficult to think of more miserable

Driving the rivers

and dangerous work. Often likened by those who did it to driving cattle, it was very similar except, of course, the logs had all the perversity of inanimate objects. Few men could handle all the types of work needed on the drive. The best men were dead water men, men who could handle a pickpole in deep and less turbulent water. The rest of the work was done with the primary tool of the trade, the cant dog. Just prior to our period, in 1858, Joseph Peavey of Stillwater, Maine, had proved himself a friend forever to the river driver by inventing the tool which bears his name. Up to that time drivers had used a long-handled tool with a dog on the end which swung freely in all directions around the handle. This rudimentary lever, usually called a "swinging bitch" or, more genteelly, a "log wrench" enabled men to move the logs if they were hung up, but with difficulty.

Although there is some controversy over this story, the most generally accepted version is that Peavey noticed this difficulty while walking over the bridge at Stillwater as the drive came down, went to his blacksmith shop, and immediately put a pick on the end of the handle and a clasp around the handle to freeze the dog in place without hampering its up and down movement when it was used as a lever to turn or pry the jammed logs. According to the story he then took the new tool back to the perspiring men who pronounced it a great success. Peavey went to manufacturing his new cant dog under the trade name of peavey or peavey stick, and by our period this tool was replacing the older instrument throughout the woods.[4] The tool continued to be manufactured almost entirely in the Bangor area.[5]

Once the ice had gone, sometimes aided by the use of dynamite, the landings were broken and the logs rolled into the water. Then the work began. Seven days a week, from dawn to dark, the men followed the logs to market. Husbanding the water, dragging logs over rocky ground, "carding the ledges" to sack in the rear, standing up to one's armpits in icy water, it was no wonder that the drivers, after eating their fourth meal of the day just after dark, would tumble in head to head in the field bunks which had been constructed at their stopping place. Even with wet clothes and all, sleep came easily. In river driving days most men wore home knit underwear, and it was said that no matter how cold and wet it became one was always warm in home knit drawers. The only change of clothes ordinarily carried was an extra pair of socks pinned to the back of the trousers with a horse blanket pin.

Once the logs had cleared the small brooks where the landings were located, usually the logs found themselves in a pond or lake.

Here the work was as difficult if perhaps not as wet. Moving the logs from one end of the lake to another involved primarily brute strength aided only by a headworks and anchor. Once the logs cleared the brooks they were brought together in a bag boom to prevent their escape. A raft was constructed with a primitive capstan or pin usually made of a large stump or butt stick into which handspikes were driven. In order to turn the pin to take up the slack in the hauling rope the boss might put a large chunk of salt pork in the bottom of the hole in which the pin was inserted. This acted as a lubricant. Men would then row ahead with a batteau, or on large lakes two, with lines passed back to the raft. After throwing the anchor over the men would walk the capstan around, warping and kedging the boom up to the anchor where the process would be repeated until the end of the lake or pond was reached. With a headwind the process was difficult if not impossible and often the work took place at night, after the wind had died down. Lanterns on the raft and bonfires on the shore provided an eerie light for the work. With a following wind, of course, the booms moved along fairly readily.

As the larger streams were reached the work was not so difficult, until finally the drives began to mingle in the main river where the corporation, or the largest owner, usually took over control of the drive to its ultimate destination. Many are the stories about the Big Heater in Ripogenus Gorge, Grand Pitch on the East Branch, Carrabassett Rips on the Kennebec, Rumford Falls on the Androscoggin, and other equally difficult places where the logs would hang up. If the logs jammed it was the duty of men to go out in batteaux, or on the jam itself, in order to clear it. A wing jam might be fairly easy to clear, but a jam in the center of the stream was always a problem. In some places, where the banks of the stream were precipitous, men might be let down with ropes from above to pick the jam, or in later days, to place dynamite to blow it out. Many lives were lost in clearing log jams as the logs might start suddenly throwing the men into the water. Perhaps the most famous logging song of all was written to celebrate such an event, "The Jam on Gerry's Rock." [6]

Moving the batteaux, (in Washington County usually a skiff) with the wangan, down behind the drive to feed the men and provide their sleeping quarters was often a difficult job for the cook and his helpers. A good cook would have his four meals a day, at five and ten a.m. and two and eight p.m., hot and ready for the men. A poor cook, or rather a poor judge of driving, might miss the place and the

Driving the rivers

men would find themselves far from the wangan when night fell. Such cooks probably did not last long on the drive.

The drive moved as rapidly as did the water. In quick water on the brooks in the spring the drive actually moved itself, and the men were concerned with clearing the logs which stayed behind on the banks, or on rocks in the stream. In slower water, in the main rivers, the drive moved at the speed the men wanted it to. They would control the speed, and on low water it was very slow. As the logs got smaller, there was less jamming, but it took longer to move a million feet than before. Good men in a big crew could always get a drive in, or nearly always. The question of driving time was thus determined more by the size of the crew, their ability to do the work, and the speed and depth of the water.

No more difficult or arduous work than river driving beckoned in the nineteenth century. The skills were many and the pride engendered was great. The physical toll was also great. Even today in Maine a hearty eater "eats like a river driver." And, for many who followed the drive to its destination the only solace was the bar and the crib. But, for those who patronized these places, there was little ill-feeling. Individuals in lumbering areas understood the trials and dangers of this work and did not begrudge the men their pleasures, elemental as they might be. As the song, "The Lumberman in Town" has it:[7]

> When the Lumberman comes down,
> Ev'ry pocket bears a crown,
> And he wa-a-anders some pretty girl to find.
> If she's not too sly, with her dark and rolling eye,
>   The lumberman is pleased in his mind.
>   The lumberman is pleased in his mind.
>
> The lumberman goes on
> Till his earnt money's all spent and gone,
> Then the landlady begins to frown
> With her dark and rolling eye, this will always be
>   her cry:
>   'Lumberman, it is time that you were gone,
>   Lumberman, it is time that you were gone.'
>
> She gives him to understand
> There's a boat to be a-manned,
> And away up the river he must go:
> Good liquor and a song, it's 'Go hitch your horses on,'
>   'Bid adieu to the girls of St. Johns,
>   Bid adieu to the girls of St. Johns.'

For the operator the spring was always a time of worry and fret. If he had judged his cut correctly the logs would all be placed on their landings just at the time when the snow began to soften to the point where it was impossible to haul. If not, the struggle to move the logs as sledding became poorer and poorer might injure the horses and oxen of the operation so that all the profit was gone. If the operator had judged correctly, and no heavy spring rain had interfered, the men might be finished with their work and even perhaps have had a week in town to await the melt. The best sort of spring provided a slow gradual melt of the winter cover to give a steady flow of water into the streams, and then the logs would come down with little trouble. Too early a spring would provide an early freshet and the logs might hang for lack of water later on. Such a year was 1874, and observers were acutely aware of the possibilities. As one said:[8]

> Good judges state that many drives of logs on small streams will hang up this spring, and those who work on small capital will be in danger of meeting with embarrassment in consequence.

This poor spring, 1874, was a hard way to start ten years of depression. The winter came late in the fall of 1873, and not much snow fell until mid-January although conditions in the extreme north woods, on the Aroostook and St. John were much better.

Operators were equally fearful in 1878. The thaw came so early that drivers on the West Branch Penobscot were caught and the men had to travel around Moosehead Lake rather than cross it on the ice, as was the usual custom. Rain came though, and by May 6th all logs in the state were expected to come in. It rained hard and steadily all the first week in June ensuring the success of the drive. In fact, so much rain came that the Bangor boom broke letting a million feet of old logs go down the river to Bucksport, to be caught and returned by raft. As it turned out, 1878 was one of the easiest driving years in the century.[9]

Among the other bad years was 1883. The winter was poor and the drought was extensive. As early as January newspapers were fearful.[10]

> The lumbermen of Maine suffer severely, the product of sawmills being reduced by some 80 million feet. The rivers of Maine are lower than for half a century and losses on account of suspended operations will reach many thousands of dollars. The mills on the Penobscot lose over one hundred thousand dollars. In southeastern Maine hundreds of farmers are compelled to resort to snow water and cattle are driven miles to ponds and brooks.

Driving the rivers

It was estimated that the lack of water on the St. Croix cost $100,000 in wages to woodsmen and sawmill hands. By the 5th of May drives were reported hung everywhere. Even though a big rainstorm came on the 22nd, it was not enough, and in 1884 fewer operators went to the woods because of the old logs still coming down from last year's cut.[11] A poor spring often cast a blight over several years to come.

Almost as bad was the possibility of a great and sudden freshet. The freshet of 1846 is still legendary in Maine, and the freshets of 1832, 1849, and 1869 are almost as well known. The 1869 freshet was a fall rise and is called "the pumpkin freshet" because the rise floated so many of the vegetables down stream. The famous freshets of 1901 and 1902 played an important role in the struggle over control of the rivers. After the terrible winter of 1886-1887 everyone looked for a spring freshet of considerable proportions. They were not surprised. In the spring it rained very hard and the waters began to rise. From April 28 to May 6 rivers all over the state went to flood conditions. Booms and logs were lost on the Kennebec and Penobscot. At St. John and Fredericton it was described as the worst disaster since the great freshet of 1854. The west wing and some of the center of the great dam at Chesuncook Lake went out with the water and although men rushed from Bangor to attempt to repair it, the water left so rapidly that it stranded about six million feet of logs on the lake shores. On the Kennebec the loss was nearly 10,000,000 feet which were swept to sea and not recovered.[12] Such bad years and dangerous freshets were always in the mind of the operators as spring came and they readied their equipment.

The farther south in the state one operated the earlier the drive got in. In 1875, 3,000,000 feet, all pine cut on the Little Ossipee River in York County, was in on April 26 even though it was the largest drive in twenty years.[13] The drives on the Saco were usually quite early. In 1878 the first drive of 1,500,000 feet was in by May 1.[14] The next year Joseph Hobson of Saco, the big operator, had his eight million feet in the boom by June 1.[15] In 1885 the head of the drive, 700,000 feet of pine, arrived by May 25, and in 1889 some logs were in the Biddeford boom by the middle of April. The total of logs in the Saco drives that year, 17,500,000 feet, had all cleared the boom to the mills by August 13. Observers thought this was excellent as it was the largest cut on the river in some time.[16]

Although this area, southern Oxford and Cumberland counties, had been heavily cut, new growth pine and poplar for the pulp mills provided logs to cut and to drive. Some of these rivers had not been driven in many years. Poplar had to be peeled to drive. This meant laying the logs over in the woods for a season. The sap then dried out of the logs making them like cork on the water. On the big rivers, for this reason, the pulp often came after the long logs as they needed less water. On the smaller rivers pulp was often the only thing being driven and this explains the early arrival dates.

In the middle eighties the demand of S. D. Warren Company for poplar opened up for lumbering the drainage basin of the Crooked River in Norway, Waterford, and Harrison. The route of the drive was down the river to the lakes, to the Songo River, and thence to Sebago Lake and through the Presumscot River to Westbrook. Some sawmills in the Portland area also utilized these drives to get pine for their mills.[17]

The Presumscot was vital to S. D. Warren. A short and deep river draining Sebago Lake, it was the source of both logs and power to the paper mill. Once the paper company was forced to send a posse of men to raise the gates to give them water. The owner of the dam, the Oriental Power Company, seized the dam back after a pitched battle, although the paper company's army of thirty men was able to retake the stronghold and hold it against all comers as long as it was necessary.[18] Later the paper company purchased the dam site and rebuilt it to their own needs. The pulp came down this river throughout the eighties and nineties. For instance, in 1884, 1,000 cords cut in Gray were driven in to the booms, and in 1885, 2,000 cords came from North Gorham. The 1886 drive of 4,000 cords, from Harrison, did not come in until August 23.[19]

Drives on the larger rivers were somewhat more regular. On the Androscoggin logs were stopped at Berlin, New Hampshire, Lewiston, and Lisbon. Later newer mills, both sawmills and pulp mills, were built on the river and the drives grew more difficult as the logs had to be sorted more often. Usually the logs going to Lewiston, Lisbon, or down river parties cleared first, and the Berlin logs later. The Lewiston Steam Mills Company generally had the drive as they were the largest operators downstream. In the sixties, before Berlin became a big sawmill town, the drives were more haphazard.[20] By the seventies with the Berlin Mills Company also getting out a big drive, 13,000,000 feet in 1871 for instance, the routine was more standard.[21] Sometimes jams still bothered. The Bearce drive hung at Gilead in 1877 and it took two and one half days to pick the logs. This was a very hard year, and the Androscoggin Water Power Company's drive

Driving the rivers 71

for Lisbon did not arrive until August 2. The mill had been closed for six weeks for lack of logs.[22]

In 1879 a new mll appeared on the river. C. H. and Z. Gilbert, who owned one mill at Berlin, constructed a new mill at Canton Point. They built a boom in the river with twenty piers which was designed to hold 100,000,000 feet of logs.[23] The new mill was just in time to participate in the Androscoggin long drive of 1880.

The Androscoggin drive for the lower river amounted to about twenty-five or thirty million feet of logs, and word came early of the difficulties ahead. Reports began to filter in of logs hung everywhere. The worst jam of all was on Rumford Falls, with more than 4,000,000 feet hung along a mile or so of shore. Crowds came for miles by buckboard and team to watch the men pick off the logs one by one. Finally the jam cleared and the rear was reported to be at Jay Bridge on August 7. The rear consisted of 15,000,000 feet though, and by now reports of the drive did not speak of driving, but dragging. The only hope was for a late fall freshet. By October 25 plans were being made to erect booms in the river wherever the logs finally stopped. Logs were reported in the river at Canton Point and south to Lewiston. Rains came at the end of October which made some difference. It was thought in mid-November that the logs would all get to the boom. It was a vain hope, however, as on November 24, the river froze, catching some six million feet in an area from the Lewiston boom up the river sixteen miles. The mill men were forced to hire men to cut the ice, drag the logs from the water, and shingle them up on shore. Eventually all but about 2,000,000 feet was safe and these finally came into the boom the next spring.[24]

The drives continued to grow in size. Over 51,000,000 feet came down to Lewiston and Lisbon in 1887, and by 1889 Bearce and Wilson had cut 20,000,000 feet for their drive alone. Six million of this was hauled from the river by donkey engines for the Otis Falls pulp mill.[25]

Some of these pulp drives were fairly large. The Forest Paper Company in Yarmouth received some of their wood by drive. Cut in four foot sticks and peeled, it was driven to Lewiston, taken from the water by donkey engines and endless chains, loaded on Grand Trunk Railroad cars and then carried to Yarmouth. Their contractor drove the first lot in 1883 from the Swift and Ellis Rivers. This was probably the first four foot wood driven in the state.[26]

These later Androscoggin drives were always long and arduous. Logs cut in the northernmost regions often took two years to get to the mills as they had to be boomed in the Rangeley Lakes over one winter. Good log drivers began to be scarce, and men went from Bangor and farther east to drive these rivers at premium wages.[27] This demand coupled with the demand on the Connecticut kept wages high all over Maine. In 1893 the Berlin Mills Company was forced to run a newspaper advertisement for river drivers.[28]

WANTED

400 River drivers. Apply in person
at our office, Berlin Mills, N. H.
Berlin Mills Co.

One of the difficulties was that the Androscoggin was the only river of any size in Maine which did not have a corporation to drive the logs. Here it remained every man for himself. Bearce and Wilson usually had the lower river drive until the old Lewiston Steam Mills burned, and then they turned to handling pulp for Livermore Falls parties. By this time the Union Water Power Company really controlled the drive by controlling the water from dams constructed near the Rangeley Lakes. The only drives left on the Androscoggin were to the few small sawmills below Livermore Falls, and to the pulp mills at Berlin, Rumford, and Livermore. The railroad was beginning to replace the drive by the end of the period on the river and would soon do so entirely.

Among the reasons for this were the usual difficulties caused by the upriver mills. In the 1890's the Berlin Mills Company ran their booms clear across the river because they did not have enough boom room for their own logs. In one year over 4,500,000 feet of spruce logs were stopped which were intended for Lewiston, Canton, and Lisbon Mills. Some were stopped for over a year, and the general drive took sixty days to get to Canton when the average had been only fifteen. Twenty million feet hung completely that year, and thirty million feet more were being waded or dragged in. Bearce and Wilson joined with the Canton Steam Mill drivers to get what they could, but most of the Lisbon and some of the Lewiston logs did not arrive on time. Suits were threatened, but as the logs got nearly if not all in, they were not consummated.[29] The future was there for anyone to see however.

That inevitable future did not come as quickly to the Kennebec River as it did to the others. It would be possible to say that it still has not come, one supposes, because it, with the Machias and very occasionally the St. John, are the only rivers with any considerable

Driving the rivers 73

amount of driving still being done. The Kennebec drive is all four foot stuff, pulp; the Machias drive is long logs, and the St. John drive is mixed, much pulp and for some recent years (before 1963) about 5,000,000 feet of pine annually. The end is near for all these drives, however.

The Kennebec was a corporation river, and the drive was always taken in by the corporation. In fact there was, in addition to the Kennebec Log Driving Association, the Moose River Log Driving Association, the Dead River Log Driving Association, and even the Dead River North Branch Log Driving Company. These corporations took the drives as they left the small brooks in late April and then moved them down river. The Dead River drive usually arrived at the Kennebec in late May, and the Moose River drive at Moosehead Lake at about the same time. The two drives were then merged and brought down to the mills together arriving in Augusta at the end of August or early September. Tables in the appendix give some idea of the size of these drives, and the time taken to drive logs on the Kennebec River.

These drives were large, employing in the early days four hundred men and at the end of our period about eight hundred men. One of the reasons was the amount of sluicing needed at various dams for power and storage on the river. In 1872, 1875, 1877, 1879, and 1880 logs were hung at various places on the river often, as in 1877 and 1880, only coming out with a fortuitous fall freshet.[30]

Carratunk Falls was the worst spot on the river. Although work was done here nearly every year by the corporation, the logs continued to give trouble and great jams were formed. One such jam occurred in 1881. By June a mile-long jam with a face a hundred feet high had formed on the falls. The drivers had a constant crowd of viewers, which on Sundays increased to over a thousand people. No matter how much work was accomplished more logs came from up river. By the use of "Atlas powder" or dynamite the jam was finally broken, but the logs did not arrive at Norridgewock until September 1st, and the rear was not at Augusta until October 1.[31]

The Kennebec drives increased in size after 1881. No year was below 100,000,000 feet, with one year reaching 188,446,509 feet. That year, 1893, was the best of the period, but as late as 1911 the log drive was still 115,626,169 feet on the river. Dynamite was used to blast out rocks and channels at Carratunk Falls and elsewhere in order to handle these increased drives, but the amount of poplar being driven to the pulp mills in addition to the saw logs made the drives slower as well as larger.

As the drives increased in size great jams continued to occur. In 1885, 12,000,000 feet hung two miles below Indian Pond Dam on the east branch of the river. Unable to move it the workers raised the dam allowing some water to come out; this only succeeded in breaking the boom and depositing another 2,000,000 feet on the jam. Finally, in desperation, the men raised the gates to the top, letting out ten feet of water, and this coupled with strategically located dynamite, broke the jam letting the half mile of logs go by the Forks all at once in one magnificent, smashing, crashing mass.[32]

The major difficulty was caused by the poplar being cut in addition to the softwood. Riding in a train near Skowhegan a Bangor *Commercial* reporter remarked on the pulp in the river and elicited the following remarks from "a prominent Kennebec Valley businessman." [33]

MAINE'S PROSPERITY

> Between Somerset Mills and Skowhegan the Kennebec seemed to be filled with pieces of wood from four to eight feet long, hundreds of cords of it. Calling attention to it our Maine man proceeded: 'That flow of stuff, you can't call it timber, illustrates the new fields, of industry which have been found here. That is poplar wood. Fifteen years ago it and the land on which it grew were worthless. The term poplar was one of contempt, but now it has, by the inventive genius of our people, acquired value. . . .'

When this poplar arrived at Fairfield, the logs were put on an endless chain and sent up an inclined plane. Upon arriving at the top the logs were thrown out in a huge pile. The wood went up the chain night and day when the drive came down.[34]

Huge jams were formed at the end of the drive as the pulpwood came in on low water, and the rear was often in Waterville until late or mid-September.[35] Other obstructions continued to be placed in the way of the drives. In the early nineties Hollingsworth and Whitney put a dam in the river at Winslow. It had a log sluice in the center of the dam, but when the 1892 drive came down the dam was not finished. After some bickering the workmen suspended their operations to let the drive through.[36] The next year more trouble came. The Moosehead Pulp and Paper Company boomed the river at Solon to catch their logs. The logs accumulated faster than they could be sorted and within a short time twenty-five million feet were jammed at Carratunk Falls. The pulp company put a crew of men on to break the

Driving the rivers

jam but with no luck. They tried dynamite and the jam still held. Finally the company rented a locomotive from the Somerset Railroad and with a wire cable were, after some days time, able to start the jam. The company estimated their booming experience cost at $10,-000. The locomotive could not, however, move the logs rapidly enough, and by mid-August the jam had grown again this time to 35,000,000 feet. Men from the Kennebec Log Driving Association were put on the other side of the river from the locomotive to sort at the pulp firm's expense. Finally late in August the jam was cleared. The drive cost so much that, coupled with the tightness of money in the depression of 1893, the men on the drive were paid in interest bearing notes of the Log Driving Company for the first time since 1862. The eventual assessment on the owners of the logs was 10¢ per M on the main river. It may have been prophetic to viewers of this expensive season to note that some of the Moosehead Lake logs were carried to Bangor on the Bangor and Aroostook Railroad. It was cheaper to bring them across the lake to the railroad and then to Bangor than it was to drive them past the pulp and paper mills, each with its sorting boom to slow the logs down.[37]

Downeast events were much more prosaic. The lumber cut was small and got smaller each year, and the pulp mills did not arrive until well after this period. The logs usually went down quickly if not easily. In 1871 the St. Croix Log Driving Company moved 55 million feet from the outlet of Chipetnicook Lake to the Baring booms in fifteen days. The drive was described as the "largest and cleanest drive ever made on the river or ever will be."[38] This was probably a correct observation. Logs often got in quite early. Six million feet arrived at Calais early in June of 1873.[39] James McReavy drove two million feet from 5th Machias Lake to the Whitneyville booms on the Machias by June 12th in 1879, the first time the drive had ever come in in one year.[40] In 1880 the St. Croix drive was at Grand Lake Stream by June 1st, as it was in 1887 when twelve million feet came down.[41] The Machias river cut of thirty-two million feet was in the boom by mid-May in 1885, and thirteen and one half million feet was at Beddington on the Narraguagus by April 15, 1888.[42] Driving wages were lower here, $1.25 to $2.25 a day most years, and this is why the men went to the Androscoggin and Connecticut where $3.00 to $5.00 was often paid to good men.[43] Only in 1891 was much trouble offered. The Narraguagus River drive was delayed some time at Beddington, and the main river drive did not come in until the end of June with a big rain storm.[44] By and large though driving conditions were not as difficult in Washington County as elsewhere. The rivers were shorter, the cut was smaller, and the pulp mills had not arrived to stretch their booms to slow down the drives to the sawmills.

On the Penobscot and the St. John Rivers, corporations usually took the drive. In the spring after the cut was known an annual meeting was called, almost invariably in Bangor for Penobscot branches, and the drive was let at auction, to the lowest bidder. The drive was an opportunity to make money for the contractor if he did not meet with low water, early freshets, or trouble with his men. The contractor would guarantee to move the logs from the corporation limits, Jellerson Boom on the Mattawamkeag, Head Chesuncook Lake on the West Branch, and so on, to the boom at Argyle.

On the Mattawamkeag there were often three drives. Sometimes they came quite quickly. The third Mattawamkeag drive left Jellerson Boom on June 6, 1874, for instance.[45] The first drive was at Winn by April 30 in 1877.[46] The Websters of Orono were big operators on the river. They often had the drive. Later they would raft and boom at Bangor, and eventually own large sawmills and an early pulp mill on the Penobscot. In 1878 James Webster took a million feet he had cut, twelve miles to the Jellerson Boom in four days with a crew of six men. He must have had high water.[47] Occasionally the Mattawamkeag did get quite high. The next year, 1879, the Skatarack Boom, two miles below Kingman, broke letting five million feet of logs go adrift. The word came to Bangor as the logs passed Mattawamkeag at nine p.m. It was early and the Argyle Boom had not been hung. A special train left Bangor for Greenbush at eleven p.m. with a crew of rafters and the boom was hung out in time to stop the logs.[48]

This river was as likely to be low as high. In 1886 the third drive of five million feet hung and was boomed on the river. Later the drivers moved the logs to Argyle by building plank-dams on the river until two or three feet of water had collected, then letting the logs go forward until the water was gone. The whole process was then repeated until the logs came in to the boom.[49]

The drive was taken often by the biggest operator, and the cost of the drive went up steadily. The table, next page, reproduces the bid cost of driving the Mattawamkeag River from Jellerson Boom.[50]

The Passadumkeag River also nearly always had a drive coming in to the Bangor Booms. J. W. Hathaway brought the drive down in 1875, with a crew of twenty men who moved the six million feet to Argyle in just under two months. His bid price was 49¢ per M which

Driving the rivers

Cost of Driving Mattawamkeag River—Selected Years, 1878-1905

Year	1st Drive	2nd Drive	3rd Drive	Extra Assessments
	Prices in cents per M			
1878	24	29		
1879	27	29		
1882	22	25 5/8		
1883	27 7/8	30 5/8		
1885	26	31		
1888	32 1/2	37 1/2		
1892	33	38		14
1893	37 1/2	42 1/2		8
1894	49 3/4			8
1895	30			6
1896	32 7/8	37 7/8	42 7/8	8
1897	29 1/2	34 1/2	39 1/2	6
1898	42	47	52	5
1899	33 1/2 plus 5¢ for each dam sluiced			
1900	35	40		7
1901	65			7
1902	68	73		8
1903	51			
1904	52 3/4	62 3/4		8
1905	The corporation drove and prorated each mark with 10¢ M being charged for dam sluicing.			

must have been profitable, although the logs were rafted down from the Passadumkeag Boom.[51]

The table, next page, shows the prices for the Passadumkeag for the years available, again showing the cost increase of river driving. For small operators, as most of those were, the increased cost often meant the difference between success or failure.[52]

Drivers from the Penobscot were always in demand in other areas, and before discussing driving and rafting on the Penobscot proper, it is appropriate to spend some time on the work of these "Bangor Tigers" as they were called on other rivers. The Connecticut River, draining nearly all of western New England, was difficult to drive, and Bangor river drivers were in demand on its drives. Perhaps the most famous of these drives concerning Maine men was that of the year 1876.

> Hartford will no longer be obliged to get his [sic] lumber by vessels or car from Maine and elsewhere, for it will be floated down the river to the city.

Cost of Driving Passadumkeag River—Selected Years, 1885-1904

Prices in Cents Per M.

Year	1st Drive	2nd Drive	3rd Drive
1885	23	37	61.5
1886	23	37	61.5
1887	22	30	55
1890	19	30	75
1892	14.5	27	57
1896	45	50	
1898	23	53	
1899	17.5	33	40
1900	25	50	52
1901	24	24	65
1903	29	50	by company
1904	50	75	98

That comment was the way the Hartford *Palladium* greeted the news of the forthcoming drive.[53] W. S. Stevens of Hartford in partnership with John Ross, of Ross and Leavitt in Bangor, had taken stumpage and bought land to the amount of 111,000 acres on the headwaters of the Connecticut. Thirty teams and fifty men from Bangor had gone in to these cuttings, in addition to New Hampshire men, and some 6,000,000 feet were waiting on the landings to be driven.

The plans were to drive the logs to Hartford, Connecticut, boom them there, raft them in to the channel and then tow the rafts by tugboat to the Little River where Stevens, Ross, and Leavitt had rented a sawmill. By the 17th of April men had begun to leave Bangor to work on the drive, seventy drivers going out in the first crew. The Connecticut was low and the three hundred and twenty miles to Hartford were long and arduous. Ross, who had the drive, took five million feet for Bellows Falls, Vermont, and five million more for Holyoke, Massachusetts mills, in addition to his own logs. With a few other smaller drives the total in the river was 20,000,000 feet. The first drive arrived in the Hartford booms about September 1, and the second rear, of 750,000 feet, went by Springfield, Massachusetts on September 19. One hundred and fifty drivers from Bangor, with eighteen Indians from the Old Town reservation for batteaux work, had taken the drive in. The drive had created quite a sensation on the river as the men estimated that 20,000 people visited the camps on the river to watch the drama of the log drive. Ross announced that they expected to drive 25,000,000 feet the next year.[54]

Driving the rivers 79

Men continued to go from Penobscot areas to the Connecticut River drive, although it was not so purely a Bangor operation after the great drives of 1876 and 1877. In 1880 two hundred and fifty went from Bangor, fifty or so in 1883, and even as late as 1888 two hundred and fifty men went to the Connecticut River drive from Maine rivers.[55]

On the Penobscot itself the West Branch was far and away the most important section of the river. Efforts to control the water flow had early been made, culminating with the construction of the dam at the outlet of Chesuncook Lake about 1840.[56] Rocks were blown in Ripogenus Gorge to aid the passage of logs as well. The dam, a quarter of a mile below the head of Chesuncook Carry, was built in a gorge with one side thirty feet high as an anchor. The other side was bolted to a ledge. The dam was five hundred feet long, with twelve gates. The log sluice was forty feet wide and two hundred feet long.[57] This was the way it looked near the end of its life. Much work had been done on it since the original construction, however. Hempstead,[58] who had access to records now lost or misplaced, describes rock blowing in 1845, 1846, 1866, 1874, and 1878. Work was done on the dam itself in 1865, 1866, 1867, 1874, 1878 and 1886.

The Penobscot Log Driving Company had begun to take logs from the head of Chesuncook in 1856, which had opened up a large area for cutting. The distance from Chesuncook to North Twin Dam, the area controlled by Chesuncook Dam, was treacherous water, falling two hundred feet in the three miles of Ripogenus Carry, and twenty-six feet in less than half a mile from Chesuncook Dam to Ripogenus Carry. In the middle of Ripogenus Gorge were two huge rocks, in appearance like old-fashioned flatirons. It is small wonder that these rocks, known as the Big and Little Heater, and their passage were so well-known to woodsmen. Here was concentrated as much driving difficulty as perhaps any other place on the river.[59]

The total distance from the head of Chesuncook Lake to Medway, where the East Branch came in was just a shade under ninety miles.[60] This was the difficult area to drive, and it was the primary responsibility of the Penobscot Log Driving Company. This was the area over which the drives were let at auction each spring. Figures of the cost of the drive, the bid price, are useful. The table, next page, reproduces them in the last twenty-five years of the dam.[61]

After the famous long drive of 1901 on the Penobscot figures were compiled from the books of the driving company to indicate

Penobscot Log Driving Company Prices
Chesuncook Lake to the Bangor Boom

Year	Head Chesuncook	Foot Chesuncook	Additional
1878	.77	.53	
1883	1.05 1/2		
1884	1.08		
1885	1.14	.89	
1886	.83	.58	.25
1887	1.00	.80	
1888	.94	.74	
1889	1.05	.85	
1890			
1891	1.02	.82	
1892	.93		
1893	.94 1/2		
1894	.93	.73	
1895	1.05		
1896	.94	.74	
1897	.92 3/8	.72 3/8	
1898	.89 15/16	.69 15/16	
1899			.25
1900	1.00	.80	
1901			
1902	Association drive		
1903	c1.00 (Association drive)		
1904	.41 (from the Great Northern Holding Grounds)		

the usual length of the drive and the amounts driven. The table, next page, reproduces his useful information for the West Branch Penobscot.[62]

The East Branch, with Hulling Machine Falls, Stair Falls, and Grand Pitch among other driving hazards, was perhaps as difficult to drive and work as the West Branch. For some reason, it did not make the impression on observers that the West Branch did. Thoreau, for instance, who came down it was apparently in such a hurry to get back to civilization that he didn't even mention any of its beauties. The best contemporary account of the East Branch, although done with some of the superciliousness which normally marked the outlander, is the little-known *Canoe and Camera*,[63] by Thomas Sedgewick Steele. From the etchings one is able to pick up both the beauty of the East Branch of the Penobscot River and its driving difficulties.

There were fewer bad years on the Penobscot in our period. One of the reasons was that men were usually plentiful to drive the

Driving the rivers

West Branch Penobscot Log Drive—1873-1900

Year	Left Chesuncook	Arrive Penobscot Boom	Amount of the drive
1873		July 12	
1874		August 4	
1875	—	—	35,460,350
1876	—	July 31	27,978,840
1877	May 21	—	17,833,370
1878	May 31	July 31	38,129,490
1879	June 9	—	no drive
1880	June 3	July 4	32,423,570
1881	June 5	July 3 and July 26	56,467,680
1882	June 14	August 12	42,960,580
1883	June 23	August 8	33,047,980
1884	June 3	July 18	28,858,780
1885	June 26	August 11	23,396,630
1886	June 10	September 18	40,457,830
1887	June 8	August 24	38,547,630
1888	June 9	August 7	36,014,230
1889	June 5	August 8	43,607,640
1890	June 6	July 31	54,038,680
1891	June 4	August 6	53,126,800
1892	June 18	August 29	50,094,450
1893	June 27	August 29	40,308,100
1894	June 16	August 7	42,348,050
1895	June 9	August 1 (about)	35,779,050
1896	June 22	August 18	54,086,740
1897	June 19	August 2	58,951,270
1898	June 12	August 2	44,026,920
1899	June 22	July 25	47,664,630
1900	June 23	August 20	32,628,790
	Total West Branch Logs 1875-1900		1,008,238,080 feet

logs, and so much down-river depended on the logs getting in on time. In 1870, one of the hottest summers on record, the drive was limited. About 50,000,000 feet hung that year on the various branches.[64] Most years were like 1874 though. Both drives were a bit late starting, with the West Branch rear clearing Ambejejus Lake on June 30. The drives (East and West Branch) joined at Medway on July 18 for the trip to the boom. When the logs arrived it was a fairly typical time. The last quarter mile into the boom was always a batteau race. Twenty-two batteaux with crews raced in, and the usual crowd went up from Bangor on the railroad to watch the fun. A cannon was brought out and a salute was fired to welcome the logs and their drivers with their full purses.[65]

The only year in our period when real trouble was had on the Penobscot was 1879. The season was late to begin with. The men did not start for the woods until April 25—fifty men to the East Branch, one hundred and twenty-five to the West Branch, one hundred men to the Mattawamkeag, and thirty for the Passadumkeag. Just as they arrived a great freshet came, carrying out two hundred feet of the south end of North Twin Dam. The news shook the lumbering community. "It will surely be a great calamity. If the whole West Branch drive is hung up it will be a severe loss for the operators, and the supply of logs will not be large enough for the mills of the Penobscot." [66]

A special meeting of the Penobscot Log Driving Corporation was called. Jasper Johnston the great dam builder of Corinna was sent for, and twenty men and their supplies were sent up river as quickly as possible. It was hoped that cribs could be sunk and emergency repairs made in order to save some of the water. Owners in Bangor waited anxiously as the water began to rise rapidly at the city. The dam was in too bad shape to repair and the men were paid off. When part of the Chesuncook Dam went out a few days later the drive was effectively lost, and in mid-June the drive was given up in Ripogenus Lake and the drivers all came to Bangor.[67] The newspapers were lugubrious when reporting the news.

> The hanging up of this large drive was a serious blow. Many of the operators whose logs are thus kept from market till another year will be likely to suffer serious loss, especially those who are working on borrowed money. If there should be an increase in the demand for lumber this summer the supply may be a little short. Owners of logs on other streams will probably get better prices. Manufacturers will have to pay more than they have been paying.

About 7,000,000 feet from other areas, mostly south of the lakes, did get into the boom, mixed with East Branch logs, and counted as such.[68]

One of the few changes noticeable in this time was that by the 1880's the river drivers, as well as the loggers, were nearly all men from the Provinces. The Penobscot men, many of them, were attracted to the Connecticut by better wages, and in 1881 handbills were printed and scattered throughout the Maritimes in hope of raising the one thousand men needed on Maine rivers. The call caused a large crowd of potential river drivers to come to Bangor.[69]

Driving the rivers 83

As the loggers began penetrating deeper and deeper into the north country, it became more difficult to get the drives to Bangor in as good season as before. Trouble was had in 1886, 1887, and 1889 although the logs arrived every year.[70]

Apparently as a result of this lateness, the Penobscot Log Driving Company voted at the annual meeting in 1890 to build a steamboat to facilitate movement of the logs, especially on Chesuncook Lake.[71] Steamboats were not new; they had been used on Moosehead for some time towing the booms to the Kennebec outlet. W. H. Strickland, John Morrison, John Ross, and Cornelius Murphy, all prominent Bangor lumbermen, were appointed as a committee to supervise the building of the new steamer. The material was taken up from Bangor over the railroad, carried across Moosehead, and the boat was built at Northeast Carry. The *John Ross* as she was named was a sidewheeler with two 25 horsepower steam boilers. After construction the boat was run down the West Branch to Chesuncook, let over Pine Stream Falls into the white water by blocks, tackles, chains, and pulleys, and then let go under her own power.[72]

In her early career the *John Ross*, built like a scow, gave trouble to her operators until a false prow was added. Apparently the first rudder was an elementary affair, and men steered her partially by the use of setting poles or long sweeps. The first year of her use she started up averaging about two million feet to a boom. The last boom of the year had four million feet and when *John Ross* moved down the lake the load was too much. She could not be steered, and her mainshaft broke under the strain. Over the winter the boat was repaired and she ran until she was scrapped and replaced in 1901.[73]

Over the winter of 1890-91 logs did not sell well. In mid-June there were sixty million feet of logs from the previous years drive still unsold. As there was little room in the boom and along the banks, many wanted the West Branch drive to be boomed in the lakes over the winter. The mill operators wanted them driven out in order to depress the price even more. Many of the men on the boom went home to New Brunswick on the assumption that no more logs would be rafted. The market began to look a bit better by the 1st of July, and the logs came down.[74]

The *John Ross*' success in towing was so spectacular that the P. L. D. voted in 1892 to build another steamer for use on the lower lakes. This one, the *F. W. Ayer*, was put in operation in 1893.[75] Then the drives began to come in when the operators wanted them. With the dams and the towboats they could boom and hold the logs for the proper time at the boom and the proper price. By 1894 the operation had become so standard that the Bangor and Aroostook railroad ran advertisements calling attention to their special trains running north to see the start of the drive, as this one in 1894.[76]

> Why not see
>
> MOUNT KATAHDIN AND THE LOG SLUICING
>
> at
>
> NORTH TWIN DAM?
>
> The Bangor and Aroostook R.R. Co. will
>   sell tickets at $1.50 each
>
> BANGOR TO NORCROSS AND RETURN
>   July 18, 19th, and 20th
>
> for all regular trains. These leave
> Exchange Street at 6:45 A.M. and 3:25 P.M.,
> and due to leave Norcross for Bangor at
> 10:49 A.M. and 4:13 P.M.
>
> TAKE A LUNCH ALONG

There were forty million feet of logs waiting for the train. The gates were raised when the train arrived and the logs were sluiced through. When the gates were raised there was a head of seventeen feet of water behind the dam.

Still, in this very same year, the future loomed for those who wished to prophesy. Barker and Thurston of Old Town received three hundred cars of logs from Greenville via the Bangor and Aroostook Railroad, and another Old Town firm was taking five to six cars a day from the same source. A newspaper commented:[77]

> This seems to be a new departure in the lumber business. Lumber that cannot be transported by water owing to the distance can be brought by rail. One would think the manner of lumbering in vogue in the far west would be a good plan for use in these inland timber tracts, taking the mill to the timber instead of *vice versa?*

This was the voice of prophecy, speaking through the writer of the "Milford Notes" in the Bangor *Commercial*. Others were beginning to think the same thing, notably Garrett Schenck, and it would be but a few years before this final move did, in fact, take place.

Driving the rivers

Once the logs were driven down the river to the mill towns something had to be done in order to hold them until they were needed and to separate them according to their owners. When the logs were placed on the landings, skilled axemen cut their owner's log mark on each end. These marks, very similar to cattle brands, were all registered, and lumbermen could tell at a glance who owned the logs. In order to assure easy handling of the logs the first boom on the Penobscot to stop the logs was erected at Argyle in 1825. In 1827 Rufus Dwinel purchased it, and when in 1832 the legislature allowed the construction of a new boom for further sorting at Pea Cove, Samuel Veazie purchased the entire boom works for $20,000 in 1832 and $20,000 more in 1833. The boom was sold in 1847 to David Pingree and others for $90,000. Great difficulty arose between the owners of the logs, and the owner of the booms, and in 1854 a new corporation was formed, the Penobscot Lumbering Association, or P. L. A., which was authorized to lease the boom from the owners, for fifteen year periods. The lessees were to maintain the boom in the correct manner, make all necessary repairs, and pay the owners of the boom 10¢ M royalty as rent. They in turn could charge tolls to cover their expenses. In 1869 when the lease was extended the new lease called for a 9¢ M royalty.[78] There were in reality four booms in the Penobscot Boom. Heading down the river the first boom met was the Nebraska Boom. Logs were stopped and sorted here. Then came Argyle Boom, where further sorting and much rafting took place. Further down the river just above Old Town where the Penobscot splits in two channels was Pea Cove Boom behind Orson Island. Much sorting and rafting took place here in its two parts, Pea Cove Boom proper, and the Mexico Boom. Temporary booms also could be and often were strung behind Freeze's Island, Thoroughfare Island, and the other islands in the river. These islands, or most of them, and the shores were owned in perpetuity by the Penobscot Indians who collected an annual rental from the owners of the boom.

At the booms the logs were sorted and separated into their various marks and rafted together to be carried to the sawmill by the operator or the purchaser of the logs. Rafts were constructed by driving hardwood wedges in the end of the logs and taking half-hitches around the wedges to hold the rafts together. Later crotched wedges were invented which took less rigging. The rafting was bid off to the low bidder who guaranteed to do the work under the supervision of the Association. Although many people speak of the Penobscot Boom as the Bangor Boom, this is a misnomer, as the Bangor Boom was a separate entity charged with moving the logs to the steam mills down the river. It also moved the great rafts of sawed lumber which filled the river in the early days. Logs which came down without marks, or with undecipherable marks, were auctioned off periodically as prize logs belonging to the Association. The men who were employed at the boom had to be experienced water men who could move the great logs from place to place with ease. Large boom houses were constructed at each boom to house and feed the men during the booming and rafting season which often ran from ice-out to ice-in.[79]

Rafting was more often than not done by the firm of Daniel Lunt and Eben Webster. Others who did rafting were Charles White, E. and J. F. Webster, and Webster and White. The Websters used their profit from rafting to increase their lumbering operations and eventually to go in to pulp mills. Occasionally the bidding was spirited as profit could be made easily in a good year. Such years were 1878, 1885, 1891, and 1893. The bids did not vary much as the number of men needed was about the same. The main variant was the cost of the wedges, which were supplied by John Treat and Son of Enfield, at about $2.00 a thousand, and rope and rigging which varied rather markedly in price. The men's food, the ubiquitous beans, biscuits, and pork of the lumbering camps, also increased in price as well.[80]

The man who did the rafting ordinarily paid rental for use of the boom; after the charter renewal of 1884, when the royalty was cut to 7½ ¢ M, this rental seems to be set at $12,000 a year.[81]

The river was usually full of logs and rafts, especially if much lumber had been wintered over. When this occurred new booms had to be hung in the river to stop the drives, and navigation was frequently obstructed. The newspapers were often used as a method to notify owners of rafts or logs of these new obstructions in order that they could plan their sawing to take advantage of the market.[82] The fees for boomage were due thirty days after the logs were rafted, and in the seventies prompt payment earned a six per cent discount on the cost.[83]

From time to time the men working on the booms gave trouble. In 1872 a crew of Frenchmen were hired to aid in the work at Boombridge Brook. The men at the Nebraska Boom, nearly all Irish, resented the French who were working for less pay. They attacked them at their work but the French beat them off. That night the Irish went up to the tents of the Frenchmen and a bad fight ensued. "The Irishmen received a bad calking from the Frenchmen." [84] Police rode the trains out of Old Town and Bangor for a few days and no new trouble

Driving the rivers 87

ensued. The men were getting 75¢ a day, or $19.50 a month, during the depression of the seventies for work which lasted from 3:30 a.m. to 8:00 or 9:00 p.m. In May of 1879 one hundred men and boys of the one hundred and thirty employed struck for $1.00 a day. Lunt and Webster paid them off and closed the boom. The next day others had taken their place and the boom opened again.[85] In 1891 when wages had gone to $1.00 a day the men struck again for $1.25. The same process repeated, the men were fired and replacements opened the boom the next day for the old wages.[86]

The biggest year in the boom's history was 1872. The crew of rafters handled 216,841,430 feet that season. In 1871 they had rafted 190 million. Twenty million were wintered in the boom, and another forty million along the river shingled up. In 1871 forty-five million had been wintered.[87]

When the booms were full, work began soon after the ice was out of the river although it was very dangerous in the high water of spring freshets. In 1873 in the depression three hundred and fifty men presented themselves for the one hundred and twenty-five jobs when the boom opened.[88] The waters were very low that season. Spruce rafts only as large as 18 M were floated. In 1872 rafts had been floated of 47 M, and in 1871 35 M.[89] This, of course, made the work that much more difficult. In 1876 the boom broke under the pressure of the ice from the Stillwater River, and about 100,000 feet went down the river. A few logs were captured as they went past Bangor but most went to sea. Very few logs had been wintered so the loss was small.[90] In good years lumber was run down to Frankfort, Bucksport, or Belfast to ship after the ice closed.[91]

Sometimes the amount of work was prodigious. On July 18, 1876 forty-one men at Pea Cove rafted 11,984 logs of 1,648,480 feet. This was described as the largest days work ever done at the boom.[92] These figures were only challenges to the men. The following letter appeared in the papers the next year.[93]

Mr. Editor:

    A crew of sixty-eight rafters, under the charge of Edson P. Chapman, of Old Town, did the following piece of work in one week's rafting at Argyle Boom, from Saturday morning, 23rd of June, to Friday night, 29th of June.
    Whole number of pieces rafted, 60,440
    Whole number of feet rafted, 7,252,930
    The largest days work was done on Thursday, June 28th, when 11,-000 pieces were put under the rigging, making 1,320,000 feet.

Mr. Chapman is an experienced and skillful workman, and what he dosen't know about rafting logs is not worth much study.

We don't think the above weeks work can be beaten. Where is Nebraska?

Penobscot.

The amount of logs rafted never again approached the great years of 1868-1872 but much was handled nevertheless. After that time, 150 or 160,000,000 feet was recorded several times, in 1881, 1887, 1889, and 1891 for instance. As the logs grew smaller in size the work involved in handling one hundred million feet was similar to that of handling two hundred million feet in the more palmy days of the boom. Although we do not have complete records as to the size of logs throughout the century, in 1833 logs averaged 343 feet per log, and as late as 1857 were still nearly 200 feet per log. In contrast the average size in 1889 was 108 feet, in 1891 114, and in 1892, only 105 feet. (Incidentally, during this time, three very big years, the boom was clear of logs August 28, September 22, and October 3.) For this reason one suspects that the work was probably more difficult and more arduous at the boom at the end of the period. As driving became somewhat easier, booming and rafting became harder, especially as the drives tended to come in later and later. Near the end of the period more and more logs were wintered over shingled up on the banks for the next spring's sawing. Bad weather conditions and a shortage of workmen (as crews would already be gone to the woods) forced this change in method by the 1880's.[94]

Under the conditions that were true near the end of the period the rafters often found themselves doing quite enormous feats of work when the drives came in. In early June 1880, for instance, the rafters at Argyle Boom rafted 24,044 logs in two days, footing up 2,792,080 feet. At this time there were about one hundred and twenty men at Argyle, and one hundred and thirty men at Pea Cove. The average day's work was 500,000 feet at Argyle, and 800,000 feet at Pea Cove.[95]

Toward the end of the decade the totals were higher. For the fifteen days ending September 5, 1888 the crew of "Red" Smith of Veazie rafted 17,400,000 feet for an average of 1,160,000 feet a day at Argyle. The two booms, Nebraska and Argyle, were handling 13,-000,000 feet a week.[96] The next year three hundred men worked at the upper two booms and averaged 2,500,000 feet a week all season. Their year's total looked like this:[97]

Driving the rivers

Rafting Penobscot Boom—1889

Pea Cove	280,821 sticks	25,400,520 feet
Argyle	781,189	82,900,240
Nebraska	340,192	41,897,180
Totals	1,402,202	151,198,540 feet

The boom work continued to be, as it had been, a staple of work for many in and around Bangor even though the work was arduous, cold, wet and the hours were long. Occasionally on holidays the cities provided entertainment—batteau races, canoe races, and log races[98]—but this was small recompense for the work demanded of men in and on the river.

More arduous, perhaps, than even this work, because of the danger involved in running rafts of logs and timber over the sluice at the Bangor Water Works Dam, was the work demanded of men employed on the Bangor Boom proper. Nearly all logs going to the mills in Bangor, Brewer, Hampden, or Orrington came down from the Penobscot Boom in rafts. Rafts of sawed lumber from the mills in Orono, Old Town, or Bradley also came down to the vessels anchored at Bangor. It was difficult and dangerous business moving the heavy rafts in the spring freshets over the rips in the Bangor Salmon Pool, and it became even more dangerous after a dam was put across the river at Treat's Falls for the Bangor Water Works. After that time rafts could be brought down to Bangor only on each high tide, twice a day. Five rafts could be brought down on each high tide. The men had to be skilful at their work, and it is no wonder that except for a scant handful of years William Connors of Bangor had charge of the rafting from 1865 to his death. His sons succeeded him until there were no more rafts to be piloted over the dam.

The average of rafting from 1875 to 1885 was just over 41 million feet of logs each year, although 1884 saw only 36 million feet. A typical year was 1885. The work began May 15 and ended November 3, with 46,250,000 feet being transported to the Bangor mills. The log rafts were taken at the Argyle Boom, and driven the sixteen miles to Bangor, passing four dams as well as the water works dam just above the Bangor Boom. Connors had thirty men moving the rafts from the Penobscot Boom, and forty more (fifteen of them boys) rafting and running from the Bangor Boom to the mills. It took an average of eight days to move the rafts from Argyle to their destinations. The drivers received $1.50 a day, the pilots of the rafts $1.75

to $2.00, and the rafters $1.00 to $1.25. They lived together in boarding houses. There were then about six mills sawing below the Bangor Dam, and each laid up about three million feet for the spring sawing. Connors estimated that he had paid out in wages $13,000 for the year 1885.[99]

Although board rafts continued to come down throughout our period, apparently the last really big year was 1878. In that year 1800 rafts came over the sluice, all but one hundred and ten of them of boards. They went to the big shippers in the Bangor area. Each of these rafts held from twenty-two to eighty M of boards, and usually a deckload of laths or shingles sawed at the up-river sawmills and ready to be loaded on the big schooners in the harbor. After 1878 the railroad took an increasing amount of the board traffic as the damage over the dam sluice was too much to gamble with in an era of falling lumber prices.[100]

By 1888 Connor's crew, which numbered always now between fifty and seventy-five men, was handling mostly log rafts. Each raft, made up of five hundred pieces, was delivered to the steam mills with the exception of those destined for the firm of Morse and Company. They came to the boom and got their own, and rafted them up the Kenduskeag Stream on high tide. Connors was using about thirty tons of three inch hawsers each year for rigging on his rafts. The rafts, just as on the Penobscot boom, were secured by wedges manufactured by John Treat and Son of Enfield. In this time about 600,000 wedges were required each year for the work.[101]

The boom had been organized the first time in 1843, and the charter extended in 1869 to the year 1890. In 1881 the tolls had been fixed at 20¢ per M. In 1889 the legislature faced the problem of renewing the charter. The boom company wanted to build more piers in the river near Veazie in order to hold about seven million feet for their work. The booms then in the river held only four million feet and were too small especially in the fall of the year. Lengthy hearings were held where the selectmen of Veazie and others protested the new piers, while others were heard who desired lower tolls. The boom company answered complaints by statements that those who were unhappy could always buy in and share the profits, and that if the new piers were not allowed the logs would be secured in the river anyway by the use of cables and buoys. The charter was eventually extended with a compromise of lowered tolls, the promise of fewer piers, and the opening of stock sales to outsiders especially the builders of new mills in the Bangor area.[102]

Driving the rivers 91

The river below Veazie was always a source of trouble. Sawdust, edgings, and refuse from the sawmills collected in the river and obstructed navigation. The mayor of Bangor called attention to the problem in his annual messages of 1866, and 1868, and the Harbourmaster tried to get better enforcement of the law prohibiting dumping of such refuse in 1867 and 1870.[103] These efforts were of little avail. Newspapers, such as the *Commercial* in 1881, lent their editorial voice to this discussion, and in 1883, William Connors said that it cost him from $10 to $12 a day to keep the booms clear of the debris. There was a jam of three acres of sawdust, edgings, and bark at the boom at one time. One man did patrol the river in a boat to keep down the nuisance, but it was not until Connors threatened to keep the river clear by force that men were hired by the city to prevent the illegal dumping of the refuse.[104] By this time the number of sawmills upriver was beginning to dwindle anyway, and the violations of the law were of less importance. Dredgers in the river as of this writing, however, still find sawdust, planks, edgings, and logs on the bottom.

The rafts of logs and lumber started to come down as soon as the ice was out. In 1871 the first one appeared in Bangor March 19, in 1872 on April 23.[105]

Everyone in Bangor was happy when the rafts commenced again. It was estimated that the lumber business gave employment in Bangor to well over four hundred men outside the sawmills in this the banner year of the century. This included thirty-five surveyors of lumber, each for two hundred days. In addition there were also about one hundred and forty overhaulers of lumber, and one hundred and forty men hauling long lumber into the vessels, each for two hundred days. Short lumber took about seventy men and the Bangor Boom another seventy-five to one hundred for the season. The average wage in Bangor was $1.00 a day. Even beyond this, of course, was the fact that all businesses were tied to the lumber industry. Iron and foundry work employed two hundred men in Bangor at this time, and nearly all the product was destined for woods work.[106]

As well as giving employment to men in Bangor the boom and the rafting gave entertainment to the young bloods of the area. Until the time that the dam was placed across Treat's Falls young men of society used to ride the trains to Old Town with their female friends and come back down the river on the rafts of freshly sawed lumber eating a picnic lunch. What better way to woo the fair sex than on a deck load of fresh spruce boards on the Penobscot River?[107] Years later small boys used to drop from the railroad bridge between Bangor and Brewer and ride the rafts of logs to Morse's Mill on the Kenduskeag, but now all of this is just a pleasant memory of days long since gone by.

There were times when the work itself became more difficult. In 1874 the boom broke and men had to scatter to Bucksport collecting the logs cast adrift.[108] All of this was child's play compared to the difficulty created when the new dam was placed across the river. The first rafts came through the sluice on April 26, 1879. The first one, consisting of 50 M long lumber and 75 M laths, had been made up the previous fall for L. and F. Strickland, and had been caught when the river froze. The raft was badly beaten up on its trip through the sluice. The rafts all suffered great damage, but the demand was so great that in the first fifteen days of rafting 2,000,000 feet of boards came down over the dam.[109] The damage reports caused engineers to come forward with plans to eliminate the difficulties.[110] The lumbermen held a special meeting and asked the city to do something, perhaps to build a rollway, or an apron to cushion the shock. Leander Strickland, the owner of the first unfortunate raft, announced that during the previous summer one hundred and twenty foot rafts were not uncommon but now the absolute limit was ninety feet and usually less.[111] Even smaller rafts were not the answer, and eventually the city built a sluice twenty-seven feet wide on the upper part, twenty-five feet on the lower, and one hundred and fifty feet long designed, as it was said, to take any common raft.[112] The first of these rafts came through on November 3, 1879, with no trouble.

After this there was relatively little trouble at the boom. The logs were made up into six to ten big drives each year for the steam mills below the dam, enough so that their mill ponds would always be full. Each one of these drives had from six to eight million feet in it. With the disappearance of the lumber rafts, or their great curtailment, the men at the boom could turn their attention to logs and log rafts.[113]

Although the figures are much higher at the end of the period than earlier (sixty-one million in 1892 and fifty-one million in 1893) these reflect the last spurt of the Bangor industry as the combination of portable mills, railroads to the woods, and upriver pulp mills would deal a death blow to sawmilling within the decade. Those that hung on longer would not survive the opening of the great Pacific Coast forests by the completion of the Panama Canal just before World War One.

On the St. John the problems created by the length of the river were early made much worse by the digging of the Telos Canal. This

Driving the rivers

canal, which connected Telos Lake with Webster Lake by a channel eight hundred feet long, diverted water from the Allagash watershed to the East Branch of the Penobscot River. Of the thirty-one miles of water on the Allagash watershed, just over twenty-five became artificially tributary to the Bangor loggers.[114] Maine people remember the Telos War [115] between interests on the Penobscot as to who would control the canal, but what is forgotten is that the real effect of the canal was to make the St. John a much more difficult river to drive. In fact, "the main" John Glazier went up with a crew soon after the building and destroyed the dam in order to get his logs into Fredericton Boom.[116] The dam was rebuilt, and after this New Brunswick lumbermen cut their logs so as to conserve the remaining waters and cursed the Penobscot lumbermen every time there was a dry driving season.

The waters in the north were difficult to drive, and there were few years in which some logs did not hang up. In 1874 the logs only came in from the Aroostook River on a fall freshet and 1878 was similar.[117] The worst year was 1887 however. A huge jam formed on the Aroostook Falls and some logs did not come in until late fall. In 1889 22,000,000 feet jammed on the falls, and although a heavy rain (the week of the Johnstown Flood) helped, dynamite was finally used to clear the jam.[118]

On the main river St. John were more difficult places, with the most notorious being the Grand Falls in New Brunswick. The entire river was difficult though, and much work was done during this period blasting out rocks and buildng storage dams. Toward the end of the century a steam driller was brought up the river to work on the rocks from Seven Islands to the Northwest Branch.[119] Below Madawaska the rule of thumb was to use as many horses as men. Spring water was so high and swift that logs were always stranded on the banks and horses were used, as on the Androscoggin, to twitch the logs into the water. Men carded these banks four and five times a year for stranded logs.[120]

It was hard work as the following exchange would indicate:[121]

    Q. You worked your men how long?
    A. All the hours they would stand.
    Q. And on Sundays?
    A. Sometimes. Some Sundays the men refused to work.

From Grand Falls to Fredericton was not too difficult. On a good pitch of water the logs might get to the boom in thirty-six hours but getting to and over the Grand Falls was the difficulty. The river was a busy place. Hay and cattle came down from Fredericton on the log rafts from the boom, and the men piloting the rafts always had to watch for the famous St. John woodboats which filled the river. From time to time steamboats operated in the river, mostly wheelbarrow boats, but after the loss of water on the Telos Canal these boats seldom operated much above Fredericton, and certainly not above Grand Falls.[122]

Originally it was every man for himself on this river; it was not until 1884 that the St. John Log Driving Company was chartered, and it drove only from Fredericton. Eventually an upper corporation was formed which drove from Grand Falls and did the rafting. These two were merged finally. It was not until the nineties that the Madawaska Log Driving Company began to take a corporation drive from Big Rapids to the Grand Falls.[123]

In the early days of every man for himself life on the river was sometimes interesting. A hard year was 1873, with logs still coming to the boom as late as mid-October. So was 1887. In 1889 it was nearly as bad, but the great rains at the end of September brought down the drives as forty million cleared the Grand Falls on a four foot fall freshet.[124]

A most interesting year was 1886 though, as the Canadian lumbermen did something more than curse the Mainers who had stolen their water. New Brunswick parties had purchased stumpage in Maine. In the spring a driving dam which they had constructed proved to be too small for the amount of water they needed, and in May their drive hung twenty miles below Churchill Lake. The master driver took a small crew of men and went up to Chamberlain Lake to the main dam, hoisted the gates and fastened them up. Water which was released drove the logs into the main river. As Cyrus Dickey, who had logs in the main river, described the sight:[125]

> I can remember the water commenced to rise here in the main river without any cause rhyme or reason. Good, warm, sunny weather, with no snow in the woods, and we had a rise of water of three or four feet, and in a day or two the logs commenced to run, and this drive sailed by us, and I remember their men hooting us. Our logs were up high and their logs were in the bed of the river, and they went by, and the men went by in batteaus, and they hooted at us. We didn't know the cause of it. . . .

This sort of river driving was not often used, but it is easy to see how much the waters diverted to the Penobscot would have meant to the north country operators.

Driving the rivers 95

By the end of our period the St. John Log Driving Company had taken over all the driving, rafting, and booming from Grand Falls to St. John. It was a big operation. Nearly four hundred men were employed at the booms in Fredericton alone.[126]

As the logs came into the Fredericton boom area they were separated by marks. Twenty-five or thirty logs would be put parallel in the water with a small pole or spar across each end with pins driven through the spar securing it to the end logs. The top end of the spar was about four inches in diameter, and the pins (hardwood) were one and three-quarter inches in diameter. Eight or twelve other logs would be rolled on to the raft thus created. This joint, as it was called on the river, would then be fastened or bracketed together with the outside logs, from four to eight of them, tailing out to form a fish-tailed raft when finally completed. Each joint was eighteen feet long, and thirty feet wide, and the finished raft would be from seventy-two to one hundred and forty-four feet long, depending on the height of the water. On good days the corporation would deliver about four hundred joints, or 800,000 feet, to the contractor below Fredericton who would tow them to the mills outside St. John. Rafting was usually over by the end of October or the first week in November. Logs which came in late were shingled up on the bank and were rafted early the next spring. Most logs on the St. John were two years from the woods to the saw.[127]

River driving and booming practices did not change much in our time nor would they later. The rivers were cleaner because of dynamite; the telephone would give some aid, but basically, here, as at no other place in the industry, the logger remained dependent on pleasant weather, high water, even water, and good men.

NOTES

[1] Poem collected by William Fraser from A. H. Chisholm, "The Boss of Wind River," printed in *The Northern*, April, 1925.

[2] The first railroad made an impact in Bangor and Old Town. Bangor *Whig and Courier*, July 12, 1853; Bangor *Journal*, February 22, 1855.

[3] Mattawamkeag Lake Dam Company records, "Dividend Book." The largest shareholder was the S. F. Hersey Estate with 10 shares. Dividends were paid in 1862, 1864, 1871, 1872, 1874, 1876, 1877, 1878, 1879, 1881, 1882, 1883, 1885, 1886, 1887, 1888, 1889, 1892, 1893, 1897, 1898, 1899, 1900, 1904, 1905, 1908. These records are presently in the hands of Ballard F. Keith of Bangor who kindly allowed me to use them. Dividends were, of course, over and above the cost of the repair of the dam which occasionally came to a

fairly large amount, as in 1875. The law incorporating this company is in *Acts and Resolves of Maine, 1860,* chapter 452.

[4] The best account of this is in F. H. Eckstorm, and Mary W. Smyth: *Minstrelsy of Maine,* Cambridge, 1927, 38. Mrs. Eckstorm interviewed the great grandson of Joseph Peavey for the family account of the invention.

[5] Advertisements appear in Bangor *Daily Commercial,* February 15, 1875. There were two big concerns who made cant dogs in the Bangor area, the Bangor Edge Tool Company who held the patent, and R. W. Kimball, in Orono. By 1878 the Bangor firm had moved to a new location as the demand was so great—150 dozen axes, in addition to their cant dogs, 2,000 for the spring drives that year. Kimball started in this year, making 100. By 1885 he was making 6,000 cant dogs each year. His metal parts were all cast steel shipped to him from Pennsylvania. He sold his tool, with rock maple stocks, for $1.50 each. The Peavey firm expanded into a considerable shipping trade. In 1887, for instance, they made 2,500 for the Connecticut Lumber Company, and nearly as many more to be shipped to Washington Territory for a Seattle firm. By 1891 this firm operated a small saw mill, to produce the 28,000 maple stocks they expected to use that year. This information is from interviews with old-time Bangor residents such as William Hilton, from discussions with employees of Snow and Nealley, in Bangor, who presently hold title to the Peavey name for their cant dog, and from stories in the Bangor *Daily Commercial,* January 31, 1878; April 10, 1885; March 25, 1887; and March 14, 1891.

[6] Much of the foregoing material came from interviews conducted at various times with Harold Noble, Topsfield, Maine; William Hilton, Bangor, Maine; and I. H. Bragg, Patten, Maine, as well as from walking many miles over lumbering country. Other sources worth mentioning are Allen Eric, "River Life—Wild Life Among the River Drivers in the Wild Woods of Northern Maine," *The Industrial Journal,* September 7, 1888; *The Loggers,* Boston, 1870, where at 48-9 is an excellent description of a headworks and boom at Grand Lake in 1869; F. H. Eckstorm, *The Penobscot Man,* second edition, Boston, 1927; *Minstrelsy of Maine,* Cambridge, 1927; Springer, *Forest Life and Forest Trees,* Boston, 1851; and F. L. Barker, *Lake and Stream as I Have Known Them,* 66-127, (Androscoggin waters in 1870). Some of the Indian names indicate the difficulty: Aboljackarmegus—no trees, all bare or smooth; Mahkoniahgok—a hole in the river, "The Gulf"; Millinocket—lake with many irregularities; Passadumkeag—falls running over a gravel bed at the head of the rips; Sysladobsis—rocky lake. These examples are from Hubbard, Moosehead Lake, *op. cit.,* 193-213.

[7] *Minstrelsy of Maine,* 96-7. This song is, of course, a variant of a very old sailor song. Not all were this way, apparently as some early accounts stress the sobriety of the riverdrivers. *Bangor Whig and Courier,* April 1, 1839; March 6, 1844.

[8] Portland *Eastern Argus,* May 20, 22, 1874. The quotation is from May 20.

[9] Bangor *Daily Commercial,* April 24, May 6, 7, 10, June 25, 1878.

[10] Portland *Eastern Argus,* January 12, 1883.

[11] Portland *Eastern Argus,* January 16, May 5, 23, 1883; April 14, 1884.

[12] Bangor *Daily Commercial,* April 28, 29; May 2, 3, 4, 5, 6, 9, 13, 17, 18, 24, 1887; Portland *Eastern Argus,* April 30, 1887.

Driving the rivers 97

¹³ Portland *Eastern Argus*, April 28, 1875. In 1879 a similar drive came in even earlier. Bangor *Daily Commercial*, March 25, 1879.

¹⁴ Portland *Eastern Argus*, May 1, 1878.

¹⁵ *Ibid.*, June 20, 1879.

¹⁶ *Ibid.*, May 26, 1885; Bangor *Daily Commercial*, April 14, August 15, 1889.

¹⁷ In 1886 one million feet of pine came to Portland firms, mainly the Star Match Company. In 1889 their drive of 1,500,000 feet of pine was brought in a boom in Sebago Lake on June 6. They moved the logs across the lake with headworks and a windlass, and the total drive, river and lake combined, only took five weeks. The logs were transferred by railroad from Sebago to Portland. See Portland *Eastern Argus*, April 26, 1883; December 2, 1885; February 19, April 24, 1886 and Bangor *Daily Commercial*, June 6, 1889 for discussion of these early mixed drives.

¹⁸ *Paper Trade Journal*, February 3, 1877 has an account. I could find nothing of this in the Warren files, nor could the company's historians.

¹⁹ *The Industrial Journal*, July 18, 1884; April 17, 1885; Portland *Eastern Argus*, August 10, 1886.

²⁰ Oxford *Democrat*, May 20, 1864, the best description of the difficulties of Rumford Falls at this early time; June 26, August 7, 1868, for this very bad year. The drive of 11,000,000 feet did not arrive in Lewiston until August 1. In 1867 the Androscoggin drives cost the lives of eight river drivers. Portland *Eastern Argus*, June 8, 1867 quoting the Maine *Farmer*.

²¹ In addition during these years about three or four million feet each year was coming to the Androscoggin Water Power Company Mill in Lisbon. The Bethel Steam Mill Company also had from five to six million feet in the river themselves but they cut and drove their own usually in advance of the main drive. Portland *Eastern Argus*, May 22, 1871; Bangor *Daily Commercial*, June 11, 18, 1872; Oxford *Democrat*, May 16, 1876 on Bethel.

²² Oxford *Democrat*, April 16, 1877; Portland *Eastern Argus*, August 2, 1877; the next year, though fifty men took only sixty days to move all 17,000,-000 feet in the drive to Lewiston and Lisbon. The logs arrived June 26. See Bangor *Daily Commercial*, June 26, 1878.

²³ Bangor *Daily Commercial*, May 19, 1879. They put in sorting gaps with room for thirty men to handle the logs. The sawmill will reappear later in this account. See below, pages 115-6.

²⁴ *Ibid.*, April 29, May 27, June 17, August 2, 7, October 18, 26, November 5, 24, 1880; Portland *Eastern Argus*, August 13, November 11, 1880.

²⁵ Portland *Eastern Argus*, June 30, 1887 for drive and *The Industrial Journal*, April 26, 1889 for Bearce and Wilson.

²⁶ *The Industrial Journal*, August 3, 1883. In 1885 one man cut and purchased 15,000 cords in and around Paris for this drive. By 1889 it took sixty men at the Lewiston boom to split and pile the pulp. Four cars or sixty cords then went each day to Yarmouth by train. Portland *Eastern Argus*, April 14, 1885; *Industrial Journal*, May 24, 1889.

²⁷ Common river drivers went on the Swift and Androscoggin drives for $2.75 and $3.00 a day. Bangor *Daily Commercial*, April 8, 1890; April 8, 1891.

²⁸ Advertisement ran for a week in *Ibid.* starting April 8, 1893. Harold

Noble's father drove the Androscoggin for several years and the Sandy River once as men from Washington County annually left home for three months to take the drive on southern Maine waters in the 1890's. Interview, Harold Noble, Topsfield, Maine, July, 1962. Farmington *Chronicle*, May 6, 1875. They didn't go as wages were too slight this year.

[29] *The Industrial Journal*, June 30, 1893 reported on the boom and the possible court case. See Bangor *Daily Commercial*, August 24, 1893 for the lateness and difficulty of the drive. Horses were used with jumpers to twitch the logs back into the water so they might go down river. This was a fairly common sight on the Androscoggin anyway.

[30] Bangor *Daily Commercial*, July 8, 1872; May 16, 1875; August 12, 1879; Portland *Eastern Argus*, March 4, 1873; August 29, October 25, 1877; June 22, October 19, 1880.

[31] Portland *Eastern Argus*, June 14, 15, 1881; Bangor *Daily Commercial*, August 27, 1881.

[32] *The Industrial Journal*, June 12, 1885; Portland *Eastern Argus*, June 2, 1885; Bangor *Daily Commercial*, June 22, 1885. When the jam broke it came with such force that it parted the Weston boom at Skowhegan, and the Somerset boom at Fairfield, allowing some 20,000,000 feet of logs to go down to Augusta. The next year was also very difficult especially on the Dead River, Phillips *Phonograph*, August 4, 1886. Wilton *Record*, October 13, 20, 1886.

[33] Bangor *Daily Commercial*, September 5, 1885.

[34] *Ibid.*, October 8, 1886.

[35] *Ibid.*, July 6, 14, August 2, September 2, 1887; October 30, 1889.

[36] *Paper Trade Journal*, October 1, 1892, has an account of these events printed with a description of the new mill construction.

[37] *The Industrial Journal*, July 14, 1893 and Bangor *Daily Commercial*, August 10, 1893 for the jam. *Ibid.*, August 22, 1893 for the method of payment, November 6, 1893 for the assessment on operators, and August 13, 14, 15, 1893 for the railroad. The following table shows the dates for leaving Moosehead and arriving at Augusta at a somewhat later period, and when most of the drive was pulpwood.

Time of Kennebec Log Drive, 1900-1912

Year	Rear leaves Moosehead	Arrives Augusta
1900	June 23	August 27
1901	July 3	August 9
1902	June 15	August 15
1903	July 5	September 9
1904	July 2	September 8
1905	July 24	August 27
1906	June 28	September 15
1907	July 5	September 4
1908	June 23	August 31
1909	July 9	August 23
1910	July 7	August 18
1911	July 21	October 2
1912	July 3	August 30

Third Annual Report, State of Maine Water Storage Commission, 1912, 54-6. This report is best located today in Volume IV, *Public Documents, Maine*, 1913. Some other small streams stopped driving as the railroad came. One such was Sandy River in Farmington, see Farmington *Chronicle*, May 2, 9, 16, September 26, December 19, 1872; May 22, 29, June 5, 1873.

[38] Portland *Eastern Argus*, July 11, 1871. This was the biggest year on the St. Croix. Calais shipped 300 million feet of timber. *Ibid.*, July 15, 1871.

[39] Bangor *Daily Commercial*, June 11, 1873.

[40] *Ibid.*, June 19, 1879.

[41] *Ibid.*, June 8, 1880; May 27, 1887.

[42] *Ibid.*, May 16, 1885; April 9, 1888. One hundred and fifty men moved the 1885 drive down the river.

[43] *Ibid.*, May 16, 1885; *The Industrial Journal*, May 3, 1889.

[44] Bangor *Daily Commercial*, June 1, 22, 1891. The rain storm was June 19-20. The East river drive was ten million feet; the main river eighteen million.

[45] Bangor *Daily Commercial*, June 6, 1874.

[46] *Ibid.*, April 30, 1877.

[47] *Ibid.*, April 16, 1878.

[48] *Ibid.*, May 3, 1879.

[49] *Ibid.*, June 25, July 14, July 31, 1886.

[50] These figures come from the reports of the annual meetings, and special meetings, of the Mattawamkeag Log Driving Company. They were often reported in the local newspapers as well. See, Bangor *Daily Commercial*, March 19, 1878; March 12, 1879; March 14, 1883; March 14, 1888; March 15, 1892; and March 21, 1893, as well as Portland *Eastern Argus*, March 12, 1885. The main source is "Record of Meetings," a log book in the company records. These records are in the custody of Ballard Keith of Bangor who kindly allowed me to use them.

[51] Bangor *Daily Commercial*, January 26, 1875. It is surprising how close to Bangor that lumber was cut and driven to the mills. This lumber which came, of course, to the booms very early, often commanded premium prices. See, e.g., accounts of such small drives in Bangor *Daily Commercial*, May 8, 1879 (300,000 feet of hemlock on the Kenduskeag stream right in Bangor); May 12, 1887 (three small drives on Great Works Stream and Pushaw Stream in Old Town. The lumber was spruce and it sold for $13 M cash which was an excellent price.) April 16, 1889. (Nearly two million of spruce on Pushaw Stream which came to the boom April 12.) In 1890 a drive of cordwood for Boston markets came down the Kenduskeag from Lake Chemo, but the loss was high from sinkage. See *ibid.*, from July-November, 1890 for these operations amounting to just over 1,000,000 feet.

[52] Bangor *Daily Commercial*, March 5, 1886; March 16, 1887; January 14, 1890; March 8, 1892; March 24, 1896; March 29, 1898; March 29, 1899; March 20, 1900; March 21, 1901; March 16, 1903; February 22, 23, 1904; Portland *Eastern Argus*, January 22, 1885; Bangor *Daily News*, March 12, 1901, all reports of annual meetings of the Passadumkeag River Log Driving Company. The Log Driving Company also handled the affairs of the Passadumkeag Boom Company, and we have detailed records for one year, 1887, which are worth producing to show the variety of lumber being handled on the rivers. Seventy

men worked for three months at the boom that year, making up and moving 826 rafts of lumber. See Bangor *Daily Commercial*, August 11, 1887.

Lumber—Passadumkeag Boom—1887

Item	Amount (feet)
Pine	1,721,100
Spruce	4,640,360
Hemlock	2,505,250
Juniper (Hackmatack)	204,880
Cedar	445,460
Ash	34,570
Spars	72,760
Total	9,714,380

[53] Quoted in Bangor *Daily Commercial*, March 21, 1876. They always had been in demand. In 1853, 100 men left Richmond for Kentucky to cut and drive on the Ohio and Mississippi. Oxford *Democrat*, November 11, 1853. Even earlier 9 Mainers left Portland for Savannah to lumber, drive and saw on the Little Ocmulgee River in Georgia, *Maine Farmer*, September 26, 1834. Both of these operations were on Maine owned lands.

[54] *Ibid.*, April 17, August 15, September 21, 1876; Portland *Eastern Argus*, April 24, 1876. For general remarks about Maine men on these waters see Robert S. Pike, *Spiked Boots*, St. Johnsbury, Vermont, 1961. Edward Ives and the author have edited the autobiography of one of these early river drivers. *Fleetwood Pride 1864-1960: The Autobiography of a Maine Woodsman*, Vol. IX, *Northeast Folklore*, 1967 [Orono, Maine], 1968.

[55] Bangor *Daily Commercial*, April 13, 16, 1880; April 21, 1883; April 23, 30, 1888. Perhaps the most famous feat of a Bangor river driver was to cross the Hell Gate to Manhattan on a log as did Edward A. Chase in 1910. See *Paper Trade Journal*, June 16, 1910.

[56] This is Hempstead's best date. Alfred Geer Hempstead, *The Penobscot Boom and the Development of the West Branch of the Penobscot River for Log Driving*, Orono, 1931, 39-43.

[57] F. S. Davenport, "Some Pioneers of Moosehead-Chesuncook and Millinocket," in *The Northern*, Vol. II, No. 6, 5-6.

[58] Hempstead, *op. cit.*, 46-9. One famous rock dynamiting occurred on the Wassataquoik Stream in 1879. So much was used that it was poled up river in batteaux. The railroads refused it. *Pittston Farm Weekly*, March 7, 1963.

[59] The best description is in Hubbard's *Moosehead Lake, op. cit.*, 70-9.

[60] Figures from *ibid.*, 78.

[61] See reports of annual meetings of P.L.D. Co., in Bangor *Daily Commercial*, February 20, 1878; February 13, 1884; February 20, 1885; February 17, 1886; September 24, 1886 (a special meeting as the prices were so low on the original bid); February 16, 1887; February 14, 15, 1889; February 11, 1891; February 17, 1892; February 13, 1894; February 12, 1895; February 11, 1896; January 29, February 9, 1897; February 8, 1898; August 4, 1899 (special meeting); February 3, 1900; February 9, 10, 1904. In 1889 a combination was supposed to have been formed to keep the price high and make a large

profit out of driving. If this was so it was thwarted by Philo Strickland who bid in the drive and then let it to Cornelius Murphy, John Ross and James L. Smart who did the most driving for the P.L.D. in the years covered by this table.

[62] These dates and figures compiled from Bangor *Daily Commercial*, July 14, 1874; Portland *Eastern Argus*, August 5, 1874; Bangor *Daily News*, October 12, 1901.

[63] Orange Judd Company, New York, 1880. This book is a companion piece to his *Paddle and Portage*, Estes and Lauriat, Boston, 1882, an account of traveling from Moosehead Lake to the Aroostook River. Both of these works deserve to be better known.

[64] Portland *Eastern Argus*, June 6, 1870. Article entitled "Severe and Protracted Drought" and dateline Bangor, June 4. It was 90 degrees on June 5, and 96 degrees on June 6 in Bangor.

[65] Bangor *Daily Commercial*, June 30, July 8, 14, 17, August 4, 1874; Portland *Eastern Argus*, August 5, 1874, an account by a reporter at the festivities.

[66] Bangor *Daily Commercial*, April 16, May 12, 13, 1879. Quotation is from May 13, 1879.

[67] *Ibid.*, May 12, 13, 22, 23, June 17, 1879. Quotation is from May 22, 1879.

[68] Portland *Eastern Argus*, July 1, 1879.

[69] Bangor *Daily Commercial*, April 6, 1881. French mill and woods workers in the Patten, Stacyville, and Sherman areas were river driving by this time. *Katahdin Kalendar*, May 1, 1880.

[70] Bangor *Daily Commercial*, August 14, September 7, 1886; July 26, August 4, 19, 25, 30, 1887; July 25, 27, 30, August 10, 1889.

[71] *Ibid.*, February 11, 1890. For the Moosehead steamers see *Bangor Historical Magazine*, VI, 240.

[72] Bangor *Daily Commercial*, March 5, 18, 1890; *The Northern*, Vol. VI, No. 9, December 1926, 3-4; Hempstead, *op. cit.*, 50.

[73] Bangor *Daily Commercial*, March 5, 18, June 4, 5, 1890. My account of her steering apparatus, and the cause of the first failure is based on a reading of the newspaper accounts. I have been unable to find anyone who remembers seeing her before the changes to verify this account.

[74] *Ibid.*, June 23, 25, 26, 30, 1891.

[75] *Ibid.*, November 30, 1892. The *F. W. Ayer* appears to be something of a mystery, as there is almost no mention of her anywhere, including Hempstead, *op. cit.*, where she is barely mentioned on page 50, or in *The Northern*, which spent more time on West Branch history than on any other thing.

[76] Bangor *Daily Commercial*, July 17, 1894.

[77] *Ibid.*, January 5, 1894.

[78] The early history of the boom is in Bangor *Whig and Courier*, September 24, 1847; July 8, 1848; March 17, 1850; Bangor *Democrat*, February 14, 1843. *Senate Document #10*, Thirty-Third Legislature, Augusta, 1854; *Statement of Facts on the Petition of Henry Campbell for a Boom at Greenbush*, Augusta, 1847.

[79] Description of the boom and its operation is derived principally from David Norton, *Sketches of Old Town*, Bangor, 1881, 23-6. Norton was clerk

of the boom from 1837 to 1864. Other sources of importance are Hempstead, *op. cit.*, 16-29. Hempstead was much less interested in the boom than his title would imply, and the chapters discussing it were probably added as an afterthought. There is a good discussion in Bangor *Daily Commercial*, January 18, 1886. I have also profited from discussions with Ballard F. Keith who was the secretary of the Penobscot Lumbering Association at its demise, and who has custody of the records from 1916 to 1960, and Harold Noble, Topsfield, Maine who worked on the St. Croix boom, in the early period, and from general conversations with people in and about Bangor who remember the boom in active days.

[80] Bangor *Daily Commercial*, March 14, 1876; March 5, 1878 (Lunt and Webster started their rafting experience in 1872, a big year on the river); March 7, 1882; March 7, 1883; (rigging advanced 3¢ a lb.); March 5, 1884; March 4, 1885; March 6, 1886; March 1, 1887; March 4, 1890 (rigging was at its highest point in 1889); March 3, 1891 (this was perhaps the liveliest bidding as John Ross made a strong attempt to take over the rafting); March 1, 1892 (the association agreed to provide wedges to the contractor at a price of $1.50 M.); and March 7, 1893 (again furious and spirited bidding between Ross and Lowe.) These are all reports of the annual meeting of the Penobscot Lumbering Association. The first rafts of lumber noticed were at Foxcroft in 1811. In Asa Loring, *History of Piscataquis County*, 1880, 176.

[81] Portland *Eastern Argus*, March 7, 1883. They sent a reporter to cover the annual meeting of the P.L.A. because of the interest in the legislative renewal of the charter.

[82] See advertisement in Bangor *Daily Commercial*, July 26, 1878.

[83] See advertisement in Bangor *Daily Commercial*, June 12, 1878.

[84] *Ibid.*, July 27, 28, 1872; quotation is from July 27.

[85] *Ibid.*, May 15, 16, 17, 1879.

[86] *Ibid.*, August 11, 1891.

[87] *Ibid.*, November 27, 1872. The largest raft I have located was one of Rufus Dwinel's in 1855. It surveyed 113,327 feet. *Drew's Rural Intelligencer*, June 2, 1855.

[88] Bangor *Daily Commercial*, April 29, 1873.

[89] *Ibid.*, August 1, 1873, account of rafting difficulties in low water.

[90] *Ibid.*, April 19, 1876.

[91] *Drew's Rural Intelligencer*, December 1, 1855.

[92] Portland *Eastern Argus*, July 21, 1876.

[93] Bangor *Daily Commercial*, July 5, 1877. I could find no evidence that the Nebraska boommen ever took up the challenge.

[94] This paragraph depends on various sources. Chief among them are Norton, *Sketches of Old Town*, 25-6 (tables); Portland *Eastern Argus*, January 22, 1885; Bangor *Daily Commercial*, January 29, March 29, May 28, 1877; March 5, November 2, 12, 1878; June 8, October 7, 1880; June 16, October 19, 1881; July 7, 1882; April 7, 11, December 3, 1884; March 4, April 4, December 20, 1885; November 11, 15, 1887; September 10, 22, October 10, 11, 22, November 15, 1888; July 10, 23, 27, August 10, 24, 26, 29, 1889; August 15, 23, 1890; September 22, 1891; October 3, 1892; November 16, 1893.

[95] Bangor *Daily Commercial*, June 8, 1880; June 16, 1881.

[96] *Ibid.*, September 10, 22, October 10, 11, 22, November 15, 1888.

Driving the rivers 103

[97] *Ibid.*, August 29, 1889.

[98] *Ibid.*, July 7, 1882, has an account of the Fourth of July races in 1882, perhaps the biggest and best year for them. The bateau races were 2½ miles long; four crews entered. Five canoes entered the ⅝ mile race on the Kenduskeag stream, and three men ventured into the log race equipped with setting poles.

[99] Bangor *Daily Commercial*, January 18, 1886, interview with Connors.

[100] Portland *Eastern Argus*, November 22, 1878, report on the season.

[101] Bangor *Daily Commercial*, June 18, 1888.

[102] *Ibid.*, January 22, 24, 25, 26, 1889. The Legislative Records for 1889 have almost nothing on this squabble; it was not until later that better records appear. It is presumed that the new mill builders were those investing in F. W. Ayer's plant in South Brewer. The records of the Bangor Boom Company have apparently disappeared; at least, I could not locate them, nor could I locate anyone who had ever seen them. It is conceivable that one of the big companies, such as Morse and Co., or the Eastern Paper Company, purchased the boom in in the last days in order to control it, and they may well have the books and papers stored away somewhere. This has universally been the fate of such records in Maine.

[103] *Annual Address of the Mayor*, 1866, 1868, *Annual Reports*, City of Bangor; *Annual Report* of the Harbourmaster, 1867, 1870, *ibid.*

[104] Bangor *Daily Commercial*, April 7, 1881, for their editorial, and August 6, 1883 for the big jam and Connors' threats.

[105] *Ibid.*, April 23 1872. Dates for ice leaving the Penobscot as well as the dates for freeze up are tabulated in the Appendix.

[106] Bangor *Daily Commercial*, May 8, 1872.

[107] The best account of these parties is in *ibid.*, June 15, 1876, but see the *Commercial* in July and August for all the seventies.

[108] *Ibid.*, June 23, 1874.

[109] Bangor *Daily Commercial*, April 16, 26, May 13, 1879.

[110] Letter of H. B. Maynard in *ibid.*, April 30, 1879 offering to remedy the situation for $1,200, and letter of James H. Sheehan, May 10, 1879, offering to do it for $1,000, and further offering to wager $100 to $500 that rafts would carry eggs over the dam if his way was followed.

[111] *Ibid.*, May 9, 20, 1879.

[112] *Ibid.*, September 13, November 3, 1879.

[113] Bangor *Daily Commercial*, June 28, October 5, November 7, 17, 1880; October 7, 9, 1881; November 17, December 18, 1882; November 1, 7, 15, 1887; November 19, 20, 22, 1888; November 12, 1889; June 10, October 19, 1891; August 6, 7, October 4, 1892; November 16, 1893.

[114] Testimony of Harold S. Boardman and Cyrus C. Babb, *St. John River Commission*, transcript of testimony, hereinafter cited as *SJRC*, and further described in bibliography, 1751-1800, and especially 1792-3; testimony of Boardman recalled, 1801-1825, especially 1801.

[115] See Report of William Parrott to Committee on Inland Waters, December 1, 1843, found in *SJRC*, 2417-2499, and "Evidence Taken Before the Committee on Inland Waters in 1846", found in *SJRC*, 2566-2690. This can also be found in the University of Maine Library, and Maine State Library, Augusta, where it is bound and catalogued under the title, *The Telos Canal*. By the

diversion of the waters it had been possible to cut, from 1842-1846, 22, 722, 859 feet above the canal, and about 10 million feet on T6R11 where the canal is located. A detailed summary of the cut appears at 2624 of *SJRC*. There is a description of the canal as it appeared in this time in Hubbard's *Woods and Lakes, op. cit.*, 223.

[116] Testimony of Richard Hand, *SJRC*, 2124-2148. Hand as a boy was with the crew which destroyed this dam.

[117] Bangor *Daily Commercial*, May 23, July 1, 1874; June 8, 1877; May 9, 1878.

[118] *Ibid.*, June 4, 8, 30, 1887; May 21, June 1, 2, 3, 14, 18, 1889. Other years were difficult as well, *Presque Isle Sunrise*, May 1, 29, July 3, 1868.

[119] A good description of the river is in Hubbard's *Moosehead Lake, op. cit.*, 126-33, especially 133 for the Grand Falls. Also see Rev. J. P. Raymond, *The River St. John*, (Fredericton, 1896), for the entire river. From the point of view of the river driver an interesting document is the testimony of Richard A. Estey, *SJRC*, 1200-17, which describes the river almost rock for rock. On the steam driller see Testimony of George S. Cushing, *SJRC*, 1283-4. Still problems occurred in 1895 and some small law suits were filed. St. John *Daily Sun*, January 30, November 26, 1896.

[120] Testimony of John Sweeney, *SJRC*, 232-98; at 254 the following occurred:

> Q. Were you a river driver?
> A. Well, I thought I was. I was tripping dog for a four horse team, using a peavey.

Also see Testimony of Jerome "Come" Cyr, *SJRC*, 1604-29.

[121] Testimony of John A. Morrison, *SJRC*, 655-714, this at 667.

[122] On the hay and cattle, see Testimony of Louis H. Bliss, *SJRC*, 1153; on the woodboats see George MacBeath, "Johnny Woodboat", *American Neptune*, XVII, No. 1, January 1957, 5-16; on the steamboats Testimony of Francis Violette, *SJRC*, 1440-1445. There is a lovely cut of the drive in "Driving Logs Down the Falls of the St. John," *The Illustrated London News*, August 28, 1858.

[123] On driving, Testimony of John Kilbourn, *SJRC*, 587-652; "Early Days of Log Driving on the St. John River," *Canada Lumberman*, September 15, 1925.

[124] Bangor *Daily Commercial*, May 31, June 11, May 17, October 2, 13, 1873; June 4, 30, 1887; July 2, 3, September 20, 26, 1889.

[125] Testimony of Joseph (John) Savage, *SJRC*, 2196-2217, (one of the men who hoisted the dam); testimony of Edward McElveney, 2249-65; and Cyrus Dickey, 2218-22; quotation is at 2221.

[126] Portland *Eastern Argus*, August 2, 1883, a detailed report. In 1909 the corporation owned the following items: a dewelling house, blacksmith shop, carpenter shop, repair shop, turning shop for rafting pins and wedges, a power plant at St. Mary's Ferry, and at Douglass a boom house, boom, eating house, sleeping house, cook house, bakery. At Mitchell's boom a boom house and foreman's house as well as a camp for crewmen. It also owned a floating scow for

Driving the rivers 105

drift drive purposes, 10 miles of log booms, 63 jam piers, 12 block piers, 12 sunken piers, 1,000 feet of sheer running boom, 2 steamers, 3 steam scows (one equipped with a pile driver), cordage, chains, small boats, a gasoline launch, $1,500 worth of boom chain, and miscellaneous small chains. The total value was $148,000. Testimony of Louis H. Bliss, *SJRC*, 1112-50, especially 1113-1115.

[127] The best description of sorting, rafting, and towing at the Fredericton boom is in the testimony of J. Fraser Gregory, *SJRC*, 187-231, and Gregory recalled, 2914-2950, especially 2921. Also see Portland *Eastern Argus*, August 2, 1883.

5

THE MANUFACTURE OF LUMBER—1860-1890

> Of course, everyone in the lumber business knows how sawmill property in this vicinity has fallen in value within the past few years, but it is doubtful if the people generally have any idea of the great change that has taken place in this respect.
>
> For instance, when the fine Palmer and Johnson mill was built and after it had been repaired a few years ago, the property was worth fifty or sixty thousand dollars. Then, from 75 to 100 men were employed there, while now it is wholly unused and has been for the past two years. Today, it would probably be difficult to find a purchaser for it at $15,000 and one of the heaviest dealers and manufacturers of lumber here recently remarked that he did not know as he should be willing to pay $12,000. The Bangor Savings Bank has a mortgage on it for three times that sum, but it is doubtful if they realize half that amount on the property for some time to come. Several years ago the property began to fall off in value, and has continued to fall ever since. This is only used as an illustration of how property of that kind has fallen. . . . Bangor . . . must for the present turn its attention to something else.
>
> Bangor Daily Commercial,
July 28, 1879

Once the logs were cut, driven down the river, and sorted out to their owners, they were then turned as rapidly as possible into boards, planks, scantling, laths, clapboards, shingles, edgings, and sawdust. The finished product was then carried away to its destination to be used for building materials. The process of manufacture was a simple and profitable one at the outset of this period. The larger mills were run by steam, but most were still operated by water power. When the water was at the correct sawing pitch and the logs were available the owners of the mills would "hist" their gates and commence their sawing. With the exception of low water times the sawing would continue until the fall freeze-up.

In the southern half of the state most of the mills were owned by the operators and they serviced local and Boston dealers. On the Kennebec the operations and the mills were large and the markets were often farther afield. On the Penobscot, where lumber was king, the mills were usually owned by large capitalists who then rented them to the operators. These mills manufactured everything, but Bangor was noted mostly for its long lumber whch went all over the world. Farther north along the coast the operations were more varied, and much of the lumber cut went to foreign markets, especially to the Caribbean. In the far north the business was dominated by a few capitalists, many of them operating in Canada with Maine capital, who sent their long lumber to England as long as it was economically feasible, and manufactured much, if not most of their Maine logs into short lumber for tariff-free re-transfer back to Bangor, Portland, or Boston dealers.[1]

From 1872 on the sawmill end of the lumber business was in a serious depression. Capitalists moved into land ownership to protect their equity, and at the end of the seventies they had begun to search for new methods of investment. Mill property declined in value and bankruptcies were frequent. South of the Kennebec fewer and fewer mills were in operation. On the Kennebec and north the mills grew larger in the downriver towns, but more and more small mills, predecessors of the portable mills of a later day, began to appear. The economic squeeze caused the mill operators to search for many different methods of protecting themselves against this pressure. Most millmen must have looked at the pulp paper revolution with sighs of relief by the turn of the century, although the revolution was not to be entirely a peaceful one, probably because so much was at stake.

Prior to the coming of the depression in 1873 the lumber business had been very much every man for himself. There were occasional meetings, as in 1861, to discuss possible tariff provisions, but by and large operators looked out for themselves and expected their competitors to do the same. After the bonanza year it was obvious that such methods would not be of much value as prices began to decline precipitously. Long before the Sprague interests went down in the fall of 1873 to trigger the depression, lumber manufacturers had called a convention to meet in Bangor. This meeting was the first real attempt at creating a lumber trade association in the northeast.

Once the lumbermen had come together they acted with considerable dispatch. Any dealer in Maine or New Brunswick was allowed to become a member of the Association upon payment of a $2.00 fee and with the concurrence of two-thirds of the present membership. Each company had one vote, but all members could debate. It was proposed to appoint one man who would have the duty of regulating the survey of all lumber sold in the territory. This man could in turn appoint an assistant who "would consult with dealers and manufacturers to establish such rules and regulations in regard to uniformity of price and survey as they deem for the interest of the association." [2]

A Committee on Sale and Survey had been appointed and it made an immediate report. The report suggested that all areas adopt a similar survey and employ a surveyor general. In addition it recommended that all lumber to be sold should be measured by the Eastern or Home survey, and that all prices be FOB wharves or mills. The Committee also proposed that the minimum price for random spruce be set at $15.00 per M.[3] This last provision was not adopted, although the first two were.

The second day of the convention J. Manchester Haynes, a large operator of Augusta, reintroduced the $15.00 per M provision, but the convention adopted the following resolution:[4]

> Resolved—that the price of merchantable and random spruce be not less than $14.00 per M. The price for seasoned hemlock not to be less than $12.00 per M. If during the season the judgment of the General Committee is that an increase can be demanded, they are to notify the several local committees and recommend an increase in price.

Some discussion of the lath business followed. A few members advocated the sale of laths by dimension rather than by the bunch, but the discussion died a desultory death. The last business of the convention was a return to the problem of enforcing the price minimum. When Hastings Strickland of Bangor wanted to know what the Association wished to do about undersellers, Haynes said that the undersale price would become the standard, and Peleg Wadsworth of Calais remarked that those who undersold the $14.00 M price would never survive long enough to make serious competition.[5] With this the convention adjourned, and the first attempt to beat the stringencies of economics was history. It apparently failed. One never hears of it again, and therefore the suspicion is that undersellers were able to profit, (at least prices dropped below $14.00 per M for random spruce), and the Association failed because of its inability to control its own members.

The next year a further attempt at organization was apparently made although almost nothing is known of it. In August most of the larger mills on the Kennebec and Penobscot shut down, and there were rumors that this was an attempt to drive prices up and wages down by the operators acting in concert.[6] After this time cooperative efforts lay fallow until the next bad years, in the early eighties. At that time a new call was issued to lumber dealers to meet in Portland to try and control prices. The new group was more limited in its membership. It only involved those shipping their lumber over the Grand Trunk, the Portland and Ogdensburg, and the Passumsic Valley railroads. A few attended as observers from Lewiston, the Kennebec region, and the upper Connecticut Valley.

The convention was called by men most of whom were members of the Connecticut River and Northern Lumber Association, which had been in operation since 1880. They met first at Berlin Falls and then issued a wider call for a meeting to be held in Portland. When they came together in Portland they immediately formed a group to be known as the Lumber Manufacturer's Protective Union, and set the minimum price on random spruce at $15.50 per M, $1.00 over the then Boston price. It was said that those who met controlled an annual average cut of 235 million feet. After a two-day discussion the price demands were revised. The final decision of the group was that all lumber ten inches and under in width and twenty-five feet and under in length would sell for $13.00 M in Portland and Biddeford, and $15.00 M at Boston. All lumber twelve to fourteen inches wide would sell for $16.00 and that over fourteen inches wide at $16.50 in the Boston market. Random spruce twelve feet long and over would go at $14.00. Planing and grooving on one side would increase the price $1.50 per M; on both sides $2.00. Planing alone would cost $1.00 for the first side, and 50¢ for the second. The water shippers were not included in the convention, but they raised their price to about this level on receipt of the news of the price-fixing.[7] This Association did not die with adjournment, and in fact held two more meetings at least, the first of which in July, offered a 5% discount to purchasers for quick payments. The second, in September, withdrew the discount and raised the price of laths 25¢ per M.[8] Again there was no enforcement of price, and the Association attempts to void the economic laws foundered on the rocks and shoals of private enterprise. Without some method of rigid enforcement and thus compliance such attempts at price fixing were of no avail. There was a violent distrust of cooperative efforts. The lumber manufacturers would not cooperate until the last extremity, in the bitter battles of the first decade of the twentieth century, and by this time the war was over. Supply and demand determined the price of random spruce in Maine and Boston markets, and this price was nearer $12.00 a

Manufacture of lumber

M, not $14.00 or $15.00. Only the large manufacturers could survive at these figures and price-fixing failed to stem the tide.[9]

With the exception of these price-fixing attempts, events remained fairly stable in the sawmill towns as each manufacturer moved along about as he pleased, until 1886 when the Knights of Labor appeared. Labor troubles away from the sawmill towns had caused the lumber manufacturing business to take a setback, and a number of lumber manufacturers met at Waterville in an effort to form an organization which would regulate the hours of labor, the output of the men, and the selling prices of the product.[10] Alas for the manufacturers, the Knights had already penetrated into Maine laboring circles and they had their own ideas about hours, wages, and output. Stevedores on the Kennebec refused to load ice because of low pay, and the news of this was quickly followed by a major strike at the Richards Paper Company in Gardiner.

The men were fired and asked to vacate their tenements. In rapid succession others went on strike. The crews at Milliken's sawmill, Lawrence Brothers, Bradstreet Brothers, and Putnam and Clossen, all of Gardiner, went out on strike, asking for the ten hour day rather than the regular working time of 11½ hours. They asked for the same pay after the cut in hours. The strike involved three hundred and twenty-five men. Sawmills at Richmond did not open at all because of the laborers' demands. Disaffection spread. Bangor stevedores circulated a paper calling for a minimum wage of 30¢ M for coastal loading, 38¢ for the West Indies, 45¢ for South America, and $1.00 for a standard of deals. In Mechanic Falls the Dennison Paper Company skipped a payday, and the Knights of Labor prevailed on the operatives not to appear for work.[11]

Some of the strikes were quite successful. On the Kennebec the ten hour day became a reality for many of the larger mills, and the men went back to work. In Gardiner the Richards Paper Company granted a small wage increase and these men also returned. The Bangor stevedores were less fortunate, although some increase was granted. In Mechanic Falls Dennison's were through although others would take over the operation. For these men the passage of the fortnightly pay law by the next legislature would be a real boon.[12]

The success of the Kennebec operatives in getting the ten hour day in the sawmills apparently emboldened the men working in the tidewater mills at Bangor, although it was some time before they made their demands, and the demands were not occasioned by a nationwide effort such as the Knights of Labor. Still the success of the

Kennebec operation was of importance in determining the results of the strike. A great sawmill strike, perhaps the largest in Bangor's history, occurred in 1889. This strike is worth studying in some detail as it sheds light on conditions in the Penobscot sawmills, and enables us to determine the relationship between employers and employees at the time of transition in the Maine lumbering industry.

On June 14, 1889 a meeting was called at the East Hampden schoolhouse of all employees of the tidewater sawmills. More than three hundred attended. The meeting was orderly, and the discussion was concerned almost entirely with the hours of labor. The working hours then involved a twelve and one half hour day made up as follows: the first whistle blew at 4:30 a.m., and the men worked until 6:00 a.m. when they were given a half hour for breakfast. Dinner came at noon for half an hour, and the day finished at 6:00 p.m. The men asked for the day to begin at six, with breakfast beforehand, and work to finish at six, with an hour at noon, making an eleven hour day. A committee of three men was appointed from each mill to consult with the owners and operators as to these demands and to find out the regular tour of duty in the up-river mills.

The demands were refused, and after a second meeting the men voted to go out on strike. Sterns and Company did grant an hour at noon, and these men went back to work after one day. At a third meeting held a week after the first the men voted to continue the strike although only about half the men voted. The mill owners asked the *Commercial* to suppress all news of the strike, and the newspaper carried a bitter editorial saying that news was news, and suppression of news did no one any good except where personal grief was concerned.[13] Public sympathy began to build up on the side of the strikers. All the men at Morse and Co., Hodgkins and Hall, and F. W. Ayer were out on strike. Only D. Sargent and Sons, and Gould and Hastings were able to continue to operate. On June 24th, Sterns and Co., and Gould and Hastings went out on strike, and the idle men went to Sargent's to discuss the strike with the workers. A good deal of harsh talk ensued but no fighting as the sheriff and his deputies broke up the meeting. This only increased the public sentiment for the strikers and a large public meeting was held in Bangor, particularly as word spread that some of the men had been paid off and attempts were being made to hire scab labor. On the 25th Sargent's closed, the Bangor Boom closed down as the men handling the logs joined in a sympathy strike, and word was received in Bangor that the men in the up-river mills were ready to walk out as well. A mass meeting was held in Bangor and $200.00 was collected to pay possible fines incurred by the strikers as some tempers began to be short. The city closed all saloons for the duration of the strike in an effort to forestall difficulties.

The *Commercial*, by this time easily the foremost Bangor newspaper, now stepped in in an effort to stop the strike while at the same time achieving the demands of the men. A reporter went to the Kennebec in an effort to find out the effect of the shortened work day in that locality. He found that most mills worked an 11½ hour day, with many having a ten hour day. In Lewiston the ten hour day was the rule. The men worked from 6:30 a.m. to 6:00 p.m., with an hour's nooning, and until 5:00 p.m. on Saturdays for the first half of the season. On August 1st the men started getting out of the mills one half hour earlier on weekdays. Ira H. Randall, president of the Augusta Lumber Company, said that since they had gone on this schedule their production had increased from 5,000 to 8,000 more feet of lumber per day than on the old 11½ hour schedule. At the Augusta Lumber Company the men were paid $4.50 a day for the head sawyer, $2.75 for the head edger, $2.00 for lath machine work, outside men $1.75 and teamsters $1.50 to $1.75. Some boys got $1.25. This was not the rule though. The average wage in Augusta was $3.00 to $4.00 for sawyers, $2.00 to $2.50 for edgers, and $1.25 to $1.75, with $1.50 being the average for regular men.[14] After discussing the Kennebec findings the *Commercial* advocated arbitration of the strike with three men from each side, the six to nominate an impartial seventh to moderate. The strikers accepted the offer and went to the mayor asking him to set up the arbitration. The mill-owners would not meet for the purpose of discussing the move.

On the 27th, with the saloons still closed, the mill-owners also met with the mayor in an effort to promote the *Commercial's* plan, but the intransigence of the mill owners defeated all efforts.

The strike continued. Saloons at Old Town and Great Works were closed as it was felt that sympathy strikes and physical difficulties might arise among the up-river employees. Vessels in the harbor began to report difficulty in filling their charters as the mills remained idle. It was not until after the 4th of July that the strike began to die out.

First to go back to work on the 8th were the men at Morse and Co., D. Sargent and Sons, and Hodgkins and Hall. They went to work at six and worked till six, with a half hour for dinner, splitting the difference in dispute. On the 9th the men at F. W. Ayer went back to work getting a ten hour day but taking a cut in wages. This was decided on the vote of the men as they were given the option of following other mills or taking the ten hour day. The Dirigo mill went back on the old hours, and the other three remained out on strike. Finally on July 15th the remainder of the men went back to work, six to six with a half hour nooning. Many of them had gone haying in the interim, and these last mills were forced to delay full operations until after the first of August. The men had won a partial victory. It was not a union victory but it was a concerted one. With a union they might have obtained their full demands but whether they would have had the aid of the newspapers and the public is debatable.[15] This was the only big strike to appear in Maine either in the woods or elsewhere throughout the period. Perhaps the most successful thing about the strike was that no violence occurred in an era when violence was the ordinary way in which labor's demands were met.

One other small strike occurred during the period involving wages rather than hours. In June of 1890 the workers at mills of H. F. Eaton and Sons, and J. S. Murchie's in Calais went out on strike. The men were getting $1.25 a day and they asked for $1.50. Murchie offered them $1.35 and promised them $1.50 if he found conditions similar after a trip to Bangor. These men went back to work. The Eatons offered nothing, fired the men, and maintained their mill during the summer using scab labor, paid by the M, under police protection.[16]

A further attempt was made by lumbermen to meet and form a trade association to regulate prices. Most of the men of the Kennebec, who were feeling the pulp competition very strongly, met to do something about the price of logs which had been driven very high as the pulp mills bought from the loggers. They proposed to pay for their logs at boom scale, not woods scale which was notoriously off the mark. The combination hoped by doing this to set maximum prices for logs. At a second meeting all but two of the millmen agreed to buy at the re-scale, but this last dying gesture of the sawmill operators did not amount to much. Rather than fight the competition most decided to join it.[17] Later that year, when it was thought that lumber might go on the free list under the new Wilson tariff, lumbermen met at Bangor to protest the move, and a delegation from Bangor, Calais, Ellsworth, and Lewiston went to Washington to issue their protests in person.[18] This was the most successful effort on the part of lumbermen to combine their forces. On the tariff all agreed; on prices, wages, or hours of labor it was still every man for himself.

Manufacture of lumber

These instances of unified action, both on the part of the employer and the employee, are indicative of the changing character of sawmilling. In the southern half of the state mills grew smaller and disappeared, and in the northern half they grew larger and more modern. Both of these changes simply reflected attempts to deal with the changes coming in the woods. Like most regimes which are faced with revolution these attempts to modify the changes were quite unsuccessful. Small mills or large, eventually the pulp grinder would overcome them all.

In the south smaller operations were the rule. Perhaps the largest firms were the Lewiston Steam Mill Company and the Berlin Mills Company, both of which have been treated briefly above. These mills sawed their usual amounts, from ten to twenty million feet a year, and shipped their output to Boston by railroad. They also suffered the usual vissicitudes of business culminating in the depression of 1873. It was not until 1876 that demand increased enough on the Androscoggin to be an item of note.[19] By 1881 demand was good, and the Lewiston firm was shipping from forty to fifty carloads of lumber a week to Boston. This business continued until 1889 when the firm suffered great fire damage which ended sawmill operations in that area. Even before, the Lewiston firm had begun to shift to the contract cutting of pulpwood, the work for which it is known in the nineties.[20]

An interesting medium size sawmill operation was the Bethel Steam Mill Company. This mill, which was started just as the Civil War broke out, suffered a blow in 1863 when it was destroyed by fire. The people of Bethel held an emergency town meeting and voted to exempt the owners from taxation for ten years if they would rebuild. After the new construction, the mill ran for many years and was quite successful. In 1866 it sawed eight million feet of long lumber, 1,250,000 shingles, and 125,000 clapboards. About three hundred men were employed in the woods obtaining 10,000,000 feet of spruce and pine logs. This was the last big year though. In 1876 the total production was only about a million feet, and in 1877 only two million. By the end of the decade, the new owner was sawing spool strips for England, two million feet, and headings, shovel handles, and shingles. They sawed a little cedar into long lumber, but most of their production was the manufacture of eight hundred cords of poplar and birch into headings and handles.[21]

The only mill of any size to be constructed during this time was that built in Canton at the end of the seventies. This firm, the Canton

Steam Mill Company, started as a small operation, about 4,000,000 feet annually, near Berlin, New Hampshire. In 1878 it began expansion south to Canton. It built a steam sawmill with the newest equipment. Dining rooms and bunk houses were built for the men; new booms were hung in the river, and the Buckfield and Canton Point railroad erected a spur to the mill. Canton exempted the builders from taxation for ten years.[22]

The mill was immediately successful, so much so that in less than two years its owners purchased another mill, timberlands, and a farm property in Coos County, New Hampshire. In 1881, their employees cut above 15,000,000 feet to supply these three mills. The mill at Canton continued to prosper. In 1885 the production was more than 6,000,000 feet of long lumber and the month of October saw 910,000 feet leave for Massachusetts points.[23] Eventually this mill would be taken over by the pulp interests, but for a brief ten year span it was quite successful.

Although this section of northwestern Maine is in the middle of the lumber country and the number of sawmills was larger, a survey of the state at this time showed nearly every town and hamlet in the state as possessing one or more mills dependent on the lumber industry. There were spool mills, stave mills, toothpick mills, shovel handle mills, casket mills, shingle mills, and lumber mills, all small but the economy of the state depended on them to a surprising degree.[24] Occasionally they were multi-purpose mills such as that in North Turner. In 1878 this mill produced 12,792 corn boxes, 2,072 boot and shoe cases, 250,000 shingles, 15,000 laths, 5,000 clapboards, 300,000 feet of long lumber (pine), and it ground 1,500 bushels of grain.[25] This type of operation was not at all unique in Maine in the 1870's and 1880's. One even has a feeling that the number of such mills increased at the time of transition. By the time that Boardman made his survey in the mid-nineties there were only four mills of any size south of the Kennebec, and only the Berlin Mills Company, with its 40,000,000 feet a year, could be rated large.[26] There were a few small mills left but most sawed only thirty to forty thousand feet a year and made up the difference in grinding grain, hoping for a quick sale to someone willing to toss money away. There was one new development. In 1893, an Ellsworth man, C. W. Morrison, put a portable mill in at Jackson, New Hampshire, and cut 1,400,000 feet of spruce.[27] This mill, right at the cutting, was the voice of the twentieth century coming early.

Manufacture of lumber 117

Further north in the state, when the deal trade[28] was a major factor in Maine lumber during the seventies, Wiscasset and other coastal towns were big sawmilling centers. The Kennebec Land and Lumber Company's big mill at Wiscasset spit out the deal for England at a steady pace, reaching 121,477 feet in one ten hour day. During that month the mill produced 1,656,000 feet, averaging 61,000 feet a day.[29] Another big mill near the tidewater was that at Bath of M. G. Shaw and Sons of Greenville. In 1884 and 1885 this mill ran about six months and produced each year about 6,000,000 feet of long lumber as well as 3,000,000 each of shingles, laths, and clapboards, in addition to 600,000 or so slats.[30]

Upriver from Augusta the big sawmill town was Fairfield. At the end of the seventies the production of mills located here was about fifteen million feet annually, and the Maine Central Railroad sent out from sixteen to seventeen hundred railroad cars of lumber each year.[31] In the next decade these mills grew even larger. The largest was run by Lawrence, Phillips, and Co., and it had some big runs. For instance, in one June week of 1887 its saws produced 400,658 feet of long lumber, 163,000 shingles, 24,000 clapboards, and 115,000 laths.[32] A new band saw in 1891 increased the output until on June 11 of that year the record was set—96,883 feet of long lumber as well as 35,000 shingles, 5,500 clapboards, and 25,000 laths.[33] The table will give some indication of the production of these large mills at the time of transition.[34]

Lawrence, Phillips, and Company Production
Fairfield, Maine—1889, 1891

Year	Long Lumber	Shingles	Clapboards	Laths	Curtain Rods
1889	10,064,548 ft.	4,622,000	481,000	not known	
1891	9,531,780	4,995,150	569,775	3,003,900	39,555

There were other big operators in the Kennebec area, among them the Bradstreets of South Gardiner and the Lawrences of Gardiner. Most of them sawed about 10,000,000 feet annually into all sorts of stock. The mills had begun to move up the river though, and the great days of sawing in Augusta, Gardiner, Richmond, Bath, and Wiscasset were about all through. Fairfield was the big town, but these towns would not fight off the challenge of the pulp grinders very long.

Smaller specially mills dotted the valley. These mills, most of them dealing in novelty products such as croquet sets, lasted throughout the period but few of them could have made much money after 1880.

One of these smaller mills was quite enterprising though and it deserves passing mention. The Sandy River Lumber Company owned 4,000 acres on Letter E township in northern Franklin County. It built a mill eleven miles from the nearest town with a daily capacity of 22,000 feet. This stump-to-market operation ran all year round, and the employers planned it to cut for ten years in the area. To facilitate the toting of supplies and carriage of their lumber production the firm constructed a wooden railroad to Phillips, with fifty-five horses doing the toting. They hauled six or seven cars a day over their road.[35] Another small mill worth mention was that of Packard and Staples in West Phillips. Here the men went in to cut and yard with oxen and hauled to their portable steam mill located on Township No. 6. After the wood was yarded the men worked at the mill. A fire delayed them briefly but for the months of January and February 1887 the woods fairly hummed with logs, and then boards, and clapboards. In July the mill closed as the men went haying but it soon opened again, sawing in its biggest day over 100,000 feet. Others were active in the same general locality. Small portable mills with a good chance were very successful.[36]

When Boardman came to make his survey in the early nineties he estimated the total possible production of Kennebec mills at 126 million feet. Even he, though, said that twenty million was idle, and in all truth the figure was probably nearer forty million.[37] The Kennebec days were coming to an end for sawmilling, although the transition on this river was nowhere nearly as sharply contested as on the other rivers of the state.

When one thinks of sawmilling in these days of the nineteenth century it is Bangor and the Penobscot which come to mind. Here was the premier sawmill town. When other towns cut their 100,000,-000 feet a year, Bangor produced 200,000,000. The scream of the saw, the hum of the steam engine, the shouts of the men moving the logs to their destination—these were the sounds of Bangor town on the Penobscot River. The mills on the Penobscot were nearly all clustered in an area from Hampden to Stillwater on the south side of the river, and from Brewer to Milford on the north, although on that bank most were in Brewer. Many of them were great barnlike structures which had been built originally in the speculation of the

Manufacture of lumber 119

1850's. They were highly susceptible to fire, and during the sixties and seventies nearly all of these old mills went up in flame and smoke. Although it is easy to say that this fire damage hastened the inevitable day when sawmilling as a way of life would be extinct, or nearly so, on the Penobscot, as a matter of fact, the men who were in this business regarded such fires as simply an ordinary hazard of the trade and went about their business building new mills. The fire danger was always present, and the decade of the seventies seemed to show a constant scene of fire damage.

The worst year for fires in this time was 1876. The losses were very heavy for all sawmillers. Early in the spring the freshet had carried away the mill of Gilman and Webster of Orono. They leased another mill, but on September 16 a tremendous fire in Orono destroyed that mill and two others. The town itself was just saved by the intervention of a heavy rainfall.[38] Later on that year the Dwinel mill in Old Town also burned, along with much planed and sawed lumber. The loss in the two fires was well over $100,000.[39]

In 1878 the fire loss was again great, with the Veazie mills with its twenty or more saws, and 200,000 feet of lumber first to burn early in June. Later in September another great mill burned.[40] There were other bad fires in 1881 and 1883. This last one, a fire in the great corporation mills in Veazie was a famous one. Bangor firemen saved the mill with quick action, thus preserving the jobs of three hundred men, and the production of 15,000,000 feet of lumber.[41] Again by the speedy action of firemen who were forced to use even old hand pumpers the town of Milford was just saved in 1890 when a tremendous fire got started in a pile of drying lumber just outside the town.[42] The fire fiend (as it was then called) was no respecter of modern mills either. In 1891 the new mill of the Bodwell Water Power Company in Milford, less than two years old, burned flat with a loss of over $75,000.[43] These fires were just an added tribulation to individuals who were rapidly being pressed to the wall by other factors, factors which affected all lumbermen not just those unfortunate enough to have their mills burned.

After the Civil War sawmilling built up to a tremendous pitch on the Penobscot culminating in the great year of 1872. The mill referred to in the epigraph of this chapter employed five hundred men in the years from 1868 to 1872, and rented to operators for $40,-000 a year. It was then appraised at half a million dollars. After 1873 the appraisal fell to $228,000 and the rent barely paid the taxes and insurance. Blunt and Hinman purchased the mill for $70,000 in

1873 but they sold it for $14,000 in 1877.[44] It is small wonder that the number of mills diminished rather rapidly after 1872.

The year before they purchased the Bangor mill, Blunt and Hinman had leased the mill which was located in East Hampden. They operated a gang of saws, a rotary, a single saw, a shingle machine, a clapboard machine, and three lath machines. Running 167 days of 11½ hours the firm had had a banner production: 12,007,003 feet of long lumber, 9,292,300 feet of laths, 235,500 fish barrel staves, 165,500 four foot pickets, 733,000 shingles, and 144,700 clapboards.[45] Another big mill on the river in 1872 was the mill at Bradley, co-owned by three Bangor firms. This mill had nearly thirty saws, including a few old muleys, and employed two hundred men. This mill fought the fire danger by wetting the outside of the building constantly with a force pump while sawing went on. The mill was partially mechanized. In the mill yard a series of small cars driven by endless chains carried the sawdust and edgings away to be burned. A fire was kept going constantly to destroy the waste, which in the one hundred and sixty days of sawing was estimated to be about 16,000 cords.[46]

In addition to these mills, in 1872 there were five water mills in the Orono area, and five steam mills on tidewater which were in operation. The largest mill on the river was the water mill at Orono, called Basin Mills. A dam eight hundred feet long had been constructed at Ayer's Falls in Orono. This dam, fifteen feet high, created a fall of water of about 1,000,000 cubic feet a minute. Below the dam, between the island and the shore, was the basin, covering about fifty acres, with the boom space for 6,000,000 feet of logs. A bridge carrying railroad tracks and a walkway ran to the island, supported by stone piles laid down in 1836. The mill, located just below the bridge, used the rapid fall of the water to drive its saws, which consisted in 1872 of four gangs, four single saws, one rotary saw, as well as six lath, four shingle, and two clapboard machines. The rotary saw had a fifty inch blade and was designed for huge logs. The sawmill was rated at just over 250,000 feet a day. All movement of logs or boards was done by power. At this time a new mill was being constructed for short lumber and a cabinet mill was contemplated.[47]

There had nearly always been a mill at this location. The first one had been started in 1835 on money advanced by the Coburns, but the owner failed in 1838, and the mill had gone into the hands of Samuel Dakin of New York City. Dakin ran the mill, and in 1842 introduced into this mill the first gang saws on the Penobscot. The

Manufacture of lumber 121

mill ran then nearly all the time on pine destined for the West Indies. Dakin sold the mill to Courtland Palmer, also of New York City, who in turn sold it to General Samuel Veazie in 1861 for $50,000. Walker rented the mill from the Veazie heirs after the Civil War at $28,000 per year, and in 1870 purchased mill, island, dam and all including three hundred acres of land for $125,000. By June of 1872 Walker said he contemplated a woolen mill below the dam and raising the dam some six feet in order to obtain more power. The depression came too soon though and the mill passed to other hands; eventually it burned and was replaced by a pulp mill.[48]

Just as the Basin Mills were the typical water mill of this time, the mill owned by E. H. and H. Rollins of Brewer was the typical steam mill. All mills below tidewater were steam mills at this time. A short description of this mill is in order to indicate the state of the Bangor business in the last great year before the depression.

The Rollins mill, used almost entirely for planing, was constructed of brick, on a granite and cement foundation. The boiler room was rated at 75 horsepower, although it could produce 100 horsepower if necessary. There were seven planers in the mill, including a new one, which could handle anything up to a 8 x 16 timber, planing all four sides in one operation. The mill also housed a molding machine and one which made wooden gutters. The mill, built in 1849, was the oldest in Bangor. It set ten feet above the wharves. The lumber was drawn up to the mill by steam power, through the mill, and down to the vessels waiting to carry the finished product away. The owners were proudest of their new clapboard machine, which consisted of two saws, self-dogging, set in an arbor, which planed, set, and butted clapboards all in one motion.[49]

The year of this description was the last big year, and the next year brought the depression. In the spring only one mill above tidewater had opened by May 7, and it was rumored that the others might not open at all. The word had been out the previous fall. After the great Boston fire the *Commercial* had offered some advice to lumbermen.[50]

> Would not our lumber dealers find it for their interest to clear their docks of all the random odds and ends and ship to Boston before the river closes? It would seem that every foot of it would be wanted there for temporary buildings and stagings in early spring.

Logs were high and demand was slow. As the economy began to drag the demand for lumber grew less reflecting the conditions in the construction and building trades. By July of 1873 all the up-river mills were shut, even though the river was at an excellent sawing pitch, and it was not until mid-September that mills began to start. Many were unemployed and times grew very hard along the Penobscot. The next spring a few perhaps answered the call of mill owners at St. John who advertised for five hundred millmen in Bangor newspapers.[51]

By 1875 the situation had worsened so much that the newspapers grew almost ecstatic over the efforts of one man to run his mill on short time.[52] Mixed with the ecstasy though were a few hard and ominous words.[53] "Professional Philanthropy is good, but we must have some lumber sawed this season, some real labor, or it will go hard with us." Still, things looked somewhat better even to the point where the new owner of the old steamboat, which had run from Old Town to Winn for years, took it from the water and used it to generate power for a sawmill.[54] The situation was not much improved though. In 1876 all the mills on the river closed down in mid-July, and they did not open till mid-October.[55]

Conditions did not much improve. Mill property came onto the market as men tried to liquidate their investments.[56] Even in the early eighties, when the economic horizon brightened somewhat, much of the lumber sawed on the river was box-boards, an item which would have produced only disdain in the late sixties. Box-boards were sapling pine, three-quarters to one inch thick, edged and sawed into shooks, and used for packing cases. From 1880 to 1883 about 15,000,000 feet each year went out of such lumber, and from then until the new century the figures of such dry pine lumber never scaled under 16,-000,000 feet and occasionally went as high as 23,000,000 feet. From 1870 to 1901 some 476,000,000 feet of this lumber was surveyed at Bangor, but 319,308,590 feet of it came in the last fifteen years.[57]

The larger mills on the Penobscot that had survived the deep depression of the seventies did fairly well in the next decade. Their competition had been frozen out and they were able to continue on shortened time. Operators like James Walker of the Basin Mills continued to mill large amounts, although never quite as large as in 1872. When he discussed the question early in the eighties with a newspaper reporter, Walker said the real future lay in pulp mills, and he cited the example of Berlin Falls, New Hampshire, as a town which was getting in on the coming boom early.[58]

Occasionally a new mill was built such as that the Bodwell interests erected in Milford in 1881.[59] Even the older mills were often run by men whose names had no echoes from earlier times. The depres-

Manufacture of lumber

sion and death had taken away most who were prominent in the sixties and seventies, and a new group, generally more interested in the main chance and more ready to experiment, had come in to take over the industry.[60]

These new men would bring such innovations as the band saw. This new saw was the wonder of the age. The first one in the Penobscot area was put into the mill of F. W. Ayer in South Brewer, easily the most modern mill in New England and perhaps in the country when it was constructed. Millmen from Maine, New Hampshire, and New Brunswick came in droves to view the new mill and the miraculous new saw. In the first eleven hours of running the new saw proved its worth. During this time the little gangs in the mill sawed 73,287 feet. The big gang sawed 83,208. The new band saw, though, sawed 128,357 feet which was called the new world's record. Only thirty-eight men were employed around the saw in this new mill, which was an economic feature which did not miss many viewers. Ayer put in another saw the first winter, and W. T. Pearson ordered one the first time he saw the Ayer saw in operation.

Ayer's mill was a marvel. He had a mile of waterfront in Brewer, with 1,000 feet of wharves. The mill employed one hundred and sixty men, and had a daily rated capacity of 160,000 feet. A spur chain brought the logs from the river to the men. Only one man was needed to push them into the sluice; they were then taken by steam feed to the three band saws (a third was added in the spring of 1890), and the three gang edgers, two clapboard and four lath machines with which the mill was equipped. The clapboard and lath machines were on a second floor, and after the lumber was sawed it was carried out a sluice to be tied up, and from there to the wharves all by machinery. The third floor held two automatic grinders, a lap grinder, and a brazer to keep the saws sharp and running. The sawdust from the saws was carried by an automatic conveyer to the furnaces which drove engines rated at 350 horsepower. The mill was protected by an automatic sprinkler system as were the separate brick planing mill and machine shop.[61]

With the introduction of these more economical methods the new era had come. Fewer and fewer mills would be operating, and those only when conditions were good. Walker at the Basin Mills was still a big operator. In 1889 he produced seventeen million feet of long lumber, ten million laths, one million box shooks, 500,000 staves, 20,000 pairs of heads for barrels, and 5,000 barrels.[62] This was an exceptional year, though. By the early nineties times were more difficult. The *Bangor Business Directory* for 1890 still listed fifteen firms interested in lumber and sawing, but many were doing very little business.[63] In the spring of 1894 the mill owned by Dole and Fogg in Bangor was sold at auction, the terms of which were announced as one-half down cash and the balance to be paid equally over six years. Everything went, machinery, scows, wharves, mill, and so on. When the auction opened only desultory bidding took place and it was finally knocked down to F. H. Clergue and F. W. Hill for only $8,800 much to the surprise of the crowd and everyone else. It had been expected that the property would bring $30,000,[64] but mill men had already read the handwriting on the wall and the auctioned value was just about the true value.

Ellsworth, Cherryfield, Machias, Calais, Woodland, and Milltown were the larger sawmill towns downeast. At Ellsworth, where the Union River comes to the sea, the largest operators throughout this period were W. F. Milliken, and Hall Brothers. These operations were always smaller than on the Penobscot but large enough so that Ellsworth remained a major sawmill town. In 1862, for instance, the town had a production of 28,000,000 feet of long lumber, although this total was probably not reached again until 1872.[65]

In the latter years Hall Brothers sawed 15,000,000 feet themselves. The firm shipped on their own vessels, lumber which their own crews had cut and driven to their own saws. They owned a retail lumber yard in Boston and in this year erected a planing mill to insure better prices for their product.[66] Ellsworth had always a varied production, and the appended table shows the extent of the production in one year during the depression.[67]

Lumber Production—Ellsworth, Maine—1877

Type of Lumber	Amount Produced
Long Lumber	19,300,000 feet
Staves	15,500,000
Shingles	11,850,000
Barrel Heads	210,000 pairs
Laths	3,450,000
Stove Wood	700 cords
Clapboards	18,000
Excelsior	210 tons
Pickets	35,000
Fish Boxes	5,000

One of the things which protected the Ellsworth industry during the period of hard times was the great boom and growth of Bar Har-

Manufacture of lumber 125

bor, then called by its summer visitors "Eden." Much, if not most, of the lumber sawed during the late seventies and throughout the eighties went to the island where it was used in the construction of the fashionable summer homes then building. Milliken's kept two small schooners going back and forth all summer with lumber for this work.[68] The Bar Harbor boom was over by 1888 or 1889. With this most of the mills disappeared. Only Hall Brothers survived, and their production was almost entirely confined to cement staves for New York.[69]

In Cherryfield,[70] on the Narraguagus River, there was no Bar Harbor to tide the mills over. Even so the mills produced their 14,-000,000 feet or so long lumber and about as much short lumber each year throughout the 1870's. By 1885, however, the port was about done. After this time what little was sawed remained in the area. The river was quite denuded of lumber, and prices and competition were starving these small ports out.[71]

The Machias river was another small river downeast. In 1863 the mills on the river sawed 28 million feet of long lumber, 7,500,000 at Whitneyville, the rest at Machias,[72] for perhaps its high point. Later statistics show the impact of the depression on these little towns which were as dependent on the lumber industry as were Bangor, Old Town, or Augusta.[73]

Lumber Production Machias River—1874-1877, 1882 (Board feet)

Year	Machias Cut	Whitneyville Cut	Total
1874	—	—	23,074,000
1875	18,045,000	4,929,000	22,974,000
1876	14,405,000	3,679,000	18,084,000
1877	13,056,000	3,227,000	16,283,000
1882	15,600,000*	5,300,000*	20,900,000

*Estimated from known cut of long lumber, laths, and shingles.

Everything in these little towns revolved about the industry. In Whitneyville, several monster piles of sawdust, edgings, and refuse wood had collected just outside the town from the years of sawing. The largest pile covered more than an acre, and was estimated to hold more than 10,000 cords of sawdust. In 1872 the winter was severe, and the ice and snow were heavy. Town officials dug trenches around the big piles and set fire to them that spring. Spectators came from miles away to see the big blaze. The piles burned for two or three days before the spectacle was diminished, and the fire danger was considerably relieved.[74]

Towns downeast were not free from the fire curse even with such methods as these. In 1891, for instance, a bad fire at the Danforth wharf in Machias destroyed 4,000,000 feet of dry pine boards and two tenement buildings.[75] The largest and most destructive fire in the period though occurred in Baring in 1875. All the mills on the American side were destroyed and the Canadian side was saved only through strenuous effort.[76]

The St. Croix lumber industry has been well treated by Davis already, and it serves little purpose to repeat it here.[77] Suffice it to say that in the period before 1872 the river produced about fifty to sixty million feet of long lumber each year although the capacity of the mills was nearer 75,000,000 feet. After the depression the cut fell off considerably and in 1884, which was a good year, the cut was about 59 million feet of long lumber, 35½ million laths, and 12 million shingles.[78] The 1885 cut was about the same.[79] In 1888 it was down to 40 million feet on the river, and by the depression of 1893 the cut was only 20 million.[80] The story was a similar one to the other small rivers in the state, although perhaps not as pronounced.

The St. Croix, like the St. John, was a border river. Maine lumber, sawed in Canadian mills, after being driven down these rivers, was admitted free of duty under the provisions of the Webster-Ashburton treaty. It was extremely difficult to differentiate Maine lumber from Canadian lumber once it reached the saw mill. There were rumors throughout the period that the lumber, once sawed, went for sale according to the market rather than according to the geographic location of the growing tree. This was especially true in the mills in St. John, New Brunswick, owned by Maine men. As one newspaper which was unhappy about the investment of Maine money into Canadian enterprises remarked:[81]

> Since 1866 [the end of reciprocity] much Maine capital has been put into lumber mills in St. Johns and St. Stephens, some of the foremost names in Bangor and Calais, with a few from Portland, being known in the enterprses which flourish under the state's special lumber privilege.

It is impossible at this distance to deal with this problem. However, in discussing the St. John River it is necessary to keep it in mind.

The largest of these Maine firms which operated in St. John was that which was known at various times as Hayford and Stetson, and Stetson, Cutler and Company. This firm's employees cut their lumber all over Aroostook County, and the logs were sawed in Carleton, N.B., and Indiantown, near St. John. Much of their product was shingles sawed from cedar cut in the swamps of northern Maine. After the great St. John fire of 1877[82] this firm dominated the Maine cut sawed in St. John. Shingles were sawed in Carleton, 1,500,000 feet of cedar going this way in 1879.[83] In 1880 the firm sawed ten million feet of spruce and three million feet of cedar, all cut on the St. John and Aroostook Rivers. By 1883 Hayford and Stetson produced ninety M long lumber, twelve M shingles, five M barrel staves, fifty M laths, three M clapboards each day in their Indiantown mill. They employed one hundred and thirty men at this operation. Three other mills located in this town produced one hundred and seventy M of long lumber, and one hundred and thirty M of laths each day.[84]

In 1884 the Maine firm was cutting one million feet of spruce a week when the bottom dropped out of the market and the mill was forced to close. A correspondent of a Maine newspaper was caustic.[85]

AN UNHAPPY CITY

St. John, N.B. going down hill at a rapid rate. Deadening effect of protection on the prosperity of the city.

A high protective tariff, a disastrous fire, and iron steamships have almost knocked the life out of this once prosperous and still attractive city. . . . The fire was an event of 1877, the tariff was made the law of the country about 1878, and iron steamships became a terror in the land about 1882. . . .

The firm owned much land in Maine and partly because of this, and partly because of better prices they were soon at work again. In the next year the firm produced 31,600,000 feet of long lumber, 33,-000,000 shingles, 20,000,000 laths, in addition to some six or seven million feet which went into headings, staves, boxes, and clapboards. Three years later the firm had risen to be the largest producer of shingles in New England and was the middleman for most shingles produced in Maine and New Brunswick. In that year the firm sold seventy million shingles.[86]

The largest sawmill operator in the States was Johnson and Phair who had a large mill at Washburn. In 1885 they turned out over 16,000,000 shingles which were then sold to Stetson and Cutler.[87] There were smaller mills as well, among them S. W. Collins and Company who are still operating, and others such as that of W. H. Card at Alder Brook were fairly well known.[88] Aroostook County was still a frontier territory though, and it would remain that way until the railroad came near the turn of the century. As a result the projected and prospective towns all wanted sawmills to provide lumber for the new settlers and money for those who had already settled in the area. And, of course, once the railroad came the town with a mill might be large enough to rate a depot which would in turn advance the fortunes greatly of the promoters. Some were quite blunt about their desires as the following newspaper article attests.[89]

MASARDIS

We are in great need of a sawmill here. The only mill that does anything is seven miles from here and that cannot supply the country. Mr. Eben Trafton offers the privilege and a house lot free to any good man who will put in a good sawmill. The dam is built, and to be kept in repair. On it at present is a starch factory and a grist mill. There is plenty of lumber that can be driven to the mill; spruce, cedar, hardwood, and some pine.

Competition was strong all over the state for new mills, although the growth had been held back somewhat by the Maine Supreme Court ruling declaring Chapter 716 of the Maine Statutes unconstitutional. This law had allowed towns to pledge their credit to private manufacturing firms. The town of Jay had attempted to attract a sawmill from Livermore Falls by lending them $10,000 at 6% interest if the company would move to Jay, put in a grist mill as well as a sawmill, and invest $12,000 of their own. Some taxpayers filed a suit to void the action, and the court ruled null on the grounds of unconstitutionality.[90] Towns could abate taxes to attract new sawmills and many did. Others were more than willing. Letters went out to known investors in Boston, Portland, and New York, and were reprinted in the *Commercial*.[91]

The mills were moving toward the railroads and the woods but most of them moving inland were portable or semi-portable mills. At the end of this period there is much evidence of this sort of operation, and in particular, one worth some study—the operation of the Webster family of Orono near Shirley, Maine. Others had been in operation for a specific reason.[92] None of these mills ranked with the almost revolutionary mill at Shirley, though.

The mill was put into place in the fall of 1887 by E. and J. F. Webster, prominent mill men and rafters from Orono. Thirty men went in to cut for them and they expected to get out two million feet the first year. By January of this year the mill was in full operation producing 30,000 feet a day from its saws. The Websters shipped twenty railroad cars a week to Bangor with the mill production.

Manufacture of lumber

In 1889 the operation really began to boom. Eighty men and twenty horses were cutting and hauling in the woods, and the mill sent out three railroad cars a day with sawed lumber to the Bangor wharves ready for the ice to leave. Two million feet of spruce logs went through the saws that winter. With the success they were having the owners decided to branch out. They purchased the land in the fall, rather than to continue to pay stumpage, and their new acquisition was described as a bonanza with from five to six million feet of good spruce in proximity to the mill. Their crews were nearly doubled that winter. The mill sawed 3,000,000 feet of lumber, and another two million feet of logs came down in the cars for their Orono mill. A special train of logs went over the railroad five days a week. By beating the ice the Websters were making an excellent profit on the early high price of the lumber.

The mill in the woods worked round the clock and averaged 60,-000 feet a day. The biggest night in 1890 was 32,500 feet. But these records were made to fall. With the rotary saw and the gang edger going constantly the mill put out 382,000 feet in one week, including 123,324 feet in one thirty-six hour stretch. By the end of April the six cars a day to Bangor had brought down 4,500,000 feet, and another 1,500,000 was on its way. The year's production was down somewhat because of a fire which destroyed some of the machinery in January, 1891. The workmen saved the mill though, and it was back sawing within the week. The men finished off their winter's work by shipping all the slabs and edgings to the firm's pulp mill in Orono.

By this time the pulp mill was beginning to take most of the family time and energy so they sold the mill to George M. Fogg of Bangor who ran it for a time on last blocks. The Websters now went out of lumber into pulp. Before they left, however, they had made a good deal of money by using the industry in its more profitable phases —rafting, driving, portable mills, the sale of pulpwood, and pulp and paper manufacture.[93] Probably this was the only way that one could make money consistently in the lumber business after the price fall which began in 1872. Whether others did or not, the Websters always saw the main chance and took it, and as a result rank as one of the more successful lumbering concerns from 1860 to 1890.

Another man who came to prominence at about this same time was Fred W. Ayer of Brewer. Ayer introduced the band saw to Maine sawmills, but he was not content with this, being involved in pulp and paper mills, modern carrying methods in the woods (The Eagle Lake

Tramway) and other peripheral areas in the business. One which was not as successful, but which provided much comment to observers, was his attempted corner of the log market in 1892. In August of that year great activity was noted in the buying and selling of logs, and observers soon noted that Ayer had purchased more than 25,000,000 feet from various parties. Others refused sale but his efforts drove the price up $1.25 per M to a season's high of $13.00. The log corner was successful enough so that some mills closed down, and the price dropped slightly. In 1893 when the attempted corner met the depression commencing in that year the price dropped to $8.00 to $10.00 per M depending on the size of the logs. This was too much though— one man, even with the help which he apparently had, could not control the whole Penobscot market, but even so Ayer got logs for his mill quite cheaply for one year at least in 1893.[94]

One last series of efforts to utilize the manufacturing end of the business ought to be noted. These were the efforts of individuals to make a use for sawdust. The Maine Compress Company, a firm owned in part by F. W. Ayer, devised a sawdust baler and compressor which would package the waste product for use in packing, not only for ice but also around breakable objects. This device created a small furore, and one of the compressors was installed in Ayer's plant but it did not amount to much.[95] Earlier some had thought to use the sawdust to make paper pulp. The Rutland, Vermont *Herald* even printed an issue on sawdust pulp newsprint in 1885. Maine newspapers were enthusiastic over the possibilities.[96]

SAWDUST PULP

A new substitute for the stock employed in the manufacture of paper.
—Why not utilize the sawdust of your mills in manufacturing Paper—
A project worthy of the attention of our business men . . .

These were the headlines. The newspaper had samples and information regarding the machine at the disposal of the budding investors. Lumbermen at Fairfield investigated the possibilities but the big bonanza did not develop.[97] The use of waste wood was some time away yet. In any case the big bonanza was not in utilizing mill waste but rather in shutting down the mill and converting it to pulp operations. These mills will be discussed immediately after a chapter on shipping and a general economic view of the period under discussion.

NOTES

[1] This is based on accounts of short lumber re-entering the United States as given in the Bangor Harbourmaster, *Annual Reports*, 1860-1875. To convert these figures which are given in terms of shingles, it was estimated that 7,000 shingles equaled 1 M of long lumber.

[2] Article 16 of Constitution of Lumber Manufacturer's Protective Association.

[3] Bangor *Daily Commercial*, February 20, 1873.

[4] *Ibid.*, February 21, 1873.

[5] *Ibid.*, February 7, 8, 20, 21, 1873.

[6] Portland *Eastern Argus*, August 26, 1874.

[7] Bangor *Daily Commercial*, April 11, 1883; Portland *Eastern Argus*, April 11, 16, 1883.

[8] Portland *Eastern Argus*, September 7, 1883.

[9] Later in 1897 a Bangor meeting of Box Shook manufacturers was called. The U.S. representatives of the Sicilian firm controlling the business was convener and they met at T. J. Stewarts. After bemoaning the tariff they disbanded. Lewiston *Weekly Journal*, May 13, 1897.

[10] Bangor *Daily Commercial*, April 30, 1886.

[11] *Ibid.*, April 20, May 4, 12, 13, 24, June 24, 1886. The strike was the last straw for the paper mill venture. Phillips *Phonograph*, May 14, 1886. Two attempts were made to open some of these mills with scabs. They both failed.

[12] This law went into effect on May 1, 1887, and although there were some attempts at non-compliance, the law did improve relationships between workers and their employers, and when coupled with a log lien law giving workers a first lien on their employers' logs when wages remained unpaid, it made conditions quite tolerable for workmen. One of the famous old logging songs, "The Good Old State of Maine," by Larry Gorman, devoted a verse to the difference between Maine and New Hampshire on this innovation. See Ives, *op. cit.*, 104-5.

> And for these sub-contractors now I've got a word to say,
> If you work for a jobber here you are apt to lose your pay;
> For there is no lien law in this state, the logs you can't retain,
> While the lumbers holding for your pay in that good old State
>   of Maine.

[13] Bangor *Daily Commercial*, June 21, 1889.

[14] *Ibid.*, June 26, 1889; an excellent article on conditions in Augusta mills.

[15] The best story of the strike is in the Bangor *Daily Commercial* where it was discussed with spirit and honesty. See the issues of June 14, 19, 20, 21, 24, 25, 26, 27, 29, July 5, 9, 15, 1889. For once there were no sententious editorials about fellow servants, and the laborer worthy of his hire. The *Commercial*, a Democratic newspaper, is often interesting to read in light of the changing character of that party in this period as laborers, populists, farmers, unions, and others become more and more an important part of the parties voting strength.

[16] Bangor *Daily Commercial*, June 4, 16, 1890. One historian has located only eight strikes in the lumber industry from 1880-1900. Charles A. Scontras,

Two Decades of Organized Labor and Labor Politics in Maine, 1880-1900, Orono, 1969, 176-180.

[17] Bangor *Daily Commercial*, February 23, 27, 1893.
[18] *Ibid.*, December 9, 1893.
[19] Portland *Eastern Argus*, October 19, 1876.
[20] Bangor *Daily Commercial*, June 7, 1881; July 6, 1889. For their pulp wood operations see below. The only operator of any size south of Lewiston was Joseph Hobson of Saco. He sawed from fifteen to twenty million feet a year in his mill. In 1878 his receipts were supposed to be about $62,000, and in 1879 when he sawed 14,000,000 feet he was supposed to have taken in $127,-000. The firm made a thousand dollars a day that fall filling late orders. *Ibid.*, October 20, 1879.
[21] Oxford *Democrat*, December 18, 1863; May, 17, 1867; Portland *Eastern Argus*, May 30, 1876; August 14, 29, 1877; Bangor *Daily Commercial*, February 4, 1878.
[22] *Ibid.*, May 19, 1879. The mill was a big one, generating 200 horsepower. The section of Canton where the mill was located is still known as Gilbertville after the owners of the new sawmill.
[23] Bangor *Daily Commercial*, February 7, 1881, November 23, 1885 quoting The Canton *Telephone*.
[24] These sentences are based on a reading of the *Maine Register*, 1868-1878, and especially 1872 and 1878. This semi-official annual publication lists nearly every enterprise in the state and its owner, and occasionally gives information of rated production. The files of the Oxford *Democrat* in this period were also helpful.
[25] Bangor *Daily Commercial*, December 30, 1878. Other small mills are described in Farmington *Chronicle*, February 27, 1873; January 1, 1874; July 22, 1875; Wilton *Record*, August 4, November 10, December 2, 1886; February 24, May 2, 1887; Phillips *Phonograph*, December 3, 10, 1886; *Katahdin Kalendar*, December 27, 1879; January 17, 1880. These made rakes, shovels, boxes, and barrels.
[26] Boardman, *op. cit., The New England States*, 1230-2.
[27] Bangor *Daily Commercial*, April 3, 1894.
[28] Deal was a plank, at least three inches thick, of spruce or pine. These planks were carefully sawed from the best lumber, and the ends were dressed. They were reckoned in great or standard hundreds, 120 single pieces, or 2,750 board feet, and were made to be re-sawed in England. Much pine in the early days went out this way, but at this time most of the trade is spruce, which had a great flurry in the 1870's. See the next chapter for a discussion.
[29] Portland *Eastern Argus*, July 4, 11, 1876; June 4, 1877. Bangor *Daily Commercial*, September 9, 1876; December 27, 1878. The big day was July 10, 1876.
[30] Portland *Eastern Argus*, April 14, 1885; Bangor *Daily Commercial*, September 19, 1884; December 20, 1886 quoting the Bath *Independent*.
[31] Bangor *Daily Commercial*, June 22, 1878; March 17, 23, 1880.
[32] *Ibid.*, June 20, 1887.
[33] *Ibid.*, June 12, 1891. The band saw accounted for 51,147 feet of the long lumber.
[34] *Ibid.*, December 17, 1889; February 16, 1892; August 3, 1893.

Manufacture of lumber

[35] Portland *Eastern Argus*, December 20, 1883, a nice story. Later the portable mills began to make their way into northern Kennebec waters. By 1890 one was operating on Mount Abram town south of Addison. It was a steam mill, and it cut 4,000,000 feet which came down the Phillips and Rangeley Lakes Railroad that winter. Another was operating on Madrid, and it produced about a million feet. Bangor *Daily Commercial*, September 19, 1890. Phillips *Phonograph*, January 15, 1886. Recently some papers of the firm which operated this mill have come to the Fogler Library. One box of these Chandler papers illustrates its smallness and ingenuity. It was successful because of the Sandy River and Rangeley Lakes narrow gauge railroad. See H. Temple Crittenden, *The Maine Scenic Route*, Parsons, West Virginia, 1966 for a good history.

[36] Phillips *Phonograph*, December 24, 1886; January 6, February 18, 25, July 22, November 4, 11, December 23, 1887; Wilton *Record*, March 10, 24, 31, April 28, May 26, September 15, November 3, 10, December 9, 1886; February 17, 1887. These mills would provide about 5-6 carloads of freight a day for the railroad, sawing between 30 and 80,000 feet on an average day.

[37] Boardman, *op. cit.*, 1233.

[38] Bangor *Daily Commercial*, May 22, September 16, 1876.

[39] *Ibid.*, October 3, 1876. Fires had always been a problem. See *Maine Farmer*, April 15, 1833 on Stillwater fires; April 18, 1834 (Ellsworth); April 22, 1836 (Old Town); *Drew's Rural Intelligencer*, November 27, 1856 (Dwinel's Great Works Mills); Presque Isle *Sunrise*, October 16, 1868 (Fort Kent fire).

[40] Bangor *Daily Commercial*, June 8, September 30, 1878.

[41] *Ibid.*, June 4, 1881; May 3, July 27, 28, 1883. Firemen were at the fire twenty-two minutes from the firehouse, and had water on the flames in fifty minutes, which would be fast even today.

[42] *Ibid.*, April 15, 16, 18, 1890. Eight million feet of lumber, valued at $95,000, went up in this fire.

[43] *Ibid.*, June 20, 1891.

[44] Bangor *Daily Commercial*, July 28, 1879.

[45] *Ibid.*, January 2, 1872.

[46] Bangor *Daily Commercial*, March 25, 1872. A muley saw had a twelve inch blade (single saws were eight inches in diameter). They were not hung in a frame, as were the single saws, but rather ran from above, and were pulled down onto the log by another rod connected and driven by a crank. The logs lay on a carriage running on a rail. At each stroke of the saw the carriage was driven forward by a pinion fixed to the axis of a ratchet wheel connected with the crank. The log was always kept up to the saw by this arrangement, and at the end the carriage was rolled back and the log either rolled over, or a new one placed on the carriage. This saw was designed for larger logs than single saws, but for smaller ones than gangs, which were nothing more than several (usually four at this time) single saws hung in a frame and designed to produce several boards at once. The muley gave more precise boards, although it produced a larger kerf. Band saws did not come in until 1889.

[47] *Ibid.*, March 27, 1872.

[48] Bangor *Daily Commercial*, March 27, May 3, 1872. Early descriptions of mills on the river are in Bangor *Democrat*, January 10, 1842; Bangor *Whig and Courier*, August 10, 1842, November 28, 1854, August 15, 1859.

[49] Bangor *Daily Commercial*, June 5, 1872. One reason for the brick construction was that Brewer was the chief manufacturing town for bricks in New England at this time.

[50] *Ibid.*, May 7, 9, 1873. Calais lumber merchants were complaining of slow demand and low freights as well. The mills on the St. Croix were still cutting, but the lumber was piling up on the wharves. See *Ibid.*, May 9, 30, 1873. The quotation is from an editorial published in *ibid.*, November 18, 1872.

[51] Bangor *Daily Commercial*, July 23, 1873. For other bleak comments on the earliest days of the depression see Portland *Eastern Argus*, August 12, September 14, 1873; *Kennebec Journal*, July 25, 1873.

[52] Bangor *Daily Commercial*, February 10, 1875.

[53] *Ibid.*, April 28, 1875, on the occasion of the mill opening.

[54] *Ibid.*, July 3, 19, 1875. The best description of Bangor in this mid-year of the depression is in *Knowles' Bangor Directory and Business Almanac*, 1875. For sawmills see 23-4, 31, 37, 42-3, 71-4.

[55] Portland *Eastern Argus*, July 18, October 18, 1876; Bangor *Daily Commercial*, October 13, 1876; April 21, 1877.

[56] Bangor *Daily Commercial*, May 4, 1876; July 20, 1878; October 2, 16, 1879; April 21, May 19, 26, July 12, 1880. These stories are all general comment on the industry, and nearly all of them concern some sale or transfer of sawmill property in the Bangor area.

[57] *Ibid.*, April 23, 1881; September 12, 1882 on box-boards. Box-boards sold as high as $12.00 M in this time of generally low prices, which was a good price for "junk" lumber. The statistics in the last few lines were compiled from newspaper accounts of survey results, and the books of the Surveyor General, Bangor.

[58] Bangor *Daily Commercial*, June 25, 1881.

[59] *Ibid.*, June 27, September 3, 1881.

[60] This paragraph is based on a reading of the names which appear in the Bangor and Portland papers of the period, and articles which appeared in Portland *Eastern Argus*, November 13, 1882, and Bangor *Daily Commercial*, April 23, 24, 1883, and April 2, 1884. The first two dates in the Bangor paper refer to the discussion of the issues in a court case involving the amount of lumber sawed in a rented sawmill, *Prentiss vs. Pearson and Pearson, SJ Court*, Penobscot County, Spring Term, 1883. Bangor *Daily Commercial*, July 30, 1884, an article entitled "Business Depression." *Ibid.*, February 20, May 23, July 30, August 8, 1885; February 22, May 9, September 17, October 29, 1887; Portland *Eastern Argus*, August 19, 1885; January 20, 1886.

[61] Bangor *Daily Commercial*, August 22, 23, 1889; May 10, 1890. A band saw is a continuous ribbon of saw blade always operating and the logs are pushed through the blades. The advantage lies in the much smaller kerf produced, and the speed with which the saws slice the logs. When mounted in gangs they literally melt logs into lumber. See Henry Disston and Sons, *The Saw in History*, Philadelphia, 1916, or almost any issue of the trade journals after 1890 for pictures and descriptions of the variations produced. Disston's were the largest saw manufacturers in the country. They produced a trade journal of their own which is as good as any for following these trends, *The Disston Crucible*.

[62] Bangor *Daily Commercial*, March 22, 1890.

Manufacture of lumber

[63] This directory is available in a few libraries, and is published in its entirety in *ibid.*, October 11, 1890. Also see, *Ibid.*, October 19, 1892; August 1, 2, 11, 16, September 16, 1893.

[64] *Ibid.*, April 10, 1894.

[65] Oxford *Democrat*, February 6, 1863 quoting the Ellsworth *American*.

[66] Bangor *Daily Commercial*, April 13, 1872.

[67] *Ibid.*, December 27, 1877 quoting the Ellsworth *American*.

[68] Bangor *Daily Commercial*, December 21, 1878; January 3, 1881; February 8, 1882; February 14, 1883. A. F. Norris, agent at the mills, issued a yearly summary of the work done in these years.

[69] *Ibid.*, December 19, 1884; December 7, 1888.

[70] Cherryfield was not the only town downeast with small mills. For instance, in 1871, there were three mills in Dennysville, that produced the following amounts:

Dennysville Mills—1871

	Long Lumber	Laths	Shingles
Allan and Sons	918,000 feet	1,500,000	250,000
Lincoln Bros.	2,900,000	3,000,000	400,000
P. E. Vose	850,000	600,000	300,000
Total	4,668,000	5,100,000	950,000

The depression, and the pulpwood revolution, eliminated such mills completely which was the point of the story in *ibid.*, December 31, 1897.

[71] Bangor *Daily Commercial*, July 30, December 3, 1883; December 5, 1884; January 20, 1886; October 8, 1889; Portland *Eastern Argus*, September 19, 1873; December 9, 1885. This was the last year of any consequence. The Cherryfield boom rafted 98,000 logs, 5,500 cedar logs for shingles, and 7,000 sticks for staves that year.

[72] Oxford *Democrat*, February 26, 1864.

[73] Bangor *Daily Commercial*, April 2, 10, 1877; January 23, 1883; Portland *Eastern Argus*, April 3, 15, 1878. There were ten mills on the river and the largest sawed just under five million feet in 1877. Two of the mills were of the eastern half of the family known in west coast lumbering as Pope and Talbot. The firm returns in this story in another connection. For the west coast see Edwin T. Coman, *Time, Tide and Timber*, Stanford, 1950.

[74] Portland *Eastern Argus*, March 11, 1873.

[75] Bangor *Daily Commercial*, October 2, 1891.

[76] *Ibid.*, September 9, 1875.

[77] Davis, *op. cit.*, Chapters X, XIV, XVIII.

[78] Portland *Eastern Argus*, January 26, 1885.

[79] *Todds of the St. Croix Valley*, 19.

[80] Davis, *op. cit.*, 274-5.

[81] Portland *Eastern Argus*, November 11, 1887. The *Argus* was commenting on the request of a Minnesota congressman for a similar tariff-free privilege. The Secretary of the Treasury had refused the request, and the Maine newspaper applauded the action.

[82] The best description of the fire is in Hannay, *op. cit.*, II, 326-8. George Stewart, *The Great Fire of St. John*, Toronto, 1877.

[83] Bangor *Daily Commercial*, May 26, 1879.

[84] *Ibid.*, January 21, 1880; July 11, 1883.

[85] *Ibid.*, August 2, 1884. When the mill closed it threw two hundred and seventy-five men out of work. The article was a diatribe against the Confederation of 1867, and it said that as a result New England and New York commerce had been diverted to Upper Canada to the detriment of the Maritime Provinces. It proves, if nothing else, that some New Brunswickers had (and have) long and apparently painful memories.

[86] Bangor *Daily Commercial*, March 18, 1886; November 23, 1887; November 26, 1888. Other large firms included Andre Cushing and Co. in St. John. They produced about 25 to 30 million feet each year throughout the period, and increased their production in 1889 when they put a band saw in their mill. This mill was burned in 1893, and replaced by one said to be the most modern in the area. See Testimony of George S. Cushing, *SJRC*, 1241-8.

[87] Bangor *Daily Commercial*, October 12, 1885. This mill had four shingle machines, each one rated at about 125,000 a week.

[88] See *ibid.*, April 2, 1884 for Card's mill, and S. W. Collins, *A Hundred Years*, Presque Isle, 1959, *passim*, for this early venture. Card produced 700,-000 feet of long lumber, 500,000 laths, and 2,500,000 staves in 1883.

[89] The quotation is a news item, not an advertisement, from Bangor *Daily Commercial*, March 31, 1886.

[90] The answers of the court to constitutional questions posed by the Legislature came down on February 14, 1871. See *58 Maine* 590 ff. or House Document #47 (1871). The *Commercial* was still unhappy about it on March 18, 1873.

[91] Printed in *ibid.*, December 2, 1889.

[92] Among them were the mill of the Sandy River Lumber Company already mentioned. Others were the mill put in near Katahdin Iron Works in 1884 and 1885 to handle the hurricane blowdowns from 1883. This mill shipped a million feet of short lumber a year until the blowdowns were gone. After the Canadian Pacific came through another portable mill ran on spool bars near Hardy Stream for a year or so. See Bangor *Daily Commercial*, April 3, 1885 and January 1, 1891 for accounts of these early portable mills.

[93] Bangor *Daily Commercial*, October 21, 1887; January 25, March 22, 1888; January 22, February 5, April 11, August 3, 12, November 30, December 27, 1889; January 11, 23, February 7, April 25, 1890; January 20, 21, February 26, June 16, 1891. The Websters owned besides the saw mill at Shirley, one at Orono, the Orono Ice Company, and they were involved in the new pulp mill of F. W. Ayer at Brewer, a hotel at Phillips Lake, and the ice houses there, as well as the Blanchard Slate Company, at this time.

[94] Bangor *Daily Commercial*, July 8, 18, August 5, 1892; October 20, 1893.

[95] *Ibid.*, January 29, July 13, 1889. Sawdust had had a boom earlier for ice shipments, Farmington *Chronicle*, January 13, 1848 describes six vessels loaded with sawdust leaving for the South.

[96] *Ibid.*, January 5, 1885.

[97] Bangor *Daily Commercial*, January 13, 1885. The Fairfield men came to look at the sample after the article appeared. Nothing ever came of it.

6

SHIPPING THE PRODUCT

Oh! It's where away is the good ship bound,
    Liverpool Town, Liverpool Town,
Then! It's merrily turn the capstan 'round,
    For we're going away to Liverpool town.

                          Oldtime sea chantey, quoted in
                          a story on Bangor's foreign
                          trade (spool, deal, shooks.)
                          Bangor *Daily Commercial*, June 15,
                          1897

    Contrary to popular belief, vessels have not done over and above well, for the reason that the railroads have taken away many hundred carloads of lumber, giving low rates and quick delivery to such points as are reached alike by rail and vessels, and thus freights have been low. When it is known that about one-third of the manufactured lumber goes out by rail, and that the general freight and passenger traffic from the east has doubled within ten years, it will be seen that the liberal improvements and increase of service lately made, are fully warranted.

                          Portland *Eastern Argus*, November
                          21, 1887 (commenting on the
                          season just finished on the
                          Penobscot River.)

The usual way of transferring lumber and lumber products from the northern spruce area to the markets was by vessel. At the sites of down river sorting booms sawmill towns such as Bath, Wiscasset, Saccarappa, Gardiner, Augusta, Bangor, Ellsworth, Calais, Cherryfield, Machias, Fredericton, and St. John grew up. Later when the railroads came other towns such as Fairfield, Lewiston, Berlin (Berlin Falls), Houlton, and Van Buren were to be large lumber shippers. The lumber then left these shipping and sawmill towns to travel to its eventual destinations—places such as Norfolk, New Orleans, Philadelphia, Baltimore, New York and Boston in the United States; and to more esoteric locations abroad, Granada, the river Plate, Turk's Island, New Providence in the Bahamas, Havana, Castellamare in Italy, Palermo in Sicily, Grennock in Scotland, and Fleetwood and

Hull in England. These large receiving ports occur over and over again as the destination listed on manifests, but in the nineteenth century nearly all ports on the eastern seaboard of the United States, in the West Indies and eastern South America, in the Mediterranean, and in the British Isles were touched fairly consistently by ships bearing cargoes of Maine lumber: spruce first, and later birch for spools, and pine for shooks.

The trade suffered the usual vicissitudes of commerce. Tariffs and other commercial restrictions frequently interrupted a seemingly flourishing business. New products changed the direction of trade. New areas came into competition, first the Baltic, and later the Pacific west coast when the Panama Canal was completed. The backbone of the trade remained, at least within the continental United States, the coastal business protected to vessels of United States registry. Even this trade, however, found itself under increasingly sharp attack from a new competitor, the railroad. A shipping and transportation revolution, similar in nature to that of the early part of the century, took place as goods went by different routes and were carried by different methods.

Eventually the railroad was to win out. The victory was aided by the shift in ownership of the woodlands to the giant paper companies, whch sometimes shipped their product by vessel, but now in the form of paper and pulp and in the twentieth century, usually in their own ships. More and more of their products went by rail ordinarily. By the first World War the only towns still shipping much were Bangor and St. John, and it was an artificial trade. The vessels took lumber away, almost in the form of ballast, after touching with their heavy cargoes of chemicals for the paper mills, coal for the cities, and cement and iron for new construction. By the end of the twenties the only cargoes into Bangor were incoming pulpwood, from the Maritimes and Scandinavia; some paper left (mostly from Bucksport as the tidewater mill came into operation), but this was the end. The shipping industry which fifty years before had been the lifeblood of many family fortunes had died within the lifetimes of the founders of some of these fortunes. The rise and fall of this business is indicative in many ways of the rapidity of industrial growth and change in the later part of the nineteenth century.

On the eve of the Civil War nearly every town on the coast was involved in lumber sawing and shipment. Years afterward a boy from Saccarappa reminisced about life in a premier sawmill town in those

Shipping the product

days. "The river was full of logs almost all the time . . . the yards were well filled with lumber although teams of six and eight yoke of oxen with a span of leaders, were hauling it day and night to Portland to be shipped."[1] Portland harbor was open for shipping the year round and lumber went out of Portland, particularly to foreign destinations, when the other river ports were closed with ice. In Saccarappa in the fall the mill owners hauled the logs out of the river into piles taller than the mills, and after the river was frozen men rolled the logs off into the water breaking the ice to provide water to saw the day's logs even during the winter season.[2]

All along the coast it was of course in everyone's interest to keep the ports as free of ice as long as possible. There were even occasional experiments in keeping the ports free of ice longer in the fall, and because of premium prices for early cargoes opening them earlier in the spring.[3] In the spring of 1873, a year which turned out to be disastrous for the lumbering business but looked in the spring to be perhaps the peak year of the century, Calais businessmen and the city fathers hired crews to clear the St. Croix river with saws and dynamite. "The borderers appreciate the money value of open ports, and subscribe liberally to the ice cutting fund."[4] The Kennebec and Penobscot rivers usually opened at about the same time;[5] the St. Croix, Machias, Union and St. John were later, the Saco, Presumscot and Androscoggin were earlier.

The cost of shipping the lumber to coastal ports depended primarily on the freight charges. Most of the vessels were privately owned, often by the captain, although it was not unusual for anyone with money to invest to have his sixty-fourth, or one hundred-twenty-eighth of a vessel. In any case the freight charges were as high as the traffic would bear. At the end of the ice-free season if lumber demands were high, as they were after the great Boston fire of 1872, freight charges were high as well. In mid-summer, with many vessels and much lumber, charges usually hit their lowest point. They accurately reflected the state of the Boston, New York and Philadelphia lumber markets, and the shipper on speculation was dependent on his understanding of those markets. As time went on and the depression of the seventies and eighties made life more difficult for the small capitalist, lumber manufacturers began to own and operate their own ships in an effort to beat out the costs of the middleman's profit. Of course, such larger firms as the Eatons of Calais had always owned their own vessels, sometimes operating a fleet of nine or ten at any one time.[6]

Most of the vessels plying up and down the coast were small schooner rigs, and usually of only a hundred or so tons burden. Some of them were in service for a very long time,[7] and some made small fortunes for their owners.[8] These vessels ordinarily went on fairly set routes, if not schedules. A typical trip would be from Bangor to Providence with 100 to 150 M of lumber, and then back to Bangor with hay, provisions, or perhaps cloth. Returning from Norfolk or Philadelphia the cargo might be coal. Cargoes in any case were heavy goods, and as railroad competition grew more intense many vessels came to the lumber ports in ballast. It was a steady and probably monotonous life as they made their three or four trips a year and then laid up for the winter. Once the ice trade became heavy in the seventies and eighties, some captains were lucky enough to work winters from the tidal ports which did not freeze, running to Philadelphia, Florida, or even on occasion to the West Indies. Ice was even more speculative than lumber though, so not as many cared to be involved in it. As well as being speculative, coasting lumber was also dangerous since vessels did not get much maintenance in this highly competitive business and many sailed into port afloat only because of the lumber aboard.

In peak years nearly every lumber dealer was involved in coastal shipping. Major Lord, a Portland lumber merchant, was perhaps typical of a small operator. Lord operated a wholesale and retail business on Fore Street in Portland.[9] He owned and operated a lumber mill in northern New Hampshire and occasionally purchased lumber from other mills in that area.[10] The lumber came to him by the Grand Trunk Railroad where he, after survey, either transferred it to his lumberyard or to the docks for shipment to New York. Like other small businessmen at this time, Lord also owned houses which he rented, and he discounted notes for others from time to time. In 1872 and 1873 his business was at its height. Spruce deal and some random spruce came in at the rate of two or three railroad cars a day and was tranferred to schooners at the wharf. In 1873 he sent out schooners to New York on April 8, June 12, and September 16, but in 1875 only one went out, on July 29.[11]

From an inspection of Lord's books, though, it is possible to say that shipping played a very small part in his business. He depended on his sale of lumber to local contractors for his profit. It took a great deal of money to be involved in coastal shipping, and commission merchants dominated the business. It was such people as the Eatons of Calais, T. J. Stewart of Bangor, and Alexander

Shipping the product

Gibson of New Brunswick who were the lumber shippers. Stewart was the only one who was not a landowner and lumber king as well. His was a specialty business. Gibson,[12] commonly known as "Boss" Gibson, operated primarily on the Nashwaak River in New Brunswick outside the scope of this work.

Captain Thomas J. Stewart was the great shipping figure in Bangor. A self-made man with independent views, he came to dominate shipping to foreign ports from northern Maine. He rose to be captain of his ship, and after being shipwrecked in 1850 he came ashore in Bangor to run a grocery store and ship chandlery. From this he went into ship brokerage and made his mark in northern Maine as a commission merchant. His views were spicy and well-known, especially on the tariff, and he ran as Democratic candidate for Congress in 1888 because of these views.[13]

Stewart, like most commission merchants, handled any cargo. He dominated the box shook[14] trade to Florida and the Mediterranean, and the spool wood trade to Scotland. Few years went by from the middle seventies to the end of the century without five or six cargoes of box shooks to Palermo or Castelamare. By the late seventies, in fact, the cutting of lumber for shooks was a big thing with some Maine lumbermen. For instance, Hawthorn, Foss and Company cut 900,000 feet in 1878 especially for this trade.[15] T. J. Stewart probably advanced the money for the operation.

In that year four vessels under commission to Stewart cleared with fruit box shooks from Bangor to Sicilian or Italian ports in the month of August. Later in the year, in addition to several vessels for the West Indian trade, another cleared for Palermo on October 14.[16] In 1879 nine vessels cleared for these ports, one leaving as late as December 6.[17] Most of these vessels carried from sixty to sixty-five M of shooks. Occasionally the loads were huge however. In 1880, one vessel had 146 M aboard, perhaps the largest shook cargo carried in a sailing vessel.[18]

Stewart had become the dominant figure in the trade by this time. In 1881 he shipped abroad shooks for 504,000 boxes and staves for 3,600 barrels. In addition, he sent to Bermuda some 128,500 onion boxes, 19,400 tomato boxes, and 15,000 feet of lumber; to the West Indies 1,500,000 feet of long lumber; and to Granada and Port-of-Spain another 680,000 feet of long lumber.[19] From 1878 to 1888 some sixty-six vessels cleared Bangor for southern Europe and fifty-three for the Caribbean with T. J. Stewart charters.[20]

Several things might be noted about this shipping trade. Nearly all vessels to the Indies were American owned and most of them carried some ice. With the other shippers ice was often the main cargo with lumber being used only as a deck cargo. Very few American owned vessels were chartered to the Mediterranean. Most of the vessels were owned by Italian companies and were probably the ones that shipped the fruit, bringing their product back to New York and then coasting to Bangor in ballast for the shooks. As the eighties wore on an occasional Norwegian or Dutch ship came into the trade as these old sailing vessels were driven into commerce that was less demanding of speed. It is indicative of the state of the United States merchant marine in this time that most of the foreign commerce went in foreign vessels.[21] Even so Stewart sometimes had to advertise for vessels as the demand was high throughout the early and mid 1880's.[22] By 1888, however, the shook business was suffering from competition. In that year only one cargo went out from Bangor as Austrian lumber came down over the Alps to crush American trade.[23] Bangor continued to compete, though, and as late as 1894 six vessels went to the Mediterranean. Seven went to the West Indies in that year, but only three of them were Stewart charters. The rest went with ice cargoes on speculation.[24] The Stewart firm was heavily involved in another trade by this time, however, the shipping of spoolwood to England.

Spoolwood went almost entirely to Scotland to the firm of J. and P. Coats at Glasgow. The Coats firm was one of the great British examples of a family owned firm which by controlling the "narrows" of an industry soon came to control most of that industry.[25] They received the spoolwood at Grennock for their mills. White birch was cut in the Maine woods, sawed into square pieces four feet long, tied into bundles and shipped for resawing into the carriers of sewing thread. Some went to Scotland from Canada, but in the late nineteenth century most of it came from Maine. The Maine industry still continued until very recently, and the American subsidiary of the firm, Coats and Clark, still obtained some spoolwood from a mill owned by them in Dixfield.[26] Spoolwood first appeared in the shipping season of 1882, although at the height of the trade it was suggested that J. and P. Coats had been buying Maine spoolwood since 1859.[27] It is conceivable that an odd cargo or two went from Calais earlier but the lush time of the trade was the middle and late eighties. Two vessels went out in 1882, and the great spool business was on. In 1883 one of the vessels which appeared in the trade was a portent of things to come as the British steamer *Creighton* cleared on November 26

Shipping the product 143

with 820 M from Stewart and 380 M more from the Canton Steam Mill Company.[28]

The trade increased steadily. By 1886 it was rated as the most valuable lumber export from Bangor.[29] When the demand grew to this point T. J. Stewart found themselves forced to advertise for men to cut the timber.[30] Maine lumbermen had a new product to cut and haul, and cut and haul they did because the best years for birch spoolwood were still ahead. Eight vessels cleared in 1887 and in 1888 over 7,000,000 board feet of birch left Bangor for the Coats' plant.[31] The demand was so great in that year that the Stewart firm advertised for vessels to carry the product away, and freights which had been quoted at 52 shillings, sixpence in 1886[32] rose to 80 shillings for a three-masted schooner. The newspapers reported that there was difficulty in getting railroad cars enough to bring the spool stock to Bangor for shipment in the vessels.[33]

One of the vessels which left in 1888 was laden with 1,400,000 feet of the spool bars.[34] The demand was so heavy in these years that vessels continued to ply the Atlantic from Maine to Scotland during the winter months, clearing from Portland.[35] By August of 1889, 5,000,000 feet had left Bangor and charters had been laid on for at least 2,000,000 more feet to come.[36]

In 1890 the market was still good and eight vessels cleared for Scotland. About five million feet went this year.[37] The peak had apparently been reached. In 1891 the Stewart firm was no longer purchasing much from other lumbermen. They were operating a small mill at Milo to manufacture their own shooks, spoolwood, and excelsior. The mill employed twenty-five men the year round.[38] Vessels continued to leave for abroad, however. Seven went out this year, but freights had dropped to fifty shillings,[39] and the margin for shipowners must have been narrow. Only four vessels went out in 1892, and one of them the Norwegian bark *Hrvat* ran aground and was wrecked.[40]

One of the ships that did clear, though, was a steamer with 1,250,000 feet on board.[41] In 1893 and 1894 clearances totaled three vessels each year, steamers all, and the trade seemed to have settled to about three million feet each year. With the lower freights the steamers had driven the sailing ships out of the trade. They could not carry the huge cargoes of the steamships, neither in spoolwood nor in the other product of this time, paper pulp, which replaced the old cargoes for a brief fling. During 1882 to 1894 when this trade was important, about sixty-five vessels left Bangor for Scotland. Probably fifty to sixty million board feet of birch went to the thread company from the Maine woods and the port of Bangor.

The Stewart firm continued to ship out cargoes, both foreign and domestic, but by the middle nineties shipping from Bangor had dwindled greatly beginning the big decline which was soon to be accelerated first by the railroads and later by the transfer of ownership of the timberlands to the big pulp companies. These developments were a few years away, however, and the Stewart interests were instrumental in forming a combination designed to weaken the railroad competition.

The lower freights abroad have already been noted, but the lowering of domestic freights had been noticeable throughout the eighties. By 1891 newspapers were reporting many vessels laid up.[42] In the same month of this notice Bangor ship owners met and formed the Bangor Navigation Company with a capital of $200,000. It was said that all vessel owners would join, entering their vessels at a fair valuation, to be managed by the new firm which would do its own insuring to save the 8% insurance costs. Most of the ship owners were represented in the new firm; E. B. Nealley was president, I. K. Stetson, treasurer, Edward Stetson, secretary, and Roland W. Stewart, son of old T. J., as agent. Other ship owners were directors. Although they announced their purpose to be building, repairing, owning, managing, chartering, and sailing vessels for both passengers and freight, there is little evidence that the firm ever did much to weaken the railroad competition.[43] The sailing vessel was on its way to hauling grain from Australia and nitrates from Chile.[44] It was a futile effort; the railroads were just too powerful. They hauled the goods more cheaply, and more importantly they went where the vessels could not go—to the source of the lumber itself.

Over the years Bangor was probably the largest shipping port in the northeast section. However, St. John, New Brunswick, was not far behind, and in many years more lumber went out from St. John than from Bangor. Although the coastal shipping was limited to vessels of American registry, the fact that Maine lumber cut on the St. John, or Aroostook rivers could be manufactured in New Brunswick mills and then transferred to United States ports without duty meant that St. John often acted like a United States coastal port. Portland also continued to ship much lumber, but more and more after the Civil War, this was Canadian lumber brought down over the railroad and consigned to South American ports. Portland's advan-

Logging Camp in Autumn

Larson Collection, University of New Brunswick

A Horse Hovel

Larson Collection. University of New Brunswick

Woodscook at Work

NAFOH, no. 56.2

Lunching Out—Mid-Morning NAFOH, no. 7

Interior of Camp NAFOH, no. 52.1

Necessary Camp Furniture

NAFOH. no. 43.1

Loggers Scrimshaw—Spruce Gum Books

NAFOH

Blacksmith at Work—Author's Grandfather—A Nova Scotia Emigrant

David Smith

Shipping the product 145

tage of being ice-free in most winters put her in competition with Canadian ports, a competition which they were unable to stem entirely even with such devices as the Inter-Colonial Railroad, New Brunswick's bribe for entering Confederation,[45] or later the Canadian Pacific Railroad which was built across Maine to connect St. John with the Canadian West.

There seem to have been several well-defined pathways of lumber commerce. From northern New Hampshire, western Maine, and eastern Quebec one led over the railroads to Portland[46] and thence to the River Plata for foreign, and to the usual nearby ports for domestic orders. The product shipped here was mostly high quality spruce deal with deckloads of laths and other short lumber. This trade continued from 1860 to the decade of the nineties with the only interruption coming when the profit margin was too slight in the foreign trade. From the Kennebec most of the traffic was domestic, running out to Boston, New York, and Philadelphia. Bangor had its dual traffic, and Calais,[47] Machias, Ellsworth, and the other coastal ports were little Bangors. The heart of the trade was domestic but there was always the foreign commerce, deals to the British Isles, then the shooks, and finally the spool stock and the paper pulp. These eastern Maine ports competed nicely with the Maritimes as long as the spruce was easy to get and was long and straight.

The Maritime provinces suffered some disadvantages. The repeal of the Navigation Acts put them into direct competition with both the United States and the Baltic nations. The shortness of the shipping season in these northern latitudes and the relatively undeveloped nature of the country meant a retardation of growth even more pronounced than in Maine. Although absolute proof is not available, one suspects that Maine lumber leaving St. John on the one hand, and jobs in the Maine woods for New Brunswick and Prince Edward Island boys on the other, were the economic salvation for these provinces often enough. It is too bad that for once political boundaries had not been drawn along natural economic lines.

After the Civil War, which throttled much of the lumber commerce both domestic and foreign, business from Portland, dependent on the railroad for its lumber, picked up slowly. Low prices ruled, and when the boom came it lasted such a short time in southern Maine (1870-1872) that Portland shippers were never to realize the pre-war business again. It was not until the end of the seventies that Portland wharves were very busy. In winter there was the incoming Liverpool boat filled with items for the ice-bound Canadians, and outgoing wheat from the Canadian prairies, as well as the United States, but lumber languished. In 1879, for instance, freights from Portland to New York were only $1.75 M and the vessel paid the stevedoring costs, although by 1880 some vessels received $3.00 with shipper paying the stevedoring. One commission merchant, J. W. Frederick Company of Belfast, reported that this latter price was the best in four to seven years.[48] It was not until the eighties that the trade to the River Plata began to pick up, (nine vessels were loading for the South American estuary at the end of October) and the four big Portland firms shipping south—R. Lewis and Company, W. and C. R. Milliken, Frank Dudley, and S. C. Dyer Company—began to look forward to some good years.[49]

In 1885 nearly 33,000,000 feet of lumber went to South America on seventy-two vessels.[50] Freights had dropped greatly though. The high figure, $28.00 M, which had been received by the brig *E. Winslow* at the end of 1872 had never again been reached. In 1884 the price was around $15.00, but it went steadily downward ranging from $12.50 at the beginning of the 1885 season to $9.12½ per M at the beginning of 1886.[51] Even so ten vessels were loading as the year opened, although as newspapers pointed out, every one of the ships was Canadian and every foot of the lumber Canadian as well.[52]

The next year was better. Portland harbor saw one hundred fifty square-rigged ships and barks, and these vessels transported 24,000,000 feet of lumber to the Plata region.[53] Competition drove freights down to between $8.50 and $10.50 in 1888 although they rose to $17.00 in early 1889.[54] There were advertisements for $20.00 freights later on in this year, but this last spurt did not last long.[55] The freights soon fell off to $8.50 per M and leveled at that price, and in 1892 when the schooner *Penobscot* took 325,000 feet from Calais to Cardenas for $4.50 per M it was described as "good business as times are now."[56]

And even though the transaction was in Spanish milled gold, this was a far cry from the halcyon days of $28.00 or even $20.00 per M. In the middle nineties the figure was $7.00 to $8.00, no higher, and as a consequence not much went out. By 1902 the total winter exports from Portland only amounted to 370,302 bundles of shooks, 41,306 pieces of deal, and 9,340 bundles of pulp.[57] Portland shipping had met the railroad, even courted her[58] and lived off her earnings for a while, but had been cast away. It is interesting to note, however, that Portland was more successful in the great battle

Shipping the product

for western produce from lumber than from grain, which had been the great hope of John A. Poor and his followers a half-century before.

The small ports along the southern shore shipped out their lumber, but aside from brief times the economics of lumber shipping after the Civil War limited most of the trade to the larger towns.

The deal sawed in the Wiscasset sawmill mentioned in chapter five gave Bath, the customs port involved, two or three good years in the mid-seventies.[59] Bath was always a fair-sized port. Even as late as 1890-1891 it was still a busy port shipping about 85,000,000 feet of all kinds of lumber during the fiscal year.[60] Bath was the only big port, outside the big three however, in this year as well as all others. For the size of the towns, such ports as Calais, Cherryfield, or Eastport shipped fair amounts, (perhaps 10 million a year) but taking it in the aggregate it did not amount to much compared to the great ports.[61]

Bangor, the largest of these ports, often had over a hundred vessels in port at a time and employed a full-time harbor master. His duties were to assign berths, negotiate tows down to Bucksport, clear foreign vessels, and generally maintain order along the wharves and in the channel. The Penobscot River at Bangor is deep enough but it does have a tricky channel, and on busy days the harbor master and his crews were extremely hard pressed.[62] The harbor master records have disappeared for the period before World War One, but an annual abstract was presented to the Mayor and it was printed in the *Annual Reports* of the city. Although it is incomplete, notably for domestic clearances since 1876, by placing the report in tabular form one can see the ups and downs of Bangor commerce very well.[63]

Commerce, Both Domestic and Foreign, Bangor, Maine
1861-1916

Year	Foreign Clearances	Domestic Clearances	Remarks
1861	126	1,652	42 for S. Europe
1862	126	1,834	50 frozen in
1863	140	2,128	
1864	190	1,626	
1865	140	1,615	
1866	135	2,088	
1867	112	2,359	
1868	112	2,307	
1869	176	1,956	
1870			
1871	64	2,650	

Year	Foreign Clearances	Domestic Clearances	Remarks
1872	84	2,774	Biggest lumber year
1873		2,556	Total for both
1874		664	Total for both
1875	72	1,753	
1876	40		
1877			
1878			
1879	46		358,000 feet pine
1880	One of the better years		imported from St.
1881	34		John in 1879.
1882	40		
1883	45		
1884	51		
1885	54		A million feet of pine
1886	41		imported in 1885,
1887	44		and about every year
1888	40		after.
1889	52		
1890	39		
1891	27		
1892	38		
1893	34		
1894	23		
1895	27		
1896	31		
1897	42		
1898	34		
1899	32		
1900	28	920 lumber cargoes	
1901	22	687 cargoes (587 lumber; 100 cooperage)	
1902		518 all told	
1903	38		
1904	23		
1905	20		
1906	9		One cargo to West Indies
1907	4		
1908	none		
1909	4		
1910	2		
1911	3		
1912	5		
1913	3		
1914	5		
1915	10		
1916	2		

Shipping the product

The typical imports into Bangor in the early period (before 1875) were molasses, salt, flour, corn, and pork. Most of it continued up the river to feed the men cutting the logs. The railroad had already begun to bother the shipping; in 1864 for instance, of the 70,000 barrels of flour, and 260,000 bushels of corn which entered Bangor,[64] nearly half came by railroad, and soon much if not most of the flour, corn, oats, and pork was entering Bangor by this method.

Other items besides lumber went out. In 1873, for instance, the Harbor Master listed exports of 3,027 tons of hay, 3,200 cords of bark, 25,000 ships knees, 71,000 knocked down barrels, 8,050,000 hoops, and 180,000 orange boxes to domestic ports. Even with all this, Mayor Augustus C. Hamlin, in his annual address of 1877, when commenting on the shipping business said:[65]

> Once she [Bangor] was the first lumber mart of this continent, if not the world. Not many years ago the sails of her commerce might have been seen on almost every sea. It is no fault of hers that her forests have vanished and her ships have disappeared. These misfortunes do not arise from want of prudence or energy of her citizens. They are due to causes over which mercantile communities have but little control. . . . .

His Honor was somewhat premature in his eulogy, but the time of which he spoke was not far off. Perhaps the biggest of the last years was 1880. In that year the domestic exports were over 123,000,000 feet of long lumber, 145,000 shooks, 100,000 fish barrels, 109,000 bundles of staves, while the foreign exports amounted to 445,740 feet of shook, nearly 2,000,000 feet of long lumber, and 226 spars.

Only three years later the new mayor, Frederick A. Cummings, when discussing the lumber trade, saw the end in close view:[66]

> Trace them all [*the different branches of the lumber industry*] to their fountainhead, and you will find that all begin and end in the spruce log, and a famine in logs means hungry mouths for all other businesses in Bangor. . . . The axe and the worm are fast reducing the crop of lumber. We have but one resource left, . . . the water power at the falls.

If he was right the city was never able to use this last resource, try as it would down through the years.

In the earlier discussions of the T. J. Stewart Company some of Bangor's speciality exports, shooks and spoolwood were touched upon. Bangor and to a greater extent St. John were more famous in a somewhat earlier period for another export, spruce and pine deal. The deal business was a big one. For instance in 1862 twenty-five ships or barks cleared the Penobscot with 17,000,000 feet of deal. Under the impress of the Confederate raiders, that trade fell off in 1863 by about a half, but in 1864 12,000,000 feet went abroad. The last year was lighter than before, but not as bad as the other war years.[67] English cotton towns had suffered during the American Civil War and building lagged.[68] One hears little about deal again until 1873, the year of the crash. During that summer the market for English deal was described as "brisk", and cargoes left nearly every week. Calais sent out 3,000,000 feet for which the merchants received $11.00 M in gold, and Bangor parties sent out several ships.[69] The trade continued to be important throughout the next few years.

In the depression the trade was welcomed.[70] The last year of any consequence was 1877. Newspaper readers in Bangor saw articles on British prices nearly every week all spring. The trade was apparently enhanced by the terrible fire which leveled much of St. John. Fifteen vessels left with deal for the British, including eight full-rigged ships. Most went to Liverpool, but one went to Gloucester, one to Drogheda, Ireland, and one to Rotterdam for the continental trade.[71] Three more went out to Drogheda in 1878, but in 1879 the *Commercial* remarked, "The deal business is very dull across the waters. It has been gradually declining for the past two years. There is almost no demand now..." [72] Ten more vessels went out loaded with deal through 1882,[73] but after this time deal only went as a deck cargo with the shooks and spoolwood. Canadian and more particularly Baltic trees furnished this sort of lumber for the British consumers. The last time that deal entered into an account of Bangor shipping is in 1907. In that year a cargo from Norway with 150,000 feet of spruce deal bound for Buenos Aires stopped for water and provisions. Under maritime law she cleared Bangor with the cargo.[74] This notation told more than some volumes about the competition and the result of it; a cargo of pulpwood from the Maritimes was landed at Bangor that year, and the two items together told it all.

To place still further emphasis on the point of competition in the shipping business some accounts of freight rates and stevedoring costs is needed. The freight rates were for the carriage from one port to another, with the owners of the lumber usually paying dockage, insurance, and loading fees. When times were hard the stevedoring costs were often the negotiable point in the contract. Whoever could exert the most pressure generally forced the other (whether shipowner or lumber owner) to assume these costs.

Shipping the product 151

The boss of the loading crew got paid by the M for the loading, and he provided the crew who were paid by the hour or by the day. These men, "mudlarks" as they were called in Bangor, worked long hours in disagreeable conditions. Wages remained low as "buckwheat eaters" from New Brunswick were nearly always available waiting for the fall logging to start. New Brunswickers nearly always kept the labor supply high enough to depress wages in this rough, hardy work. The old rates, before the Civil War, had varied with the tonnage of the vessel to be worked. The going rates in 1856, for example, were $1.00 M for vessels fifty tons and under, $2.25 M for vessels of 75 to 100 tons burden, and $4.00 M for vessels which were rated at from 150 to 200 tons.[75] The difference probably occurred because of the height of the lift and the length of the carry. All this lumber of course was moved from place to place by hand.

This may well have been the high point of wages before the Civil War. With the introduction of some hoisting machinery, prices dropped off. In 1880 the price was about 20¢ M with the high point at the very end of the season 24¢. The next year the stevedores asked for a raise to 25¢ minimum, and the dealers granted it.[76] In 1882, a good year, stevedores averaged $2.50 to $3.00 a day and green hand boys got $1.50 to $1.75. Well they might; sailors wages had risen from $20.00 a month, described as "the highest wages in years," to $25.00 for ordinary seamen. On one occasion they rose higher, to $30.00 but that was an extraordinary year. At that time, newspapers reported recruiting going on in the back country.[77]

Portland stevedores were always better off. They didn't have the "buckwheat eaters" for competition, although when they threatened to strike this same year, the companies told them to go ahead for they would import Frenchmen from Quebec to do the work. Even with this threat a Longshoreman's Society of Portland was organized and the men received 5¢ an hour pay boost from the previous 25¢ by day and 30¢ by night. They worked a ten hour day, and a spokesman claimed there was more money in 30¢ an hour than there was in 30¢ M. He did say they were expected to handle spruce "all out". An attempt to organize the Bangor laborers was not successful, but they got the 30¢ an hour on spruce anyway.[78] Even as late as 1894 Portland stevedores were paid only $3.00 a day, with $3.50 for night work. They got $4.50 for Sundays, 35¢ an hour for 12 hours. They went out on strike for 50¢ an hour on Sundays pointing out with some reason that New York stevedores were getting 60¢ for Sunday work. They got it.[79]

Stevedoring was rugged, dirty, heavy, exhausting, miserable work. Any wage increase granted to the "mudlarks" would have only been due them. Still, from the point of view of the lumber operator, this was only one more cost and, as will be shown, few businesses in the late nineteenth century were in such an economic squeeze. The temptation to go west must have been almost irresistible both for laborers and capitalists alike and, of course, many went.

Freight costs were another big factor in determining lumber profit or loss. Probably the standard figure for freights on domestic lumber after the Civil War was $2.00 M to Boston, $2.25 to Sound ports, and $2.50 to New York. These prices are from Bangor. In the depression of the seventies prices fell off sharply, later rising to $1.75 to Boston and others in proportion. The Boston price rose again to $2.00 early in the 1880's and as high as $2.50 in the small boom of the 1880's, but after this $1.75 was probably the top price as the railroad competition intensified. Under difficult conditions, or with a boom market at the end of a season, they might rise very high, or conversely, in mid-summer fall off sharply. We have a fairly clear record from 1876 to 1894. The table which follows will indicate the shifts.[80]

Lumber Freight Prices, Bangor to Boston—1876-1894

In dollars by the M

Year	High	Low	Average	Remarks
1876			$1.00	
1877			1.25	
1878			1.37½	
1879			1.25	Some competition in port
1880	$2.25	$1.25	1.75	
1881	2.50		2.00	Great competition
1882	2.50	2.25	2.50	
1883	2.25	2.00	2.25	
1884			2.00	Trade "dull"
1885	1.75	1.62	1.75	
1886			1.62	
1887			1.87	
1888	2.25	2.00	2.12½	New York freights higher than usual
1889	2.25	2.00	2.12½	
1890			1.75	
1891			2.00	
1892		1.62½	1.75	
1893	2.00	1.50	1.75	Trade "brisk" at the $1.75 average both years
1894	2.00	1.50	1.75	

Shipping the product

With the higher prices for wages and all the other costs of running a vessel it is no wonder that the sailing ships were driven from the seas. It was easier for an owner, particularly a skipper-owner, to lay out one voyage, and then two, and finally to use his vessel to haul stone across Eggemoggin Reach or perhaps to sail with the extremely dangerous cargoes of lime casks from Rockland. It was the big four, five, or six masted schooners that carried some of the lumber trade. The ubiquitous railroad got the remainder.[81]

The St. John River and its two large lumbering towns—Fredericton with its sawmills and boom, and St. John,—itself a major saw mill and shipping town—stood in friendly competition with the Penobscot and Bangor. Much of the lumber shipped from St. John had been cut in Maine and driven down the river to be sawed. Under the Webster-Ashburton Treaty this could be done without reference to tariffs or other customs fees both here and on the St. Croix, the other border river. The St. John however, also benefited from certain colonial preferences with the English as long as they lasted. After this change, which occurs just about at the start of this period, the competition was stiffer and sometimes was conducted with less friendship.

The Canadians increasingly found themselves tied to American fortunes and this (for them) unpleasant situation continued until the days of pulpwood. The United States Civil War had an effect on Canadians. The *Acadian Reporter* described 1862 as the smallest shipping year in a decade, and said the recession was due to the blockade of southern ports and the consequent slowdown in the English cotton towns.[82] In the next year shipping and shipbuilding increased north of the border[83] because of the commercial destruction at sea, and as the fortunes of the North increased the business of Canada began to grow until in 1866, the last year of reciprocity, there was a great Canadian boom. After the end of reciprocity, however, it was a different story. The Canadians, finding themselves in competition with the Americans, attempted to become dominant. Only in the shipping of deal to England and the Continent were they successful. This was done partially through non-compliance with the law. The English had had a law for years which prohibited deck loads in the North Atlantic lumber trade between September 1 and May 1. By enforcing the provisions of the law on the Americans, and winking at it in the Canadians, the Empire vessels were benefited. That is, they were benefited until the terrible fall and winter of 1872-3 when fifty-seven vessels were wrecked and abandoned in the North Atlantic between provincial ports and the Continent. This resulted in a tighter law passed by the Canadian Parliament which specified no deck loads after October 1st, although this law was later changed to allow deals to the height of three feet to be carried on deck.[84]

The other way to compete with the Americans was to take away Portland's position as Canada's winter port, a situation which had come about with the construction of the Grand Trunk Railroad. This was part of the continuing nineteenth century battle for control of western produce, although fought on a somewhat smaller scale than in New York and Boston. The Provincemen had little chance until June of 1889 when the Canadian Pacific short route across Maine was completed and the lease connecting Mattawamkeag and Vanceboro insured the route. After this the citizens of St. John really tried to beat out the Maineites. The Portlanders had a mail boat all winter from Liverpool; it was a lucrative thing, and St. John's citizens wanted it. They, in short, wanted to become Canada's winter port. They built a grain elevator for the CPRR at a cost of $40,000 but it wasn't enough. The Liverpool-Portland route was shorter and the Maine port retained the mail steamer with its return cargoes of grain and lumber. The citizens of St. John then laid out a million dollars in improvements to their wharves and docking facilities and, with the energetic efforts of J. D. Haxen and John A. Chesley in the Dominion Parliament, were able to get a $25,000 subsidy from Ottawa for a weekly boat in the season of 1895-1896. In the twenty-two sailings that winter to Liverpool were shipped 272,910 bushels of wheat, and 6,521 standards of deal. The Portland mail steamer survived only until 1897. The St. John effort had been successful. By 1908 their exports amounted to $30 million, of which $10 million came from the States.[85]

St. John had always been a busy port. After recovering from the initial shock of losing reciprocity advantages, shipping from the port gained until in 1872 it was the fourth largest port in the British Empire. Only Liverpool, London, and Glasgow had more shipping.[86] Much if not most of this shipping was deal across the Atlantic. Great amounts went out even during the years of lumber depression elsewhere. Four hundred and twenty vessels left in 1874, for instance, carrying 219,000,000 feet of deal. Nearly all this came from the St. John valley.[87] By 1876 a rise in freights had wiped out the American competition and the trade really began to hum. The trade continued to be good, upwards of 200,000,000 feet each year, with 1883 the second largest year in the history of the province. Four hundred and

Shipping the product 155

twenty-seven million feet left the Province, 181,500,000 from St. John itself.[88] The table shows these New Brunswick exports in detail.[89]

New Brunswick and St. John Shipments to Transatlantic Ports
1870-1900

Year	St. John Deal (feet)	Birch Timber (tons)	Pine Timber (tons)	Provincial Total
1870				244,750,000 (deal)
1875				275,000,000 (deal)
1878	188,168,610	7989	2493	
1879	153,279,357	11548	3237	
1880	215,485,000	16035	2441	330,000,000 (deal)
1881	210,281,730	5134	1734	
1882	201,413,717	7576	3332	
1883	181,517,932	11778	3883	427,000,000 (deal)
1884	164,829,825	14006	3836	333,000,000 (deal)
1885	152,543,026	13769	3686	292,000,000 (total)
1886	138,934,392	7354	4313	276,000,000 (total)
1887	118,450,590	5197	1587	250,000,000 (total)
1888	153,184,187	4721	457	277,000,000 (total)
1889	180,167,488	7221	487	369,000,000 (total)
1890	132,608,516	1311	4317	293,000,000 (total)
1891	122,242,682	5004		253,000,000 (total)
1892	146,529,309	10200		325,000,000 (total)
1893	156,653,334	5294		312,000,000 (total)
1895				311,750,000 (deal)
1900				419,750,000 (deal)

Once the American competition was finished in the mid-seventies freight charges did not vary much,[90] and the Canadians had nearly a clear field with only the Baltic exports to trouble them. By the time it had driven the Canadians from the British ports, however, the demand by American newspapers for wood pulp paper had driven most of the sawmill owners into that manufacture or into retirement. Change did come later on the St. John than it did in Maine though.

It is difficult to determine how much of the St. John lumber was actually cut in Maine. The figures on lumbering cuts on these northern rivers which drain into Provincial waters are for the later period, that is after 1885.[91] It is obvious that once lumber is sawed it is very difficult to determine where it was grown. Much lumber from New Brunswick must have come in duty-free as Maine timber, and at the appropriate time much Maine lumber must have gone to Britain under the protection of the Navigation Acts. Before 1885 no one cared, especially those who hoped to see the United States purchase the Maritimes, a dream which did not die until the Bangor and Aroostook railroad made its way to Van Buren, Madawaska, and Fort Kent by which time St. John lumbering, except for the pulp mills, was nearly gone by. Scattered figures show some of the trade.[92]

Maine Lumber Arriving From Provincial Sawmills
1866-9; 1885-7

Year	Shingles	Clapboards	Total (in feet)
1866	41,000,000	834,000	
1867	15,000,000	655,000	
1868	25,000,000	500,000	
1869	28,000,000	261,575	
1885			71,500,000
1886			110,300,000
1887	25,000,000		96,000,000

By this time some Maine mill owners were beginning to be very unhappy about the competition. As a result in 1891 and 1892 a close watch was made of Maine lumber entering New Brunswick to be sawed. With the customs inspectors enforcing the laws, other means had to be found to deal with the United States duty—$2.00 M on sawed boards. This led to an exciting and little known method of lumber transport in the Northeast: the rafting of logs over the Atlantic to United States coastal ports. Logs which entered in the round were not subject to the $2.00 M tariff so enterprising individuals decided to raft their logs to New York City, not only saving the tariff costs but the freight and handling charges as well. It was not the first time this had been tried. As early as 1841 a big raft of squared timbers in the form of a ship's hull and ship-rigged had tried to go from Bath (Swan's Island) to England. The crew was unhappy with the unorthodox vessel and only sailed after a small schooner was hoisted on deck. Their fears proved to be an ill omen for the success of the voyage and the timber ship was abandoned off the coast of Nova Scotia.[93] Coasting the route from the Maritimes to New York City was somewhat more successful though.

Big rafts started leaving for the States in the early 1880's. Some of them experienced severe problems. The first, seventy smaller rafts chained together, went from port to port. Towed by two tugs, it spent its third night at sea in Boothbay Harbor after some of the chains

Shipping the product 157

parted. It finally arrived in Portland on the seventh night, and after this it spent time in Provincetown and Vineyard Haven. Even so some of the timbers were lost and the owners were happy when it arrived at New York City, nineteen days out from St. John.[94]

The success of these pioneers apparently emboldened the Provincial lumbermen as reports came thick and fast of new attempts. Men at Joggins, Nova Scotia, under the direction of H. H. Robertson of St. John, began a truly enormous venture. They cut long logs, two and one half million feet and built on ways a tremendous vessel held together with fifty-four tons of chains to be launched into the Bay of Fundy and towed to New York City. This huge vessel, four hundred and ten feet long, fifty feet wide and thirty-five feet high, broke down the ways from its weight on the day of its launch, July 31, 1886, and finally set half in the water, half out.[95] Maine newspapers greeted this news with glee.[96]

The speculators did not give up however. More logs were put on the monster and it was raised on new ways. By November 15 of the next year it was reported to be five hundred and eighty-five feet long, sixty-two feet wide, and thirty-seven feet deep. It drew nineteen and one-half feet of water. The 2,750,000 feet of lumber was estimated to weigh 120,000 tons. When it was finally launched it traveled the 1,200 feet of the ways in thirty-four seconds and the friction set the ways on fire. Tugs began to haul it toward New York but in heavy seas off Nantucket it parted from its tow and broke up into constituent logs three hundred miles off shore in the coastal shipping lanes. Vessels reported drifting logs for six months until they finally grounded off the African coast.[97] A Bangor newspaper expressed the standard Maine view. "Bangor lumbermen and vessel owners are very much pleased at the loss of the raft. They understood that if this experiment proved a success it would deal a heavy blow to the lumber business and coasting trade." [98]

Robertson was not daunted by his failures. As his rafts were unsuccessful he now turned to timber ships similar to the older Bath experiments. He laid down ways for a six-masted vessel, six hundred and fifty feet long. The *Commercial*, when it reported the news, was cool.[99] "However he may lack in judgment, there can be no doubt that the builder has the courage of his convictions."

This vessel was another monster. It took 25,000 spruce sticks of an average length of thirty-eight feet, with the minimum top diameter of six inches, to construct this vessel. Her six masts were square-rigged with fore and aft try sail, and she was designed to carry a crew of twenty.[100] After it was launched, on July 25, the owner J. D. Leary of Brooklyn sent two tugs down east to haul her to his yards. She was insured against loss for $30,000.[101] Even though Maine newspapers were not sanguine of her success she made it to New York.

Others followed in 1890 and 1891. The 1890 venture, a raft of logs this time, had a difficult trip. It partially broke down and had to put in to Rockland harbor for repairs after earlier repairs had been made in Rockport. When it arrived in Portland it was in very bad shape and had to be towed to Peak's Island for extensive repairs, with logs reported to be scattered all up and down the Maine coast.[102] Maine newspapers were still unhappy about these ventures.[103]

> This venture of Canadian lumbermen furnishes employment, logs, and something to look at to many dwellers along the Maine coast, but the steamboat men, to whom the logs are a great bother, are not pleased with the unique visitor, and it doesn't look as if the raft would be of much profit to its owners unless a way can be devised to keep it together.

The big raft, designed to be delivered to the Astor family, to be used in construction of new wharves, finally left Portland on July 28.[104] The next night it was reported off Highland Light, Massachusetts, and finally after putting in at Vineyard Haven because of heavy weather and for more repairs, it arrived in New York August 5.[105] Two others went that year, both via the deep sea route, but it was costly. The second lost two of its ten sections; and the third eight of its fourteen. This part drifted on shore off Rockland and the logs were salvaged by the coastal residents.[106] One other huge raft, the largest of the lot, went to New York in 1891. It consisted of three and one half million feet of lumber in logs, built in sixteen sections banded together with iron chains. Each section was sixty feet long, forty feet wide, and nineteen feet deep. Two tugs towed the monster, each with two nine inch hawsers one hundred and fifty fathoms long. Even these precautions did not help much. Bad weather off Mt. Desert Island caused thirteen sections to break away. The tugs put into Bar Harbor, and chased and caught the sections. After repairs it sailed again breaking up once more off Massachusetts. Repairs were made at Vineyard Haven and the big raft finally arrived at its destination.[107] Robertson soon took his talents to the Pacific Coast, and his departure was hailed.[108]

> It is said that the idea of towing large quantities of logs in cigar-shaped rafts is becoming very popular on the Pacific Coast, and is giving, generally, success. The inventor of the Joggins system, H. H.

Shipping the product 159

> Robertson of St. John, N. B., is constructing a large raft of timber . . . which when completed will be transported to San Francisco. . . . If Mr. Robertson will confine his timber-towing operations to the Pacific Coast, the vessel owners and masters of the Atlantic side of the Continent will wish him every possible success.

The shipping of less well-known but important forest products was also an important part of this business. This includes Christmas trees, pulpwood, wood pulp, sawdust, and short lumber for shoring ice cargoes. Christmas trees started going to the eastern cities in the eighties. It was a big trade. These small firs were cut on pasture land, tied up and shipped to centers for the holiday trade by schooner or mostly by rail. In 1883 one contractor had orders for fifteen carloads for New York City. Each car carried from 1,000 to 2,500 trees. He received from 50¢ to $5.00 a tree.[109] By 1889 trees were worth $5.00 M on the stump and farmers cut and hauled to the wharves in Orland many thousand.[110] Most were carried by train though; there was less damage than by sea. One week of 1894 saw sixteen flat cars go to Boston alone.[111]

Sawdust, which had been used primarily to bank up houses against the Maine winters, was used to hold ice cargoes in place after the big boom, and it sold for $1.50 a cord at times.[112] When the Hudson River didn't freeze suitable ice, both sawdust and short lumber for deck loads disappeared at high prices. The big ice boom of February 1890 cleared the docks along the coasts with random spruce selling for $19.00 M for deckloads at the height of this boom.[113] Everything, except the tops and foliage, went to sea at one time or another.

Of more importance than Christmas trees or sawdust though was the shipping of pulpwood and on occasion wood pulp. At first the pulpwood went out; after the turn of the century it came in. Most of this was poplar designed for the soda mills which did not use spruce. The trade started early. Some came down over the Grand Trunk in 1879 from Canada, and in that same year Holyoke, Massachusetts, mills purchased 1,000 cords cut at Princeton, Maine.[114] The next year more was cut at East Machias, shipped from Machias by vessel to New Haven, and then over the railroad to Holyoke.[115] Large amounts were shipped from Brunswick, Windham, and Benton to California for paper manufacture.[116]

The big shippers were E. and J. Fred Webster of Orono. The first big year of pulpwood shipments seems to have been 1884. The Webster firm sent out 5,500 cords to Rhode Island and Pennsylvania pulp mills, and 1,200 to Baltimore parties.[117] The next year they set up a portable steam engine and saw at High Head wharf in Bangor and cut small spruce logs into pulpwood lengths, two feet long. The wood, driven directly to the saw, thus saving railroad rates, was shipped in great quantities to Providence and Havre de Grace. The firm shipped 250 cords a week at the height of the season. Some wood came down by railroad and most of it was cut in and around their portable mill near Blanchard.[118] This market, although a good one, did not last long. The mills were too far from the forest, and as these mills made their trek toward the north country pulp wood no longer left these ports. In fact, it soon began to come to the new mills.

In the years from 1907 to 1921 the International Paper Company imported pulpwood from New Brunswick to Portland; the 1908 importation was nearly 50,000 cords.[119] By the first World War it was coming at a steady flow. A fleet of four or five small steamers (mostly Norwegian registry) brought in 2,500 or so cords a trip. The tables indicate amounts and provincial locations.[120]

Pulpwood Imports to Portland—Selected Years[121]

Year	Cords of Pulpwood	
1908	50,000	
1909	51,113	
1910	43,629	
1911	51,001	
1912	93,873	with four cargoes to come
1913	—	
1914	59,397	
1920	85,000	most since 1914

Lumber Shipments to U.S. From Province Ports[122]
1912-1913

Port	1912	1913
Campbellton	35,152,226	54,351,943
Dalhousie	15,909,671	15,000,000
Bathurst	7,615,378	—
Newcastle	33,494,529	22,103,758
Chatham	45,633,040	52,384,955
Buctouche	287,556	—
Richibucto	—	1,594,990
Sackville	13,925,843	8,527,876
Albert	9,421,345	15,669,408
Shediac	2,328,131	3,792,069

Not all of it was of foreign origin. Much came from smaller Maine coastal towns such as Steuben, Brooksville, and Bar Harbor.

Hauling With Ox-Team

NAFOH, no. 6

Handloading Birch Bolts on Logging Sleds

Larson Collection, University of New Brunswick

Lombard and Crew, Grindstone, mid-1920's

NAFOH, no. 18

Tractor Accident

NAFOH, no. 21.2

Tractor With 25 Sleds—Little St. John Pond, February 16, 1949

Great Northern Paper Company

Brook Drive—Carding the Ledges

Larson Collection, University of New Brunswick

High Water—Slewgundy Heater, Mattawamkeag River, 1917

NAFOH, no. 166

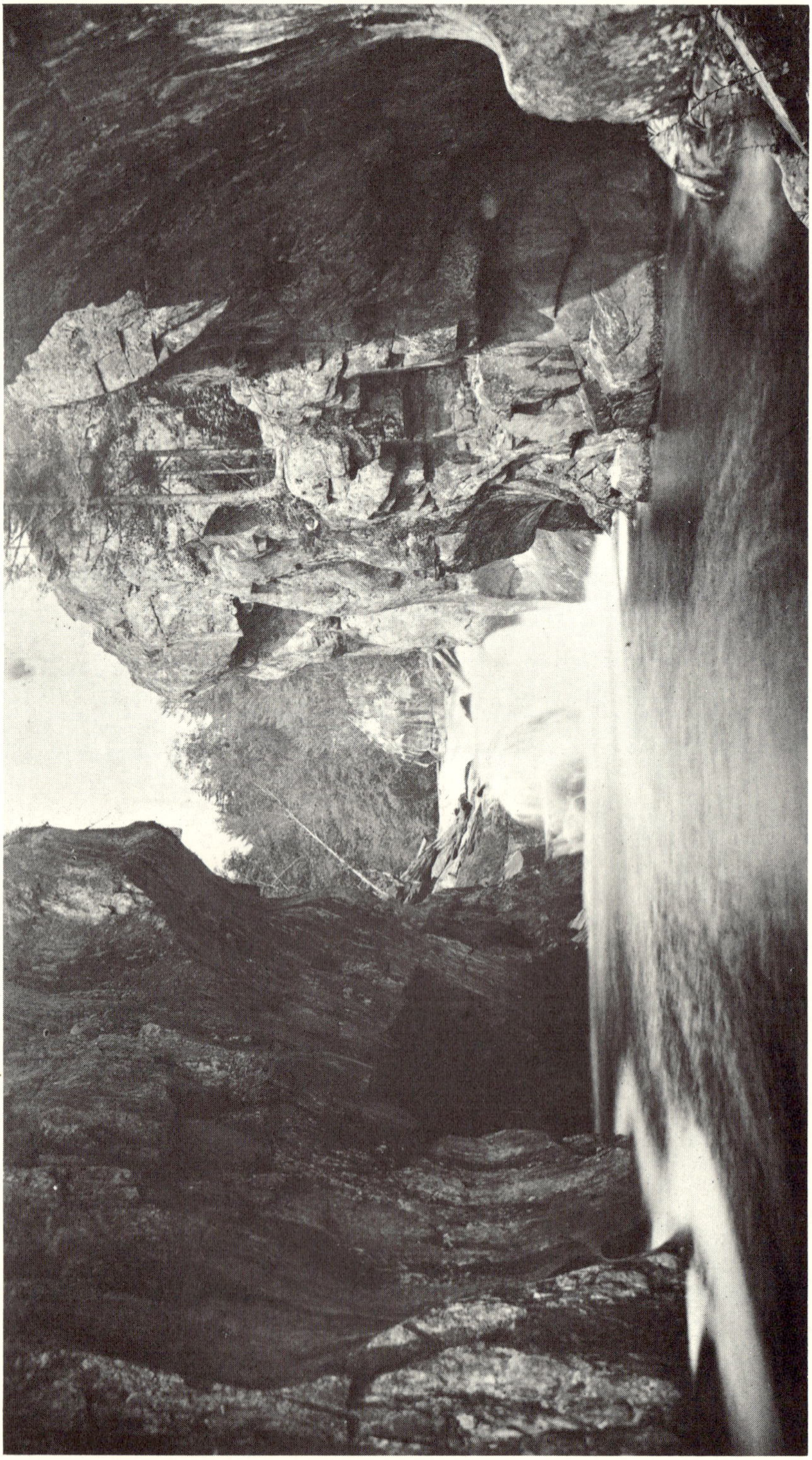

Shipping the product 161

In 1914 five hundred and fifty cords came to Bangor,[123] and in 1917 eight hundred and fifty-three cords.[124] By 1927 the amount of pulpwood imported to Bangor was nearly 25,000 cords and the amount climbed steadily.[125] For 1928 the records are detailed as the daily log of the port has been saved. Two kinds of vessels were coming to Bangor by this time. Small schooners with twenty-five to thirty-five cords per load were coming from Penobscot Bay, and larger steamers were arriving from Nova Scotia about every week with five to six hundred cords aboard. All of this went to the Eastern Manufacturing Company in Brewer, 30,543 cords all told.[126] In 1929 the total was nearly 50,000 cords, some of which arrived in very large vessels. In 1930, for instance, the steamer *Dansborg*, arrived from Archangel, U.S.S.R., with 3,100 cords of spruce pulpwood. This was the largest vessel to navigate the Penobscot to that time.[127] Protests about the arrival of pulpwood from a country not recognized by the United States soon cut off that source of supply. Even with this setback the pulpwood continued to come in throughout the thirties both for the Eastern firm and the new St. Regis mill at Bucksport. In 1939 foreign imports totaled 30,832 cords and the domestic supply only 8,514.[128] This was about the smallest year of the decade. It had been as high as 85,000 cords in 1937.

The last item of importance in the shipping of lumber products was the further refinement of the product, wood pulp. At first, as with the other products, the pulp went out. The Androscoggin Pulp Company of Brunswick shipped wood pulp to London for the *Times*.[129] It was well liked, but not too much was shipped. It was not until the nineties that the trade was of any consequence.

Then F. E. Clergue and his associates of the Moosehead Pulp and Paper Company with mills at Veazie and Solon attempted to ship wood pulp to Europe. Quite an amount went out, although 1893 and 1894 were the only years of the miniature boom. The pulp left from both Bangor and Portland, all on big ships, each taking eighty railroad cars of semi-dry pulp to a load.[130] The trade continued winter and summer, but Norwegian producers lowered their prices to the English purchasers, and by the summer of 1894 the trade was all over. Some paper left after that, and once Englishmen came to look at the mills, but little came of either venture.[131] All that remained was the Canadian pulpwood as the American demands were more voracious than any possible output.[132]

These last few items of shipment make little impression as one views the totality of lumber product shipments from Maine. The largest item through the years was sawed lumber, and it went out from the Maine ports by the million board feet. Gradually in the eighties and nineties the railroads began to replace the seacoast towns as the point of origin. For instance, two-thirds of the box-board trade of 10,000,000 feet in Bangor in 1880 went out by rail.[133] This was for the domestic shook market, but long lumber went out this way too.

Lawrence, Newhall, and Page sent two hundred and fifty-five cars of long lumber from Shawmut station in July, 1893.[134] Even before this in 1889 the Maine Central reported hauling 20,000 carloads of lumber along its line, and shipments ran over 2,500 cars a month leaving the state in the fall of 1891.[135] Once the Bangor and Aroostook made its way into the north country it would be all over for the river ports. The table shows what was possible even by indirect connections.[136] By the twentieth century the mechanical revolution was so encompassing that the railroad and the truck even replaced the river drive of logs.

NOTES

[1] "The mills at Saccarappa in 1858", excerpt from a letter from E. V. Haskell to John A. Warren, 1877, in S. D. Warren Company files.

[2] *Ibid.*

[3] The Boston steamers, which were made of iron, frequently kept the Penobscot open for a week by acting as icebreakers in the eighties and nineties.

[4] Bangor *Daily Commercial,* March 10, 11, 1873. Bangor had experimented with their method in the late spring of 1836. *Maine Farmer,* April 15, 1836.

[5] The Penobscot openings and closings are well documented, the Kennebec less so. See the appendix of this work for dates for the Penobscot and St. John.

[6] Louis Woodbury Eaton in his *Pork, Molasses, and Timber,* New York, 1954, Chapter IV, describes going to their timberlands with the captain of one of their ships and crew of choppers to pick out the timber for a vessel to be used in their trade. For early discussions, *Penobscot Journal,* August 9, 1831; Bangor *Whig and Courier,* February 20, April 20, 1834, January 10, March 31, 1842; June 9, 1847.

[7] The *Polly* was probably the oldest and most famous. She was launched in 1804, rebuilt in 1814, and was still in service in 1905. A photograph of her in Portland harbor in 1905 is in Lincoln Colcord, *Sailing Days on the Old Penobscot: The River and Bay as they were in the Old Days,* Salem, 1932, between pages 192 and 193. The *Polly* was reputed to be the oldest vessel in continuous service in the country. After a career as a privateer in the war of 1812 she was re-rigged and was operated as a coaster for most of her life. She cleared Bangor July 29, 1889 and arrived in Providence on August 10 so her age did not diminish her speed much. See Bangor *Daily Commercial,* August 13, 1889, and June 1, 1893 for stories. Another famous schooner was the *Hiram,* 64 tons n.r.

Shipping the product 163

A topsail schooner with a square rig, she was laid down in Biddeford in 1819, and worked the West Indies and Caribbean for years. By 1891 she was coasting Calais to Boston. *Ibid.*, September 19, 1891.

[8] For instance, the *Charles Cooper*, 86 tons n.r., laid down in 1872, rebuilt in 1879 at a cost of $2,800, was claimed to return a net profit of $3,500 in 1887, and was estimated to have five such years ahead of her. *Ibid.*, March 23, 1888. Earlier in the 1830's many vessels were sloop rigged. I am at work on an article discussing the 1830's coastal trade from the Kennebec.

[9] Six account books and some miscellaneous papers are in the Maine Historical Society in Portland. Following is derived from these records.

[10] In Northumberland and Milan, New Hampshire.

[11] The railroad cars averaged around 7,500 board feet per car. From survey book for 1872-3 showing number of pieces and scalage for each car. It cost him $ .25 a M to have the lumber scaled when his own men were not available. Survey bill dated May 1, 1876. In 1873, his big year, he operated almost entirely on credit furnished by J. H. Hamlin of Portland who also handled his insurance account. Hamlin transferred money to Northumberland to pay his sawyers, supplied him with running capital, and shipped his own timber under Lord's charters. There is a complete yearly statement for 1873 in these papers from Hamlin. The 1875 vessel went on his own account. The *Olive Elizabeth* took 123 M of spruce to New York at freightage of $2.00 M. Voucher signed by master and Lord dated May 29, 1875.

[12] The best short account of Gibson's life is by Fred Phillips, St. John (N.B.) *Telegraph-Journal*, March 15, 1937. I am indebted to Mr. Phillips for a copy of this and other articles on "New Brunswick Lumber Kings".

[13] Most information comes from Stewart's obituary, Bangor *Daily Commercial*, March 7, 1890. His views on the tariff are prominently placed in letters and speeches printed in *ibid.*, throughout 1887-1888. Stewart's papers seem not to have survived, or at least, I have been unable to locate them.

[14] Shooks were pine boards, cut ordinarily from "sapling pine", about five-eighths of an inch thick. The amount necessary to make a box to ship lemons, oranges, or other citrus fruits, were tied together with string and sent to the ports shipping out this fruit. In the West Indies onions, and similar vegetables, were also shipped this way until about 1875 when rough bags of jute were found to be cheaper. The American dominance of the West Indian trade ended with this discovery.

[15] Bangor *Daily Commercial*, August 12, 1878. This was cut in LaGrange, just outside Bangor.

[16] This information comes from a column entitled "Commercial Marine Register" which appeared every day *ibid.*, listing all vessels arriving or clearing Bangor that date with other marine information of note. This appears to be the only real source about Bangor shipping now as the Harbormaster records only go back to about World War One. After the great shipping days the office became a sinecure and the records disappeared.

[17] Vessels left on June 3, 17, September 11, and November 24 for Palermo; August 16, September 27, October 20 for Naples, and on November 8 for Castellamare as well.

[18] She cleared September 15. The brig *Lapland* cleared November 9, that same year with another big cargo, 127 M of shooks.

[19] Bangor *Daily Commercial*, December 24, 1881.

[20] Compiled from the Marine Register for the decade. Others involved in this trade, especially to the West Indies, were Pierre McConville, Henry Lord, and Cutler and Eddy. To these three firms though the lumber actually came as an adjunct to their ice business. The best year was 1885 when twenty-four vessels left, the earliest another famous schooner, the *Mary Ann McCann*, leaving Bucksport for Bermuda before the ice went out. The first Bangor clearance was another Stewart charter, the Italian bark *Emilia*, for Sorrento on April 27. By the end of this period the cargoes to the Caribbean were mostly lumber, not shooks. Typical 1887 cargoes were the *Mary Ann McCann* which cleared Bangor November 23, with 144,000 feet of pine boards, 20,000 spruce shingles, 105,000 cedar shingles, 5,000 clapboards, and 54 pieces of oars for Granada, or the British brig *San Jacinto*, September 30, with 431,859 feet of white pine to Granada. Mediterranean cargoes continued to be mostly shooks. On August 16, 1887, the Italian barkentine *Emilia* went to Palermo with 12,236 bundles of shooks and 1,280 knocked down barrels, and the Italian bark *Andrea Lo Vica* left November 19 with 25,886 bundles of shooks for the same destination, for example.

[21] On this point see John G. B. Hutchins, *The American Maritime Industries and Public Policy 1789-1914: An Economic History*, Cambridge, 1941.

[22] Bangor *Daily Commercial*, November 25, 1885.

[23] *Ibid.*, October 26, 1888.

[24] Marine Register for 1894, *passim*.

[25] The best account of the firm is in J. H. Clapham, *An Economic History of Modern Britain*, Cambridge, England, 1926-38, especially III, *passim*.

[26] N. P. Wood (Coats and Clark, Inc., New York) to the author, April 5, 1963. Plastics are ending the trade, however.

[27] Bangor *Daily Commercial*, December 10, 1888.

[28] The *Creighton* was worried about being iced in. She arrived November 21, and cleared the 26th, after working stevedores day and night to load her cargo. *Ibid.*, November 21, 26, 1883.

[29] Customs officials valued the spool cargoes at $52,963.40. Shooks that year were worth $28,939, and the lumber exports $25,757. See *Ibid.*, December 11, 1886, and Harbormaster's *Report*, City of Bangor, 1886.

[30] Bangor *Daily Commercial*, November 12, 1883.

[31] *Ibid.*, October 26, November 7, 1888.

[32] Bangor *Daily Commercial*, July 14, 1886, reporting the charters of bark *Charles Stewart*, British bark *Mary Hogarth*, and British bark *Maplute*. The advertisement and the 80 shilling quotation is in *ibid.*, May 5, 1888.

[33] *Ibid.*, November 7, 1888.

[34] On October 5, the *SS Crawford*, (British).

[35] *Ibid.*, December 2, 1889, reporting on the winter's prospects.

[36] *Ibid.*, August 19, 1889. The Norwegian bark *Ronda* was among the last vessels to leave the Penobscot that season, clearing on December 4; it is easy to imagine a fairly stormy passage for her in the wintry Atlantic. She arrived in Grennock on January 24, fifty-one days out. Her owners probably didn't mind as her charter was for eighty shillings. See *ibid.*, December 4, 1889; January 27, 1890.

[37] Marine Register, 1890.

Shipping the product

[38] Bangor *Daily Commercial*, November 13, 1891.

[39] *Ibid.*, September 4, 1891.

[40] The cargo of the *Hrvat*, some 384,000 feet, was salvaged by coastal residents and purchased back by the Stewart firm for reshipment at 60¢ M. *Ibid.*, May 25, June 30, 1892.

[41] The *SS Barden Tower*, (British), September 1, 1892.

[42] Bangor *Daily Commercial*, August 31, 1891.

[43] *Ibid.*, August 18, 1891 for the main story on the new development.

[44] There were a few dying flurries. In 1901 eleven vessels left for the Mediterranean, nine from Bangor. Four steamers cleared with spool stock. Other vessels left, one for each port, to Bermuda, St. Pierre and Miquelon, Curacoa, and Turk's Island. These last two went because the salt trade to the Caribbean died as hard as any other. The shook cargo, 3,249,000 boxes, was valued at $114,242.31. From Marine Register 1901, and Bangor *Daily News*, December 7, 1901.

[45] New Brunswick residents could see what Confederation would do to them easily enough. Those in eastern New Brunswick, and in the St. John river valley, who depended on Maine lumber, or St. John shipping for their livelihood were particularly vehement against it. At least one provincial government fell over the issue, and for long years after the economic ties were as strong as the political ones. The relevant articles dealing with the immediate question of Confederation are George E. Wilson, "New Brunswick's Entrance into Confederation", *Canadian Historical Review*, IX, 1928, 4-24; C. P. Stacey, "Fenianism and the Rise of National Feeling in Canada at the Time of Confederation", *ibid.*, XII, 1931, 238-261; Alfred G. Bailey, "The Basis and Persistence of Opposition to Confederation in New Brunswick", *ibid.*, XXIII, 1942, 374-397; and Alfred G. Bailey, "Railroads and the Confederation Issue in New Brunswick, 1863-1865", *ibid.*, XXI, 1940, 367-383. See especially 370, 380-2. As the *True Humorist*, (Saint John, New Brunswick) said on June 22, 1867, "Died—at her late residence in the City of Fredericton, on the 20th day of May last, from the effects of an accident which she received in April, 1866, and which she bore with a patient resignation to the will of Providence, the Province of New Brunswick, in the 83rd year of her age." Quoted in Bailey, "Railroads and the Confederation", *op. cit.*, 383.

[46] Before the railroad such lumber went by the Erie Canal. See Franklin *Register*, August 17, 1843 for a disgruntled comment. In 1855 about 50 million went out and by 1873 the trade was close to 150 million. *Drew's Rural Intelligencer*, August 25, 1855; Farmington *Chronicle*, February 5, 1874.

[47] The Calais trade was steady. In 1843 20,369,000 feet went out (½ spruce) along with laths, shingles, clapboards, sugar boxes, and even 6,000 cords of hardwood for fuel. In 1856 the deal trade was domestic. Franklin *Register*, March 14, 1844; *Drew's Rural Intelligencer*, November 11, 1856; Presque Isle *Sunrise*, June 26, 1868, both the last two quote the Calais *Advertiser*.

[48] Bangor *Daily Commercial*, February 21, 1880. This was a story commenting on southern Maine shipping matters. My remarks on Portland shipping are based on comparison of the amount of shipping news carried in the largest selling daily in each city. Very little work has been done on Maine economic history.

[49] *Ibid.*, October 30, 1883. These nine vessels were chartered for 4,000,000 feet.

[50] *Ibid.*, January 12, 1886. These vessels included one ship, fifty barks, nine brigs, and twelve schooners. The next year 62 vessels took 32 million feet. Phillips *Phonograph*, January 14, 1887.

[51] Bangor *Daily Commercial*, January 8, 1884; January 12, 1886.

[52] *Ibid.*, January 1, 25, 1886.

[53] *Ibid.*, November 2, 1887. Thirty-six of these vessels went out for Ryan and Kelsey who were handling the bulk of the Canadian lumber by this time.

[54] *Ibid.*, January 9, April 17, 22, 1889. In this year the *Governor Ames*, a five masted schooner, sailed with the largest load noted in this time. The vessel had a crew of 15 men, and was rated at a tonnage of 1689 tons n.r. When she cleared Portland Head her deckload of spruce and pine towered eight and one-half feet high. The total load, 1,896,239 board feet of lumber, was destined for the River Plata. *Ibid.*, April 22, 1889; Portland *Eastern Argus*, April 16, 21, 1889.

[55] Bangor *Daily Commercial*, July 22, 1889. This was for a vessel which would load 400 to 600 M for the Plata. Also on freights see *ibid.*, July 13, October 26, 1889.

[56] *Ibid.*, October 30, 1891, and for the quotation, April 1, 1892.

[57] Portland *Eastern Argus*, January 13, 1903. For the 1890's see the stories in Bangor *Daily Commercial*, April 25, 1893; March 30, November 8, 1894 in particular.

[58] For this story see Edward C. Kirkland, *Men, Cities, and Transportation*, 2 volumes, Harvard, 1948, and Alfred D. Chandler, *Henry Varnum Poor, Business Editor, Analyst, and Reformer*, Harvard, 1956, especially the bibliography.

[59] These years were 1876 and 1877. In the latter year 1533 vessels cleared Bath, perhaps half with lumber. In the first six months of these two years the surveyors scaled over 40,000,000 feet each year. Portland *Eastern Argus*, July 4, 11, 1876; June 4, 7, 1877; Bangor *Daily Commercial*, January 9, 1878.

[60] Bangor *Daily Commercial*, August 3, 1891.

[61] For Bath shipping see Parker McL. Reed, *History of Bath* (Portland, 1894). For the other ports see Bangor *Daily Commercial*, December 3, 1883; October 8, 1889 (Cherryfield); August 3, 7, 1891 (general stories on all downeast ports); Portland *Eastern Argus*, October 31, 1874 (Calais); and Davis, *International Community, op. cit.*, 316-321. The towns and mills which really felt the squeeze of profits in the latter part of this period were the tidewater mills and towns of the lower Kennebec. The big steam mills drove them out of business early in the seventies. However, labor troubles and a construction boom in Massachusetts caused a little spurt in 1887. Eleven little mills in Winneganse, Arrowsic, and Mill Island operated that year. They sawed and shipped 11,000,000 feet of spruce. The lumber went out in their own coasters to Boston. There is some evidence that these mills were a blind for M. G. Shaw and Sons of Bath who were troubled by the Knights of Labor at their big steam mills. On this whole business see Portland *Eastern Argus*, November 30, 1887.

[62] Wasson and Colcord, *op. cit.*, have much to say about the difficulties of navigating "Bangor River."

[63] Table is compiled from *Annual Report of the Harbormaster*, printed in *Annual Reports*, City of Bangor, various publishers, 1860–present. The only

Shipping the product

clear run of these is in the Bangor Public Library. The extant Harbormaster records (1916 or so to date) are in the City Clerk's Office in Bangor. These were saved, or that is, those to 1950, only by someone clearing an attic. Such records, which consisted mostly of seaman-type logs, were kept by the custodian of the office, and never were city property. I am indebted to the City Officials of Bangor for allowing me to rummage in the basement of City Hall for more. Early accounts are in Bangor *Whig and Courier*, January 1, 12, October 25, 1842; July 17, 1847 all of which discuss amounts. In the last "There is actually shipped from this city more lumber . . . than from any other part in the whole world."

[64] Harbormaster's *Annual Report*, Bangor, 1864.

[65] These addresses are printed in the *Annual Reports*, City of Bangor.

[66] *Ibid.*, 1883. The figures on imports and exports here, unless otherwise cited, come from the *Annual Reports* of the Harbormaster, for the years indicated.

[67] On the war see the Harbormaster, *Annual Reports*.

[68] See Clapham, *op. cit.*, II.

[69] Bangor *Daily Commercial*, May 21, 1873 quoting the Calais *Times*, and June 6, 1873.

[70] Bangor *Daily Commercial*, May 19, 1874.

[71] *Ibid.*, May 28, June 25, 27, July 10, 19, 27, 1877.

[72] *Ibid.*, March 25, 1879.

[73] One to Drogheda (1879), seven to Rio De Janiero (5 in 1879, and 2 in 1880), two to Liverpool (1882).

[74] Harbormaster, *Annual Report*, 1907. Although a cargo of clapboards had arrived from St. John in Bangor as early as 1849. Farmington *Chronicle*, August 23, 1849.

[75] Bangor *Daily Commercial*, April 16, 1890 citing a poster from 1856 found in their files.

[76] *Ibid.*, April 23, 1881.

[77] *Ibid.*, November 14, 1880, May 7 and October 11, 1882.

[78] *Ibid.*, October 9, 1882.

[79] *Ibid.*, April 16, 1894.

[80] The table is based on a study of both Portland and Bangor newspapers, with special reliance on the following news stories which actually listed prices. Portland *Eastern Argus*, November 5, 1875; December 9, 1887; *The Industrial Journal*, February 24, 1882; Bangor *Daily Commercial*, September 16, 1876; May 21, 1878; August 20, 1879; April 30, May 17, October 28, 1880; April 30, May 12, September 24, October 9, December 24, 1881; April 19, May 17, October 11, November 1, 1882; April 28, May 12, August 25, November 12, December 8, 1883; April 19, 1884; May 22, November 25, 1885; December 2, 1886; June 9, August 6, 13, October 27, 1887; May 26, June 16, October 26, November 16, 1888; October 11, 12, 29, November 9, 12, 25, 1889; January 23, March 13, August 13, 1891; April 1, 20, 25, May 3, August 30, October 5, 1892; April 11, 20, June 6, 1893; April 3, October 2, November 6, 1894.

[81] On the decline of the shipping told in general terms see Wasson and Colcord, *The Old Penobscot, op. cit.*, Hutchins, *The Merchant Marine and Public Policy, op. cit.*, and William Hutchinson Rowe, *The Maritime History*

of Maine: Three Centuries of Shipbuilding and Seafaring, New York, 1948, especially Chapters 7, 8, 11, 12, 13.

[82] *Acadian Recorder*, January 31, 1863, quoted in A.R.M. Lower, et. al., *The North American Assault on the Canadian Forest*, Toronto, 1938, 135.

[83] Toronto *Globe*, January 1, 1864, quoted in Lower, *op. cit.*, 135. See also, for further comments, 136.

[84] Frederick William Wallace, *In the Wake of the Windships*, George Sully and Company, New York, 1927, 121-3.

[85] James Hannay, *History of New Brunswick, op. cit.*, II, 355-9. I shall speak more of the CPRR in connection with the opening of the north country. See *Saint John, New Brunswick, Canada: Canada's Winter Shipping Port*, Common Council, Saint John, 1914, esp. 26-33.

[86] Hannay, *op. cit.*, II, 61; Bangor *Daily Commercial*, May 18, 1875.

[87] St. John *Telegraph*, November 15, 1874, quoted in Bangor *Daily Commercial*, November 16, 1874. Their business was good enough so that in May of 1874 St. John millmen had advertised for workers in Maine newspapers. See Bangor *Daily Commercial*, May 16, 1874. On the 1874 trade see *ibid.*, May 18, 1875.

[88] *Ibid.*, June 26, 1876; October 2, 1880; September 1, December 13, 1883; January 10, 1884; August 3, 1886; November 3, 1892; March 5, 1894. The best year for the province was 1877; 441,000,000 feet was shipped from provincial ports.

[89] This table is compiled from *Special Consular Reports*, "American Lumber in Foreign Markets", XI, Bureau of Statistics, Department of State, 53rd Congress, 3rd Session, Misc. Doc. #92, Washington, D.C., 1896, 217, and S. A. Saunders, "Forest Industries in the Maritime Provinces" in Lower, *North American Assault, op. cit.*, 348, 350. The month of July, 1897 was the last really huge month for St. John port itself. In the first seventeen days of the month 38 million feet of deal, 15 million laths, 5 million shingles, and 500 pieces of piling left the harbor, all on steamers, each handling three to five million feet. Much of this went the week of the 20th, twenty-seven million feet of the deal, in fact, with twenty-one million going to the British Isles. See the Bangor *Daily Commercial*, July 20, 1897, and St. John *Sun*, July 18, 1897 for a discussion of St. John port and the other provincial ports in this busiest of months.

[90] Liverpool freights ranged from 50 shillings to 67 shillings on deal, and about 30 shillings on ton timber. Bordeaux freights were about 5 shillings higher. Bangor *Daily Commercial*, July 14, September 1, 1883; February 14, 1884; January 26, 1886.

[91] Some figures appear in the appendix. Throughout the period from 1840 to 1890 immense amounts of kilnwood came to Maine's lime ports from the provinces. The only treatment is Roger Grindle, "A History of the Lime Industry of Maine," unpublished Ph.D., Orono, 1971.

[92] *Annual Report*, Bangor Harbormaster 1866, 1867, 1868, 1869; Bangor *Daily Commercial*, November 25, 1887. Newspapers were beginning to be disturbed by the competition by this time, and a special report was made by their special correspondent in St. John from the files of the American consul at St. John. For 1891 and 1892 an even more detailed special report appears at *ibid.*, December 10, 1891, and March 17, 1892.

Shipping the product 169

⁹³ The best account of this early attempt is in Bangor *Daily Commercial*, February 6, 1888 when the story was recalled as background to the new attempts.

⁹⁴ *Ibid.*, August 28, 1883 has a running account of their adventures. A second one went over the same route this year. This raft, 800 feet long and 30 feet wide, had 5,500 sticks of 65 foot piling on it. It was said that the tug rental was less than the freightage would have been. See Portland *Eastern Argus*, October 5, 1883.

⁹⁵ *The Industrial Journal*, July 16, 1886; August 6, 1886. Wilton *Record*, August 4, 1886.

⁹⁶ *The Industrial Journal*, November 19, 1886.

⁹⁷ Bangor *Daily Commercial*, November 17, 1887; *The Industrial Journal*, November 18, 1887; December 23, 30, 1887.

⁹⁸ Bangor *Daily Commercial*, December 21, 1887.

⁹⁹ *Ibid.*, March 10, 1888.

¹⁰⁰ *Ibid.*, April 10, 1888.

¹⁰¹ *Ibid.*, July 27, 1888.

¹⁰² *Ibid.*, June 28, 1890; July 5, 8, 15, 1890.

¹⁰³ *Ibid.*, July 15, 1890.

¹⁰⁴ *Ibid.*, July 26, 28, 1890.

¹⁰⁵ *Ibid.*, July 29, August 5, 1890.

¹⁰⁶ *Ibid.*, August 7, September 12, 15, 1890.

¹⁰⁷ *Ibid.*, August 16, 1891.

¹⁰⁸ *The Industrial Journal*, March 18, 1892. For The Pacific see Kramer A. Adams, "Blue Water Rafting. . . .", *Forest History*, 15, No. 2, [1971], 16-27.

¹⁰⁹ Bangor *Daily Commercial*, December 10, 1882.

¹¹⁰ *Ibid.*, November 25, 1889.

¹¹¹ *Ibid.*, December 7, 1894.

¹¹² *Ibid.*, December 2, 1881.

¹¹³ *Ibid.*, March 1, 3, 1890.

¹¹⁴ *Paper Trade Journal*, June 14, 1879 quoting the Toronto *Globe*, and August 16, 1879.

¹¹⁵ Bangor *Daily Commercial*, May 6, 1880.

¹¹⁶ *Paper Trade Journal*, October 22, 1881. A good trade was set up round the Horn. See my *History of the U.S. Paper Industry*, N. Y., 1971, for a brief account.

¹¹⁷ *The Industrial Journal*, February 8, March 28, 1884. For an historical account see Boston *Transcript*, November 26, 1920; *Paper Trade Journal*, December 2, 1920.

¹¹⁸ *The Industrial Journal*, August 28, October 10, November 1, 1884; August 24, August 7, 1885; June 11, 1886; August 3, October 14, 1887; Bangor *Daily Commercial*, October 9, 1886; August 6, 1888.

¹¹⁹ *The Industrial Journal*, November, 1907. This was not the first shipment into the state though. The S. D. Warren Company had imported 80 cords of poplar to Portland from Merigonish, Nova Scotia, on August 7, 1885, which was the first one I have located. See *ibid.*, August 7, 1885.

¹²⁰ In addition to the sources in the next two notes *Paper Trade Journal*, May 14, 28, 1908; July 7, 1910; April 20, May 4, August 31, September 14, October 5, 19, 1911; September 26, 1912.

[121] *Paper Trade Journal*, January 26, May 4, November 16, 23, December 7, 1911; January 11, December 5, 1912; November 5, 1914; January 6, 1921.

[122] *Paper Trade Journal*, February 5, 1914. No shipments from St. John are included in the table. Nearly all of this is pulpwood, even though the figures were in board feet.

[123] Bangor *Daily News*, December 28, 1914. 14,562 cords came to Bangor in 1908, however. *Pulp and Paper Investigation*, 53rd Congress, Vol. 6, 1909.

[124] Harbor Master, *Report*, 1917. Companies like the Eastern Corporation sold their schooners when they closed their sawmills and depended on charters for their pulpwood. *Paper Trade Journal*, March 30, 1916.

[125] Harbor Masters *Report*, 1927.

[126] *Daily Log*, 1928. This is contained in a ledger listing all port entries 1928-1941. The ledger is located in the City Clerk's Office, Bangor, Maine.

[127] Harbor Master, *Report*, 1929, and for the *Dansborg*, Bangor *Daily News*, July 9, 1930. The *News* reported that this wood had been cut by slave labor at a cost of 10 cents per cord.

[128] Harbor Master, *Report*, 1939, and in general throughout the decade.

[129] *Paper Trade Journal*, August 19, October 28, 1876; Portland *Eastern Argus*, October 23, 24, 1876.

[130] Bangor *Daily Commercial*, August 2, November 3, 1893; *The Industrial Journal*, September 29, November 17, 1893.

[131] *Paper Trade Journal*, February 10, 1894; *The Industrial Journal*, June 1, June 15, 1894; December 3, 1897 (for the visit of the British industrialists.)

[132] *Ibid.*, March, 1911. For this story in particular see L. Ethan Ellis, *Pulp Paper Pendulum*, New Brunswick, 1958, *passim*, and my *History of the U.S. Paper Industry*, N. Y., 1971.

[133] Bangor *Daily Commercial*, April 23, 1881; September 12, 1882.

[134] *Ibid.*, August 3, 1893.

[135] *Ibid.*, October 14, 1889; November 30, 1891; July 26, 1892. Actually though the revolution almost sneaked up on the lumbermen. For instance, tucked in the bottom of a detailed report of the year 1877 in the Maine woods was the following table:

Maine Central Exports of Lumber, 1876-1877

(up to Dec. 13)

Year	Long Lumber (cars)	Short Lumber (cars)
1876	2984	1902
1877	2393	1814

This was food for thought for anyone who noticed the item when it appeared. Portland *Eastern Argus*, December 25, 1877 quoting the Kennebec *Journal*.

Shipping the product

[136] Bangor *Daily Commercial*, January 14, 1889.

Items Shipped on New Brunswick Railroad for South of Bangor Destinations—1879-1887

Year	Pota-toes	Shingles (M)	Starch (Lbs.)	Hay (Lbs.)	Bark (Cord)	RR Ties
1879	135,304	53,997	6,000,000	3,000,000	***	500
1880	270,911	59,710	8,250,000	12,500,000	1,131	3,600
1881	219,949	65,806	6,000,000	6,500,000	3,679	3,505
1882	260,945	111,877	7,000,000	10,000,000	2,214	108,971
1883	126,168	149,947	10,000,000	19,000,000	3,622	65,257
1884	143,464	171,457	11,000,000	18,000,000	12,599	208,016
1885	238,027	179,823	10,000,000	21,000,000	12,701	166,977
1886	411,738	229,630	6,250,000	22,500,000	10,327	154,236
1887	593,116	224,483	***	12,500,000	13,997	179,199

*** unavailable—Some figures are rounded. The figures for potatoes are for pounds shipped August to December each year only. Floods in May of 1887 knocked the road out of business until July of that year which cut down some shipments. Nearly all of these shipments were of goods raised or produced in Aroostook County, shipped into New Brunswick, and then into the States again via Vanceboro. The increase points up not only the effect of a railroad on an area economy, but also the increased population in this frontier area.

7

LUMBERING AND LAND SALES—1860-1890

> It has been well known in the past that the lumber business has been something like a lottery. One year people have made a number of thousand dollars on successful operations only to lose it all on the next. A gentleman who has been in the lumber business all his life said recently, "You can't tell how you are coming out on a winter's operation. One year I was offered $15,000 profit on my year's work. I thought I could do better than that but I kept my logs just one day too long. There was a change in the market and I lost money. I might have made $15,000 on that winter's work, but I didn't. I tell you this business is all a lottery."
>
> From an interview with an
anonymous lumberman published
in Bangor Daily Commercial,
August 18, 1883.

Lumbering was a lottery in the State of Maine throughout the period from 1860 to 1890.[1] Beset on one side by falling prices in the economic debacle of the seventies and eighties, the lumberman found himself under increasing attack from the citizens of the state for his use of the land, his methods of obtaining the land if he were a landowner, and demands to utilize the state's lands for settlement purposes rather than logging. In addition the state faced constant competition from western states which had large quantities of land to sell; the federal government, also in the land disposal business; and the western railroads, who were enticing the citizens of the east to their great bonanzas of land. The competition within the state was met by attempts to transfer the state land to prospective farmers first and later to a giant railroad scheme. The lumberman however could not meet his economic pressures as easily, and those pressures resulted in much lower profits and bankruptcies for many. Those who survived did so only to succumb to the burgeoning pulp paper industry which overcame all resistance to its growth ending finally with ownership of the land, control of the rivers, and domination of the politics and economics of the state.

The disposal of the Maine public domain was an extremely complicated business. Originally all or nearly all of the area of the state then owned by Massachusetts was thought to be a vast Garden of Eden, to be sold to prospective farmers in order to pay Massachusetts debts and to finance new growth within that Commonwealth. Huge land grants were made and revoked. Land was sold. Land was given away to Revolutionary War veterans until the area now known as the State of Maine became free of Massachusetts in 1820. From that time on until the middle 1850's the public domain of the state was owned by both Massachusetts and Maine, and it was disposed of in part to opportunists in the great land speculation of 1834 and afterward to lumbermen and proprietors who wished to utilize the land for its timber values. After refusing the opportunity to purchase the lands from Massachusetts several times, Maine finally took sole possession of the lands, except those already under grant by Massachusetts, in 1854. Now that Maine owned the remaining land, about 28% of the total or 5,400,000 acres, she too began to regard her holdings as a source of income to pay debts and finance growth. As a result the state was in the market to dispose of its lands, but unfortunately this disposal was to occur at the same time that other giant tracts of lands came into the market. Maine found itself in the unenviable position of having property to dispose of in a buyer's market.[2]

The state followed the practice of granting and selling land to prospective farmers. Certain portions were put up for sale after the land was surveyed and either disposed of at public auction or in across-the-board cash sales. By 1858 twenty-three townships, all in the first six ranges, had been surveyed and named for sale. Of this land (494,805 acres) 297,037 acres still remained to be sold in 1858 including practically the entire townships of GR1, GR2, IR2, 13R3, 4R4, 9R4, 12R4, and 3R5.[3]

The state lands had not moved very quickly, and the hoped-for agricultural utopia in the Maine woods was still in the future. The state decided to move its lands somewhat more quickly into market in the hope of achieving its dream of growth. First the state set aside all or part of fifteen townships to be sold, with the proceeds to go to public schools. These lands moved fairly rapidly. By 1869 most had been disposed of with the purchasers being nearly all prominent lumbermen. The land went cheaply too—from 12¢ an acre up and at an average price of only 30¢ an acre.[4] The state then announced that it was putting up for sale its settling lands, one hundred and seven townships in the first seven ranges. Unfortunately very few of the townships so denominated as settling lands were actually available

Lumbering and land sales

to prospective settlers. Much of the state's lands had in fact been sold before 1855. The big purchasers had been lumbermen and large amounts had gone. A perusal of the deed books shows that of the new settling lands fifty-two townships were already in private hands.[5] Much confusion resulted from this and other conflicting attempts to move the lands. To a great extent the confusion over the disposition of the state lands derives initially from the fact that on the eve of the Civil War the state itself did not know what lands it had sold, or what remained to be sold.

In any event the state was willing to dispose of its lands. As one Land Agent said in response to a request to buy stumpage from the state:[6]

> [*The*] object [*of the state*] is to realize a large value of money either by sale of the lumber or by sale of the land; therefore the more you will agree to haul, that much easier it will be to facilitate an application for stumpage.

According to the law of the state those lands sold for settling purposes under the Maine version of a Homestead Law had to be cleared for farming and the timber could not be sold until the settling duties were met. The state was constantly admonishing its scalers and agents to watch out for such illegal sales by those attempting to get the value of the lumber and then perhaps to clear out for western states with the grubstake so obtained. The instructions were always clear, as in this example:[7]

> In scaling for operations on 5R3 it is expected and you will be careful to see that lots in actual occupancy, but where the settling duties have not all been complied with, that the lumber shall remain on such lots.

The lands moved quickly. Through the sixties Ebenezer Coe, Pingree and Coe, Eaton Brothers, and other large lumbermen and land owners continued to add to their holdings. The state occasionally acted as agent for institutions to which it had granted land.[8] The Land Office was a big business. The sales from 1860 to the eve of the slump in 1873 can be followed in the appended table.[9] (next page)

In addition to regular grants and sales the state also continued to grant land to educational institutions. Such grants during the Civil War period totaled 140,856.5 acres.[10]

Throughout the period 1850-1870 state officials and others interested in the lands seemed to assume that the amount of land to be granted was almost unlimited until the large grant was made to the European and North American in 1869. By that time the competi-

Land Sold and Granted by State of Maine—1860-1872

Year	Amount (acres)
1860	39,712
1861	9,967
1862	21,857
1863	145,336½
1864	21,344
1865	3,124
1866	119,634
1867	130,655
1868	23,872
1869	*
1870	
1871	11,781
1872	1,333
Total	528,615½

* 734,942 acres granted to European and North American Railway—see below.

tion inherent in the Federal Homestead Act had made Maine people willing and anxious, almost frantically so, to get the last of the state lands into private hands. By doing this all at one time it was felt by many that not only would the incubus of the public lands be lifted but also that the "Garden of the North", as Aroostook County was called, would finally realize its potential with railroad connections to the outside world.

The state had taken one flyer of a similar nature before. In 1862 individuals living in Princeton had organized a company to build a turnpike from Princeton to Milford on the Penobscot River. In Princeton the road would connect with the Lewey's Island Railroad which ran from Princeton to Calais. After some initial attempts at swamping out the right of way the company fell on hard times and in 1866 the legislature appropriated $30,000 from the receipts of public land sales to be used in building the road. Under the provision of the grant when the company had spent $13,000 of its own money the state agreed to provide an additional $10,000. The company took out a charter under the name The Granger Turnpike Company (named for one of the promoters) and began to work. In 1870 another appeal was made to the legislature and that body authorized the expenditure of $2,000 of state funds for each $2,000 expended by the company. The road was grubbed out three or four miles to the west of what is now Grand Lake Stream Plantation before financial and other difficulties beset the promoters. Finally, in 1876, after failing to attract more aid the

Lumbering and land sales 177

remaining money was returned to the state and the project abandoned.[11]

The history of the European and North American Railroad grant is in some ways even more obscure. Only the published records have survived. The records of the state land office are missing from 1868 to 1890.[12] It has been said that these records were deliberately removed from the land agent's office although there is no real evidence to this point. Maine, like most states, has its share of skeletons in its official closets, but the one which is revived most often, especially during political battles, is the bones of the "Great Land Steal." Percival Baxter while he was governor from 1920 to 1925 used this with great effect in his battles with the paper companies, and others have also rattled these bones from time to time. Whether or not there was a conspiracy to relieve the state of the remainder of its public lands, whether some individuals saw the main chance in the failure of the road to do as it was supposed, or whether the situation just fell out with no guidance from interested observers, it is difficult to say. In any event the lands were transferred and apparently with very little opposition at the time.

One of the dreams of Maine promoters after 1850 was the construction of a railroad to connect Bangor—and through Bangor by steamship and railroad, Boston and New York—to St. John, New Brunswick. This line once built would throw open much land for agricultural settlement and would enhance the economic ties between the Maritimes and northern New England. Henry Varnum Poor the railroad enthusiast, editor and promoter, who was a native of Portland, had always advocated this road, which he called the "short route to Europe." In the fifties he had organized a trip through the region to interest prospectve investors.[13] Poor's attempts failed, but his dream survived.

The state made a survey and investigated the possibility of granting land to the railroad in order to finance the construction.[14] Little came of the attempts until the Civil War, primarily because the land title was not cleared by Maine's purchase of Massachusetts land until the fifties. The desires of the promoters were revived during the Civil War as some felt that the road would be an aid in defending the Maine frontier. The bloodless Aroostook War and the Fenian disturbances were never too far from the minds of those on the border. After the war the road was discussed as a possible aid to international good will in light of the strained relations between Great Britain and the United States. In 1864, ten townships or about 230,000 acres were granted to the putative railroad, but nothing came of it. Not until the Confederation idea began to take root in Canada was the railroad seriously considered. Many in the lower St. John River valley wished to cement their ties with New England by building a westward extension of the road to connect with the Maine side. Some in fact used this as a lever to hold up Confederation until 1867.[15] In 1863 the New Brunswick Parliament gave a subsidy of $10,000 a mile to build the road (eighty-eight miles to the Maine border with a twenty-two mile branch to Fredericton from St. John). In addition the New Brunswick government took stock worth $300,000 at par, and both St. John and Fredericton gave money to the venture.[16] By the terms of the ten township grant Maine transferred old claims against the United States dating from the Webster-Ashburton settlement to the railroad. The promoters then began to work on the Massachusetts legislature to get the Commonwealth to transfer their claims as well. In 1865 the claims were given over to the road.[17] The government in Washington did make a very small grant in 1868 but it and the ten townships did not seem to be enough to insure the road's completion. Maine's legislature now came to the aid of the promoters again. The original grant was repealed, and was replaced with one which conveyed most of the state's remaining public lands to the railroad.

Certain lands were held out, among them ten townships amounting to 242,366 acres, which had been set aside for public school townships,[18] as well as the lands previously set apart for actual settlers.[19] When the land finally went over, the deed conveyed land in ninety-three separate townships and gores to the railroad.[20] About the only restrictions on the use of the land were that they must be sold in accordance with the state's settling law, rather like the Homestead principle, and they were granted under the condition that the proceeds from the sale of land should be applied to the construction of the main line from Bangor to the Maine boundary or of branches into Aroostook or Piscataquis Counties.

The promoters now attempted to raise money to construct the road by using the lands for collateral on loans. The Maine Central Railroad loaned $20,000 to the road in 1868 and 1869.[21] Bangor took out a bond issue for nearly a million dollars to finance the construction between Bangor and Winn with that part of the road itself as their collateral. Other bondholders were in New York, Boston, and abroad. However sanguine its promoters had been though, the road was almost immediately in difficulty. As a result, in 1872, the halves

Lumbering and land sales

of the road in Maine and New Brunswick were consolidated and a new $5,000,000 mortgage was taken out with the lands as security, and bonds were sold to individuals to provide the refunding. This new bond issue of six million dollars was to cover the consolidated mortgage, three million on the Maine half, two million on the New Brunswick portion, and another million to be used for current debts and to purchase new rolling stock.[22]

Of this authorization only a million and a half dollars worth was printed and of that only one million was posted for sale. These bonds of forty years duration were to pay 6% interest in gold. The bonds were secured with a first mortgage on the line from Bangor to St. John and a second mortgage on the land grant. Only $6,000 of the bonds was sold; the rest was used as collateral security for loans.

After the panic of 1873 the road paid no interest, and the investors began to be very restive especially the city of Bangor public officials who had put out $814,000 in loans in 1869, 1871, and 1874, in addition to another $100,000 covered only by the unsold bonds. The road like many others of its time was too sanguine of its future. It had leased the Bangor and Piscataquis Railroad in 1873, and the Bangor and Bucksport in 1874, hoping to consolidate them and eliminate competition, but these hopes were not fulfilled as the deadening weight of the 1870's fell on all commerce. The railroad was forced to default on its bonds, and a trusteeship was set up to operate the road under Benjamin F. Smith of Bangor.

The road ran for a year under this arrangement until in July, 1875, the interest due Bangor was defaulted again. G. K. Jewett, president of the road, had pledged the credit of several firms with which he was connected to payment of the road's obligations, among them Jewett and Pitcher of Boston, E. D. Jewett and Co. of Bangor, and E. J. Dunn and Co. of St. John. Something now had to be done. These firms and the railroad had taken to paying their employees and their lumber operators in shin-plaster I.O.U.'s which were scattered all over Aroostook County and the St. John River valley. On the news of the latest default the shin-plasters fell to 50¢ on the dollar and no takers.[23] A meeting of the creditors was called for Bangor. At that meeting it was announced that the debt of the road amounted to $6,226,749 of which only $4,765,000 was funded. To balance this the assets were announced at two hundred six miles of road bed and equipment worth $668,216, real estate in Bangor and St. John worth $215,897, and the land grant now totaling 600,000 acres and other assets of approximately $123,000. In addition there was the Bangor and Piscataquis Railroad. The debt had increased by $85,000 in the past six months, but it was hoped that the rise in the debt was in the process of being stemmed. The gross receipts of the road for 1874, including the B. and P., had amounted to $724,000. Jewett announced at the meeting that the road could meet wages and other regular bills but no more.[24]

A committee consisting of leading businessmen from Bangor, St. John, Thomaston, Rockland, Portland, Belfast, and Boston, all of whom had apparently been investors, was appointed to investigate the possibilities open to the road and its creditors. Although the future was not as bright as it had been thought to be when the road was first promoted, local newspapers were heartened by the reports and counseled moderation in their editorials. As the *Commercial* said,[25]

> The sacrifice on the part of the creditors will not be great. An extension of time and a liberal policy by them will prove a boon to the community, and will probably result in placing the road on an ultimate sound financial basis.

The land grant seemed to provide the way out for the investors. An excursion was held along the route of the train in an effort to interest prospective buyers. The proprietors expected to sell 1,500 tickets for the excursion, but only 900 were sold and some of those went to owners of the road's securities trying to drum up business. A correspondent attempted to place the blame for the failure of the excursion on rainy weather and the fact that the train was late. Realizing that these excuses were lame he went on:[26]

> If it [the railroad] has not been successful, it is by no means owing to its management, for in no section of the United States have these hard times told so powerfully as on the Penobscot River.
> Almost the entire business of the people is lumbering, and with no sale of lumber our whole business is paralyzed.

Whatever the reason the lands did not sell. The railroad continued to limp along making a small surplus in 1875 so that the treasurer of the road could announce that a regular payday would be met in January, 1876, something which had not been done in nearly a year.[27] This was not enough. In mid-December the city of Bangor foreclosed its mortgage on that part of the road between Bangor and Winn in an effort to recoup part of its loss.[28] In October 1876 the stockholders of the Bangor and Piscataquis took over that line from the E. and N.A.

Lumbering and land sales

These stockholders were primarily the municipalities of Dover and Bangor who had pledged credit to the road.[29] The New Brunswick section was also split off in 1876, and the trustees of the land grant bonds, Hannibal Hamlin and William B. Hayford, took over complete control of the road. They caused the gauge of the road to be changed from five feet six inches to the standard gauge during the summer of 1877.[30] The road began to look more prosperous.[31] In 1880 a new company was formed under the old name with an exchange of stock taking place and new bonds being sold to refund the old debt. Only the Bangor mortgage remained from the initial debts after the $2,000,000 bond sale was complete.

The road apparently had made some effort to entice settlers on the lands, not only of the road but also of the state. A pamphlet was published in 1871 with maps showing the locations of the settling lands on the first seven ranges in the state and giving information for immigrants on how these lands might be obtained.[32] The pamphlet was quite explicit about the benefits of pioneering in northern Maine, saying:

> It is an exceedingly healthy climate, fevers and other diseases of a malarious nature are unknown; and other acute diseases are not of common occurrence. A surgeon who resided for some time at Fort Kent, reports that he not only never saw a case of consumption in the country, but some persons who, when they came to reside there, and who had suspicious symptoms when they came into the country, recovered from them entirely. Children enjoy the best of health and are never afflicted with any of those complaints, so common in warm climates.

The company was directed by the grant to divide its settleable lands into 160 acre plots for which immigrants would pay $1.00 an acre as a minimum price. The settlers also had to agree to live on the land for five continuous years and pay the usual state settling duties which amounted to 50¢ an acre, payable in one, two, or three years in road labor in the township. This method of settling the land would have been effective if other land had not been available. As it was the only way that it would work was to have a number of prospective settlers in an area at the same time so that they could pool their efforts both in home building and road construction. In addition, the fact that the timber on the lands could not be sold until the lands were paid for denied the prospective settler his first real cash crop.

Although the deed was fairly specific about what lands belonged to the railroad the proprietors apparently were either not sure of the provisions or perhaps were willing to attempt to get even more lands. Whatever the case, in 1870 the railroad granted stumpage permission to E. G. Dunn on T11R3. When the Land Agent found that Dunn was cutting on this land previously reserved by the state, he notified him of his trespass. Dunn continued to cut and on January 8, 1872 the sheriff of Aroostook County seized his teams, supplies, logs, and equipment and took them to Presque Isle. All but the logs were sold at auction; they were sold conditionally and driven to the Fredericton boom where they were claimed by Dunn. Court cases resulted in which the state was confirmed in its title to the lands, the railroad was given the stumpage amounting to about $2,000, and Dunn was allowed to sell his logs.[33]

By the time of this decision it was obvious that very little land was going to pass into the hands of small holders and many in the state began to be incensed at the railroad's failure to live up to expectations. Governor Nelson Dingley described the state's land policy in 1874 as wasteful,[34] and in 1875 had further remarks to make to the legislature.[35]

> Whatever disposition may be made of the vacant state lands suitable for settlement, it seems to me highly important that nothing should be done to retard their conversion into cultivated farms. Indeed, it is desirable that you should inquire whether any measures can be devised to further encourage the settlement of wild lands held by proprietors, of which about 9,000,000 acres are owned by individuals, and about 734,000 acres are owned by the European and North American Railroad Company.

The legislature did not take up the cue offered by the Chief Executive and it was until 1878 that the lands again came into the public eye. As the depression deepened and Maine continued to lose its citizens to the siren song of the west, newspapers began to agitate the old question of settlement in the north country. And finally in September the *Argus*, the leading Democratic newspaper in Maine, and perhaps in New England, began to turn its attention to the lands. It described the lands in detail as to their potential, sent individuals to investigate, began to feature news of Aroostook County, and called attention to the fact that settlement was discouraged by the lumbermen because of the danger of fire. In one editorial it reopened the old sore with this blast.[36]

Lumbering and land sales

> Let then immediate steps be taken to open its lands to settlement. Let the State demand the lands granted to the E. and N.A. RR. Co. be surveyed and offered on favorable terms to actual settlers. . . . Let proprietary lands be assessed at something like their real value. Let provision be made for opening roads so that settlers can get to the new lands. . . .

The editorial continued:

> The land policy of our state needs to be reversed and remodeled as much as does that of the federal administration; both have been conducted in the interests of land speculators. It is high time to have them conducted in the interests of the people.

The legislature did not wait long with this challenge offered them. A bill was offered charging the road to survey its land grant into settling lots of 160 acres at a maximum price of $1.00 an acre. The bill became law with little discussion.[37] The railroad ignored the state's request and in 1881 the old issue was revived. A new bill was offered naming the land to be surveyed and setting out the conditions of settlement. The land agent under this bill was charged with conducting the survey and sale with the costs of his labors to be paid by the railroad. If the road defaulted, title to the lands was to pass back to the state. The bill was reported to the Joint Committee on Lands of the Legislature, and on March 17, 1881 it was postponed indefinitely.[38] There the situation remained until the next summer. The railroad had come to its end and it only remained to clear up the wreckage.

Rumors began to float through the state that the Maine Central Railroad was going to take over the road. At first it was thought that the Maine Central would take over the entire road to St. John, but early in August it was announced that the New Brunswick Railroad had assumed responsibility for the Maritime section but that the Maine Central had in fact consummated a lease for the E. and N.A. within the state boundaries.[39] The Maine Central was to take over the old road on April 1, 1882. It was to pay the city of Bangor $60,000 annually on its mortgage with the loan to be paid off in 1894. The old stockholders, mostly English investors, were to receive gold bonds paying 5% on their older investment most of which had been picked up at from 30 to 70 cents on the dollar.[40] The stockholders of the Maine Central agreed quickly by a vote of 29,953 to 1,514 to lease the road for 999 years, paying 5% on the old stock, or $125,000 a year as rental.[41] The new board of directors after the reorganization was a who's who of Maine finance and lumbering at this time with such individuals as Lysander Strickland of Bangor, Abner Coburn,

Arthur Sewall of Bath, T. J. Stewart, and Thomas W. Hyde of Bath as members.[42]

Those interested in the railroad's fate had perhaps noticed even earlier that something was afoot. An advertisement had appeared early in May announcing the auction sale of the remainder of the land grant of the E. and N.A.[43] In July some of the excitement had been taken from the sale by the announcement that David Pingree's heirs had acquired all the lands where only the soil was involved, having already purchased the timber rights from the state in the fifties.[44] The remaining land was sold on September 21, 1882 in the Bangor City Hall. The sale was well attended even though the Maine Central had reserved the right of one bid on each lot up for auction. Most of the larger lumbermen in the state came to Bangor to observe this auction and hopefully to pick up some bargains. About 260,000 acres changed hands on that day, most for low prices, because much of the land had been "permitted" (that is, the stumpage had been sold) by the state earlier.[45] The sales are recapitulated in the following table. (next page)

Bangor in its capacity as partial owner through its mortgage had received a rebate on the state excise tax until the Maine Central leased the road. The state refused the rebate then, and the city claimed that it should be free of the tax just as it had been earlier. Because of this a slight furore was stirred up culminating in an attempt to investigate the land sale.[46] The attempt came to little. Many however were unhappy with the sale and in fact with the entire disposition of the state's public domain, and from time to time attempts were made by those who were unhappy to rattle these old bones for political purposes.

An investigation was made in 1887 which held that the company was not at fault in its failure to attract settlers to its lands. In fact, the inquiry held that the state was at fault for not designating the lands more specifically.[47] In 1895, Peter Keegan, a Democratic lawyer and representative from Van Buren, resurrected the old claims again. After discussion and investigation the majority of the investigating committee held that the "legislation had been inexpedient." A minority, consisting of Keegan, asked the legislature to enforce the terms of the original grant. The minority view lost by a vote of 16-63 in the House, and the House accepted the majority report. The Senate did not discuss it at all.[48] Here the issue lay, occasionally being discussed but not for long, until Baxter would use it in his land battles in the 1920's.

Lumbering and land sales

Auction Sale of European and North American Railway
Land Grant—September 21, 1882

Location	Acreage	Purchaser	Price	Remarks
18R2	5,688	S. J. Adams	44¢	
9R3	23,000 app.	M. C. RR	15¢	soil only
N½17R4	6,892	M. C. RR	25¢	
S½8R6	11,500 app.	M. C. RR	30¢	soil only
S½5R7	11,500 app.	J. F. Webber	31¢	
13R7 less 4000	18,000 app.	Jewett & Co.	30¢	some permits
½N½17R8 undivided	—withdrawn from sale—			
17R9	15,000 app.	M. C. RR	40¢	less grants
½17R10 undivided	11,600	E. L. Jewett	30½¢	
½12R11 undivided	11,600	M. C. RR	15¢	less stumpage
½W½12R14	5,000	S. H. Blake	34¢	
4R18	23,000 app.	M. C. RR	50¢	less permits
7R18	23,000 app.	S. J. Adams	51¢	less permits
9R18	23,000 app.	W. J. Engel	61¢	less public lots
5R19	23,000 app.	M. C. RR	30¢	less permits
5R20	23,000 app.	S. H. Blake	41¢	less public lots
3R5	23,000 app.	M. C. RR	40¢	less public lots
½E½3R3 NBKP	5,000 app.	M. C. RR	15¢	less permits
Gore 6R3 NBKP		Bernard Pol	20¢	
Gore 6R4 NBKP		S. H. Blake	63¢	
AR8&9 (Long A)		Caleb Holyoke	10½¢	less permits

4R18, 7R18 and 5R19 were permitted only until Jan. 1, 1884 as school lands. The other permits were until the areas became organized towns, or in the case of most of these lands, in practical perpetuity, as they have not yet been organized into towns, nor are they likely to be.

Although it had seemed to many that the public lands were a source of almost inexhaustible wealth it had become obvious after the passage of the Federal Homestead Act in 1862 that these lands, located as they were in a desolate area with a severe climate, were in fact little more than a white elephant. The land grant to the railroad had settled some of the problems by relieving the state of the responsibility for the lands, but the attraction of the west and the depression of the seventies meant that the problem of settling the Aroostook frontier still remained. Lands ungranted went begging until in desperation the state auctioned off the remaining acreage to lumbermen. The population frontier even began to recede as the more fertile lands of the west tolled family after family away.

There was little the state could do to oppose the drift to the west. Nearly every newspaper was filled with advertisements from railroads and others who wished to sell lands or from those who wished to attract capital to invest in the mortgages.[49] Some appeals were addressed specifically to Maine lumbermen.[50]

Newspapers and private individuals within the state attempted to hold back the westward drift by any method available. These ranged from editorial comment to descriptions of the giant yields available on the Maine farms for those who would work. Sometimes only a wife was necessary to fill the void of companionship in the far north.[51] The editorialist said, "Stay at Home" [52] and [53]

STAND BY THE STATE

> .... Farmers! Be of good courage. Study and practice the economy of your fathers. Maine's blighted interests and industries shall yet bloom again. Her happy and prosperous homes of the past, shall yet be prosperous and happy in the future.

After all why should one go west to a land filled with sharpers and confidence men, to a land where the morals of the Puritanical easterner would be shocked by practices foreign to his nature?[54]

> The Lewiston Journal says that an examination of some feed lately brought to that market from the mills of Missouri showed the following proportions: to one peck of bran, sixty nails and a piece of glass. The nails are headless, or slender and sharp-pointed refuse, or such others as doubtless were dropped while heading up flour barrels. Will not the use of such fodder lessen to a considerable degree the consumption of hay? Our dealers are not to blame, for the quality of this feed, but let stock-raisers beware. Such an iron tonic is likely to be too bracing for New England cattle, to say nothing of the dirt that comes with it from the sweeping of the western mills.

By staying in Maine one could hire farm labor at $20.00 a month and board, or by 1876, $15.00 a month although by the early eighties such labor got $2.00 a day at least during the haying season.[55] Maine farms yielded fortunes to those who persevered, or so the newspapers said. Some got five hundred bushels of potatoes to the acre, others one thousand bushels.[56] Great yields were particularly noted in the "Garden of the North" as Aroostook County had begun to be known. Van Buren, Maysville, Presque Isle, all these towns were prosperous. Even in southern Maine fortunes could be made.[57] All it took was hard work. Still the wagon trains went west, from the Provinces, from Aroostook County, from Lewiston and Farmington, all going to the new El Dorado on the homestead lands of the fertile prairie.[58]

Lumbering and land sales

As the trek west continued the newspapers shifted their attacks to others, those who had control of the state's destiny from their vantage point in Augusta and Washington. The following editorial discussing Aroostook County after an "exploration" trip by Boston capitalists in 1886 is classic in analyzing the difficulties of Maine during these times:[59]

> .... They wanted a wise public land policy, that would make the acquisition of farms easy as possible to actual settlers, and prevent the sale of large tracts of land to speculators. They got the reverse of this. . . . All the unimproved lands which the state once owned are now the property of capitalists who do not live in the county and who oppose rather than favor settlements. . . .

All aspects of the state's agricultural policy were unwise, unfounded, and wasteful of money according to this editorialist, who continued his attack:

> But in addition to the hindrancy in the way of Aroostook's progress in consequence of the unwise land policy of the state, as indicated above, and as also proved in the case of the lands granted to the states for an agricultural college . . . , its refusal to aid in opening direct railroad communications with this great country—a part of our own state—has produced broadly undesirable results. . .

Of course, an editorialist dealing with Aroostook could hardly ignore the fact that without the railroad the timber cut of the northern country, which he estimated at one hundred millions of feet each year, was driven to St. John and manufactured in foreign mills. Even the tariff was a product of minds who could not understand ordinary economic facts. However, success and good fortune were possible. Finishing his attack with a great flourish the editorialist pointed the wide road ahead for those who were not blind.[60]

> We have pointed out some of the hindrances to Aroostook's progress and with a purpose . . . . in that unreasoning craze, ... which prevails among our young men, that it is only in the great west that fortunes are successfully to be achieved. A greater error could not be. . . . We believe Aroostook is a better, more desirable field for honest industry and enterprise today than any part of the cyclone or otherwise afflicted states of the Great West.

Although this editorial writer and many others were unhappy over the disposition of the Maine public lands, given the feeling of most Maine citizens that the lands should be an immediate source of wealth both by sale and through taxation, it is difficult to attack the state for its action in disposing of the land. After all the disposal of the Maine public domain was not entirely a giant giveaway. The state made honest and earnest attempts to attract settlers to its lands, not the least of which was the interesting colony of New Sweden.

New Sweden was the brainchild and almost the sole responsibility of William Widgery Thomas. Thomas (1839-1927) was born and raised in Portland. He was educated at Bowdoin, graduating in 1860. For his services to the new Republican party he was rewarded with several small diplomatic posts abroad eventually being posted as consul at Gothenberg, Sweden. After the Civil War he resigned the post, and returned to the states to finish his legal education at Harvard. In 1869 he was appointed Commissioner of Public Lands for Maine where he became interested in the possibility of attracting foreigners, particularly Swedes, to Maine's public lands. He served on the state's Commission on Immigration from 1870 to 1873, following this with terms in the Maine House of Representatives from 1873 to 1875 and a term in the Senate in 1879. He continued his interest in Sweden and in Swedish in which he had become fluent by translating several Swedish works of literature into English. In 1883 he was appointed Minister to Sweden and Norway, holding this position during Republican presidential tenure 1883-1885, 1889-1894 and from 1897 to 1905. He married a Swedish noblewoman Dagmar Tornebladh, October 11, 1887. In 1892 his giant work *Sweden and the Swedes* was published in English in America and in Swedish in Sweden. His ability to speak Swedish led him to be used as a Republican stump speaker in Swedish communities throughout the west. His first wife having died, he married a second time, also to a Swedish woman Aina Tornebladh on June 2, 1915. To the day he died he remained perhaps the foremost friend of Sweden and Swedish immigrants in the United States, especially in his own colony New Sweden.[61]

At Thomas' behest the state legislature had passed in 1870 a law "to promote immigration and to facilitate the settlement of the public lands." Under this law a township, T15R3, was set aside for potential Swedish immigrants. The State Land Agent appointed a surveyor to run out lots and locate roads on the township which was an unbroken wilderness. Log cabins were built, five acres for each lot cleared, and a stove and kitchen utensils were provided for the homes of the prospective settlers. Thomas went to Gothenberg where he began to campaign in earnest for individuals to come to Maine. He sent agents to the northern section of the country, advertised in the local papers,[62] and within a short time had fifty prospective

Lumbering and land sales

colonists whom he had brought to Maine himself by way of Halifax and St. John. At the end of the first year one hundred and two individuals were living in New Sweden, and by the end of the decade over eight hundred people had immigrated to this town and its neighbor T14R3-4 called Stockholm. Thomas continued his advertising and speaking in Sweden in behalf of the new colony.

The colony was a success, so much so that some mourned the fact that land had been granted to the E. and N.A. Railroad Company which could have been put to this use.[63] The success of the colony was a thorn in the sides of those who felt the lands had been distributed badly. The smallest news item was the occasion for an attack on the policy as the following shows:[64]

> Mr. Stadig, (from Jempland, Sweden), is perfectly satisfied (after a trip home to visit his father) that New Jempland in Township 15 Range 3 Aroostook County, Maine is the Jempland for him. He says many Swedes would come to New Sweden were they sure of obtaining wild land at a fair price per acre. Where is the railroad land that was to be put in to the market at $1.00 an acre?

Thomas continued to call for further immigration to Maine. In 1886 he visited his colony and said that 10,000 Swedes would come if only the land were available. He blamed the proprietors for the failure to sell land for this worthwhile purpose.[65] Many were still interested. The Aroostook *North Star*, a weekly published at Presque Isle, carried a column of news printed in Swedish until well into the nineties. Few more came, however. Thomas was right. The land was not available. The state had none and the proprietors would not sell their holdings. New Sweden and Stockholm remain, New Jempland [Jempland] no longer exists except as a section of a town. As the traveler comes to these towns today he is struck by the Swedish surnames and the fair complexions and blond hair of the inhabitants, visible proofs of the efforts of one man to solve the problems raised by the immense tracts of land in the Maine public domain.[66]

Even after the large grant to the railroad and the attempts to attract Swedes to Maine, much land remained unsold. Throughout the decade from 1870 to 1880 the state found itself in the position of disposing of these lands, most of which finally went the way of the auction block as the state apparently wished to wash its hands completely of the land business. A good deal of the land remained in the state's hands as the decade opened. There were 246,843.38 acres of land which had been denominated as settling lands still unsold; 126,-843 acres contracted to settlers which had not yet been deeded, and

152,427 acres of timberlands which had not been sold as yet.[67] They went quickly though. By 1874 the townships on the railroad land grant on which the timber and grass stumpage had been reserved were all contracted for. The state received nearly $55,000 for educational purposes from the sale of this stumpage.[68] Before discussing the auctions where most of the lands were disposed it is worthwhile to break down the remaining sales by years to see in general the disposal.[69]

Land Sales—Maine Public Domain
1873-1881

Year	Amount Sold (in acres)
1873	46,052.4
1874	114,338.31
1875	141,769.02
1876	9,721.03
1877	
1878	1,029.16
1879	3,178.05
1880	765.41
1881	2,437.46
Total	319,290.84

This total of 319,290 acres when added to the 529,000 acres sold between 1860 and 1872 and the nearly 735,000 acres granted to the European and North American Railway, as well as the close to 141,000 acres granted to educational institutions, shows that the State of Maine disposed of just under 1,725,000 acres of land in the period from 1860 to 1881. The settling lands and the Swedish settlements are not included in this total, and if these lands were figured in it would mean that nearly 2,000,000 acres of land went into private hands in the east at the same time the federal government was also disposing of its lands in the west. This record helps to explain why it was necessary for Maine to dispose of its lands in large amounts. The other alternative would have been to hold the lands under the supervision of the land agent for some far-off day when their value would rise. No one ever put this plan forward as a viable solution to the problem of the lands. Disposition was the solution sought by all—lumbermen, land agents, and citizens alike.

Most of the large lumbermen had filled out their holdings well before the last disposal of the Maine lands. Pingree and Coe, Ebenezer Coe, the Hinckley and Egery interests, and Llewellyn Powers all

Lumbering and land sales

bought land from the state in the period after 1872, but most of these purchases were only to round out the areas which they already controlled. In 1874 and 1875 Pingree and Coe bought parts of fifteen separate townships all in the western part of the state, about eight thousand acres, for $3,895.20.[70] The Hinckley-Egery interests bought land in fifteen different townships, 2,000 or so acres for just over $1,100, all this in 1874.[71] Coe bought land in 1875 filling in parts of eleven townships already controlled.[72] Powers, from Houlton, soon to be a member of Congress and governor of the state, began his series of purchases at about this same time. In 1874 he bought T6R18, 22,883 acres for $14,873.95. Three weeks later he bought half of T8R5, 11,429 acres for $8,572[73] and he continued to buy odd lots from time to time as well.

Some of these purchases came from the auction block, which was where the state disposed of this land. There were three big auctions and one small one to clean up the residue in the period from 1874 to 1878. They were big affairs, well advertised and conducted in the presence of the Governor and his council in the confines of the Bangor City Hall. Nearly every lumberman of any importance in the state attended. The lands were well known as the prospective purchasers sent out explorers to ascertain the amounts of merchantable timber available on each township. The prices as a result varied greatly, anywhere from 5¢ an acre to as high as $1.30. The following table shows the variation in prices in the largest sale, that of September 23, 1874.[74] As an observer described it while drawing a moral lesson from the auction:[75]

> The bidding was lively, and the prices obtained indicate that the lands which the state gave the European and North American Railroad

Auction of Land and Prices Paid: Bangor

September 23, 1874

Location	County	Acreage	Purchaser	Price per Acre
1R5	Oxford	20,227	E. S. Coe	.35
5R5	Oxford	3,624	M. Giddings	.45
4R8	Franklin	8,848	L. Cowan	1.30
3R9	Piscataquis	22,043	F. A. Reed	.42
17R16	Aroostook	11,521	A. D. Manson	.57
Reserved lots	18 towns	12,070	9 purchasers from	.27 to 1.75
Right to cut till January 1, 1884				
7R16	Somerset	23,745	C. P. Stetson	.31½
8R16	Somerset	23,746	C. P. Stetson	.29½
9R16	Somerset	23,542	C. P. Stetson	.24¾

some years since, would now bring at least $600,000. Probably if the state had properly husbanded the lands which were parted with for a song twenty, thirty, and forty years since, they would now be worth many millions of dollars.

The next big auction was in the fall of 1875. At first the auctioneer disposed of the state's stumpage rights on the public lots and reserved lands in thirty-one townships and gores, about 22,000 acres. Most of these sales were for the stumpage rights until the area became a town, when the rights would revert to the new town. For most of these purchases this was an excellent buy as the townships are predominantly timberland and quite unfit for settlement. Prices ranged from 5¢ to 34¢ an acre, with the average about 11¢. The state then sold nine parcels of land in the first two ranges, the largest 10,700 acres to J. C. Madigan for 55¢ an acre. The total of these parcels came to 19,000 acres. The right to cut timber till 1884 on 4R18, 30,826 acres, went to E. W. Shaw for 17¢. Finally the auctioneer disposed of twenty-nine more lots of land all from the settling lands in the first seven ranges which had been declared to be unfit for settling by the legislature. These plots, amounting to 64,526 acres with the largest being 11,429 acres in 8R5, went for prices which ranged from 8¢ to 90¢ with the average price probably being near 30¢.[76]

All told 105,000 acres went under the auctioneer's hammer that day if one counts the stumpage on state lots where no towns had yet been built. Not all went in the sale however. The land agent held a further sale at his office on November 29 disposing of 4,150 more acres in 11R1, 15R3, and 17R7 to William Engel and I. R. Clark, for 50¢, 30¢, and 10¢ an acre. E. S. Coe also bought the stumpage on the reserved lots on N½1R8 WBKP and W½6R2 NBKP, all told 387 acres for 15¢.[77]

It had been thought that this auction would eliminate all the state's lands, and the legislature had ordered the land office in Bangor closed as of January 1, 1875. The office was still not closed at the end of February and the agent said that he had 55,000 acres or so left to be disposed of to actual settlers at 35¢ an acre, to be paid off in two years by road labor. The settlers had also to take up residence, clear fifteen acres to grass, and build "a comfortable dwelling house" within the two year period.[78] The office was finally closed during the summer and then moved to Augusta. The legislature meeting that next January did away with the office as being unnecessary, but when they found that lands still remained to be sold it was hastily brought back into existence. One more auction was held, No-

Lumbering and land sales

vember 21, 1878, in the city hall in Bangor. Eighteen parcels passed under the auctioneer's hammer, 9,246.29 acres in all. In addition the state also auctioned off stumpage permits to 1884 on some of the old European and North American grant, 343 acres on 7R18 for 3¢ and 15,947 acres on 5R19 for 2½¢ an acre[79] to Daniel Lord.

Very little land remained. Some individuals had not proved up their settling claims and this business hung on until about 1890. A few islands, mostly in Penobscot Bay, were sold from time to time. At the end of the eighties it was found that a few warrants for land which had been granted to Revolutionary War veterans still remained outstanding, and the land agent spent considerable energy tracing these down and granting the lands called for. These grants, making about 5,000 acres, were finally settled by April of 1889.[80] The land agents were left with the duties of administering the state owned public lots, a haphazard business at best. In 1891 the office of land agent was abolished and his duties were turned over to the newly created Forest Commissioner's office, where they still remain.

Now it was time for those who felt the lands had gone for too little money to make their appearance. The Belfast *Age* combined an attack on the tariff with one on the state's land policy.[81]

> The timberlands in all our older counties is exhausted, and upon our streams numerous sites of saw mills long since gone to decay are seen. The people have to depend on building lumber almost entirely on the fast receding forests in the northern half of the state. These forests, which twenty-five years ago were mostly owned by the state, are now owned by speculators, and this duty upon lumber goes almost wholly into their pockets. The point is well-stated, that the duty is a premium for the more rapid destruction of our forests, and instead of discouraging the importation of foreign lumber by putting a duty on it, we ought to encourage it by taking duties off. The proposed bill is simply in the interests of those few wealthy speculators who have clutched all our timberlands, and now manipulate the lumber market, and push up prices of Canadian or foreign lumber with the duty and expenses added.

Some called for a revaluation of timberlands which it was felt were not paying their share of schools, roads, and bridge construction,[82] and even an occasional politician called for a revamping of the state Board of Assessors.[83] The next year the board called for legislation which would keep a continuous running account of the ownership of the wild lands in the Augusta offices, as well as in the local Register of Deeds offices, and suggested that some method which would determine valuation more equitably be ascertained as well.[84] Wild lands by this time amounted to 9,233,575 acres, which were valued at $17,103,317 for tax purposes, or not quite $2.00 an acre.

Wild Lands—Location and Value—Maine—1893[85]

County	Wild Lands	Value	Public Lands	Value
Aroostook	2,810,578	$ 4,726,706	111,431	$124,704
Franklin	538,245	1,137,936	20,835	34,156
Hancock	362,893	612,209	15,360	17,760
Oxford	358,654	866,776	13,744	26,744
Penobscot	825,420	1,340,512	32,710	36,403
Piscataquis	1,989,473	4,287,713	83,424	126,779
Somerset	1,732,137	3,049,389	91,043	92,231
Washington	616,175	1,082,076	25,233	29,183
Totals	9,233,575	$17,103,317	397,780	$487,960

It can be seen from the table that the state still retained some sort of possession to nearly 400,000 acres of land. Much of this though had been sold, or at least the stumpage rights had been and those rights applied until towns were erected, so the lands did not belong to the state at all. The state had given away, granted, and sold its lands, nearly ten million acres, in the nineteenth century, two million of these acres in the period from 1860 to 1890. How much money the state received is difficult to ascertain, but it was probably close to 40¢ an acre for lands sold, but only about 25¢ an acre for the total.[86] Although this seems a very small amount, in those times this amount of money was a fairly good return especially considering the competition offered by better land elsewhere in the country.

Land continued to change hands throughout the period as the large landholders filled out their purchases from the state, or as earlier investors strove to cut their losses incurred through unwise investment in the railroads or in the lumbering business. Several large landowners went bankrupt or nearly so and their holdings went under the auctioneer's hammer. All told, the sale of lands and their prices indicate something of the economy of the Maine woods during this period.

Some small sawmill owners purchased land for their log supply. For instance, in 1866 Alexander McL. Seeley of St. John, New Brunswick sold the W½KR2 to Samuel W. Collins of Lyndon, Maine. For this acreage, about 11,520 acres, Collins paid $5,000.[87] This was a fairly typical agreement. If the purchaser did not pay cash, stumpage

Lumbering and land sales 195

was charged. If he did pay cash oftentimes the price of the land rose slightly to compensate for the loss. Collins was to parlay those holdings and others into his sawmill business which is still being operated under that name by his heirs. This was a fairly typical price paid in the years before the depression, although the Lawrence Brothers of Fairfield paid $20,000 cash for half of Pierce Pond Township the same year.[88]

Most transactions were of this nature, at least until 1872. Persons bought part of a township, a full township or a few lots for their operations. In 1872 though a land sale occurred which created much comment. Abner and Philander Coburn, the great land owners of Skowhegan, sold all their lands, their boom property, and everything else connected with the lumber business. The price of the land was $3.00 an acre. The 450,000 acres, 410,000 of it on the Kennebec River went for $1,350,000. The purchasers—Ira Sturgis of Old Town who had an eighth, Orville D. Lambert of Augusta who had an eighth, and Amasa Sprague and Company the Rhode Island textile manufacturers who took the rest—paid $300,000 down. Observers estimated that the Coburns had made a profit of $1 million besides the lumber which they had cut themselves.[89] The lands were sold to a firm styled the Coburn Land Company, and Abner and Philander Coburn held a first mortgage on the lands until the sale was consummated. This sale stimulated sales elsewhere in the state and newspapers remarked on the prosperity now coming to the state. Timberlands were said to have risen a third in value as a result of the Coburn sales. Skillins and Company operating the Bethel Steam Mills paid $50,000 for Usher town on the Magalloway, and Androscoggin lands were said to be worth about $5.00 an acre. In 1865-6 the same lands had sold for $1.50 an acre.[90] It was a false prosperity however. The Sprague interests, the largest textile manufacturers in New England, went down in the late summer of 1873, the first big victim of the depression of 1873, and when they did they carried Maine lumbering with them. In 1873 the crews which had gone into the woods were recalled and in 1874 the Coburns foreclosed their mortgage. After the death of the Coburn brothers their heirs continued to operate their lands in a small fashion, but wrangling over the estate limited operations. In fact the estate began to dispose of some of the land. The Lawrence brothers bought Lowelltown, and Coburns' interest in the Somerset booms, a store, a dwelling house, and the rights to the water power there as well as the Somerset mills, all for $80,000 in 1877.[91] Later the big pulp companies were to purchase most of the Coburn lands.

The Coburn sale marked the beginning of the big disposals, many of them caused by the failure of their owners in other businesses. When a great amount of land was put up for sale in this way the usual method of disposal was to do as the state did in disposing of its lands. An exact description of the location of the lands was published in the newspapers for six months or so in advance in order to let the prospective purchasers cruise the land and estimate its value. These advertisements appeared in all newspapers read by lumbermen. When the time had elapsed and the prospective buyers had had a chance to spy out the land the auction was held, usually in Bangor which was the center for most timberland owners and investors.

During the spring of 1872 the first of these large sales took place when the heirs of Samuel H. Dale decided to dispose of his estate by the auction method. Advertisements appeared early in 1872, and the auctions took place selling all but the timberland properties on May 22 and those properties on June 13. The bidding was spirited and lively and the lands went for a good price. Terms were one third cash and the balance to be paid in three annual installments with interest, with the estate to retain a first mortgage on the lands. Many varied items went in the first sale[92] which returned $63,655. The land sale like the earlier property sale was held at the Bangor City Hall, and it was remarkably well attended. A few odd shares of stock (mostly in the European and North American) were sold and then the land commanded the attention of the auctioneer. To give some idea of typical holdings before the depression, and the prices which land commanded, the following table (next page) has been prepared to show the auction results in tabular form. The heirs realized $117,159.35 from the sale.[93] It was very successful; after the panic others would not be.

Prices were much lower after the panic, especially for lands which were sold because of the failure of the owning firms. Such was the case in 1877 when the lands of the Dana R. Stockwell Company went onto the block. Originally some 68,000 acres were put up for sale, but 43,000 acres of this went in private sale to the owners of the townships involved. The rest of the land went at smaller prices at the auction itself.[94]

Some of the land owners had large holdings by this time. The Hinckley and Egery Company held or had held interests in seventy-seven townships in the northern half of the state by 1875. Some of

Lumbering and land sales 197

Results of Land Auction of S. H. Dale Holdings,
Bangor, Maine—June 13, 1872

Location	Amount	Purchaser	Price
Burlington	200		20¢
	160		.30
	100		1.60
	100	all J. W. Porter	1.75
T8R5	948	J. P. Wellington	.75
	639	W. C. McCrillis	.80
	168	J. P. Wellington	.25
T9R6	4949	Andre Cushing & Co.	.65
	334	Samuel Willard	.33
½W½T16R3	5260	W. C. McCrillis	2.37½
Tl N. D.	9100	A. Webb & Co.	1.72½
T6R7	21720	J. P. Webber	.41
E½T3R8	11140	F. R. Webber	.48
SE¼T6R8	6010	S. H. Blake	1.67½
½ undivided T3R9	4143	J. P. Webber	.51
SW¼ & E½T7R9	17982	W. C. Averill	1.50
½ undivided T9R10	12493.5	H. F. Eaton	1.15½
½ undivided T1R4 BKP	11520	A. D. Manson	1.42
Totals	106,966.5 acres	just over $1.00 average	

their lands had been sold to Dunn and Jewett in 1874 but most still remained in the hands of the family at this date. It took money though to be in this business. When the founders of the iron foundry in Bangor died in the sixties and early seventies, the heirs continued to manage the lands together. At the time of the depression they apparently consolidated their lands somewhat, selling thirteen separate lots to Dunn and Jewett, receiving in exchange all of T18R4 and scattered lots on other towns. They also sold land to other northern proprietors and exchanged some of their land with the state for most of T18R3. All told though they still owned acreage in sixty-six townships in 1875 with a total of 91,330 acres.[95]

Other large land owners continued to buy and sell land. Ezra Totman and Company paid $30,000 for one half of T3 on the Dead River in 1878, which was not a bad price.[96] Other auctions of land owners who had gone bankrupt occurred throughout the decade from 1875 to 1885 as these firms went down before the depression. Eleven thousand five hundred acres of S. H. Barton went under the hammer in June of 1878.[97] In August J. F. Slater bought 46,000 acres undivided in various towns for 70¢ an acre.[98]

No sale, however, was as large as the auction for the assignees of the E. D. Jewett Company in September of 1878. The lands were all sold in the traditional manner at the Bangor City Hall. To show how the lands had begun to consolidate in the hands of a few owners a table has been prepared indicating the extent of the large purchases.[99] The Jewett lands, amounting to 399,475 acres in all, went under the hammer on September 18, 1878. Although the bidding was brisk the sale was orderly, and the land was all sold by 11:30 A.M. two and one half hours after the auctioneer had called for the first bid.

Abstract of Sales—E. D. Jewett Co. Land

Bangor, Maine—September 18, 1878

Purchaser	Amount of Land	Remarks
N. C. Ayer	102,915.72	land in 16 towns
E. S. Coe	70,555.45	land in 16 towns
E. Sutton	50,585.70	land in 7 towns
S. H. Barker	40,980.79	land in 5 towns
W. H. McCrillis	32,086.29	land in 5 towns
Peter Dunn	25,167.42	land in 4 towns
Chas. Hamlin	18,831.75	land in 3 towns
H. F. Eaton	14,468.5	land in 2 towns
N. C. McCausland	11,000.0	land in 1 town
W. S. Dennett	9,565.0	land in 1 town
C. H. Sawyer	9,193.0	land in 1 town
Chas. D. Bryant	6,589.04	land in 3 towns
Mark Barker	4,234.67	land in 1 town
J. P. Bass	1,966.83	land in 1 town
Wm. F. Engel	1,329.33	land in 2 towns
Total	399,475.49	land in 68 towns

The average price of this land was much lower than had been the case in land sales only five or six years earlier. Only nine lots went as high as $1.00 an acre and many went for as low as 30¢ to 35¢. Owning land as he did in sixty-eight separate townships Jewett must have been one of the two or three largest land holders in the state at the time of the disposal of the estate. The Jewett family of course had been very active in the promotion of the European and North American Railroad, and its failure was the initial factor in bringing down

Lumbering and land sales

the firm. To own land in this amount one needed a tremendous capital such as had been possessed by the Coburns or by David Pingree. Eben Coe who was a big purchaser at this auction was Pingree's agent and limited partner. The Jewett auction was the largest sale of lands in the history of the state conducted in this form. Later some of the pulp and paper companies consummated some large transactions but these were all private sales.

Occasionally land sold for much higher, prices such as when the Bradstreets paid $3.00 an acre for land on the Moose River that same fall,[100] but by and large the prices obtained by the Jewett auctioneer were the top prices paid for timberlands in this period. Oftentimes the prices were much less[101] and whole townships went for $20,000 and less in the late seventies and early eighties.[102] When auction sales were held it was the big timberland investors who bought; no one else dared take the risk.[103]

As more and more people began to dream of a railroad which would connect the northernmost part of the state with Bangor, timberlands in that area became increasingly more valuable. The Jewett lands came into the market again. Two of the biggest purchasers, N. C. Ayer, and E. Sutton, had been acting as agents for Bangor banks, the Second National and Merchant's respectively. These lands and some others which had been foreclosed, amounting to 181,840 acres in all, were sold by the banks to Hayford and Stetson, big lumber merchants of Bangor and St. John. This sale really triggered the big cutting in the upper St. John river valley as Hayford and Stetson no longer had to pay stumpage costs. The firm rapidly grew to be one of the largest lumber operators on the river. Newspapers regarded this sale as an omen for Maine's future.[104]

> The railroads that have been built, and those that are being constructed in the northern part of Maine, are opening up the facilities for transportation and increasing the value of lands. While the depletion of forests in other parts of the country has an added influence in swelling the worth of timberlands in our own state.

The millenium was not to come so quickly. In fact the depression still hung like a pall over the Maine woods and over capitalists who had invested in these woods. Still, when auctions were held now the prices seemed to be somewhat higher. Take the sale conducted in Bangor on October 1, 1881 as an example.[105]

Abstract of Auction Sale Held in Bangor, Maine

October 1, 1881

Location	Acreage	Purchaser	Price per Acre
T11R17	22,270	J. G. Clarke	.74
NE¼T11R16	5,785.5	E. C. Burleigh	.81
¼ of S½T4R7	2,880	J. P. Bass et al	.85
¼T14R8	5,758.5	J. P. Bass et al	.91
¼T14R12	5,795.5	J. P. Bass et al	.91
½T12R10	12,316	H. F. Eaton	.60
½T13R10	11,997	H. F. Eaton	.60
SE½T6R17	5,000	E. C. Burleigh	.54
¼T16R9	5,585	J. C. Clarke	.75
NE¼T14R7	1,440	J. P. Bass et al	.85
NW ¼T14R7	1,030	J. P. Bass et al	
Total	79,857.5		

Even land in close proximity to Bangor seemed to rise in value somewhat.[106] Lands in Aroostook County, near the line of the proposed railroad, fluctuated markedly in value from around fifty cents an acre to nearly $3.00 depending on whether the road seemed any closer or not.[107]

By the nineties the railroad was obviously coming, and as it did land values went up and up.[108] At the end of our period some of the prices were very high. In 1892 Senator Hale and ex-governor Daniel F. Davis, along with William Engel, paid $64,000 for 11,000 acres in the south half of Edinburg, for instance. Even though this was old growth hemlock and spruce the price was very high.[109]

Reviewing the history of private land sales in the period from 1860 to 1890[110] there are certain trends which are apparent. As long as the state was in the land disposal business, prices for lands ruled fairly low and the only factor driving the price up was the nearness to a sawmill or a location on a more southern river such as the Androscoggin. With the first big sale, that of the Coburns, prices seemed to rise rather markedly and a few of the early investors disposed of their holdings. The Coburn sale was of such magnitude though that it aided in the bankruptcy of the purchaser, the Amasa Sprague interests in Rhode Island. When the firm fell the depression of the 1870's was on. Land prices dropped markedly, and the decade of the seventies was one of forced disposal as many firms found themselves victims of their over-speculation. In the beginning of the next decade as talk of the railroad in northern Maine became more prominent, prices took a small rise especially in the Aroostook area.

Small holders took advantage of the price rise to dispose of their lands mostly to larger investors able to withstand the economic pressure. When the railroad came lands went to fairly high figures, but this was to last only a short time as the economics of the lumber situation forced a major retrenchment. Eventually most of these lands were to come into the hands of the giant paper companies. Although there is an air of speculation over land transactions in this last part of the nineteenth century, nowhere does it take the form of the famous 1830's land frenzy. One suspects that those who could weather the storm clouds of the seventies probably made money from their lands either through stumpage sales or by sales to the railroads or the paper companies.

NOTES

[1] For earlier views of the speculative nature of lumbering, Bangor *Whig and Courier*, December 23, 1834; February 1, May 4, 1837; for even growth, without speculation, as a good thing, April 4, 1834; June 27, 1835; Bangor *Democrat*, December 7, 1841; June 16, 1845.

[2] I have dealt with this in my "Maine and its Public Domain," *op. cit.* This disposal worked out like this:

Disposal—Maine Public Domain

Sold and Granted Prior to 1783	3,785,000 a.	19.4%
Granted by Mass. after 1783	1,686,712	8.7
Sold by Mass. after 1783	6,752,987	34.6
Sold by Maine & Mass. in Common	1,750,605	8.9
Sold by Maine	3,573,323	18.3
Granted by Maine	1,968,285	10.1
Total	19,516,912	100.0

These are the figures of land granted and sold up to 1880. After that time only scattered parcels, to holders of Revolutionary War warrants, or to purchasers of small islands on the coast, were disposed of. Since 1880 the state has also reobtained title to lands, most notably in the area of Baxter State Park.

This table appears in a slightly modified form in the *History of Wild Lands*, published in 1923, and *The Northern*, II, no. 9. At the time that it was compiled the land owners were fighting a considerable battle with Governor Percival Baxter and the compilation was prepared to prove that the land disposal was honest.

[3] *Land Agent's Report*, 1858.

[4] Land Agent, *Report, 1857-1858*; House Document #90, *Legislative Documents, 1869*, "Report on Disposition of School Lands Set Aside, April 13, 1857."

[5] Land Office Records, "Records of Deeds, Timber and Reserved Lands", 2 Volumes; Book of Advertisements and Bills Paid January 1, 1834-1847, and Letter Books, especially 1832-1850. The biggest purchasers were the Coburn Brothers who owned 26½ townships in the settling lands, Henry E. Prentiss, 17¾ towns; Ebenezer S. Coe, 21+ towns; and Boynton and Bradley 15+ towns. The Coburns also owned much land not in these lands mostly on the Kennebec River—25¾ towns by initial purchase. See Deed Books, pages 51-66, 156, 167-170, 172-3, 195-6, 197½-8, 223, 248, Vol. I; 98-9, 153, Vol. II. The last deed is dated November 23, 1854. These townships were the normal six mile square towns except where the surveyors made errors, or the declination tables were in error. As a result, most of the towns were of 23,040 acres.

[6] Isaac Clarke (State Land Agent) to W. C. Creasey (?), Seven Islands, New Brunswick, May 2, 1867. Clarke was trying to blackmail Creasey (?) into taking a larger amount. "If the quantity [proposed to be cut] is one or more millions . . . . it will facilitate the application for permit."

[7] Clarke to John C. Carpenter, Linneaus, December ??, 1867. Also see Clarke to Enos Bishop, Presque Isle, December 12, 1867; to Luther Gowen, Bangor, December 13, 1867. Gowen was responsible for the scaling on 3R5, 4R5, both NBKP, and 4R18, 5R19, 5R20.

[8] See Clarke to Governor December 10, 1867 confirming the sale of E½ 9R3 to S. F. Hersey of Bangor for $4,000 for timber and lumber. This town was previously granted to the Westbrook Female Seminary and College. These letters are all in a letter book 1866-7 in the Land Office records.

[9] Land Agent's *Reports*, 1859-1872.

[10] In all 992,924 acres were granted to educational institutions in Maine; 47,710 acres of this was public lots owned by Maine and Massachusetts and the proceeds went to educational facilities. Even so, this left 945,214 acres for institutions, as well as two and one-half townships on which just the timber was granted. See House Document #100, 1879 for an analysis of these grants. Also see "Grants of Lands to Academies in Massachusetts and Maine," H. W. Marr in Essex Institute *Historical Collections*, 88, (1952), 28-47 and Ava Chadbourne, *History of Education in Maine*, Lancaster, Pennsylvania, 1936.

[11] This story is best found in Frederick J. Wood, *The Turnpikes of New England. . . .* , Marshall-Jones Company, Boston, 1919, 211-2 and Atkinson, *Grand Lake Stream*, op. cit., 29-30.

[12] It is odd that these are the particular records which are missing. The records prior to 1868 are quite complete, and after 1890 very complete. I searched the Augusta archives for a week one time, and for three days another, without success, and the then secretary to the present Forest Commissioner, Mrs. Lilian Tschamler, who had worked in the office for over twenty years, said that she had never seen them, nor had her predecessor. Writing in the *Pine Tree Magazine*, in 1906, Liberty Dennett, who was associated with the Land Office during some of this time, made a series of veiled insinuations which have never been answered, to my knowledge, concerning their disposition. See *ibid.*, Vol. VI, no. 6, Vol. VII, Nos. 1-6, Vol. VIII, no. 1 (January to August, 1907), "Maine Wild Lands and Wild Landers," especially Vol. VI, no. 6, 543-5.

[13] For Poor's role in railroad promotion see Edward C. Kirkland, *Men,*

Lumbering and land sales

Cities, and Transportation, 1948, 2 vols., and Alfred D. Chandler, *Henry Varnum Poor, Business Editor, Analyst, and Reformer,* 1956.

[14] *Proceedings of the Thirty-First Legislature of the State of Maine in Relation to the European and North American Railway,* Portland (Foster and Gerrish), 1852; *Report on the Survey of the E. and N. A. Railway*: Made under the authority of the State of Maine, A. C. Morton, C. E. Portland (Harmon and Williams), 1851.

[15] Alfred G. Bailey, "Railroads and the Confederation Issue in New Brunswick, 1863-1865", *Canadian Historical Review,* Vol. XXI, no. 4, 367-383, especially 372, 380-2.

[16] Hannay, *op. cit.*, II, 281.

[17] The *Argus* greeted its passage with the following editorial, May 17, 1865.

Aid to the European and North American Railway

The bill which was before the Massachusetts legislature, granting aid to the European and North American railway, has become a law. This is a very important measure, and will give this enterprise a substantial lift. Now if Congress will do the just thing its success will be assured, and its early completion be made a certainty.

On the Congress and that lobbying, see Kirkland, I, 470-4, and Chandler, *op. cit.*, 230 and note.

[18] The lands set aside for the public schools were T7-8-9 R16, 5R17, 4R18, 6-7R18, 5-6-7R19. Legislative Resolve, March 4, 1868.

[19] Lands set aside previously for settlers included T11, B, C, D in the Plymouth Grant, East Half; TE-F-GR1; TF-G-H in the Eaton Grant; T1-L-MR2; S½T2R3, T5R3, T11-12-13-14-15R3, T18R3; T4R4, SW¼T6R4, T12R4, T14R4; E½T2-3-4R5, T6R5, T8R5, T10R5, T11R5, T18R5; T5R6, T9R6, T11R6, N½17R6; T18R6; T16-17-18R7. See Land Agent's *Report,* 1868, 1869.

[20] The easiest place to locate the deed is in House Document 68, State of Maine, 1872, which has an exact copy. On the actual recording see Portland *Eastern Argus,* April 17, 1869. The lands totalled 266,732 acres in Aroostook County, 65,095 acres in Penobscot County, 77,032 acres in Piscataquis County, 308,583 acres in Somerset County, and 17,500 estimated acres in the gores and fractional townships north of Bingham's Kennebec Purchase. The total was 734,943 acres. See Senate Document 57, Maine Legislative Documents, 1869, and the accompanying map for the exact lands conveyed by the deed. The stumpage for the timber and grass on T7-8-9R16, 5R17, 4R18, 6-7R18, 5-6-7R19 all W.E.L.S., was reserved for the school fund, and was sold to other purchasers than the railroad until January 1, 1884. These townships were all in the extreme western part of the state and had not been put up for settling lands as yet. This action is still the most difficult part of the state to reach, and the only area of possible virgin timber might be in this area just north of Little St. John Pond in T7R18, T7R19, and T5R20.

[21] Drummond, "Maine Central Railroad," 80. Much discussion of the road and its route is in Presque Isle *Sunrise,* 1867-1869. They thought that it didn't do its job as it by-passed Aroostook County.

[22] Railroad Commissioner, *Annual Report,* 1875, *Public Documents,* Maine, 1876, II, 12.

[23] Portland *Eastern Argus*, June 19, 26, 1875.
[24] Bangor *Daily Commercial*, June 25, 1875.
[25] *Ibid.*, June 26, 1875.
[26] Bangor *Daily Commercial*, August 23, 1875, correspondent from Lincoln.
[27] Portland *Eastern Argus*, December 2, 1875.
[28] *Ibid.*, December 13, 1875.
[29] Railroad Commissioner's Annual *Report, Public Documents*, Maine, 1877, II.
[30] *Ibid.*, and Portland *Eastern Argus*, September 17, 1877. The gauge shift was completed on September 15.
[31] One of the reasons the road took so long to pay dividends was the amount of money it had paid out to start operations. Daniel Hinckley, for example, conveyed sixty feet of waterfront property on Washington St. in Bangor with the wharves and buildings for 999 years for $5,500. The railroad also assumed all taxes. See deed in Hinckley papers, Bangor Historical Society, December 21, 1863.
[32] *Situation, Character, and Value of the Settling Lands in the State of Maine, published for the Information of Immigrants*, J. A. Purinton, Immigration Agent, European and North American Railway, Bangor, 1871.
[33] This story is best found in the Land Agent's *Report* for 1872, 1873, 1874. See Bangor *Daily Commercial*, March 28, 1875; also January 22, 1872 for a story on the seizures. The operative laws were *Laws of Maine*, Ch. 25, 1850; Ch. 401, 1864; Ch. 326, 1864; Ch. 604, 1868; Ch. 9, 1870.
[34] Annual Message of the Governor, 1874, *Public Documents*, Vol. I.
[35] *Ibid.*, 1875, Public Documents, Vol. I.
[36] See especially the issue of September 23, 1878 from which the following editorial quotation is taken.
[37] Senate Document 21, 1879, *Laws of Maine*, Ch. 126, 1879.
[38] House Document 168, 1881, *Senate Journal*, 1881, 425.
[39] Portland *Eastern Argus*, July 24, August 8, 1882.
[40] The road had become a fairly large concern by this time. It had 114 miles of track, 55 of which was newly laid steel, and the rolling stock included 15 engines, 28 coaches and cars, as well as 420 freight cars. See on the stock exchange Portland *Eastern Argus*, August 10, 1882, and on the road generally Bangor *Daily Commercial*, Sept. 14, 1882.
[41] Bangor *Daily Commercial*, September 14, 1882.
[42] *Railroad Commissioner's Report 1882, Public Documents*, Maine 1883, 40.
[43] Bangor *Daily Commercial*, May 2, 1882, first advertisement.
[44] *Ibid.*, July 8, 1882; also see August 24, 29; September 9, 11, 13, 1882 for comment on the lands and the land sale.
[45] See Portland *Eastern Argus*, September 25, 1882, and Bangor *Daily Commercial*, September 21, 1882 for accounts of the sale. Most of the lumbermen present did not bid, as the railroad's control of the upset had taken most of the good lands from the market. The land sold totaled about 262,000 acres.
[46] Bangor *Daily Commercial*, January 20 (report of City Council meeting); February 2 (letter from an observer, "XXX" obviously a lawyer); February 11,

Lumbering and land sales

1887 (letter from City Treasurer, John L. Crosby, defending the comment of "XXX").

[47] Senate Document, 77, Maine, 1887. The committee of ten men consisted of at least six who were clearly identifiable as lumbermen. Only one Democrat was on the committee and he was a lumberman as well. See Portland *Eastern Argus*, March 8, 1887 for comment.

[48] Senate Document 36, Maine, 1895; *House Journal*, March 22, 1895, 655-6; *Senate Journal*, March 26, 1895, 752-3. The House had twelve lumbermen and landowners out of one hundred and fifty-one members, including the speaker, Llewellyn Powers of Houlton. The Senate figures were seven of thirty-one. Only five of the one hundred and eighty-two were Democrats. The Judiciary Committee who held the hearings was also made up of individuals who though lawyers were people who derived much of their practice from lumbermen and the lumber business. Keegan was a member of the committee. This analysis is based on knowledge of the men involved and their own biographies printed on the opening day of the legislature by the Kennebec *Journal*.

[49] Most of these advertisements ran for three months in a prominent place in the newspaper. Typical advertisements are in Portland *Eastern Argus*, March 21, 1876, and September 1, 1876. But also see for other examples *ibid.*, February 25, 1871 (Union Pacific Railroad); June 30, 1871 (Chicago, Rock Island, and Pacific Railroad); September 23, 1871 (Union Pacific Railroad); April 29, 1872 (Burlington and Missouri River Railroad Company); April 29, 1872 (The Union Pacific again, this time as "The Garden of the West"); April 17, 1873 (Union Pacific and Iowa Railroad Company); September 1, 1876 (Kansas Pacific Railroad Company); and Bangor *Daily Commercial*, March 13, 1873 (The Union Pacific Railroad Company). There were, of course, many, many more advertisements, but these were the ones which seemed to be most representative of the types of lures spread to catch wary Maine prospects.

[50] Portland *Eastern Argus*, July 11, 1871 (Lake Superior and Mississippi Railroad Company).

[51] Portland *Eastern Argus*, July 20, 1865 carried the following letter from a young man in Aroostook County.

> I am eighteen years old, have a good set of teeth, and believe in Andy Johnson, the star-spangled banner, and the Fourth of July. I have taken up a state lot, cleared up eighteen acres of it last fall, and seeded ten of it down. My buckwheat looks first rate, and the oats and potatoes are bully. I have got nine sheep, a two-year-old heifer and two bulls, besides a house and barn. I want to get married. I want to buy bread and butter, hoop shirts, and waterfalls for some person of the female persuasion during life. That is what is the matter with me. But I don't know how to do it.

[52] Bangor *Daily Commercial*, January 8, 1872, an editorial unhappy about the great numbers of persons who had left Maine since 1868.

[53] *Ibid.*, February 5, 1872. Also see Portland *Eastern Argus*, July 24, 1886 for an editorial discussing the hard work involved in the west.

[54] *Ibid.*, January 11, 1872.

[55] The price of farm labor in Paris, Maine in the years cited. Portland *Eastern Argus*, April 26, 1876; July 24, 1882.

56 These farms were in Ludlow; see *ibid.*, September 27, 1876.

57 The farm in Van Buren was owned by John W. Brown; see Bangor *Daily Commercial*, September 5, 1878; for Maysville, *ibid.*, October 19, 1878 quoting the Kennebec *Journal*; for Presque Isle, *ibid.*, October 19, 1878, and January 1, 1885; Portland *Eastern Argus*, November 9, 1882, for Buxton, near Portland.

58 *Ibid.*, April 26, 1877 for three wagons from the Provinces; July 25, 1876 for an emigrant train of three families, twenty persons, from Aroostook County, February 10, 1880 for general comments and the Lewiston migration. "The annual spring western fever has set in!" and March 30, 1883 for an emigrant train from West Farmington and Industry of twenty-seven people.

59 *Ibid.*, August 9, 1886.

60 Aroostook County was always the solution to Maine problems, at least in those days. Of course, this newspaper was the leading Democratic newspaper in the area, and although their views were honestly arrived at, the intensity undoubtedly was heightened by the chance to attack the opposition party in an election year. My "Toward a Theory of Maine History— . . . ", *op. cit.*, deals with Maine's efforts to "Keep the Boys Home!"

61 "William Widgery Thomas", G. M. S. (George M. Stephenson), *D. A. B.*, XVIII, 447; Portland *Press Herald*, May 26, 1927 (obituary); Michael U. Norberg, *The Story of New Sweden, Portland*, 1895; W. W. Thomas, "Swedish Colonization in Maine and New England", *The New England States, op. cit.*, III, 1244-8; also Land Agent's Report, 1870; Report of Commissioner and Board of Immigration, 1870 (both Public Documents, Maine, 1871); Governor's Annual Address, 1870, 1871; Report on the Decennial Celebration of New Sweden, 1881; E. W. Elwell, *Aroostook . . . with some account of New Sweden*, 1882.

62 Advertisements appeared in *America* (Gothenberg, Sweden), June 30, 1870; June 14, 1871; *Oresunds-Posten* (Oresunds, Sweden), March 20, April 14, 1871; *Hemlandet* (Chicago), March 11, 1873; and *Nya Verlden* (Gothenberg), April 3, 1873 (cited in *D. A. B., op. cit.*)

63 Portland *Eastern Argus*, December 9, 1871.

64 Bangor *Daily Commercial*, August 1, 1882.

65 *Ibid.*, January 2, 1886.

66 For a good contemporary history and account of the colony see *ibid.*, July 19, 1902.

67 Senate Document 67, 1870. The timberlands were S½11R16; west part 11R17; S½2R4 WBKP; 3R5, 3R6, 2 R7, 4R5; 5R6; 4R6; all WBKP; 3R1 NBKP; 3R9 and 8R18 WELS.

68 Land Agent's *Reports*, 1874, 1880. W. H. McCrillis paid $7,479.67 for the stumpage, 23,745 a. on 7R16; E. D. Jewett 4,739.19 for 23118 a. of 8R16; George Stetson 5,767.79 for 23542 a. of 9R16; John Morrison 3,790.08 for 21056 a. of 5R17; Daniel F. Davis 4,538.20 for 22691 a. of 6R18; 5175 for 25876 a. of 6R19; William K. Lancey 5,485.50 for 23343 a. of 7R18, 2,875.32 for 15794 a. 5R19; Samuel H. Blake 8,698.12 for 23195 a. of 7R19 (all dated Sept. 23, 1874) and Thomas A. Phair for the 3/5ths undivided of 6R9 (25876 a.) paid $300 for the time remaining (dated Jan. 5, 1880).

69 The table is compiled from the Land Agent's *Reports* 1860-1890, and the *Deed Books* before cited, located in the Forest ommissioner's Office.

Lumbering and land sales

[70] *Deed Book*, p. 174, Sept. 23, 1874, 205½, Nov. 11, 1875, 206½, Nov. 11, 1875.

[71] *Deed Book*, II, 175, Sept. 23, 1874.

[72] *Deed Book*, II, 176, Sept. 23, 1874, 210, Nov. 11, 1875 (these were all land which was auctioned).

[73] Land Agent's *Reports*, 1874-1875; *Deed Book*, II, 187, 209½, 220, October 9, 1874, November 11, November 1, 1875. Land Agent's *Report*, 1880.

[74] From Bangor *Daily Commercial*, September 23, 1874.

[75] Oxford *Democrat*, September 29, 1874.

[76] Bangor *Daily Commercial*, October 28, 1875 where the sale is discussed in great detail.

[77] Bangor *Daily Commercial*, November 29, 1875.

[78] *Ibid.*, January 1, February 24, 1875.

[79] *Ibid.*, November 21, 1878.

[80] Land Agent's *Report*, 1890, recapitulating the veteran's warrants disposal.

[81] Quoted in the Bangor *Daily Commercial*, January 5, 1883 with approval.

[82] *Ibid.*, March 19, 1889 quoting a long editorial to this point from the Lewiston *Journal*. The *Journal* said farmers paid an average of $5.00 taxes on fifty acres, while the wild land proprietor paid only 13¾ ¢ on fifty acres.

[83] Annual Address of the Governor, January 5, 1893, *Public Documents, Maine, 1894*, I, 17.

[84] State Assessor's *Report*, 1894, *Public Documents, Maine, 1895*, I, 263.

[85] *Ibid.*, passim, the calculations of totals is mine. For slightly different figures see Second Annual *Report*, Forest Commissioner, 1894, *Public Documents, 1895*, I, 10, total wild land acreage 9,666,727 acres, valued at $18,210,894.

[86] This is my estimate. One which agrees fairly closely with it, although not covering the total area is in //Lillian Techamler//, *Report on Public Reserved Lots*, Chapter 76, Resolves of 1961, prepared by the State Forestry Department, Augusta, 1963, 14.

[87] Articles of Agreement, dated October 8, 1866, between the Honorable Alexander McL. Seeley and Samuel W. Collins. This document is in the New Brunswick Museum, St. John, New Brunswick.

[88] *The New England States*, III, 1356, sketch of Edward J. Lawrence.

[89] Portland *Eastern Argus*, March 26, 1868 on land holdings; Bangor *Daily Commercial*, October 8, 1872 on the purchase.

[90] Portland *Eastern Argus*, December 13, 1872 on rising land values. Oxford *Democrat*, December 24, 1872 on the effect on Androscoggin lands.

[91] Bangor *Daily Commercial*, November 12, 1873; October 15, 1874; November 16, 1877 on the land sale. Also Portland *Eastern Argus*, October 13, 1874 on foreclosure.

[92] These were typical lumberman holdings at that time. Included, for instance, besides house lots all over eastern Maine, were stores in Bangor, wharves on the river, storehouses on the Kenduskeag stream, and interests in several vessels. Most proprietors were forced to invest in the shipping end of the business in order to insure that their log property went at a fair price.

93 Bangor *Daily Commercial*, February 12, 1872 (the advertisements begin); April 22, 1872 (advertisements for ship and wharf property begin); May 22, June 14, August 8, 1872 for accounts of the sale. Portland *Eastern Argus*, June 14, 1872 for an account of the sale. The average price realized was $1.16 an acre on the timberlands, which does not include the land in Burlington.

94 Portland *Eastern Argus*, October 3, 1877.

95 The foregoing is derived from a study of the extant Hinckley and Egery records in the Bangor Public Library, consisting of various record books from 1860 to 1868, and a file of letters in the Bangor Historical Society running from 1828 to 1869. I have in my possession a large map of northern Maine, five feet by six feet, with the Hinckley and Egery holdings in early 1875 outlined carefully in red, and the exact acreage displayed. Some of the information derives from a study of this map.

96 Bangor *Daily Commercial*, April 26, 1878.

97 *Ibid.*, June 6, 1878; the prices ranged from 12.5¢ to 75¢ an acre.

98 *Ibid.*, August 1, 1878.

99 *Ibid.*, September 18, 1878. This is an analytical and detailed account of this great land auction, the biggest of the century, and probably the biggest ever held in Maine history.

100 Bangor *Daily Commercial*, October 16, 1878.

101 For the remains of the Stockwell holdings (mostly wharf property) see *ibid.*, January 1, 1879. This sale was held at the Penobscot Exchange Hotel. The estate of M. S. Drummond is the one primarily referred to. This land was first put up for sale on June 18, 1879, but the bids were so low the land was withdrawn, and put up again on August 27, 1879. Some 90,000 acres were sold at an average price of less than 50¢ per acre. See *ibid.*, September 25, 1879, and Portland *Eastern Argus*, September 30, 1879.

102 *Ibid.*, October 15, 1879 for accounts of sales to Sturgis, Bodwell, and Company of Augusta of Holeb town and Cold Stream Town. See also Bangor *Daily Commercial*, October 14, 1879 for a similar story.

103 For instance at the auction held April 21, 1880 in Bangor Thomas N. Egery, one of the Hinckley and Egery heirs did most of the purchasing, besides many houses and lots in Bangor, some 62,000 acres of timberlands. See *ibid.*, April 21, 1880.

104 Bangor *Daily Commercial*, August 12, 1881. On this also see Portland *Eastern Argus*, August 15, 1881.

105 Report of auction in Bangor *Daily Commercial*, October 1, 1881. The first advertisement appeared September 3, 1881.

106 *Ibid.*, October 3, 1881, report of an auction of lands in Woodville, Winn, and Mattawamkeag—18,193 acres.

107 *Ibid.*, November 21, 1882, July 17, 1885, August 19, 1886, March 16, 1886; March 17, 1890, and Portland *Eastern Argus*, December 19, 1882. The highest figure comes from the sale of the undivided half of T1R5. This was mostly old growth hemlock and spruce though; see Bangor *Daily Commercial*, June 28, 1890.

108 Bangor *Daily Commercial*, January 2, 1891.

109 *Ibid.*, June 20, 1892.

110 It is perhaps appropriate here to discuss the last big public auctions held in Maine of timberlands. The first of these was the lands of A. D. Manson,

Lumbering and land sales

onetime Mayor of Bangor. The terms were traditional, and when the sale opened nearly every lumberman in the state was present. All told 56,610.5 acres changed hands for $62,744.08 or just over $1.00 an acre. See *The Industrial Journal*, September 3, 1897; Bangor *Daily Commercial*, May 16, 1897 (first advertisement), and September 1, 1897. The next year an attempt was made to auction off some 27,500 acres in the vicinity of Macwahoc. When it came time for the auction it was during the tense early days of the Spanish war, and no bids were made. The lands were withdrawn and eventually went to the paper companies a year or so later. See *ibid.*, April 12, 1898 and May 27, 1898.

8

THE LUMBERING ECONOMY—1860-1890

As the year of the depression opened the Bangor *Daily Commercial* greeted its readers with an editorial which promised that a better world was coming. "Let our people then have good pluck. Practice all reasonable economy. Pay off all debts as fast as possible. Avoid contracting new ones, especially for extravagances and luxuries. The embargo of snow-drifts and small-pox will soon be over, and then we shall forget our hard times in a reign of prosperity." [1] It was difficult to take exception to sentiments such as these, but even with this highly moral behavior, Maine lumbermen and those dependent on the forest industries would find the next few years a very trying time. In fact within a few years the remarks of this and other newspapers were much more pungent in their description of affairs, as the following excerpt indicates: [2]

LUMBERMEN OF MAINE

You are called upon to respond to the same leading question. [*Are you satisfied with the existing business outlook of the country?*] Are you content with your business prospects?— Content to behold timberlands a drug—lumbering operations dead—sawmills silent as the grave—and logs rotting by the millions in the rivers of your state? *Answer at the polls!*

The depression was a calamity, a calamity which was felt all the more severely in a state in which one industry and its offshoots so dominated the economy. The economic aspects of the industry will highlight that domination.

The sale of standing timber to the operators was and is termed "stumpage". Operators paid the owners of the land a price set by contract for each 1000 board feet (M) of logs cut on the land owner's property. Usually the contracts were for only one winter, and in our period few other conditions were specified. In a later time the contracts would often note the amount to be cut, the size of the tree to be cut, and the type of operation to be conducted. The amount of timber cut was determined by the count, or scale, totaled each week or so by a scaler paid half by the landowner and half by the operator.

Some skillful landowners used to specify that the scale would be determined by their scaler, and then the scale would often run much larger than the actual cut. Trespass, or cutting without permit, usually involved paying double or triple stumpage especially if any sort of intent was proved. Prices for stumpage were usually quoted for spruce, pine, cedar, hemlock, juniper (hackmatack or tamerack—the North American larch) and for special operations the various hardwoods. Poplar stumpage was seldom quoted until fairly late in the period as the main source for the paper mills was farm woodlots. Stumpage prices probably deviated less from the norm than other cost factors in a lumbering operation.

Stumpage payments were usually due the next spring at the various booms. Sometimes logs and stumpage were used as collateral for the loans[3] especially when the landowner had made the original loan to set up an operation. At the time of the Civil War many paid a straight fee for stumpage on a given lot, rather than by the M, but this practice was discontinued fairly quickly.[4] During the transition period from 1860 to 1867 the state often granted stumpage in both ways. But by the fall of 1866 nearly all of the stumpage was granted by the M or by the ton. Standard prices just after the Civil War were $1.25 M for spruce, $4.00 M for pine, $2.00 M for cedar, and $2.00 a ton for juniper. Some cedar went for 50¢ M.[5] To protect itself the state demanded a deposit of $50.00 for each team which would operate on state-owned lands. These prices for pine and juniper were lower than on the St. John,[6] but by and large these state prices were the standard prices for all landowners. Occasionally on superior timberlands the prices went up to $4.00 M for juniper, for instance.[7] The state land agent estimated that these figures left the state about seven-eighths of the stumpage price. The scalage costs ran about 25¢ M on spruce.[8] The state scalers were expected to work for their money though. When the land agent instructed one scaler to go to the upper St. John valley and scale all timber on 8R18, 9R15, and 9R16, he followed these instructions with other remarks:[9]

> Attend to the lines enclosing the state lands in that vicinity, scale all ——— (down?) lumber, and be particular to see that no operations shall be made upon state lands except permitted as above. After getting the van (?) of all teams and scaling up the lumber you will take a tour to Fort Kent and post up the office about operations thereabouts *in* the state.

The lumbering economy

During the gold boom leading to the crash on Black Friday some groups of landowners made money because their contracts had called for payment in gold. The premium on the gold price came to nearly as much as the stumpage payment in at least one case. In any event the large timber holders were able to count on their lands to bring in a fairly steady and oftentimes quite large income.[10]

By 1871 and 1872 stumpage prices near the more settled areas had reached their peak. Hardwood (white maple, yellow birch, white birch, and beech) went for $3.25 M, and rock maple for $3.75. Spruce was available at $6.00, and pine went for $10.00. This was exceptionally good land with valuable timber but even hardwood used for heading was selling for $2.00 M in western Oxford County.[11] The next year when Congressman William F. Frye made a speech against the tariff he said that stumpage averaged $1.50 M for spruce and hemlock, but most Maine men would have been happy if these were the real prices. Frye, of course, was trying to prove how poor Maine men were and how the tariff on sawed timber was causing the poverty.[12] If Frye had made his speech six months later he might have been right. In the St. John that winter some teams went to the woods for spruce with stumpage at $1.00 M. The next year of course, the depression drove the prices down and fewer and fewer men went to the woods at all. By 1874 $1.50 M for spruce was heralded over the state as a good price. Logs sold for $8.00 M at the Fredericton boom, and Aroostook parties cut seventy-five million. The rest of the state was still in the throes of depression.[13]

When the state granted the ten year permits on good timberland which had been granted to the European and North American Railway the price was from 20¢ to 35¢ an acre, with the average price under 25¢. This of course was after the depression was well under way and lumbering had become more of a risky business than was ordinarily true.[14] The depression worked havoc on small operators. In some cases the logs did not sell for enough money to pay off the advances, much less the stumpage involved.[15] In 1878 the price had not risen at all. Spruce was still $1.50 M on the St. John and Aroostook, although some got $1.75 on the Mednuxnekeag.[16] For the season of 1880 things began to look better. The newspapers told of great numbers of men on their way to the woods and coincidentally of a rise in stumpage prices. In Calais spruce was sold for $3.50 a M, with pine a dollar higher.[17] When Palmer moved into the area of the Gulf in 1882 stumpage all around him went from $2.50 to

$3.50 a M depending on the severity of the conditions. He got his for 75¢ M and that was one reason the operation was successful.[18] By 1884 though this little boomlet was over. The average price on spruce stumpage was said to be $2.00 and some went for $1.00. Observers said that the reason for the low cuts was not the stumpage cost which had gone so low but rather the price of labor, but as we shall see this cost had also gone down considerably.[19] By 1890 prices had steadied, and they were low. Pine and spruce went for $2.00, cedar for $2.50 and $3.00, and even rock maple was worth only $2.00 M.[20] This was the low point although spruce continued to be quoted at $2.00.[21] Occasionally it went higher but not much. Cedar and spruce went for $2.50, hemlock for $2.00, pine for $3.00 and $4.00, and railroad ties for 4¢ each in the middle of the nineties.[22] This was on the state land, but as the land agent said, "Our price is the same as the proprietors of surrounding lands." [23] The landowner found himself in a difficult position. Investments which had paid good profits in the halcyon days before 1873 now barely cleared enough to pay the taxes. The forest industries were dwindling slowly but surely throughout the period, as the examples of the once prominent industries of tanning and shipbuilding will show.

Of the minor industries which depended on forest products for their survival the two which suffered most from the economic decline were shipbuilding and tanning. The suffering was more intense in these two industries because of other influences, but they both mirror the fate of those who depended too heavily on the forest industries for their way of life. Shipbuilding, that is, of wooden ships, was already vulnerable when the Civil War broke out. Iron ships and steam were beginning to make their way in all except the most heavy trades and of course in the coastal trade where Americans had a monopoly. Confederate raiders then dealt a blow to shipping from which it never really recovered. One way that Maine shipowners met the challenge was to build larger and larger iron sailing ships to be used primarily in the grain and lumber trades.[24] Another way was to cut down on building in most of the smaller towns where wooden vessels had been constructed. Either alternative was a blow to the forest industry. Farmers had always supplemented their incomes by digging out hackmatack roots to be used for ships knees and this source of income dried up steadily after 1860.[25] What wood was used came from elsewhere, and Maine shipbuilders and others who had been dependent on the forests were ever more susceptible to the call of the west and the lure of the cities.

The lumbering economy 215

The Maine shipbuilders did not succumb all at once. Unemployed shipbuilders in Calais formed a cooperative company to build a ship in order to make employment in 1877. They contributed their labor and launched a vessel that spring in which all were part owners.[26] This sort of activity was of relatively small importance though. Newspapers continued to talk of the difficulties in the shipbuilding trade.[27] Ship masters and ship carpenters all were hoping for an increase in freights which would stimulate the coasting business. One great difficulty was in obtaining investment with freights so uniformly low. It was not entirely the decline in freights which caused capitalists to avoid the shipping business. The demand for wooden vessels diminished, and although they continued to be constructed, the true state of the trade was told in the fact that the *Polly* and the *Hiram* already by 1880 the oldest vessels on American registry continued to sail the Gulf of Maine in search for cargoes until well into the twentieth century. As vessels grew old they moved into the stone or lime trade, or were sent to the knackers, and few new ones took their place. The following tables listing the tonnage constructed in Maine cutoms districts during the period 1875 to 1895 shows the decline in building.[28] All figures are tonnage.

Shipbuilding in Maine Customs Districts—1875-1895

District	1875	1876	1877	1879
Passamaquoddy	2343	4182	2979	82
Machias	6457	4033	4421	1573
Frenchman's Bay	1872	553	1045	283
Castine	2265	1884	860	688
Bangor	3315	711	2645	160
Belfast	8358	7311	4881	2598
Waldoboro	11608	6712	14719	5475
Wiscasset	2648	282	186	41
Portland	10674	8890	6238	7200
Bath	22246	31923	30472	7383
Kennebunk	3867	7068	5882	0
Saco	0	17	294	0
York	1686	0	0	0

District	1882	1883	1889	1890	1895
Passamaquoddy	9	884	0	500	24
Machias	0	0	3861	3380	1368
Frenchman's Bay	246	598	50	742	310
Castine	764	838	114	933	51
Bangor	1442	1887	401	1423	64
Belfast	6679	8140	4600	8178	21
Waldoboro	0	0	3482	10569	901
Wiscasset	1556	3650	1057	1383	105
Portland	5450	3669	1296	985	160
Bath	42870	40192	24586	34790	9887
Kennebunk	731	949	0	970	183
Saco	0	0	0	0	0
York	0	0	99	8	0

Tonnage of Shipbuilding Built—District of Maine, 1875-95*

Year	Total in tons	Year	Total in tons
1875	75,060	1884	46,502
1876	75,573	1885	23,053
1877	76,308	1886	15,095
1878		1887	
1879	35,416	1888	
1880		1889	39,546
1881		1890	64,361
1882	59,747	1895	13,074
1883	60,807		

*same sources as note #28

Several facts are obvious from these tables. Bath became increasingly important as a shipbuilding center. This reflected the importance of the Todd, Sewall, Hyde interests, and it also reflected a good deal of building for the United States Navy. Much of this construction, if not most, was of iron and steel hulls. The smaller ports declined rather precipitously, and many of the figures are of only the

The lumbering economy 217

one or two vessels built in that year. What is somewhat less obvious is the change in type of vessel. Schooners were the most important vessel being constructed as sailing ships at the end of the period. By the early eighties it was very rare for a brig, brigantine, or bark to be built. The last brig built in Bangor, the *Telos*, went down the ways in 1883. In 1886, of all the construction—some fifty-six vessels, thirty-five of them were schooners, and five of these were four masters of over one thousand tons. In 1889, of the eighty-four vessels, sixty-eight were schooners, four were steamers, two ships, six were brigs, and there was one barkentine and one sloop built. Of the one hundred and nineteen constructed in 1890, eighty-two were schooners, and five were steamers. Vessels were getting larger; they were more and more of a fore and aft rig, and they were being built in the yards of only one or two ports. The old days had gone by. In 1891 all eastern Maine was thrilled by the news that James Swett of Newport had found eight mast pines, the largest eighty-two feet long and the smallest seventy-four. They were all about three feet in diameter, and six men and twelve oxen hauled them to Bangor for masts.[29] In a few years such trees would go for boards or be ground into pulp. Another forest industry had been passed by.

A second Maine industry which was peripheral to lumbering but which offered much employment in the woods and elsewhere was the tanning of hides. The principal tanning agent was an essence of tannin which was easily obtainable from the bark of hemlock trees. These trees were felled and the bark stripped off, after which the essence was extracted through a boiling process. The trees had to be peeled when the sap was running, that is from early May until late July. Great gangs of men performed this work as the bark season was short and the demands of the tanneries for the essence was very large. Many small towns depended almost entirely on the tanneries for the maintenance of their economy. The movement of the leather industries away from New England would deal the death blow to the Maine tanning industry, but when that blow came the industry was already reeling from the effects of a financial failure of one of the largest firms engaged in this business.

In 1872 there were nineteen tanneries in the state, nearly all located between the Penobscot River and the New Brunswick border. These tanneries used 110,000 cords of hemlock bark a year to make their 11,000 tons of leather. The Maine legislature had protected the industry by passing a law which forbade taxation of hides brought in for processing from out of state. Some of the tanneries were quite large. Henry Poor and Sons of Boston operated two tanneries in Medway and Winn, both built in 1872. Between them they operated six hundred and fifty vats using seventy cords of bark a day. It took five hundred men to cut and peel the bark each season and it was all obtained in the vicinity. They produced 180,000 tanned hides a year for Boston firms.[30]

French Canadian and Swedish immigrants were employed to do the cutting and peeling, and even to work in the tanneries themselves. They were docile, cheap help and the industries were booming.[31] The coming of the depression did not seem to bother these mills. In 1874 and 1875 tanneries ran all winter at full time and men went to the woods in the spring to obtain the bark. In the area of Burlington over five thousand cords of hemlock bark were prepared to ship to nearby tanneries in the spring of 1876. Many who had gone for logs in earlier times now peeled bark and hauled it to the tanneries. The success of the business was a godsend to some of these towns in the seventies.[32] With logs cheap and hemlock logs cheaper still, oftentimes the only item used was the bark; the logs remained in the woods to rot. It is not until the eighties that logs were hauled out and driven to mills to be sawed. In part this was because the peeling season was late enough so that the logs had to remain in the woods another winter anyway and the wood tended to get dozy and did not make good boards.

The big firm in the state was F. Shaw and Brothers, a Boston firm which dominated the Maine tannery business. They had large tanneries at Kingman, Jackson Brook [now Brookton], Vanceboro, Forest City, and Grand Lake Stream. In 1878 the firm used 36,000 cords of bark, employed four hundred men in the tanneries, and used one thousand men for cutting, peeling, and hauling in the woods.[33] The three brothers in the firm, William, Fayette, and Thackster Shaw, had learned the business from their father who had a small tannery in Cummington, Massachusetts. In the summer of 1870 the three of them started their operations in Maine after a trip to Hinckleytown or, as it is now known, Grand Lake Stream Plantation. This town was admirably located as a new road had been swamped out to it; there was a chance for a railroad connection with both Bangor and Calais, and lake transportation from the hemlock forests which were abundant in the area was easily available.

Shaw Brothers bought the township that fall for $35,000 and during the winter workers from Montreal came to begin work. They dug a canal parallel to the stream for the tannery. Houses, camps,

The lumbering economy

and a sawmill were constructed with the machinery brought in from Princeton by a steamer. By 1876 six hundred vats were in operation. Locks were constructed between the two upper lakes, Sysladobsis and Pocumpass, to regulate the flow of water and to provide access to the mill. Bark once peeled was loaded onto scows which were towed by a streamer to the canal where they were brought to the tannery itself. Roads were built into the woods to facilitate the work of the men, and some of these roads were twenty-five miles long. Over four hundred teams of horses were employed in hauling the bark out to the lakes for transport to the tannery. Logs from the operation were driven and towed to the lumber mills at Milltown, New Brunswick. This was a big business—1200 hides a day at the height of the operation—but of course it was a peripheral operation always because of the distance and effort involved in transporting the tanned hides to shoe factories in Massachusetts. The company built and maintained a hotel for the single employees who came from the Maritimes and also from the growing immigration from southern Europe. Eventually houses were built for the married men employed by the company. The Grand Lake Stream tannery was just one of many. Shaw Brothers owned thirty-nine tanneries and extract works when they failed in July, 1883.[34]

Shaw Brothers often set up extract operations in the woods. One was located in Topsfield and another in Sherman. Bark was peeled and the extract was obtained in the woods, then transported to the tanneries. The extract mill in Sherman was typical of these establishments. Five five-ton steam boilers provided the heat and power for the huge extract pans, twelve feet across, in which the bark was boiled. These pans cost close to $7,000 each and the machinery in this mill weighed nearly one hundred tons.[35]

The investment in the immense Shaw operations was too much and the firm failed in the summer of 1883. Apparently the brothers raised some money to meet their obligations by skirting the truth in their statements to their banks. The debt of the firm was over $7,500,000 and their assets were just over $5,000,000. Attachments on the property were filed all over northern Maine and New Brunswick. Creditors formed a nine man committee to investigate their finances, most of them from banking and financial interests. Lumbermen were represented by E. H. Todd of St. Stephen, New Brunswick.[36] That fall and winter the cut in the north was much less and the Shaw bankruptcy was blamed for this in good part. It was a staggering blow to the economy of the northeast and one from which there was little chance of recovery.[37] The business limped on. Trustees ran the tanneries until the nineties, selling off land to pay off part of the debts.[38] Eventually the firm paid about 40¢ on the dollar. In the late nineties the remaining property was sold to the International Leather Company which dismantled the mills and shipped the usable machinery to Boston.[39] Eventually the tannery and the waterpower came into the hands of the St. Croix Paper Company and were used for log-driving purposes. The freight boats and scows were used by their subsidiary the Eastern Pulpwood Company to move pulpwood until into the 1920's. Today there is little evidence of what was once a thriving business, and in other areas where the Shaw Brothers operated they are but the dimmest of memories.[40]

Other small tanneries outside the Shaw combine continued to operate until into the nineties. One located at New Limerick turned out six hundred tons of leather a year and the lumbermen of Haynesville welcomed the chance to peel the three or four thousand cords of bark that were needed. The logs were floated to Bangor where they were sawed into boards.[41] Six thousand five hundred cords of bark a year were peeled in Amherst and Mariaville for use by the Amherst tannery, and the economy of these towns was floated by the business.[42] "This keeps a good amount of money in circulation in those towns and employs many of the inhabitants who would otherwise be obliged to leave for other parts."

By the end of the eighties the Grand Lake Stream tannery, now run by the trustees, was getting its bark from the Machias River area, Whitneyville and Northfield producing four thousand cords in 1889.[43] This was the end of the business though. The failure of the Shaw Brothers was simply a symptom of the things to come. The manufacture of artificial extract, the movement of the industry from Massachusetts, the destruction of the hemlock forests in Washington County, and the heavy cost of transportation of the raw hides to the tannery and of the tanned hides to the shoe factories militated against the continued success of the business. Another forest industry had succumbed to the economic pressures of the period 1860 to 1890.

Not all the Maine economy was as badly off as shipbuilding and tanneries. During the time under discussion railroads began slowly to penetrate the north woods. Part of this push was created by the battle for traffic between Portland and St. John, but some of the impetus came from those who saw the northern part of Maine as a great "Garden of the North". These roads—the European and North American designed to connect New Brunswick with Maine ports, and the

The lumbering economy 221

Bangor and Aroostook constructed to open up the north—are part of the economic and lumbering history of this period. The roads came slowly however and it was not until 1913 that the network was completed with the swinging of a bridge across the St. John River at Van Buren. To some extent both roads fulfilled their promise although neither reached the heights which their promoters hoped and even expected. Northern and northeastern Maine remained a hinterland even though it did become something other than the barren howling wilderness of the 1840 to 1875 period.

The battle for Canadian traffic was between St. John and Portland. Probably Portland was a more reasonable economic choice because the old Portland and Ogdensburg Railroad was already built, but by the early eighties the demands of Canadian politics meant that St. John must win the battle. In 1885 the state legislature chartered the International Railroad of Maine, a firm identical to a Canadian company of the same name, and after procurement of the charter of the older and abortive Penobscot and Lake Megantic Railroad this firm began to make surveys from Lake Megantic in Quebec to Mattawamkeag in Maine for the purpose of building a road. The Canadian government provided considerable monetary aid to this venture. In 1886 the firm broke up and sold all its assets to a road called the Atlantic and Northwestern Railroad which immediately leased its holdings to the Canadian Pacific Railroad, then concerned with opening up western Canada, and work on the road commenced in earnest in 1888. The road was open to Mattawamkeag in 1889. The Canadian Pacific leased the New Brunswick railroad, the heirs to the St. John and Maine, in 1890 and the Canadian Pacific and the Maine Central then made a contract for joint use of the tracks running from Mattawamkeag to Vanceboro.[44] The road to the west was open and St. John was the victor. It was an empty victory however as it came too late; other ports had already claimed the major proportion of the commerce and eventually the Great Lakes route via the St. Lawrence River would provide an easier and cheaper route at least for most of the year.

The time of construction of the road across Maine proved to be an exciting one for Maine people who envisaged benefits for the state. Most of the construction workers were Italians brought in from Boston though and few local people worked on the road. Camps were constructed along the right of way from Sherbrooke, Quebec all the way to Mattawamkeag and were staffed not only with Italians but other immigrants from Europe. In late 1887 men were in the camps from Italy, Finland, Norway, Poland, and Austria. Four thousand laborers worked for over a year to complete the one hundred and twenty-eight miles of track. Brownville experienced a boom as it was the central headquarters for supplies and the road hospital was located there.[45] The Canadian Pacific still runs across the state, but one suspects that it has never been profitable except in the sense that it helped to bind the unwilling Maritimes to the rest of Canada. The completion of the road meant the effective end of secession ideas by the Maritimes and stilled the New Englanders who had hoped for so long to purchase the Maritimes from either Britain or Canada itself.

The other hope for railroad promoters and those who saw a bountiful future for northern Maine was a railroad running north from Bangor to the Maine border. Two attempts to build such a road had been unsuccessful before the Civil War. There the situation lay until the construction of the Bangor and Piscataquis railroad in the seventies. The relative success of this road caused the promoters of Aroostook's wealth to begin again to call for a northern connection with Bangor. Reports began to fill Maine newspapers of the great amounts of freight being transported over the new road.[46] The railroad did cut freight costs. To carry lumber from Shirley to Dexter, a distance of thirty-five miles, had cost $5.00 M for freight. The same lumber carried by way of Bangor was $2.50 M, and when the extension was built in 1890 the cost was of course much less.[47] Such figures indicated the worth of a railroad not only to the promoters but also to the inhabitants of the area which the road would service.

There was no question of the success of the road. At the height of the promotional activity for the Bangor and Aroostook the following table showing earnings of the Bangor and Piscataquis from 1877 to 1890 appeared in Maine newspapers. The first year is thirteen months, that is, from December 1, 1876 to January 1, 1878.[48]

Bangor and Piscataquis Earnings—1877-1890

Year	Gross Earnings	Net Earnings
1877	$ 71,383	$ 31,075
1878	72,703	31,398
1879	74,254	27,399
1880	89,955	36,651
1881	104,524	43,712
1882	123,664	60,501
1883	122,318	67,629

The lumbering economy 223

	Extension to Moosehead Lake completed this year	
1884	130,892	66,535
1885	139,272	52,900
1886	147,368	69,785
1887	197,146	98,799
1888	228,958	74,339
1889	189,931	78,265
1890	183,896	65,428

This table was not the only shot in the campaign for a new road to the north. There was a constant reiteration of the amount of freight which was being shipped from northern Maine towns across the border and then over the New Brunswick railroad back into the states. In 1882 for instance, Houlton shipped over two thousand cars of freight, nearly all forest products.[49]

However, the northern towns complained that there were never enough cars; the produce had to wait too long and the freight had to be handled too many times. A railroad would solve these problems and make northern Maine the great area that it was supposed to be.[50] The campaign was effective. In 1892 the Bangor and Aroostook Railroad was chartered, and after getting underway by leasing the Bangor and Piscataquis and the Katahdin Iron Works lines, it began building to connect the north with Bangor. The road was complete to Houlton by December 20, 1893 and regular service was in effect by the first of the year. The railroad then made its way north through a process of gradual extension arriving at Fort Fairfield November 19, 1894.[51] Here a controversy developed as to how the road should proceed from this point. County aid was asked for an extension to Ashland, and the voters agreed to help by a vote of 2230 to 1724.[52] Fort Fairfield and Caribou voted against the Ashland extension and much bad feeling resulted. That spring when the Dunn Brothers drove their logs down the Aroostook with two hundred men, they paid them off in Presque Isle rather than in Fort Fairfield as a way of showing Ashland's feelings for those who did not wish them to share the coming prosperity.[53]

The railroad became the pride of the state, and the subject of editorials. The future looked bright to all for northern Maine.[54] The Ashland branch was in operation by January 6, 1896, but the area near the river valley itself was not to be tapped until into the twentieth century.[55]

Much pulpwood and sawed lumber began to come to Bangor as a result of the new road. The most important subsidiary development though was the construction of a giant band sawmill with a capacity of 30,000,000 board feet a year on the Little Machias River just outside Ashland. The mill, financed in part by money invested by the promoters of the Bangor and Aroostook Railroad, opened some practically virgin spruce forest to the lumbermen and resulted in a minor boom in the production of English deal. For a year or two ships left Bangor for British ports with great cargoes of deal, something which had not happened since the 1870 period. As a result 1896 and 1897 were fairly big years in Bangor's commerce. It did not last too long. The price went down in Liverpool by August 1897 as a consequence of the American exports, and 1898 was the last year of any major exports to this market. The Ashland mill then settled down to produce lumber for the domestic trade.[56]

In addition to this brief stimulation of Maine's foreign lumber trade railroads continued to engross an increasingly larger portion of the internal trade as well. From October of 1888 to October of 1889 the Maine Central hauled 20,000 railroad cars of lumber along its line.[57] In some months the outgoing traffic was very large. The following table shows shipments out of the state via the Maine Central in September and October 1891.[58]

Railroad Shipments Outside Maine

Maine Central Railroad, September & October, 1891

	September (in railroad cars)	October (in railroad cars)
Lumber	2,491	2,826
Bark	78	51
Hay	387	605
Woodpulp	638	747

There was no question that a large percentage of Maine forest products were leaving the state by railroad and that the railroads were profiting by the opening up of the north country during this time. The Maine Central was in a virtual monopoly position. When in 1892 the Maine Central cut the amount of boards loaded in each car, at the same time that western roads were increasing the number of boards per car, the competition from western timber states drove the hemlock log prices for a short time to $5.00 M.[59] Lumbermen were unhappy but they were the ones who had asked for the roads so their protests were not very strong. Lumber exports continued to increase

The lumbering economy 225

over the road. Of a total freight tonnage of 2,170,538 in 1894, 584,-297 tons were lumber, and in 1895 the figures were 2,476,337 and 666,441. Of this last, 103,586 tons had had their origin in other carriers mainly the Bangor and Aroostook.[60] In the new century of course most of these exports were woodpulp, pulpwood, and paper.

Even though lumbering was made profitable in some areas by the extension of the railroad, for the great bulk of lumbermen the business became diminishingly successful after 1872. As lumbering profits dropped throughout the next twenty-five years wages were cut by lumbermen in an effort to sustain a profit level. Larger operators were able to survive and so the period was one of consolidation.

Prices[61] fell steadily after the high of 1871 and 1872 until by 1890 nearly a fifty per cent fall had taken place[62] on the items used by lumbermen. This fall was probably the main reason that lumbering remained the biggest industry in the Maine-New Brunswick area.

The price of logs at the Bangor boom indicates something of the impact of the depression. The price in 1870 was about $10.00 per M for spruce sawlogs. By the spring of 1874 that price had fallen to just over $7.00. The rest of the period to 1895 was one of very gradually rising prices to a high of just over $12.00, but it was 1882 before the price went much above $10.00. By that time many of the smaller operators had been frozen out.

Price movements in random spruce boards at the Boston market indicate what happened to those who survived. Just prior to the depression the price was $14.00 M but by 1878 it had fallen to $11.00. When the price of logs hit a low point, just at the time that wages reached their low in 1878-1880, the price of boards responded by rising slightly. This rise did not last long; the $11.00 low was again touched, and from 1882 to 1892 the price fluctuated from $12.00 to $14.00 per M.

Wages paid to men in the woods follow a similar pattern, but they never again reached the high of the 1872 period. The same thing is true generally for the wages paid to river drivers although here the wages were always higher than for nearly any other work, a reflection of the difficulty involved. It seems certain from the study of these prices that as far as the cost of lumbering was concerned fixed costs dwindled somewhat but not very much, as stumpage prices remained about the same and wages did not actually fall as far. When one considers the added cost of penetration into the deeper woods, and the somewhat higher costs because of diet changes, it is probable that lumbering cost about as much in 1892 as it did in 1872 if not a bit more. It is worth noting that driving costs rose steadily throughout the period, and shipping charges although diminished still formed a major burden in the industry.

This study of prices and wages points out what most lumbermen thought during this period. Lumbering was a lottery, a gamble for most operators. Smaller men were forced to the wall, and those who survived the earlier days of the depression usually were only too happy to sell out to the pulp and paper companies when that opportunity arose. In fact, one wonders how anyone could maintain economic equilibrium in the Maine woods under the impact of changing times and diminishing profits.[63] The profit margin must have been very slim most of the time.

It is little wonder that the reports from the period echo such sentiments as "Business here is very dull and money scarce" [64] and, "It is a season of depression in the lumber trade".[65] It is only before the depression that cheery words come from the woods.[66]

> There is a good lot of snow in the woods; the loggers have just commenced drawing timber to the river. We are feasting on camp fare, baked beans every morning, bread that would make a king smile, light as a feather, with plenty of molasses. We shall be in good working order by spring.

Not all were as sanguine even then:[67]

> Just snow enough in the woods for getting out wood, sleepers, and small timber, but hardly enough for heavy logging. The principle [sic] lumbering business with us here, is getting out R.R. ties. At the present price (oak $6 a cord) and such good weather, that pays fair; but it is most ruinous business in the long run for the owners of timber, it is so destructive of young growth; at the present rate of destruction, our forests will be denuded of spruce and hemlock, so that good timber will be like gold and silver now, very scarce.

As long as the money was available things were good, but by 1873 the money was not available and times were bad.[68] They grew worse and by the time they were better a brand new world had come to the Maine woods.

NOTES

[1] Bangor *Daily Commercial*, January 4, 1873.

[2] *Ibid.*, September 7, 1876, election day editorial which was addressed not only to lumbermen, but to all economic groups.

[3] Rufus Dwinel to Hinckley and Egery, August 26, 1857. Dwinel was requesting the renewal of a loan, and offered as security the Dwinel House mills, ¼ 14R12, and the W½6R12. On September 15 he wrote again offering to add

The lumbering economy

his three gang saws, and other machinery to the collateral. These letters are in the Bangor Historical Society.

[4] See *Deed Book*, State Forest Commissioner, Vol. II, 91-95, 97-99, 124. These are all for stumpage granted in 1861-2: p. 124 for 1867. Stumpage ranged from $50 to $150 for state lots of 1000 acres on various townships.

[5] Isaac Clarke to Darius Getchell, October 10, 1866; to Joel Valley, October 24, 1866; to Dwinel Brothers, Caribou, November 8, 1866; to Captain D. Randall, December 11, 1866; to John A. River, December 12, 1866; to Sherman W. Young, November 30, 1867.

[6] Clarke to E. F. Bradford, Patten, October 4, 1866.

[7] Clarke to John A. River, December 12, 1866. River had apparently complained of the stumpage on pine and juniper. The state was not always that difficult. "The state won't be particular if you obtain stumpage of superior quality." Clarke to Dwinel Brothers, November 8, 1866.

[8] Clarke to Sherman W. Young, November 30, 1867.

[9] Clarke to R. H. Stuart, December 17, 1867. Errors were made by the state as well. W. H. McCrillis wrote a bitter letter to the Land Agent complaining that he had paid stumpage twice on the same reserved lots in T15R9. W. H. McCrillis to Land Agent, June 19, 1878, attached to cancelled deed, *Deed Book*, II, 104.

[10] The firm was Hinckley and Egery. Total stumpage receipts October 24, 1868 to March 22, 1869 was $8,112.76. They made $5,003.11 premium in gold check drawn on the Bank of New Brunswick. See Daily Journal, December 16, 1862—June 2, 1868 (the book does not have 1867); Monthly Cash Book December 1862-April 4, 1869; both in the Bangor Public Library; and a sheaf of telegrams from Boston 1868 and 1869 from Daniel Hinckley with the gold prices and advice on dispatching their checks. There are in the Bangor Historical Society.

[11] Oxford *Democrat*, January 20, 1871; article citing contracts for the winter made in Stoneham, Lovell, Waterford, and Norway.

[12] *Congressional Globe*, 45, pt. 4, (42nd Congress, 2nd Session), May 15, 1872, general discussion of lumber industry in the United States, 3449-56, with Frye's remarks at 3451. The discussion took place in a debate over the tariff. The best Maine newspaper account is in Oxford *Democrat*, June 11, 1872.

[13] *Ibid.*, December 24, 1872, for the St. John prices in the winter 1872-3; the 1874 prices and cut are described in Bangor *Daily Commercial*, July 1, 1874 and Portland *Eastern Argus*, July 10, 1874.

[14] Land Agent's *Report*, 1873, report of sales and discussion of the auction where they were sold *supra*, and *Commercial*, September 23, 1874. One township, 4R18, was not sold until October 1875; it went for 17¢ an acre. *Ibid.*, October 28, 1875; Land Agent's *Report*, 1875.

[15] Case decided in Penobscot County Court, January 21, 1876. D. W. Garland *vs* John A. Rames. Garland furnished $3,541.62 for supplies for the operation in 1873. The logs returned $1,159.40. The court awarded $2,006.44 to the plaintiff, so he still lost money. See *Commercial*, January 22, 1876.

[16] *Ibid.*, February 29, 1878.

[17] Portland *Eastern Argus*, December 3, 1879 quoting Calais *Advertiser*.

[18] Bangor *Daily Cammercial*, April 25, 1882.

[19] *Ibid.*, January 23, 1884 editorial; and January 28, 1884, letter from

"R.G.R." agreeing. Some was higher, $4.00 M in West Phillips in 1888, Phillips *Phonograph*, November 9, 1888.

[20] Contracts: State to E. & S. J. Moores, (T6R4), December 12, 1889; to Llewellyn Powers (Dyer Brook), October 6, 1891; to Bangor Edge Tool Company, (S½ Westfield) November 7, 1891 (located in Box 211 Forest Commissioner Files).

[21] *The Industrial Journal*, April 27, 1894, story on the industry.

[22] Contracts: State to J. J. McCormick (Letter AR4), June 19, 1895; to W. Thayer (T1R3—Reed Plt.); and to Allen E. Hammond, (Letter IR2—Cyr Plt.) all the same date. These are all in Box 211, Forest Commissioner's Files. For similar prices see Bangor *Daily Commercial*, September 22, 1894.

[23] State Land Agent to E. D. & J. O. Tuel, October 31, 1895 commenting on prices asked on T14 E. D.—This is also in Box 211.

[24] Hutchins, *op. cit.*, is the best account.

[25] Ships knees provided some income all through the period. Two men in Newport dug out 380 in 1888 for instance. The account said how few were now used. Phillips *Phonograph*, December 28, 1888.

[26] Bangor *Daily Commercial*, January 18, 1877.

[27] *Ibid.*, September 28, 1877 quoting the Machias *Union*.

[28] *Ibid.*, January 18, September 28, 1877; January 5, 8, 1878; January 8, 1880; November 8, 1883; December 30, 1885; January 11, 1887; November 6, December 31, 1889; November 25, 1890; January 13, 1896. Also see *The Industrial Journal*, February 7, 1891 for an issue devoted to the shipbuilding industry, as well as Hutchins, Colcord, and Rowe, all *op. cit.*, and *passim*. A series of M.A. Theses at the University of Maine in recent years has treated shipbuilding in such towns as Blue Hill, and Damariscotta.

[29] Bangor *Daily Commercial*, January 31, 1891.

[30] *Ibid.*, January 29, 1872 citing the Dexter *Gazette*; February 14, 1872, "The Tanneries of Maine"; March 28, 1872 on the Poor operations. Between 75 and 100 men were employed full-time in these two establishments.

[31] *Ibid.*, July 2, 1872.

[32] *Ibid.*, April 14, 1875 unsigned letter from Medway; June 10, 1876, the three tanneries concerned were in Grand Lake Stream, 1000 cords, Winn 500 cords, and Lowell, about 4500 cords. Men normally employed as lumbermen on the Passadumkeag did this work, such as Levi Moor, and J. W. Porter and Sons.

[33] *Ibid.*, April 7, 1878 on the Shaw operations as well as April 9, 1879; for other firms April 26, 1878. The steamer *John A. Peters* was hauling bark up the Penobscot to the Winn mill from Passadumkeag. There is an excellent discussion of this tannery run by the Poors in *ibid.*, May 25, 1878, a letter signed Katahdin.

[34] Description of these works comes from Minnie Lee Atkinson, *Hinckley Township or Grand Lake Stream Plantation*, Newburyport, Massachusetts, 1921 (privately printed), 31-43, 48, 55. Another good description of tanneries is in Miller, "Report on New Brunswick Lumbering", 100-1. The railroad carried some bark to the extract works, 5000 cords in the winter of 1880-1881, for instance, Katahdin *Kalendar*, February 26, 1881.

[35] Bangor *Daily Commercial*, February 23, 1880 for the Sherman plant; interview Harold Noble, Topsfield, Maine, July, 1962 for Topsfield.

The lumbering economy

[36] *Ibid.*, July 31, 1883; August 1-15, 1883; August 16, 1883 for a complete statement of the firm at the time of collapse; Atkinson, *op. cit.*, 55-57.

[37] Portland *Eastern Argus*, December 8, 1883, ". . . hard times are coming and business in northern Maine is slow and dull."

[38] Bangor *Daily Commercial*, July 7, 1887 on sale of Kingman tannery and land to Casco Bank and Trust Company of Portland for $50,000; $20,000 for tannery.

[39] Atkinson, *op. cit.*, 57-8.

[40] *Ibid.*, 77-78, and personal observation.

[41] Bangor *Daily Commercial*, February 9, 1884; July 3, 1888. A tannery operated in Chesterville and took bark from Franklin County parties at $6.00 cord. Wilton *Record*, December 16, 1886.

[42] Bangor *Daily Commercial*, August 3, 1888.

[43] *Ibid.*, July 12, 1889. The logs came from T29 and 41, M.D. as well.

[44] This story is best found in Edward Clark, *Maine Railroads*, Portland, 1920, 94-5 and in Harold A. Innis, A *History of the Canadian Pacific Railway*, London, 1923, as well as in J. M. Gibbon, *Steel of Empire*, New York, 1934.

[45] For the construction see Bangor *Daily Commercial*, Oct. 1, 1887 on the crews and their recruitment and working conditions, and Oct. 19, 1887 for a detailed discussion of the route and the difficulties of building the road through the forest wilderness. The road always remained somewhat isolated from the rest of Maine. In this century school teachers hired by the State to teach the children of section hands had to go in and out by train which they flagged down after traveling by buckboard from Greenville.

[46] Bangor *Daily Commercial*, January 29, 1879.

[47] *Ibid.*, July 16, 1888.

[48] *Ibid.*, March 18, 1891. These figures are interesting because the road had some competition. After the Canadian Pacific was built it handled a portion of lumbermen's supplies by its direct route from Mattawamkeag. The Maine Central also tapped the area by an extension to Foxcroft, and the Katahdin Iron Works Railroad, by this time leased to the Bangor and Piscataquis, was a major monetary drain.

[49] *Ibid.*, January 27, 1883.

[50] *Ibid.*, March 1, 12, November 4, 1878; January 27, December 12, 1879; April 15, November 6, 1880; April 15, 1885; June 8, 1888; July 26, 1890.

[51] Chase, *Maine Railroads*, 95 especially, and for the whole story, 95-138; Bangor *Daily Commercial*, December 20, 1893; November 20, 1894.

[52] *Ibid.*, April 24, 1895; a town by town breakdown of the vote is in April 26, 1895.

[53] *Ibid.*, May 3, 1895. The politicking in this vote was intense, and feelings ran very high.

[54] *Ibid.*, April 2, and August 11, 1895.

[55] *Ibid.*, January 2, 1896. The first hearing on the extension of the B. and A. RR to Caribou and Van Buren was held May 17, 1899. Three hundred attended, including many from New Sweden who wanted the road to go through that town. The older route, surveyed originally in 1892, was more direct, going through Cyr Plantation, and TKR2, or a distance of about twenty-five miles. The new road was to go through Woodland, New Sweden, Stockholm, T16R3, T17R3, and then to Van Buren, or a distance of thirty-one miles. The two most

telling arguments for the change were the facts that the Swedes had been promised a railroad when they came to settle, and that the county still had $115,000 to aid in the building. At $4,000 a mile, which had been their subsidy before, this figured out to twenty-nine miles. The railroad commissioners decided in favor of the new route, and it was started in 1900-1901. See Bangor *Daily Commercial*, May 19, 24 (the stenographic account of the public hearing is printed here), 25, 1899. A brief history of the firm was published in 1966 and an account of the building appears in *Maine Line*, Vol. 14, No. 2, "Genesis of A Railroad."

[56] Bangor *Daily Commercial*, February 2, 1895 on pulpwood shipments to Bangor; April 1, May 20, 1896, May 8, June 26, 1897 on the Ashland mill. They produced ten million feet every seventy days. The edgings and so on were carried to the pulp mills on the Kennebec. See *ibid.*, August 2, 1897, August 3, 10, 29, September 7, 28, December 17, 1896; May 11, 14, June 15, August 18, 1897 (quoting *Farnsworth's and Jardine's Circular* and *Timber Trades Journal*, August 7, 1897 on the Liverpool auction); March 30, December 10, 1898 on the deal trade. The trade consisted of the following amounts in these years:

    1896— 2,500,000 feet
    1897—10,997,731 feet
    1898—19,300,290 feet

In the last year nearly all of this was sawed in F. W. Ayer's mill in South Brewer rather than at Ashland. Twenty-two foreign vessels cleared Bangor port with deal, spool wood, shooks, and pulp this year though, which made it one of the better years of the century. On the maritime trade generally, see *Commercial*, December 10, 1898 which is a good review of the year's business in Bangor harbor.

[57] Bangor *Daily Commercial*, October 14, 1889.

[58] *Ibid.*, November 30, 1891, also see August 5, 1898.

[59] *Ibid.*, July 26, 1892.

[60] *Annual Report*, Maine Central Railroad, 1895; Bangor *Daily Commercial*, Oct. 22, 1895. The story of the Maine railroads also appears in part in the *Annual Reports*, Maine Railroad Commission, annually in *Public Documents*, State of Maine, Augusta.

[61] Statements on prices are based on a careful reading of the "Wholesale Prices Current" published each week in Portland *Eastern Argus*, and sometimes more often, from 1865 to 1892. Prices for the following items were studied: beans, codfish, flour, oats, hay, lumber (pine, spruce and hemlock), cordwood, molasses, mess beef, mess pork, eggs, potatoes, rice, salt, tea, sugar, and lard. Other wholesale markets such as Bangor, Bath, and Rockland were used when the evidence was available. Much work needs to be done on prices throughout the nineteenth century. Such work should shed light on population movements, as well as industrial and economic changes.

[62] Baled hay went from $31 per ton in 1872 (it ranged from $14 to $19 in the years immediately following the Civil War) to a low of $11.50 per ton in 1879, 1880, 1884, and 1890. Oats went from a high of 88¢ per bushel in 1868 to lows of 34¢ in 1879 and 35¢ in 1890. The intermediate high was 68¢ in the spring of 1882. Western flours ranged from $6.80 per barrel to $3.80 for super grades. Beans went from $6.00 per bushel at their high to various lows near $1.60. The other ubiquitous logging food, pork, which was

The lumbering economy 231

as high as $26.75 per barrel during the period, went as low as $11.50, and finished the period at between $13 and $14. Comparable figures could be adduced for everything from salt to grindstones.

[63] It is perhaps not entirely clear that if a person could find steady work or employment he might survive quite well during the time of price decline. The discussion above concerns not the employee so much as the operator who hired his men, bought his stumpage, and hoped for an economic miracle. Of course, to survive one needed work, and we know little about unemployment during these periods, except to surmise that it was high in the peripheral industries. How the price decline did affect ordinary workmen is shown by this table which depicts the cost of living for a family in Lewiston. A husband, wife, an eighteen year old son, and a six year old daughter make up the family. Income was $9 weekly from the father, and $6 from the son. (From *Industrial and Labor Statistics, Maine*, 1887, 89.)

Cost of Living—Lewiston, 1877-1887

Item	July, 1877	July, 1882	July, 1887
Fresh meat	$ 3.34	$ 3.34	$ 2.90
Fruits & vegetables	3.84	3.84	3.84
Butter & eggs	.96	.96	.96
Sugar	3.57	3.14	2.00
Tea	.60	.60	.60
Lard	.84	.90	.60
Cheese	.47	.47	.47
Beans	.40	.32	.32
Rice	.18	.18	.18
Molasses	.30	.25	.25
Oil	.40	.30	.30
Soap	.24	.21	.15
Flour	4.00	3.25	2.00
Sundries	1.20	1.10	1.00
Totals	$20.34	$18.86	$15.57

(for one month)

[64] Bangor *Daily Commercial*, January 6, 1879.

[65] *Ibid.*, November 13, 1883, and for the quotation *The Industrial Journal*, October 31, 1884.

[66] Oxford *Democrat*, December 18, 1869, "A letter from a logger."

[67] *Ibid.*, January 28, 1870, letter to the editor, unsigned, from Mason, Maine.

[68] For an appreciation of the value of capital to lumbering success, see Hannay, *op. cit.*, 433, appendix "Lumbering in New Brunswick".

9

WOOD PULP PAPER COMES TO THE NORTHEAST, 1865-1900

Although the period 1860 to 1900 was a time of tremendous development in the United States in manufacturing, it seems safe to say that few industries grew with quite the rapidity of the wood pulp and paper industry. Initially at least much of this growth took place in northern New England. The wood pulp paper revolution created some new problems for the industry along with the profits for the first comers. The first of these involved wood supply and this accounted for the move to the northeast. The second problem involved the amounts of money necessary in the newly reconstituted industry and this eventually led to consolidation throughout the industry in the years before World War One.

The great cost of rags and the manufacture of rag paper coupled with the increased demand for paper triggered a massive price rise, and with it a world wide search for a substitute for rags. This search led to the development of a successful process for the manufacture of paper from wood by a variety of processes. Those areas of the United States covered with trees benefited very strongly from the new discoveries which took the world by storm in the decade of the 1870's. Nearly every town fancied itself a great manufacturing metropolis with fortunes to be made for those prescient enough to invest in the new wonder material. Maine lumbering, caught in the economic squeeze, seized upon pulpwood as salvation which in fact it was for the state's economy. Individuals were saved if they were fortuitous in locating or in selling out at the proper time. The Maine woods continued to be exploited albeit in a different fashion.[1]

In the early days of the pulp paper revolution poplar wood was used exclusively, but the great success of the new methods soon created a demand which far outran the local supply of poplar wood. By 1871 the wood used in the western Massachusetts mills was becoming expensive, and the process spread to other areas where poplar was prevalent. Chester County, Pennsylvania experienced a boom but soon the mills found themselves searching as far as Maine for their supply.[2] Since wood pulp paper was much cheaper the new processes could have revolutionized the paper business even more quickly, but the location of the first mills so far from the major supply of wood actually slowed growth. The American Wood Paper Company did locate a mill in Providence which was supplied with wood from Bangor and the Penobscot River but this too was uneconomical. It is curious, but apparently the firm believed that to locate further north was of no great use. As their agent in Maine said when reviewing the early years of that company,[3]

> During the two winters I traveled over the timberlands of Maine, New Brunswick, and Nova Scotia in the interests of this company, (nominally—in fact, in my own interest, and at the company's expense); to look up the best chances to get wood to ship to Manayunk, Pennsylvania and Roger's Ford up the Schuykill, where their large mills were located. The idea of travelling all over Uncle Sam's and Queen Vic's territory to find wood had a ludicrous side to it and the next time I had a chance to talk with Buffum (the Treasurer) I said to him, 'Why in the name of common sense don't you locate your mills in the State of Maine, either on the Kennebec or Penobscot rivers, where you can get your wood at a merely nominal price, instead of going all over the country collecting wood to be sent to Pennsylvania?' His reply was characteristic of the general opinion held concerning the State of Maine by New Yorkers and Providence men, 'Oh, you are frozen up nine months in the year down in Maine,' he replied. I could not help smiling, 'Nevertheless,' I replied, 'if there is ever any money to be made in the business the mills must be located near where the wood grows.'

This comment expressed the matter succinctly and laid the basis for the rest of the history of the industry. The processes had been developed; now it simply remained to locate the industry in the most advantageous spots. Maine and New Hampshire were where "the wood grows" and they would be where the mills would grow at least at first. The proximity of the forest would determine the location, and the paper mills already clustered in the Northeast would move farther north and east in their search for the new sources of supply.

The first notice of a wood pulp mill in Maine was in 1864 but nothing came of the venture. The first mills operated in Maine were at Norway and at Topsham. The Norway mill will be treated briefly later. The Topsham mill began in the basement of a sawmill run by Charles D. Brown and E. B. Denison, utilizing grinders from the machine shop of the Bath Iron Works. The mill produced one ton of pulp a day with Denison running the grinder and Brown the wet machine. Denison kept the books and Brown was the sales agent. The

Pulp and paper come to the northeast

poplar which they used was sawed into one-foot lengths, the bark was shaved off, the wood was then split and knots and other blemishes cut out. The wood was then pressed against a revolving millstone by a large iron weight. Water played constantly on the stone carrying the pulp away. It proceeded through a series of sieves and rollers coming out at the end of the room in sheets of thick, rough drawing paper. The mill was quite successful, enough so that Brown and Denison branched out with other mills. Over-expansion, trouble with the Voelter patent holders and a relatively poor location caused the firm finally to go out of business in the mid-seventies.[4]

Many other small mills were started in the Northeast during the early period but few of them were successful. Lack of capital, overambitious promotion, poor location, a shortage of technically able workers, and the ordinary business dangers of fire and depression proved their undoing.[5] A number of Maine firms did begin to operate on the periphery of the industry however manufacturing paper machines, felts, and other necessaries. One, the Knox Woolen Mills, is still in business. Later with the introduction of the sulphite process the Maine industry really began to boom. The new process came to Maine when the Penobscot Chemical Fibre Company built a mill at Old Town. When it began operations March 15, 1883 this large mill employed 90 men. By the summer of 1884 it was producing 18 tons of pulp a day.[6]

Even this mill used mostly poplar. The spruce forests of northern New England were drawing the pulp mills to them, yet spruce was not the principal pulpwood species until late in the 1880's and did not replace poplar generally until around World War I. Drives of poplar were still coming to Maine mills quite late. The mills took whatever came to hand and, as long as their chief source of supply was farmers, poplar predominated. When more formal methods of procurement were adopted spruce came rather quickly to the fore.[7]

Most of these early mills were not successful but the industry did come to the Northeast. For a discussion of that world it is necessary to look more closely at two firms involved in the growth period.

The first of the Maine companies to grow beyond small beginnings was the complex of firms controlled by Adna C. Denison. Denison saw the future in wood pulp paper and deserves much credit as a pioneer, but his vision was more elastic than his capital. He started out with Asa Danforth operating a small mill in Norway in 1848. By 1851 Denison had expanded to nearby Mechanic Falls and was the sole owner of the business. During the Civil War he experimented with making paper out of rye and oat straw and influenced by the high prices at the time bought a second mill at Mechanic Falls. Eventually he owned or operated mills in Poland, Mechanic Falls, Canton, Norway, and Brunswick, Maine.[8]

1866 was the first year of expansion for the Denison mills at Mechanic Falls. In that year the company built a new mill and erected a stone dam across the river. The three mills together produced five tons of paper a day, and the new mill was described as the largest in the state. A fourth mill was added in 1870 increasing production to eight tons and employment to 130. At Norway, Denison opened a pulp mill in 1869 that produced one ton of pulp and one half ton of paper daily until it was closed down in 1875 at the height of the depression.

Denison manufactured book and magazine papers. All of the paper used in James G. Blaine's memoirs, *Twenty Years in Congress*, was made from poplar stock produced at Mechanic Falls. In addition, Denison manufactured a considerable amount of paper under federal government contracts. These contracts kept the company alive during expansion and depression alike. By 1878 the firm produced about 175 tons of book paper monthly.[9]

Denison's business ventures became increasingly varied. He invested in mills in Yarmouth, Gorham, and South Paris and was involved with his son in a groundwood mill at Berlin Falls. One of his associates at Norway, G. F. Evans, invented a rifle which he felt would be accepted by the U. S. Army and Denison agreed to back him. A rifle factory was built near the paper mills at Mechanic Falls in 1873-1874. When the rifle was not accepted by the Army the arms factory foundered and, coinciding as this event did with the onset of depression, it proved to be the final item in the increasing Denison load.[10]

The firm limped through the depression, suffering a costly fire in 1876, and thereafter kept going by means of layoffs and pay cuts. When government contracts were not renewed in 1878 the Denison company passed into receivership. On June 30, 1879 the firm's liabilities amounted to $325,000 and its assets only $60,000.[11]

Among other difficulties the firm had been forced to purchase its pulp from outside sources once the mill at Norway closed down in 1875. Realizing this the mortgage holders, W. H. Parsons and Company, decided to finance a new pulp mill at Gilbertville (Canton),

Maine. A tax exemption was secured, the mill was built, and by the end of 1880 it was producing five tons of pulp per day. New government contracts came in and Denison was back on top. W. H. Parsons and Company was paid off in full by 1882 and its connection with the business was severed.

For a brief period this mill was the answer to Denison's problems. Production climbed to 200 tons of pulp per month in 1882. All of the pulp went to the mills at Mechanic Falls which were running 24 hours a day, employing 200 people and producing 175 tons of paper a month. A railroad was built from the mill to Canton to handle a growing production and by 1885 the capacity of the mill doubled to where it could handle 18,000 cords of wood annually.

But this growth was deceptive. Part of the expansion cost was borne by the workers who were constantly owed back pay. In 1885 the men went out on strike for their wages which in some cases amounted to as much as $700. The men found them increasingly tied to the company by the extension of credit at the company store. In December, 1886 another missed payday brought on a short strike. When the company failed to meet its payroll in February, 1887, the men refused to work and the mills closed for good under Denison management. The firm's liabilities amounted to $415,000 when it went under the auctioneer's hammer on May 5, 1887. Most of the debt represented the expansion at Canton and the modernization of the paper mills at Mechanic Falls. The firm was purchased for $100,000 by Charles Milliken of Portland, Maine and reorganized as the Poland Paper Company which is in operation today.[12]

The Denison interests tell a good deal about the early days of the pulp and paper business after wood pulp had triumphed. There were many ups and downs in the fledgling industry, and those entrepreneurs who were not reasonably conservative in their approach went under. In 1882, 68 new mills were built in the United States and 37 more were building. Howard Lockwood, founder and editor of the *Paper Trade Journal*, thought the expansion too rapid. He addressed questions on this point to about 40 manufacturers. All thought that there was too much expansion and many predicted failure for the new mills. None was more pessimistic than Adna C. Denison though he conspicuously failed to heed his own advice.[13] A poem was circulating in the trade during this period of expansion which told the story well.

> Once of money I had plenty,
> and, friends, too, by the score.
> I went into the paper 'biz'—
> Alas! It is no more.
>
> Men told me when my mill was built,
> and all my profits spent:
> 'Profits will be a mill a year—'
> but *ten* can't make a cent.

One firm which held to a more cautious approach was the S. D. Warren Company. The early story of the company and its success is largely the story of the founder Samuel Dennis Warren. His career in the paper industry began at the age of 15 when he started working for his uncle's firm, Grant and Daniell, Boston paper dealers. Warren was admitted to junior partnership at the age of 21 in 1838 and by 1851 he was abroad purchasing rags for the company. Grant and Daniell was one of the first to import rags and was for some years the largest importer. In 1853 the firm branched out into paper manufacture and Warren and Daniell purchased a mill and its water privilege in 1854 at Congin Falls, Maine for $28,000. After Warren purchased Daniell's interest in 1855, the Boston firm became known as Grant, Warren and Company until 1867 when it became simply the S. D. Warren Company. Warren remained active in determining the policies of the firm until his death in 1888.[14]

The firm grew steadily in size employing 50 people in 1854 and 990 in 1888. During these years S. D. Warren Co. purchased a mill at Gardiner and a pulp mill at Yarmouth in Maine, and also increased the size and capacity of the original mills at Congin Falls, or as the town would later be called, Saccarappa, and then, Westbrook. In 1873 Warren recognized that rag pulp was rapidly being superseded by wood pulp and put his firm to experimenting with the manufacture of paper made from the new source. As a result of this investigation the Forest Fibre Company in Yarmouth was purchased. The mill produced three tons of pulp a day. By 1875 the firm's importation of foreign rags on its own account was about over. Warren had caught the new revolution at just the right time. He made his decisions with intelligence and foresight yet moved slowly enough so that the firm could meet any adversity.

The Cumberland Mills complex at Westbrook occupied an enviable position on a site where the Presumscot River fell 20 feet. In 1870 it was estimated that the annual production was worth over a million dollars. From that time the mills grew in size especially as the

Pulp and paper come to the northeast

new wood pulp paper lowered prices and increased demand. In 1875 the water power owners on the river petitioned the legislature for rights to increase water storage in Lake Sebago by increasing the size and height of the dam owned by the Cumberland and Oxford Canal.

The enlarged dam was completed in the fall of 1878 making it 300 feet long, 24 feet high and providing power enough for immediate needs. But the solution of the company's power problem was short-lived and final solution was not found until the shift to electricity in 1887.[15]

During the business depression of the 1870's the firm stayed prosperous. By 1877 there were seven paper machines running at Cumberland Mills and production was 16 tons a day. Even with this growth employees were forced to take two pay reductions, but apparently their steady employment prevented any labor problems. The steady employment was achieved by winning good markets and maintaining them. During this period Warren manufactured paper for *Century Magazine, Atlantic Monthly*, and *The Youth's Companion*.[16]

Once the depression of the 1870's was safely weathered S. D. Warren Company began to grow even more extensively. The mill shifted over completely to steam power and by 1880 was producing enough to be described as the largest paper mill in the world. Growth was limited by a fire in 1882, but by 1885 the 700 employees produced some 30 tons of book paper each day. The plant covered 120 acres of land and was one of the largest firms of any type in the state. Machinery was used wherever possible. A narrow-gauge railroad was constructed in the mill yards for the movement of materials and it soon was connected with the railroads in town. The firm changed to the eight-hour day with three shifts in 1887. As the decade closed new buildings were growing up.[17]

Growth was evident in other ways. In the same year that the eight-hour day was installed S. D. Warren began to change over to hydroelectric power. The dam development which created this possibility was complete by 1890, but the firm did not change over entirely until after a serious flood in the spring of 1895. When it did come into use it marked a new era. The voltage generated was among the largest in the world at that time.[18]

The Yarmouth mill was also rebuilt during this decade. In 1883 it employed 150 men and annually produced nearly 10,000 tons of pulp, shipping it all in 789 railroad cars. The mill consumed 10,000 cords of poplar wood, 6,000 tons of coal, and 6,000 cords of edgings and slabs from nearby sawmills.[19]

The company continued to increase in size throughout the 1890's in spite of heavy losses suffered during the 1895 flood. The Forest Fibre Company (Yarmouth) expanded its production by 40 per cent to nearly 40 tons a day in 1891. In the next year the firm grew even larger with the purchase of the Richards Paper Company of Gardiner and Skowhegan. At the end of the century a modest land purchase program was underway, and the company had survived with ease the early and volatile years of this industry.[20] The secret of its success was slow, sustained growth, yet growth in advance of the rest of the industry. Another factor in its success was enlightened treatment of its employees. Employee relations were an important key to success, and for that reason some discussion of this is relevant to the history of these northeastern paper mills.

Warren was public spirited and he paid fair wages for his time, 75¢ a day in 1854 and $1 by the Civil War. In 1869 the firm gave a building site to a local church and contributed $5,000 to the building fund. This sort of charitable activity was to be part of the company's relationships with the town and by extension with its employees from that time on. Workers did find themselves under some restrictions. They were forced to live in company housing (boarding houses) unless they lived at home. Failure to abide by the regulations in the boarding houses was cause for dismissal.[21]

In the early 1880's the employees banded together to form a Mutual Relief Society. Its object was "to aid and benefit such of its members as are, by sickness or accident, unable to work." All regular employees were members. An admission fee of 50¢ was charged and on the death of a member an additional assessment was made of 50¢. The society paid death benefits of $200; unemployment benefits amounted to $5 a week for males, and half of that for females. The benefits did not start until the second week and were payable only for 26 weeks in any one year or under one illness.[22]

On the whole there were few problems with employees. Women employed as rag sorters went on a five-day strike in 1880 but apparently no other work stoppages took place. Most of the employees lived in company housing which by 1881 was furnished with running water. There was in addition a library and public reading room paid for by the company. In 1883 Warren testified before a Senate Labor Committee that the firm owned 150 houses with rents from $75 to $200 a year. Electricity was provided in some at $35 a year; water cost $10. A public hall was erected in Westbrook, with part of the money coming from the company. In 1889 the firm began work on a sewerage system for its property. S. D. Warren loaned money to employees for purchase of homes, payment of bills, or even for the education of their children.[23]

Newspapers constantly held this company up as an example to others during the troubled times of strikes in the seventies and eighties. The following comment was typical of many:[24]

> Here [*Cumberland Mills*] friction between capital and labor is unknown, affording the best practical example of the true solution to the labor question; and would that it might be more generally followed. Among the other evidences of the good will of the company is a fine popular library to which free access is had every week. Thus do they aim for the mental and moral improvement of the people.

The tradition of better living conditions and better pay continued. In the 1880's women received 83¢ a day and men $1 to start. Foremen's wages were as high as $3. In the early 1890's men started at $1.25. This was, of course, for the eight-hour day which was in itself uncommon.

Wages were increased also by a profit participation plan. The plan started in 1891 and provided that those employees who worked a minimum of 75 per cent of the working days would get a dividend credit determined by dividing a percentage of the net earnings in ratio to the total earnings of the employee. The company made it plain where they stood on the plan urging the men to use economy in their work and not to fear the introduction of machinery. As a letter said,[25]

> Under the plan now proposed, it will be for the interest of each man to work as he would on his own account; and a just regard for his own interest will make it right for him to point out to the management any failure in duty or inefficiency on the part of others.

The plan remained in operation throughout the nineties for the entire firm. Payments increased wages between 2½ and 7½ per cent. Other paper companies in the Northeast were forced to adopt similar plans. Though wages themselves did not increase much in the period the dividends raised them above the average.[26]

Although S. D. Warren had begun a modest land purchase policy by the end of the period, most firms relied on other methods to procure their wood. As has already been pointed out, until the middle eighties the major source of wood was poplar. Farmers ordinarily cut the wood, peeled it, and hauled it by team to the nearest railroad station or even to the mill itself.[27] Occasionally a drive came down on the rear of the regular river drive of logs. It was not until mills began to take 2,000 to 3,000 cords of wood a year (roughly 1,000,000 to 1,500,000 board feet) that more formal methods of procurement had to be found. S. D. Warren employed a man specifically charged with purchase of the wood supply. He eventually also acted for the Oxford Paper Company and the Penobscot Chemical Fibre Company. It was his practice to send out postcards to his farmer contacts stating the price for the year. They returned estimates of their cut and received their pay in quarterly installments after peeling, yarding, piling, and delivery.[28]

Sometimes firms advertised for pulpwood, but ordinarily people cut it when they came to it in their regular operations and sent it to the nearby mills. As the local supply dwindled more and more wood came in by railroad and some by ship. A few lumbermen moved into this sort of supply work. The demand forced stumpage prices higher as the following announcement indicates.[29]

> C. B. and B. F. Smith of Denmark have another large contract this year, to furnish poplar for the Cumberland Mills, and pay $3.50 at the Brownfield Depot. They have had large lots peeled in Denmark, Bridgeton, Fryeburg, and Brownfield. Poplar four years ago could be had for 25¢ a cord stumpage, and now five times that price is paid.

Increased amounts came down in the drive and the Warren firm began to receive much of its wood this way. Drives came in on the Presumscot on July 18, 1884, May 1, 1885, and August 10, 1886. The river was last driven in 1906.[30]

Other mills used a good deal of waste wood from nearby sawmills especially on the Kennebec and Penobscot rivers. Some of the later pulp drives were immense. In 1890 a Lewiston firm took a contract to drive 36 million feet on the Androscoggin.[31] By this time though consolidation had begun in the industry and the pioneer days were over.[32] Indeed the "pulp craze" as it was termed by some newspapers triggered in Maine a conservation movement, as in 1888 and 1889 the tonnage in the state doubled each year. The consolidation cost money though and that meant a different kind of industry than the one that S. D. Warren and the other pioneers had experienced.

NOTES

[1] This story is told in my "Wood Pulp and Newspapers, 1869-1900", *Business History Review*, [Autumn, 1963], and my *History of the U.S. Paper Industry, 1690-1970*, 1971. This chapter appears in a different form in that work, and in a still different form in *Forest History*, [Spring, 1966].

[2] Maine lumbermen soon found that the pulpwood business was a grati-

fying one. J. Fred Webster, operator of the Penobscot boom, took contracts which sent 6,700 cords in 1884. The price for the first wood shipped on the schooner *Post Boy* was $3.75 a cord delivered in Pawtucket. By 1885 they had set up a portable sawmill in Bangor and drove the wood directly to the saw. In 1886 the amounts were 250 cords a week, although by 1888 the Websters only planned to ship 2,000 cords total. Other Maine firms shipped pulpwood to Argentina and later to California. It soon proved to be more profitable to ship wood pulp though. See *The Industrial Journal,* (Bangor, Maine) February 8, March 28, August 28, October 10, November 1, 1884; July 24, August 7, 1885; July 11, 1886; August 3, 1888; Bangor *Daily Commercial,* August 6, 1888; *Paper Trade Journal,* December 2, 1920 quoting the *Boston Transcript,* November 26, 1920 an excellent story on the movement of Maine pulpwood prices from 1884 to 1920.

[3] This valuable account is by H. A. Morrell, agent of the American Wood Paper Company. It was entitled "Wood Pulp" and was first printed in the Pittsfield, Maine *Advertiser,* where he had retired, and then reprinted in *The Industrial Journal,* December 25, 1885. Also see *Paper Trade Journal,* October 16, 1897 for some comments on this early company.

[4] Hugh J. Chisholm, "History of Papermaking in Maine, and the Future of the Industry," *Twentieth Annual Report of the Bureau of Industrial and Labor Statistics For the State of Maine; 1906,* (Augusta, 1907), 161-9; Portland *Eastern Argus,* March 23, July 2, 1869; January 3, March 1, May 24, 1871; May 6, 15, 1875; May 23, 1876; Farmington (Maine) *Chronicle,* September 26, 1872 quoting the Brunswick *Telegraph.*

[5] Among the mills started, or promoted, which I have located in this period were ventures at South Paris, Gardiner, Belfast, Yarmouth, Augusta, East Turner, Gorham, Camden, Orono, Kezar Falls, Farmington Falls, East Dover, Houlton, and Bangor. The first Maritime Provinces mill was at Ellershouse, Nova Scotia. It burned in 1875 or 1876. A typical editorial appeared in the Bangor *Daily Commercial,* November 8, 1880 calling for a pulp mill to use the chemical process on sawmill waste. As they said "This has been so far demonstrated that it is no longer an experiment, and it is to be hoped that our capitalists will give the matter attention and that labor may be employed and they may receive good returns." Towns in Maine beginning to lose their population to the west or to eastern cities also thought of the pulp mill as a partial answer to their problems. See the interesting exchange of letters in the Phillips (Maine) *Phonograph,* December 31, 1886; January 1, 1887; May 20 1887. N. C. B., a former Phillips man living in Harpers Ferry, Virginia, thought they should promote a mill as did the editor, but J. E. Ladd of Augusta thought it couldn't be done as enough power didn't exist. In the event Ladd was right.

[6] *Paper Trade Journal,* February 11, 1882; March 24, 1883; May 26, 1884; *The Industrial Journal,* October 6, 1882; March 23, August 17, October 12, 1883; April 18, 1884; Bangor *Daily Commercial,* July 14, 1884; October 31, November 30, 1885.

[7] See Portland *Eastern Argus,* April 26, 1883; April 14, December 2, 1885; and August 10, 1886 for accounts of popular drives. On wood supply generally see *The Industrial Journal,* for June 12 and September 4, 1885 for a description of the Somerset Fiber Company which used both edgings from a nearby

sawmill and driven wood. The company had an endless chain rigged to haul four sticks of pulpwood from the river and pile them on the bank at the rate of 15-20 a minute. A reporter sent to look at the machine interviewed a local lumberman who described the river as "That flow of stuff, you can't call it timber, illustrates the new fields of industry which have been found here. That is poplar wood. Fifteen years ago it and the land on which it grew were worthless..." Bangor *Daily Commercial*, September 5, 1885 for the quotation.

[8] The best sources for his life are *The Industrial Journal*, April 8, 1887, an interview about his career; and *Paper Trade Journal*, September 3, 1897, an obituary notice.

[9] Oxford (County) Maine, *Democrat*, July 31, October 16, 1868; April 9, May 14, July 23, September 10, 1869; January 1, August 5, November 25, 1870; May 9, June 13, August 1, 20, 1871; January 16, February 13, 1872; August 5, 1873; May 30, 1875; Portland *Eastern Argus*, June 17, December 12, 1874.

[10] *Ibid.*, December 25, 1872; Oxford *Democrat*, March 4, 1873 (the rifle company). *Paper Trade Journal*, February 15, 1874, (good on the mills).

[11] Portland *Eastern Argus*, January 24, 26, July 26 (quoting the Mechanic Falls *Herald*), September 22, November 21, 1876; February 25, 1878; July 3, 16, 1879; *Paper Trade Journal*, July 12 (the best discussion of the failure), July 19, 1879.

[12] Portland *Eastern Argus*, February 19, 1880; July 18, 1885 (on the first strike); February 10, 11, 17 (detailed account of the firm's business, liabilities, and assets), May 5, 10, 11, 1887. *The Industrial Journal*, January 13, November 17, 24, 1882; January 5, 12, April 27, 1883; February 1, 1884; January 16, June 5, 1885; Febuary 11, 1887. Bangor *Daily Commercial*, March 28, 1882 (quoting from the Lewiston *Journal*). *Paper Trade Journal*, July 24, October 9, 16, December 11, 1880.

[13] *Paper Trade Journal*, March 31, 1883, "Production and Consumption—Are There Too Many Paper Mills in America?" Denison's reply was dated March 13, 1883. The poem which follows appeared in *Ibid.*, January 19, 1884.

[14] The major sources for Warren's life are *A History of the S. D. Warren Company* (Westbrook, Maine) 1954; *Samuel Dennis Warren... A Tribute...* (Cambridge, 1888); Westbrook *Chronicle-Gazette*, January 26, 1906; *Paper Trade Journal*, May 19, 1888, and most of the cited materials in the next few notes.

[15] Portland *Eastern Argus*, February 17, September 20, 1870; May 21, December 5, 1872; Bangor *Daily Commercial*, October 28, 1878. The Warren firm had trouble with a power company over the extent of water flow which led to a pitched battle in 1877. An account appears in *Paper Trade Journal*, February 3, 1877. The payroll and some other miscellaneous documents from the dam construction remain in the Warren archives.

[16] Portland *Eastern Argus*, September 8, November 13, 1877; March 25, July 15, September 28, 1878. On wood procurement for the Yarmouth mill see *Paper Trade Journal*, September 1, 1872 waste and edgings from a steam saw mill at Bethel.

[17] *Paper Trade Journal*, October 9, 23, December 11, 1880; February 12, 1881. *The Industrial Journal*, December 3, 1880; August 22, September 20,

1884; September 24, 1886. Portland *Eastern Argus,* May 6, 1882. Bangor *Daily Commercial,* April 19, 1886, May 18, 1887. *Warren Monthly,* Vol. IV, no. 3. Undated clipping c. 1927, S. D. Warren Co. scrapbooks (life forty years before), and *Timebooks,* 1880's, especially 1883-1886, in files of the S. D. Warren Co.

[18] *The Industrial Journal,* March 15, 1880 (an editorial calling for electricity in pulp mills); July 4, 1890 (editorial on the development). Bangor *Daily Commercial,* July 1, 1887. *Paper Trade Journal,* May 18, 25, 1889; December 28, 1895. Undated clipping c. 1932, S. D. Warren Co. scrapbooks (on retirement of William R. Bragdon who had engineered the electricity changeover).

[19] Portland *Eastern Argus,* January 24, 1881. *The Industrial Journal,* October 19, 1883, October 15, 1886.

[20] *Paper Trade Journal,* August 15, 1891, April 25, 1896, and May 29, 1897. *The Industrial Journal,* October 21, 1892. S. D. Warren Co. "Cash Report, February, 1891." This is apparently the only survivor of a monthly account of this period. The monthly payroll by this time was $35,000 for day labor, and freight bills amounted to $7,000 a month. For the Richards Paper Company see Henry Richards, *Ninety Years On,* (Augusta, Maine, privately printed, 1940), the autobiography of the founder and Laura E. Richards, *Stepping Westward,* (New York, 1932), the autobiography of his wife. Both contain delightful vignettes of the early days of the wood pulp paper business. Incidentally when they sold to the Warren's the other possibility that they investigated was a move into the south.

[21] *A History of S. D. Warren, op. cit.,* Part I. Portland *Eastern Argus,* February 17, 1870. Letter, George W. Hammond, Agent, to employees, March 1, 1875, S. D. Warren Co. Archives.

[22] *Constitution and Bylaws of Cumberland Mills Mutual Relief Association,* (Westbrook, Maine, 1882). Portland *Eastern Argus,* December 12, 1881. Portland *Evening Express and Advertiser,* January 6, 1922 (a summary of the first 40 years of the society.)

[23] Bangor *Daily Commercial,* January 28, February 3, 1880 (on the strike). *Paper Trade Journal,* April 16, 1881. S. D. Warren Co. ledger, "Rents, Hall Block and Saccarappa Property," 1890-1906. *The Industrial Journal,* October 26, 1883 on Warren's testimony which said, "If all employers were like Mr. Warren, there would soon be an end to all strikes and of the warfare between labor and capital." Also, Bangor *Daily Commercial,* November 25, 1889 (on the sewer system). For the loan policies the best source is a series of ledgers entitled "Pre-paid Labor," which run from the middle 1880's to 1929, and which detail all requests, the disposition of the request, and the payment plans. A few letters also exist on this subject. Most of the loans were paid by payroll deduction.

[24] Portland *Eastern Argus,* February 27, 1886, an article by "VIDI" entitled "S. D. Warren."

[25] S. D. Warren Co., "Time Books," 1878-1900, especially 1883-1886. *The Industrial Journal,* January 31, 1890. Undated clipping (c. 1923) in the S. D. Warren scrapbooks on earlier days (40 years previous); S. D. Warren Co. to John E. Warren, March 12, 1891 a letter outlining the profit participation plan in detail, and from which the quotation is taken.

[26] "Time Book," January 1, 1895-July 1, 1895; "Time Book, Women,"

1897. *The Industrial Journal*, February 17, 1893; February 16, 1894. *Paper Trade Journal*, February 28, 1892, February 4, 18, 1893; February 10, 1894; February 20, 1897. These last are all reports of dividend distribution.

[27] Bangor *Daily Commercial*, January 29, 1872. Here is their comment, "The farmers of Brunswick and Topsham find a ready market at the pulp mill for their poplar wood..."

[28] Remarks of C. P. Prince, in Paper Perfection Institute, *Proceedings*, March 24, 1915. This was a Warren house publication. Prince was purchasing 50,000 cords a year when the combine broke up in 1913, apparently because of the Clayton Anti-Trust Law. It has been said that this was the move which caused some Maine firms to hire professional foresters.

[29] Portland *Eastern Argus*, June 1, 1872. "WANTED—1,000 cords of spruce wood—Apply to Forest Paper Company—Yarmouth, Maine." *Ibid.*, July 17, 1877, March 12, November 19, 1879 on an operation at Woodstock, Maine for the Berlin, New Hampshire mills; May 15, July 6, 24, 1882 on the Smith operations. The quotation is from July 6. Also Oxford *Democrat*, December 31, 1872, February 24, 1874, March 16, 1875. *Paper Trade Journal*, September 1, 1872, September 1, 1875, January 7, 1882, February 24, 1883 (on the first land purchases in New Hampshire, this last three citations). *The Industrial Journal* August 7, 1885 recounts a vessel with 80 cords of pulpwood from Merigonish, Nova Scotia to Portland for S. D. Warren. This experiment would be an important source twenty-five years later.

[30] Portland *Eastern Argus*, April 26, 1883, April 14, December 2, 1885, August 10, 1886. *The Industrial Journal*, August 3, 1883, July 18, 1884, April 17, 1885.

[31] *Ibid.*, April 24, 1882, June 12, September 4, 1885. Bangor *Daily Commercial*, September 5, 1885, October 8, 1886, December 5, 1890, and April 18, 1893. For a drive on the Royal River to Yarmouth of 20,000 cords see *The Industrial Journal*, May 24, 1889. Professor Edward Ives and myself have edited the biography of one man who worked on some of these early pulp drives. *Fleetwood Pride 1864-1960: The Autobiography of a Maine Woodsman, Northeast Folklore*, Vol. IX, (Orono, Maine), 1967.

[32] Two excellent stories which summarize this very well are in Bangor *Daily Commercial*, May 3, 1888 and November 23, 1889. The last was entitled "The Pulp Craze". Both of them detail the new construction. The reports of the Maine Board of Industrial and Labor Statistics, various titles, especially 1873, 1883, 1885, 1886, 1887, 1894, and 1895 are also very useful.

10

GROWING PAINS AND PROGRESS IN PULP AND PAPER

The pulp and paper industry had come to the Northeast by the mid-nineties. It only remained to be seen how large the industry would grow and what form it would take. Observers were exultant over the soon-to-be reached future, and for good reason. In 1889 there were located or building in Maine six soda pulp mills rated at 92 tons daily capacity, six sulphite pulp mills rated at 90 tons daily capacity, and 13 groundwood pulp mills rated at 157 tons daily capacity. In the next five years, the growth was even more phenomenal. In that year the state conducted a census of the mills in the state, and it found that the industry now had a rated capacity of 1,036 tons of pulp and 508 tons of paper each day. When that part of the Androscoggin River in New Hampshire was added, the rated capacity climbed to 1,261 tons of pulp and 608 tons of paper. Now came the major consolidation of some of these smaller units into the International Paper Company. After this the only major change in the northeastern picture would be the construction of giant mills and companies in the wilderness like that of the Great Northern Paper Company at Millinocket.[1]

Consolidation had to come. There was overproduction and falling prices. Groundwood pulp prices dropped from four cents per pound during the 1870's to as low as .6 cent per pound around the time of the industry's consolidation. Some mills closed, others went bankrupt, still others suffered from poor construction and poor management. By 1897 or 1898 many mills were either losing money or barely breaking even. As prices drifted lower marginal mills were forced to the wall or into the hands of their competitors. The possible consequences of the savage competition led owners into attempts at controlling the market and eventually to formation of the International Paper Company. One other attempt to aid in the market stabilization deserves brief attention and this was the shipment of New England wood pulp abroad. The British in particular were suffering from the same fiber scarcity that has been discussed earlier in this work as occurring in this country. As a result soon after the discovery of workable groundwood pulp some was shipped for use by The *Times* of London, and apparently a small market was built up. Later when the market in the United States fell to severe lows during the panic of 1893 more shipments of wood pulp went to Europe. The little boom lasted from late summer 1893 to about mid-June 1894. Perhaps ten vessels left Bangor and Portland with pulp from the Moosehead Pulp and Paper Company of Solon. The market broke soon with an influx of pulp from Scandinava and the boom stopped. The Solon firm soon went bankrupt. All in all perhaps some 25,000 tons of dry wood pulp went abroad in this little boomlet.[2]

It was about this same time that the rumors of combination began to fly through the industry. In fact when in February, 1895 new rumors of meetings of industry leaders were heard there had been so many such in the previous twelve months that *Paper Trade Journal* dismissed the new rumors simply as trade talk. Yet the meetings continued with some regularity and in secret. It was known that the meetings were concerned with the questions of whether there should be a new firm to control production or a selling agency to control prices. After two months of discussion even the *Paper Trade Journal* was now willing to say something was about to be consummated. Throughout July and August of 1895 the meetings continued nearly every week at New York, Boston, and Saratoga Springs. The matters discussed now seemed to deal mostly with who would be in the big combine, how it would be financed, and what would be the scale of payment to previous owners. Prices rose slowly under the impact of the consolidation rumors. Apparently the plans failed as the result of disputes over the valuation of member mills. By January, 1896 the "giant new combine," as it was usually called, was for the time being quiescent.

When trade slackened again as it did in the late summer, the consolidation talks revived. After the passage of the Dingley tariff in 1897 the rather desultory talks began to have more life still. In September 1897 the *Paper Trade Journal* reported that a committee of paper manufacturers had been created to set a formal valuation on the mills. All was silence after the committee was formed and some became disturbed at the prospect of higher paper prices. Editorials like the following struck a note of anxiety.[3]

> It is claimed, however, that there is no intention of forcing up prices unduly or even materially. A large part of the benefit claimed for proposed consolidation, it is said, will result from greater economy, particularly in selling the product. Some advance in prices is gen-

erally predicted, and in fact there is a firmer market at present, but exceptionally high prices, it is asserted, are not aimed at.

At the end of November the *Paper Trade Journal* revealed that a new firm was to be set up with a capitalization of $25,000,000. In January, 1898 more stories appeared that raised the capitalization to $35,000,000.[4]

Finally on January 31, 1898 the International Paper Company was born representing the union of 20 mills in Maine, New Hampshire, Massachusetts, Vermont, and New York. It was a successful marriage of the Fourdrinier and the counting house; the future of the industry seemed to lie in this direction. The actual capitalization, when the merger was completed, amounted to $45,000,000—$25,000,000 in preferred stock and the rest in common. Only 100,000 shares of the 450,000 authorized were issued immediately. In addition, a $10,000,000 mortgage was secured on the company's equity in exchange for six per cent gold bonds payable beginning in April, 1898. International Paper Company was reported to own 1,000,000 acres of timberland in the United States and 1,600,000 acres in Canada. During the first year its mills consumed 500,000 cords of wood.

In his first annual report President Hugh J. Chisholm announced that International Paper controlled 90 per cent of the newsprint production in the East and that "competition was not of a serious nature." It was a profitable first year with over $2,000,000 in dividends and $870,828 in undivided surplus. The emphasis was on growth. In 1900 the company invested nearly $1,000,000 in a paper bag mill at Rumford, Maine. A. N. Burbank, treasurer of the company, remarked at the opening of the mill, "It has been the policy of the directors to purchase woodlands and to place all our surplus earnings in betterments, so that we will begin to obtain larger profits in a few years." The era of the huge firm had arrived.[5]

Maine and New Hampshire paper manufacturing were represented in the new company by a complex of mills on the Androscoggin River primarily and mainly in the vicinity of Livermore Falls and Rumford. During the 1880's five new mills had been built along this river. The large investors were Hugh J. Chisholm, C. A. and C. D. Brown, and W. A. Russell. Small pulp mills and a leatherboard mill built in the area stirred interest and created a small boom in the towns locally. Even though some accidents occurred as these early mills were not remarkable for their safety the boom continued and local newspapers were constantly printing stories about new machines, new buildings, or other evidences of growth. The final buildings at Otis Falls, constructed in 1887-88, were of stout construction; moreover, it was the largest in Maine at the time and the third largest in the country. This mill produced 150 tons of newsprint per day at the time of the consolidation. The Otis Falls mill was tied to a sulphite mill at Rumford and a groundwood mill at Riley, also part of Chisholm's holdings. His development of Rumford, as well as Otis Falls,—both the town and the mills were created out of the wilderness—played a significant part in the formation of both International Paper and Great Northern Paper Companies.[6]

The falls at Rumford on the Androscoggin River are among the most picturesque in the Northeast. It was obvious that with the falls' steep pitch and rapid flow they would be, when harnessed, one of the great sources of hydroelectric power in New England. Hugh J. Chisholm[7] and C. A. Brown of Portland, Maine were well aware of the possibilities. As early as 1882 they began to purchase land in the area. It was gigantic undertaking, but by 1890 the Chisholm-Brown interests had surveyed lots and connecting canals, and laid plans for a railroad to Canton 40 miles away. In 1891 they organized the Rumford Falls Power Company to develop the town and began work on the $500,000 Rumford Falls Paper Company which would have a daily capacity of 50 tons of pulp and 60 tons of paper. Other investors in the paper company included Daniel F. Emery, Jr., George F. Perkins, Edward H. Haskell, C. H. Haywood, and Edward L. Stanwood. Garrett Schenck became general manager.

The new mill went up quickly in 1892 and Rumford Falls took on all the character of a boom town. Buildings were being erected every day and rents skyrocketed. There was a minor downturn during the panic of 1893 but by and large only success greeted the rising city. Early in 1894, 100 tons of freight per day were passing over the railroad to and from Rumford Falls and the monthly payroll in the town amounted to $25,000. Electric lights lit the main streets and most homes, two miles of water main and one-half mile of sewers had been laid, three bridges built, and schoolhouses and churches were in operation.[8]

The Rumford investors were instrumental in forming the International Paper Company. Hugh Chisholm became president and Garrett Schenck sat on the board of directors. Indirectly then Rumford was also responsible for the creation of another large firm in the northeast, Great Northern Paper Company. This firm created in

Big companies come 251

partial response to International Paper Company's success has had a greater impact on the Maine economy than that of International Paper, and it ranks in the top firms in its impact on the United States economy since that time.

Though it was founded only about seventy years ago much of Great Northern's early history remains obscure. Possibly this is because the real beginning of the firm dates from the fall of 1889 rather than the usual founding date of 1899. In the earlier year Isaac M. Weston, one-time Maine man and state chairman of the Democratic party in Michigan, returned to his home state on an extended visit to look over the timber supply preparatory to setting up a pulp mill. Weston was the advance agent for several wealthy persons closely associated with Grover Cleveland and the national Democratic party who turned up in Maine in November, 1889 inspecting the sites which Weston had pinpointed. These men included Henry C. Payne, H. C. Twombly, former Secretary of the Navy H. C. Whitney, Daniel Lamont, Pierpont Morgan, George N. Fletcher, and Don M. Dickinson. With Cleveland they founded the Manufacturer's Investment Company of New Jersey which was capitalized at $1,000,000. After visiting numerous sites on the Androscoggin, Kennebec, and Penobscot rivers they decided to build their new mill at Madison, on the Kennebec.[9]

Although Maine newspapers welcomed the new addition and reported construction details as rapidly as they became known the trade journals felt the company began on shaky ground. When the Madison mill began operations *Paper Trade Journal* saluted it but commented in addition:[10]

> It is a matter of remark in the trade that the capitalization of this company is excessive and that any ordinary individual or firm could have put up either plant for about one-third of the money which they cost. . .

The trade magazine also went on to say that the office of the new firm was on Broad Street in New York and this was suggestive of the speculative atmosphere surrounding the new firm.

The Madison mill was soon in difficulty and press notices reflected the situation. The buildings were badly constructed, pulpwood proved harder to obtain than planned, and the new Mitscherlich sulphite process did not work out as well as had been expected. In addition the firm had difficulty with the town of Madison over tax abatements. In June 1893 the plant was shut down for repairs. The Appleton, Wisconsin plant of Manufacturer's Investment was also closed. Its manager was brought east to run the Madison mill, and the firm went through a reorganization. Ironically enough the first Madison manager had been Frederick Taylor of the famed Taylor scientific method of industrial management. The difficulty did not seem to have been his fault however as the building nearly collapsed once and had to be shored up. There were many difficulties. Not until June 1894 did the company resume manufacturing. The owners purchased additional woodlands and the mill was then operated continuously until it went into receivership in 1899.[11] The Manufacturer's Investment Company died as a result but it was reincarnated as part of the Great Northern Paper Company. That story is the story of Garrett Schenck, the forceful and imaginative builder of both Great Northern and the town of Millinocket, Maine.

He was born in Trenton, New Jersey in 1854 and like S. D. Warren began his paper industry career with an uncle who operated a rag business in Massachusetts. By 1886 he was the resident manager of the Penobscot Chemical Fibre Company in Old Town, Maine. He also served as agent for a mill in Lincoln. As patentee and part owner of a process for reclaiming liquids from the digester, Schenck became associated with J. Fred Webster, J. A. Kimberley, and Edward Ames in the National Sulphite Boiler and Fibre Company. The process was used all over the country, and Schenck often supervised the building of new mills as he did the mill of the Willamette Pulp and Paper Company in Oregon City in 1889 and 1890. At one time he owned part of a pulp mill at Orono and was an early investor in the development which became the Eastern Corporation in Brewer, Maine. When the Rumford development started he moved there as manager and as part-owner of the Rumford Falls Paper Company. He subsequently became a member of International Paper Company's first board of Directors until August, 1898 when he resigned to set up the Great Northern Paper Company.[12]

Dissatisfaction with International Paper's virtual monopoly of newsprint production in the East had led to reports that some competing organization would be established. In 1898 it was rumored that Schenck, D. F. Emery, and Joseph Pulitzer were negotiating with such a competitor who would construct a plant at Millinocket for some $2,000,000. The instrument for this purpose became the Northern Development Company, an obscure corporation chartered to purchase and develop lands around Millinocket along the lines of the successful Chisholm development at Rumford. The firm was mostly made up of Bangor lumbering men and included representatives of the Bass, Appleton, Prentiss, Rice and Mullen families, all famous in Bangor logging history. The company never amounted to much, primarily because of financial limitations, but it did purchase important lands in the Millinocket area.[13]

In 1898 Garrett Schenck replaced Henry Prentiss on the board of Northern Development Company, and further reorganization took place in the spring of 1899. The cost of developing Millinocket proved much greater than anticipated. As the price of Wall Street support most of the Bangor men dropped out and their places were taken by New York money men such as Oliver Payne, A. G. Paine, E. H. Haskell, R. H. Hayes, A. H. Paget, and representatives of the Wall Street firm of Grant and Schley.[14] The new organization moved rapidly during 1899-1900 to purchase 252,060 acres of timberland at an average price of $4.14 per acre.[15]

While this was happening the Manufacturer's Investment Company was ending its days. The firm's liabilities were now $2,000,000 more than its assets and Oliver Payne was the chief creditor. Payne had been one of the biggest investors in the Millinocket development as well. Now when additional capital was needed for that enterprise he simply offered his share of the Madison mill as payment and as the price of continued support. The directors of Northern Development had to accept the offer. They forced the firm into receivership, paid off the other creditors and incorporated the mill into the Great Northern Paper Company. In retrospect it seems that Oliver Payne performed a clever piece of financial surgery at a time when such operations were commonplace.[16]

Maine welcomed the new firm whatever its antecedents. Though some thought newsprint would drop in price, others thought this was Bangor's big chance to overcome the loss of its great lumbering and shipping business. New wealth would come to its citizens, and the retail and wholesale trade would both benefit. In the words of one observer, "Isn't that a pleasant outlook for citizens of eastern Maine?" [17]

Once the financial arrangements and the land purchases were underway the company began preparations for the manufacture of paper which was scheduled to start approximately September 1, 1900. The promoters of the new firm met with representatives of the Maine railroads to determine competitive rates to and from Millinocket. To accommodate the builders the Bangor and Aroostook relocated its depot. Men went in to start clearing the land in February, and large crews were on hand by the end of April. John N. Merrill of Bangor, who supervised the construction, went to Boston to secure Italian contract labor for the building. Immigrant Poles, Finns, and Hungarians were also employed. By June 1, 1900, 500 men were at work in the Millinocket area.[18]

A large steam shovel was brought in to do the bulk of the excavation. Lodging was primitive at best as 45 men slept in a room and paid $3.50 a week for the privilege. Many others slept in jury-rig tents outdoors. A newspaper commented "there is a Bohemian air about everything here which cannot be described. There really is no town, but there is an army of people working and camping."

At the end of August the Bangor and Aroostook Railroad planned a train excursion for the curious to see what was being done. Three hundred from Bangor area went on the trip. By this time 1,000 men, two locomotives, a rock crusher, and two steam shovels were hard at work. At Madison the firm employed 250 men adding new mill machinery and strengthening the old masonry. The traffic to Millinocket was so heavy that the county commissioners were forced to construct a road from Medway. It was open for foot travel in June and for horses and wagons by October.

One of the most interesting features was the impact of all the building on the port of Bangor. By the time the river was closed with ice in 1899 more than 100 vessels had been involved in the new trade. Eighty cargoes of bricks and 20 cargoes of cement arrived that fall. Bangor's "mudlarks", as the stevedores were called, had a difficult time with the huge machinery for the new mill. The city fathers rebuilt and strengthened their piers and wharves because of the business involved.[19]

By June 1900, Millinocket was a community of 100 homes and many business establishments. According to the land agent, no house could be erected for less than $750, and no liquor was to be used in the city. A thousand men were working, 400 of them Italians who lived across the river in shacks in what was known as "Little Italy." These, the visitor was assured, were to be torn down when the construction was complete. The Italians were paid $1.50 a day and board which consisted of bread, potatoes, macaroni, beans, and rice. The company also announced its payroll—$1.50 a day for laborers, $2.00 to $2.50 for skilled workers, and $3.00 to $5.00 for machine men.[20]

In this summer more strides were made. Forty new houses were built, streets were constructed, and a combination school and inter-

Big companies come 255

denominational church was in operation. A good hotel was going up and a town sewer and water supply was underway. On the eve of the mill opening 2,000 people lived in "The magic city of the North." The Great Northern Paper Company had truly wrestled a town from the wilderness.[21]

The plants themselves were huge as well. The Madison plant produced 55 tons of newsprint, 50 tons of sulphite fiber, and 40 tons of groundwood pulp a day, while the Millinocket mill produced 240 tons of newsprint, 120 tons of sulphite fiber, and 240 tons of groundwood pulp each day. To create this paper, in addition to the winter's cut, the Northern purchased 275 cords of pulpwood a day which arrived by railroad. More expansion was planned.[22]

The new century created an atmosphere of more expansion in these great mills. A seventh paper machine was installed at Millinocket. More expansion was in the offing. In June the stockholders voted to issue $3 million worth of 5% gold twenty-five year bonds. Stockholders were the main subscribers to the new loan. The money was to be used for timberland purchase, retirement of the old debt, and to erect a new mill at Madison. The new flotation made the capital of the firm $8,000,000. The plants together then produced three hundred tons a day; the new extension would increase this to at least five hundred tons of paper each twenty-four hours.[23] The expansion of the dam system on the West Branch began in 1908 and an increase in capital stock from 5 to 6 million dollars in 1909 allowed expansion to the twentieth century levels of 1500 tons per day production.[24]

With the establishment of Great Northern the structure of Maine's pulp and paper industry was nearly complete. Other large firms were talked about, but only a few were built like the St. Regis plant at Bucksport or the large St. Croix Paper Company mill in Washington County.[25]

Working conditions in the pulp and paper mills compared favorably with other industries during this time. The plants had to be cleaner, neater, and more sanitary. The industry ran twenty-four hours a day so that the eight hour day movement had more success than in most industries. During most of the period it was a time of boom so wages moved upward generally throughout the period. There was relatively little unrest.[26]

Life was perhaps not so good for everyone, but by this time the eight hour day was making great inroads in the industry.[27] There had been occasional strikes for shorter hours or higher wages. The Knights of Labor had knocked a few plants out of production in the summer of 1886, but relatively little interest was generated in the movement.[28] Most workers as shown by a questionnaire and its returns in 1886 were reasonably well off. The average worker in the state worked 10¼ hours for $1.93. He lost fifty-four days of work for shutdown, and his average yearly earnings were $417.61. Other members of his family made $117.35, and the cost of living was $391.43. Some owned their homes, many had savings accounts and benefit associations in their place of employment.[29]

In the nineties as competition increased labor unrest increased with it. Conditions were about the same in 1895 when the State Bureau took another census.[30] Major improvements had come in working conditions which were now quite decent.[31] Occasional strikes occurred but most of them were settled quickly as the owners preferred to keep the plants in production.[32]

The first decade of the twentieth century was marked by union gains. In retrospect they came quite easily in comparison with other industries. Profits were great and the prospects of shutdown were too much for the owners to deny some requests. Union recognition, collective bargaining rights, higher wages, some fringe benefits, better working conditions—all these were achieved in some degree during this time. By the eve of World War One strikes did not even seem to be necessary to achieve gains, and since that time few industries have been as free from unrest as this one. What difficulties there were do not take long in the telling.

In 1903 when the International Paper Company employees went out on strike over recognition the company locked out the union members and many sought jobs elsewhere.[33] Hugh J. Chisholm president of the International Paper Company offered some advice on strikes and strikers,[34] counselling newspapers to offer publicity and praise the companies. He thought the workers followed "unwise leaders" and committed "intemperate acts."

Whether or not this was a correct prescription, union members continued to demand more from the International Paper Company than the firm was willing to grant. As a result the year 1908 was a year in which severe labor troubles plagued the industry. Initially this involved a jurisdictional dispute within the union, and the A.F. of L. was forced to rule as to which union was the legal one. The losers continued to agitate however and the company, using the panic of 1907 as an excuse, declared a wage cut effective when the contract

Big companies come

was over. Meetings were held but little came of them except the desire to strike, and in August most workers did. In September the union members in the Great Northern mills also went out, partially in sympathy, partially because the Northern was supposedly making paper for the International, and partially over the older jurisdictional dispute. This last was the aftermath of a strike-lock out at Millinocket in 1907 which had resulted in victory for the company.

In late September a contract was signed with the International by the union president, but the locals repudiated it as it gave in on all the points at dispute. A convention of the union was called for in Albany, New York, to deal with this problem. In the meantime the manufacturers decided to go it alone and sent telegrams to the plants offering the old jobs at their conditions, with no union recognition, at least of these unions. Strikebreakers began to operate the mills, and by the end of October the mills were on partial production. On November 7th the unions called off the strike. They had lost and big firms had become "open shops."

The Oxford Paper Company followed this strike with an attempt to go back to the twelve hour day which caused their employees also to go out on strike. The Commissioner of Labor for the State of Maine became irate at this development. He regarded the Oxford attempt as almost barbaric. The twelve hour movement did not last long, and although most shops remained "open" the eight hour day was saved.[35]

In 1910 new strikes agitated the industry. One at the St. Croix Paper Company mill at Woodland apparently led to a major strike throughout the East but primarily in the International Paper Company plants. The State Commissioner offered his arbitration services, proposed that the firms drop the use of the "yellow-dog" contract and let an outside observer evaluate the employee demands for a wage increase. Several meetings were held under these conditions, and with the help of New York state arbitrators the strike was terminated. Grievance committees, recognition, eight hour days, a wage increase—all were in the new contract.[36]

With this the industry had solved most of its problems. Growth was the word, and growth it was. In 1912, 7,821 individuals worked in Maine pulp and paper plants; in 1914 the figure was 10,588; 9,698 men and 890 women. There were thirty-nine plants on the eve of World War One, and all seemed serene.[37] In 1914 when Great Northern signed a contract calling for the six day week, eight hour day, and increases of around 9% for each worker other firms in the state followed suit quickly.[38]

One more development worth mention was the establishment, in 1913, of a pulp and paper technology program at the University of Maine. These courses were apparently the first in the United States although some had begun abroad. The first professor, Ralph H. McKee, was a promoter with an expansive personality who missed few opportunities to publicize his program. The American Chemical Society met in Bangor in the first fall, for instance, where the program and its promise was the chief topic of discussion. A small paper mill was installed, students did practical work in local mills, popular lectures were given in chapel and elsewhere, a student newspaper for the neophyte chemists was circulated, papers were read at national organizations, and all in all, the course was a success from the first days. By 1916, 160 students were enrolled, half of them in a degree program, and the success of the Maine work had begun to spawn several imitators and competitors. Distinguished new faculty such as C. A. Brautlecht kept the program alive throughout the twenties when much of the university was not as successful. Some of the success came from the installation of short and summer courses.[39] The revolution had come to the Maine woods and to the Maine economy.

NOTES

[1] Bangor *Daily Commercial*, May 3, 1888; November 23, 1889. *Industrial and Labor Statistics For the State of Maine*, Eighth Report, 1894, 111-123.

[2] *Paper Trade Journal*, August 19, October 28, 1876 (they were promoting foreign trade to combat the depression all year); February 10, 1894; Portland *Eastern Argus*, October 23, 24, 1876; Bangor *Daily Commercial*, August 2, November 3, 1893; *The Industrial Journal*, September 29, November 17, 1893; June 1, 15, 1894.

[3] *Paper Trade Journal*, March 31, 1894; February 23, March 2, 30, July 27, August 10, 1895; January 25, August 8, October 17, November 7, December 5, 1896; August 21, September 4, 11, 1897; Bangor *Daily Commercial*, January 1, 16, 1897; *The Industrial Journal*, October 8, 1897 from which the quotation is taken.

[4] *Paper Trade Journal*, November 27, December 4, 11, 1897; January 1, February 5, 1898; *The Industrial Journal*, January 21, 1898. See also L. Ethan Ellis, *Print Paper Pendulum*, (New Brunswick, New Jersey, 1962), 19-24.

[5] *The Industrial Journal*, February 4, 1898; August 11, 1899 quoting from the statement. Portland *Eastern Argus*, January 6, 1901 quoting Burbank on the occasion of the opening of the Rumford paper bag mill. International Paper

Company's share of the total U. S. newsprint market was 60%. By 1913 it had dropped to 26%. *International Paper After 50 years*, (New York, 1948), 17. It was fairly obvious to others that I. P. was successful as one man said to another firm in a letter thanking them for their dividend checks, "I have just come from Palm Beach where I visited a good deal with Mr. Chisholm of the International Paper Company and got considerable information about the paper business and I judge there is a good business in sight and at remunerative prices." Ltr., B. L. Taylor to Willamette Pulp and Paper Co., New York, April 8, 1900, filed in Notebook, "Willamette and Crown to 1928—Addendum 2-B", Crown Zellerbach archives. These archives were opened for me by that company and I am grateful for this opportunity.

[6] *Paper Trade Journal*, August 18, November 24, 1877 (on related mills being erected on the Androscoggin at Berlin Falls), October 29, 1877, and on the general boom in Maine August 4, 18, 25, September 1, 8, and especially December 15, 1888; *The Industrial Journal*, August 10, 1883 quoting the Livermore Falls *News*; Wilton, (Maine), *Record*, June 30, (on a bad accident), July 28, October 27, November 10, 17, 1886; December 15, 1887 (especially good on Otis Falls Construction); Portland *Eastern Argus*, December 28, 1887, also very good on construction details at Otis Falls. Not all was peaches and cream on the Androscoggin. At the time of the International Paper Company consolidation one of the earliest mills on the river was closed and finally six months later it was sold. A newspaper remarked that this might help that part of the river out of its lethargy as compared to Rumford and the Chisholm interests generally. Lewiston *Weekly Journal*, May 13, 1897.

[7] Hugh J. Chisholm, "A Man and the Paper Industry—Hugh J. Chisholm (1847-1912)," a Newcomen address by his son at Portland December 12, 1952 printed in *TAPPI*, Vol. 38, no. 10 October, 1955. Chisholm was a Canadian who worked with Edison selling newspapers on trains as a boy. He and his brother got a monopoly on the Grand Trunk and divided the business with the brother taking Canada. From this Chisholm went into the publishing of guide books and into wood pulp via a wooden ware patent and the Somerset Fiber Company in Maine. When this burned and was rebuilt he began to move into groundwood and the Androscoggin development in 1881. See his obituary in *Paper Trade Journal*, July 11, 1912.

[8] Hugh J. Chisholm, "History of Papermaking in Maine and the Future of the Industry", XXth *Annual Report*, Bureau of Industrial and Labor Statistics, (August, 1903). There is an account of the start up at Rumford Falls and its difficulties by a participant in James O'Hara, "Reminiscences of A Paper Maker," *Makin' Paper*, Vol. 1, no. 10 [April, 1919]. This was a house organ of the Crown-Willamette Paper Company.

[9] *The Industrial Journal*, November 22, 1889; Bangor *Daily Commercial*, November 25, 26, December 14, 16, 1889. The new firm also operated and constructed a mill at Appleton, Wisconsin. Both were in operation by the spring of 1891.

[10] *Paper Trade Journal*, March 21, 1891.

[11] *Paper Trade Journal*, April 25, May 2, August 15, 29, 1891; June 25, 1892; June 3, July 22, 30, August 5, 12, 1893; February 3, 17, 24, May 19, 1894; November 23, 1895; April 25, October 24, 1896; April 17, October 30,

1897. *The Industrial Journal* April 21, December 29, 1893; July 16, 1897. Bangor *Daily Commercial*, March 28, 1891; July 21, 28, October 27, November 17, 1893; May 10, September 5, 1894.

[12] The best source for his life is *The Northern*, Vol. VII, no. 11 (a memorial issue). Also see *The Industrial Journal*, May 14, 1886; December 9, 1887. Bangor *Daily Commercial*, January 26, 29, February 7, May 18, 1887; August 2, September 26, 1890; December 2, 1891. *Paper Trade Journal*, August 6, 1898. For his West Coast activities see Directors Meetings, Willamette Pulp and Paper Company, March 18, 1889; April 29, November 29, 1890; November 25, 1891, and agreements National Sulphite Boiler and Fiber Co., February 16, March 5, 1889, all in Crown Zellerbach archives. I have also benefited from conversations with John McLeod, Calais, Maine, who is writing a Great Northern history. He has read this chapter and commented on it. I have read, in addition, the first chapters of his unpublished GNP history.

[13] *Private and Special Laws*, Maine, 1897, ch. 458; *Millinocket, Maine 50th Anniversary, 1901-1951*, n.d., n.p., copy in Bangor Public Library. Bangor *Daily Commercial*, October 18, 1898.

[14] *Paper Trade Journal*, March 12, 1898; February 11, 1899; *The Industrial Journal*, July 22, 1898; Bangor *Daily Commercial*, February 11, March 4, 1899.

[15] Figures from a book entitled "Millinocket Purchases and Sales," in files of Great Northern Paper Company.

[16] This is a story still current in Maine. For comment see *Paper Trade Journal*, February 25, 1899, and especially Bangor *Daily Commercial*, March 6, 1899 quoting and commenting on an article in *The Paper Mill*.

[17] *The Industrial Journal*, March 17, 1899. Bangor *Daily Commercial*, March 4, 1899, editorial entitled, "What This Means."

[18] From speech before Taxation Committee, Maine State Legislature, January 22, 1901. See Kennebec *Journal*, January 23, 1901. Bangor *Daily Commercial*, March 4, April 29, May 1, 10, 1899. *The Industrial Journal*, May 5, 1899 (interview with Merrill). *Paper Trade Journal*, April 29, 1899.

[19] Bangor *Daily Commercial*, May 24, June 1, July 13, 19, 22 (in this issue is a nice discussion of the plant building, with a map taken from the Chief Engineer's floor plans, as well as a map of the townsite, mill, dam and pond), August 26, September 6, 11, October 31, 1899; June 3, 1900. *The Industrial Journal*, December 8, 1899 does the best job on the effect on Bangor.

[20] Bangor *Daily Commercial*, June 2, 1900, two full pages entitled, "Millinocket—The Magic City of Maine."

[21] Bangor *Daily Commercial*, August 18, October 13, November 28, 1900. The mill opened October 17. Also see *Annual Report*, Commissioner of Industrial and Labor Statistics, Maine, XIII, 1899, 43-5; XVII, 1903, "The Development of Millinocket," 150-173.

[22] Bangor *Daily Commercial*, February 14, 1901; Kennebec *Journal*, March 16, 1901.

[23] Bangor *Daily Commercial*, February 14, 1901. Kennebec *Journal*, March 16, 1901. Bangor *Daily Commercial*, June 18, 1902. *Paper Trade Journal*, off and on all spring for the flotation.

[24] *Paper Trade Journal*, June 4, 1908; May 20, December 2, 1909.

Big companies come 261

[25] Stories continued to circulate about big enterprises yet to be born similar to Rumford and Millinocket. See, e.g., Bangor *Daily Commercial*, March 16, 1900 for plans for the Fish River, and Portland *Eastern Argus*, January 30, 1903 for a Rumford type development for Gilead. When they came, the new developments, at least in newsprint, were Canadian. Plants like the Fraser Company in Edmundston, New Brunswick would be the new frontier of this industry in the northeast. The Bangor *Daily Commercial*, May 13, 1899, predicted this in a long story with the headlines—"ON TO CANADA. THAT'S THE CRY OF THE PAPERMAKER. . . ."

[26] Bureau of Labor and Industrial Statistics, *1st Annual Report*, 1887, 107.

[27] On the shorter day and Sunday work see the great number of letters in *Paper Trade Journal* for 1882, especially September 2, 16, 1882.

[28] On a strike at the Richards Paper Company, Gardiner, see *ibid.*, May 22, June 26, 1886. On the Knights and the eight hour day movement see *ibid.*, May 1, 8, 15, 22, 1886 for letters. For an official account of the strike see *Wealth and Industry of the State of Maine, 1886*, Augusta, 1887, 95-103, both on the Richards strike and one at the Somerset Fiber Company in Fairfield.

[29] Bureau of Industrial and Labor Statistics, *1st Annual Report*, 1887, 64-75 for the census and its results, also 78 for general average of all pulp and paper mills in the state.

[30] Industrial and Labor Statistics, *Ninth Report*, 1895, 8-11, 62-67.

[31] *Report of the Inspector of Factories, Workshops, Mines and Quarries*, State of Maine 1895, 199ff, especially 210-211, and summary remarks at 221.

[32] *Paper Trade Journal*, December 21, 1895, strike at the Poland Paper Company; *The Industrial Journal*, December 15, 1899, strike at Emden.

[33] Kennebec *Journal*, April 6, 1903.

[34] Chisholm, "The Future of Pulp and Paper," *op. cit.*, 168-9.

[35] See any newspapers for an account. The best history is in the Bureau of Industrial and Labor Statistics, *XXII Annual Report*, 1908, 416-29, and on the Oxford Paper Company, 430-5. *Paper Trade Journal*, November 14, December 5, 1907; January 16, June 11, 1908 for various labor stories in Maine mills.

[36] The best account is in Bureau of Industrial and Labor Statistics, *XXIV Annual Report*, 1910, 397-8 on the St. Croix, 401-7 on the International Paper Company.

[37] Department of Labor and Industry, *Second Annual Report*, 1913-1914, 1, for the employment figures. A census is abstracted firm by firm from 2-130 *passim*.

[38] *Paper Trade Journal*, April 2, 9, 1914.

[39] *Paper Trade Journal*, May 22, October 2, 23, November 13, December 4, 11, 1913; January 15, February 19, also in the same issue, J. P. Flanagan, "Pulp and Paper Course at Maine University", 26, Ralph H. McKee, "The Training of Young Men For The Manufacture of Pulp and Paper", March 5, July 9, 1914; February 24, J. Newall Stephenson, "America's Pulp and Paper School. . . ."; June 3, 1915; February 17, 1916, R. W. Farnum, "Technical Training for Paper Makers,"; April 14, 1921, John P. Flanagan, "Pulp and Paper Making at the University of Maine,"; June 16, 1921; June 15, 1922; *Paper* May 13, 1914, Ralph H. McKee, "Course of Study in Pulp and Paper Science. . . ."

11

THE COMPANIES GAIN CONTROL OF THE RIVERS

PART ONE

WEST BRANCH DRIVE HITS TOWN ON SNOWSHOES

> Headline, "Old Town Notes"
> Bangor *Daily News*,
> November 14, 1901.

> Held: That privity of contract is not essential to the maintenance of the action. By accepting the legislative act (P. and S. Laws 1901, ch. 293), as it did by accepting and undertaking to drive the logs, the defendant came under a duty to the public, including the plaintiffs, to drive the logs in accordance with the contract. . . . Unusual climactic conditions occur so frequently that in important affairs, they must be anticipated and guarded against. . . .
>
> *Herbert W. Marsh et al v. Great Northern Paper Company*, 101 Maine 489-506, especially 489-90.
> Opinion handed down July 24, 1906.

The rivers of Maine had historically served primarily as arteries of commerce, and since 1860 and the coming of the railroads they had served almost entirely for the transportation of logs downstream from the cutting to the sawmill. Lumbermen had formed in many places mutual companies to drive the logs, repair the dams, and generally control this traffic. Rivers without such log driving concerns had by 1890 either seen them formed by the interested lumbermen, or control of these rivers had fallen into the hands of people interested in using the water for manufacturing, not log driving.

Manufacturing was mainly possible through water storage for the purpose of maintaining an even flow in order to generate hydroelectric power or steam for power. In those places where the logs were still driven downriver to sawmills these contradictory designs on the water of rivers led to protracted and sometimes violent conflict.

On the waters of western Maine relatively little trouble ensued. Upriver mills on the Androscoggin were driving the downriver mills out of business long before this time and control slipped away quickly. The slippage was enhanced by the fact that there was no mutual company on this river. On the Kennebec the situation was similar even though there was a log driving company. The situation was aided here by the growth of many pulp mills and the demise of sawmills in the 1880's. It was not until long after this period that the Kennebec was to figure in a war over water control and the reasons then were much different.

The arena for the most protracted struggle was the waterways of eastern Maine. On the Penobscot a new way of life threatened the old, and a delaying action was fought for some years before the upriver company was triumphant. On the St. John the situation was worse because the battle involved men of two different nations. The outcome was the same, however.

Geographic position was all-important in struggles of this sort. The central problem was, of course, whether or not people located near the headwaters of the river had the right to dam the water, divert it, or otherwise control its flow. There seemed to be no clear doctrine where just damming and control were involved. In 1874 the courts had issued an injunction against the Robinson Manufacturing Company of Oxford, a textile mill, which forbade it from holding back the waters of Thompson Pond and thus denying the Denison Paper Manufacturing Company of Mechanic Falls the right to usage.[1] In 1880 though, the Gardiner Water Power Company had been formed by the Richards Paper Company of Gardiner to control via reservoir dams the Cobbossee Stream.[2] However, upriver mills that interfered with the operation of downriver mills by throwing waste in the water were penalized.[3]

S. D. Warren pursued a policy of gradually gaining control of the Presumscot and although one battle was fought with their opponents the remainder gave in gracefully, and the paper company controlled water flowage from Lake Sebago completely.[4]

On the larger rivers where there were more mills and more power potential the situation was more muddled. The first real attempt to insure that water flowage would be satisfactory to the owners of mills was the formation on the Androscoggin River of the Union Water Power Company. This firm was established in 1878 and it consisted of six textile firms in Lewiston who owned the canal and water rights in that city. They purchased the upriver dams and flowage rights as well as some land from the Pingree estate. There were four dams in Lewiston and the water was controlled by seven sluiceways, a main canal sixty-four feet wide and twelve feet deep, and several cross canals. The companies involved were charged proportionately for their use of the water. This system drained one hundred and forty-eight lakes with two hundred and fifteen square miles of water surface. Engineers said six hundred and sixty-nine streams fed the main river and the annual flowage was in excess of 135,000,000,000 cubic feet. Their dam at Richardson Lake stored twenty-four billion cubic feet.[5]

The inevitable happened. In 1879, 1880, and 1882 the drives of the Lewiston Steam Mill Company hung and suit was filed claiming the river to be a public highway thus necessitating damages to be paid to the plaintiff because his drive was delayed. The court sat for thirteen days on this important case and finally awarded a verdict for plaintiff, of $5,125.[6] The State Supreme Court upheld the decision upon appeal.[7] The lesson was learned. Water could be stored, but on the Androscoggin, at least, it had to be used for log driving if logs were driven. Another dam was constructed by the company at the end of the eighties to ensure compliance, and by the eve of World War One the company owned four big gate dams with log sluices and flashboarding for nighttime storage, and a fifth was under construction (in 1910) on the Magalloway River.[8]

As mills were built in Berlin and Rumford where logs had to be sorted when the drive came down, the difficulty of the drives increased especially in low water seasons. Water was always a problem.[9] In 1893 the drives were late because of sorting at the boom of the Berlin Mills Company. Suits were threatened but if any materialized, damages were arrived at out of court.[10]

Two years later some enterprising people realized that this sort of difficulty might arise at Rumford. They asked the legislature to charter a concern to be called the Rumford Falls Sluice and Improvement Company. This firm was actually a subsidiary of the water power company, and it was designed to save water and to control the drives by sending the pulpwood at a different time than the timber and charging a toll on all of it. The bill was indefinitely postponed in the Senate with the President of the Senate casting a tie-breaking vote to ensure the delay.[11]

The old company, the Rumford Falls Boom Company, remained in control. Under its charter, granted in 1893, the company could charge up to 10 per cent of the value of its capital stock as tolls in each year. In 1893 two-thirds of the logs which went through belonged to the Rumford Falls Paper Company; in 1894 and 1895 all belonged to the paper company, and in 1896 ninety percent of them were owned by the paper company. No tolls were paid to the Boom Company on the grounds that the paper company was the sole user of the facilities, and therefore the Boom Company had nothing to do to earn its money. In the suit which followed the court held the paper company responsible up to the limits of the boom company's charter, and tolls were paid.[12]

With this case though the lower Androscoggin was finished as a log-driving river. There were no down river mills left of any consequence, and the International Paper Company gained effective control of the river when the corporation was created. The court had held that the Union Water Power Company had to drive logs and it did, but more and more moved by railroad and truck. The upriver men had won this fight with relatively little difficulty.

The situation on the Penobscot remained about the same as it had in 1846 when the Penobscot Log Driving Company was chartered, at least until the Great Northern Paper Company built its mill at Millinocket. Here the courts had held that the Penobscot Log Driving Company was the sole carrier to drive logs and that it was a public carrier.[13] Those who changed the river's natural course could be held responsible for damages,[14] but apparently not if the change only involved the dumping of pollution into the river.[15] Rafting companies had to raft no matter the conditions, they were responsible for freaks in the weather and were forced to pay the log owners for damages thus incurred.[16] On the other hand the companies did not need to provide water for individuals if they failed to get their logs into the driving place at the correct time.[17] Responsibility thus sometimes was on the shoulders of the log owner.

There were times however when the log driving company was not responsible for weather changes, as on the Mattawamkeag River in 1900. Logs were designated for two different drives on this river at two different toll prices. High water came suddenly and mingled the drives, but the courts ruled that the drives did not lose their identity under these circumstances, and the owners of the logs were liable for the payment of tolls originally set even though they differed for each drive.[18] Here the situation rested in 1901. Public carriers had been chartered with definite responsibilities, many of which had been determined by the courts. It was possible though for some of the responsibilities to fall on those who owned the logs and who employed the log driving companies.

The pulp business growing up at this time was very profitable but it was also dependent on the water supply. Each summer and fall was an anxious time. Would the mills be forced to shut down? Or would they go on short time? Low water always was a trial even long before the Great Northern came to make its demands on the river, as the following editorial excerpt shows.[19]

> While the drought is quite serious in all parts of the state, the mills on the Penobscot, though the water is lower than has ever been known before, are running day and night. *This is the season of harvest,* for every pound of pulp which they can now make finds a ready sale, and there are vague whispers of a little better figures than there have been previously known for some time. So there is a compensation for low water.

With the coming of the paper company though the battlelines were clearly drawn. At the annual meeting of the Penobscot Log Driving Company in 1899 an initial skirmish took place when the chair was asked to rule on the validity of proxy voting. John Ross, the president, ruled that they were illegal. When pressed again from the floor, this time for the reasons behind his ruling, he asked for the aid of an attorney and the meeting was adjourned until a legal ruling was obtained. The ruling proclaimed that proxy votes were all right, but only if they were filed ahead of time and with the appropriate special papers.[20] If the new company was going to vote all its lumber and horses it would do them in person at least, as a result of this ruling.

The next year was the first year that the new company had much interest in the water. A compromise had been worked out on the river. The Northern had taken some water from the lower lakes while the drive was forming, on the condition that while the drive was coming into the main river the gates would remain up. As soon as the drive passed Millinocket, however, the gates were lowered and the water grew very short for the drive. The river was jammed with logs being sorted at Mattawamkeag and Montague. Rumors and murmurings of lawsuits over control of the water began to circulate. It rained heavily in mid-October and the drives came in and mills started up.[21] Trouble had been averted but not by much.

Trouble came again early in 1901, and it wasn't averted this time. As soon as the legislature convened in Augusta a bill was presented to incorporate the West Branch Driving and Reservoir Dam Company. This company was to replace the Penobscot Log Driving Company, was to make one drive a year, and was to be paid certain rates for its driving service. The crux of the bill was the new company's board of directors, however. It read like a list of the Great Northern Paper Company's founders—Frederick H. Appleton, Joseph

C. Bass, J. Fred Webster, Fred A. Gilbert, A. Ledyard Smith, Jr., J. Sanford Barnes Jr., Payne Whitney, R. Somers Hayes, and Garrett Schenck.[22] The situation was obvious. The paper company needed to insure control of the water and the only way this could be done was to control the drives.

The bill was referred immediately to the Joint Committee on Interior Waters which included three senators, one a manufacturer of lumber, and one an owner of a sulphite mill. The House members, seven in all, included three lumbermen and two mill owners. The Legislature as a whole reflected about this breakdown. Of the one hundred and fifty-one members of the House, eight were lumbermen, and three were lawyers dependent on the lumber business; of the thirty-one senators, one was a paper maker, one a pulp maker, one speculated in timberlands and one manufactured lumber.[23] For obvious reasons the initial bill really stood little chance of passage. In the course of the legislative history of the bill thirty separate remonstrances against it were received by the legislature, with some 2,600 names appended, representing nearly every downriver town and such groups as the Passadumkeag Boom Company, "laborers on the Penobscot River," Bangor business houses, Maine pulp and lumber manufacturers, as well as river towns themselves.[24]

The remonstrance campaign did not begin immediately however. As a matter of fact the first response was quite tentative. The first to speak was William Engel, a Bangor politician and lumberman, and he was most cautious. If the Great Northern wanted to dam to save for the dry season, all right; if it wanted to divert the water, that was a different matter.[25] Newspapers were also cautious. The Bangor *Commercial* quoted an anonymous individual as saying it was the duty of the state to control the waters for the use of all. The anonymous source cited the annual messages of Governor Chamberlain in 1867 and 1868, as well as that of Nelson Dingley in 1874 as evidence for this point of view.[26] The Bangor *News* also used an anonymous source, in their case through a letter. It was a telling letter though,[27] which read as follows:

> It is a matter of record that when E. S. Coe of this city, sold his dams and privileges on the Androscoggin to the city of Lewiston for the water supply of that city, he reserved enough water to drive such logs as might be needed at Lewiston sawmills. It is also a matter of record that no logs are now driven to Lewiston, the mills were burned or abandoned, and the pulp mills on the Upper Androscoggin have absorbed all the logs that were cut, so that the reservation for the use

of that water is not needed; and the same thing is liable to happen to the city of Bangor and all west branch logs if this company secures control of West Branch waters.

The *Commercial* ventured their first comments the next day, in an editorial entitled "The Penobscot of the Future." While advocating trust in the good faith of the legislature, the writer said that storage was important and that the bill was not as horrible as its opponents thought.[28] The *News* came down on the other side citing among others Llewellyn Powers of Houlton who was convinced that the rights of lumbermen "should not . . . in any way [be] made to depend upon the will of a great corporation." [29]

The newspapers kept up their interest in the campaign, but their positions were fairly well set.[30] Protest meetings began. The Penobscot Log Driving Company, the Bangor Boom Company, the First National Bank, the Eastern Trust and Banking Company—all these called meetings, elected representatives to lobby against the bill and generally bemoaned their fate.[31] Newspapers were filled with letters. Edward Stetson, a prominent Bangor lumberman, and Patrick H. Gillin attorney for the downriver men throughout these years were in the fray.[32]

The bitterest attack of all though came from S. H. Prentiss, at a meeting of the Bangor Board of Trade. In a most prescient speech he said the ultimate effect of the bill would be to get all pulp and lumber produced above the Penobscot boom area, to restrict sales to the market provided by the railroad, thus to control stumpage prices, and ultimately to force the sale of lands. Prentiss, who was a very minor stockholder in the Northern at this time, was listened to with much respect. Others seconded his remarks, and a committee of twenty was chosen to lobby with "every possible means" to prevent this bill which "would mean the end of Bangor shipping and sawing." [33]

All, however, was really in abeyance until the Penobscot Log Driving Company held its annual meeting. The two Bangor newspapers restated their positions one more time before this meeting.[34] The *News* also printed some letters attacking the bill,[35] and the public waited with bated breath to find out what would happen when the two forces met face to face at the annual meeting of the P.L.D.

The meeting was as lively as expected. When it was called to order the downriver men immediately moved for adjournment. After a debate over the legality of proxy votes the motion was withdrawn. F. W. Ayer, speaking for the downriver forces, and James W. McNulty, for the Northern, exchanged remarks.

At this juncture John Ross accused McNulty of mismanaging the water at North Twin Dam in 1900, and a heated discussion followed which dealt mostly with personalities. S. H. Prentiss then moved adjournment on the grounds that the legislative action might be inimical to the log driving company, and thus they should await the action of the solons. Debate followed which was very noisy, and the chairman had difficulty in maintaining order until the previous question was moved—by a downriver man. The motion to adjourn was again withdrawn. A lengthy discussion followed as to the eligibility of voters and votes, and many comments were thrown with much acrimony at the Northern who had moved many horses into the woods in recent weeks. (Votes were apportioned one to each landowner, and one for each four horses at work in the woods.) The paper company answered in the person of Fred Gilbert who was incensed at the aspersions cast by the debaters. McNulty and the company then agreed to vote only their working horses.

When a vote finally came the issue under contest was the makeup of the new Board of Directors. Two slates were presented, one for each of the great antagonists:

Rival Slates—Penobscot Log Driving Company

Board of Directors Meeting of February 12, 1901

Great Northern Paper Company	Downriver Men
James H. McNulty	John Ross
James Rice	James Rice
J. F. Mullen	C. A. Strickland
Charles J. McLeod	H. B. Morrison
Fred Gilbert	Fred J. Ayer

A committee of five was chosen to count the ballots. Some were challenged and the committee ruled several invalid. The meeting went into recess while the committee made its count. McNulty was asked to leave the polling area at one time, and eventually the committee locked itself up to prevent interference with the count. They were closeted for an hour and a half, and the members were waiting with great impatience when they returned at 1:40 p.m. Two hundred twenty-three and one half votes had been ruled valid, which meant that one hundred and twelve were needed for election. Rice of course had been elected as he appeared on both ballots. The others elected were Ross (112), Strickland (114½), Ayer (114½), Gilbert (121½), and

McNulty (121½). Counting Rice as a downriver man as he proved to be, these forces had won a narrow but extremely important victory by a four to two vote.

The new directors met and immediately consolidated their victory. Ayer was elected president, and Rice clerk and treasurer. The directors then passed a resolution to send Judge Louis Sterns to Augusta to lobby against the proposed West Branch bill. The last piece of business, and it had great reverberations, was to appoint Fred Gilbert as master driver of the West Branch drive.[36]

Meanwhile a compromise had been effected at the legislature. The obnoxious bill was to be withdrawn and the Great Northern was to drive the logs. The company was however to be guided by a three man committee consisting of F. W. Ayer, F. A. Gilbert, and James W. Sewall (a civil engineer from Old Town), as to when the drive ought to leave Chesuncook Dam. A two year contract to this point was negotiated between the two groups. The old bill was thrown out and the new one substituted for it. With this a two-year moratorium had been declared and a truce signed. "Compromise is always fair," said the *Commercial*. The other newspapers simply reported the events as they occurred in Augusta.[37]

Would the compromise work? Apparently some people thought so; both the paper company and the downriver men would be happy— one with the steady supply of water, and the other with the quick delivery of their logs. It only remained to be seen how well the drive would come down.

Initially at least a good driving season was expected. There had been much snow over the winter (twenty-two storms) and it had been cold. Pessimists however pointed out that a late fall rain (the one which had brought the 1900 drive in) had meant that the ground was not frozen deeply enough. If it warmed early the snow would leave very rapidly making the drive difficult.[38]

High water came to Bangor the first week in April, rising to within an inch of the all-time record high set in 1867. Booms broke and some logs went to sea. High water was reported on the Union, Narraguagus, St. Croix, Kennebec, Androscoggin, and Machias Rivers as well. Log prices began to rise. Many began to assume that drives would be late and logs scarce as a result of the early melt. John Ross, very sanguine one day, sold his logs the next (7,000,000 feet to Fred Ayer for $15.00 per M). The drives were in trouble, and the rollways barely broken out. The news continued to worsen through the month of April. Warm days came and freshets were reported on both main branches as well as elsewhere. As one paper said wistfully, "The driving above Chesuncook must be all right, although no definite word has yet been received." This refrain or a variation of it continued to be heard throughout Bangor lumber circles during the summer.[39] The drive was always coming, it was always safe. The readers may have been reassured, but the logs did not arrive.

Water was low everywhere in the river. The East Branch came very slowly, all in one drive. The Mattawamkeag was split and part boomed off in order to bring as much as possible in. Dams upriver (on the East Branch) were closed in early June in the hopes of conserving what water there was. Downriver men became disturbed at the prospects, and a delegation went to ascertain the East Branch prospects. Only the Piscataquis drive came in in season, it was being sorted at the Howland gap in mid-June. On the East Branch, men began to drag down the logs from Grindstone, when a rain on June 21st gave hope to the log owners. Still the gates at Grand Lake Dam were not opened as the Allagash-Chamberlain lake logs were not ready to sluice. Headwinds were bad and the booms remained in place until early July.

Word finally came from Chesuncook. The South Branch Penobscot logs had not arrived until June 13. This was doleful news. When July 1st arrived and the logs had not yet cleared Chesuncook Lake it was obvious to lumbermen that only late fall rains would bring them to Bangor. The last week in June and the first week in July were terribly hot with temperatures in the nineties every day. By the tenth of July all mills on the river were closed. There were no logs in the boom. On that day however some logs from the Piscataquis sorting began to run in, and a few mills began to make plans to reopen. Rumors flooded Exchange Street bars and businesses about the West Branch drive. Some said that it would never come in. Fred Gilbert, answering to the rumors, announced that the rear was at Ripogenus and the forward end was sluicing at North Twin. Even this assurance did nothing to dampen the rumors, and newspapers described Bangor lumbermen as "panicky."

The East Branch was a bit better off. On July 16, the hottest day in Bangor in twenty years, the forward end was beginning to traverse the North Lincoln sorting gap. On the 25th it began to clear past the Passadumkeag, and when on July 30th it rained again, many breathed a sigh of relief. However as the drive came down, men for the next winter's work were beginning to head for the woods, and wages were announced as high because the usual labor supply was

The companies control the rivers

still riding the river. Sawmills began to start up on August 1, and by the 6th rafting was on in full swing. Ominously enough though, the rear was still at Five Islands and the drivers said there was not enough water in the river to move it until the West Branch water came in. By the 21st, logs were all sorted at Lincoln and were waiting, but news then came down the river that the West Branch had not yet cleared the Lower Lakes. Another shutdown occurred at the Bangor Boom, but not for too long as a rain on August 26-27 brought a few logs in from the East Branch rear.

Gilbert came to town on the 28th and stopped long enough to report that the logs were scheduled to leave North Twin that week. He said that the drive was "going ahead satisfactorily."

Gilbert was premature. The logs did not clear North Twin Dam until September 2. James McNulty went to observe and when he came back he told a reporter that the men had had three weeks of headwinds and fifty million feet of logs were piled clear to Seboomook but that the drivers were "all set to push now." Even from here the news was premature. The drive was still at Quakish Lake in mid-September and did not in fact clear until September 15th.

Some of the East Branch rear had been sacked in a boom at Mohawk Lake and, upon receipt of the news of the West Branch, men were hired to go and get it in the hopes of rafting the logs before the West Branch came down. Water was still low, and suddenly on the 16th of September the paper company shut the gates at Millinocket and stopped sorting in order to get more water. The sorting did not finish and the remaining logs were not turned into the river until September 20. By this time veteran rafting men said that the water was so low that it would prevent the logs being handled even if they came into the boom. Forty-four million feet had been rafted, none from the West Branch, when this lugubrious prediction was made.[40]

Apparently the men had gone to get the East Branch rear at the urging of Fred Ayer and others. Ayer claimed later that he made many phone calls to the Penobscot Lumbering Association urging that these logs be rafted. On September 25, 1901 he was granted a writ by Penobscot County Court directing Samuel Sterns and the P.L.A. to put more men on as rafters. The writ said, "It is easy to obtain sufficient number of men by paying a fair rate of wages, . . ." The writ also stated that unless a full crew of rafters was put on, the Eastern Manufacturing Company would close down at damages of

$1,000 a day for which the Penobscot Lumbering Association would be held responsible.[41]

The P.L.A. had already advertised for a hundred men to come to work apparently as a result of this legal threat. On that day there were just six sticks to be sorted, and the crew was "drizzling them into Argyle Boom", all twenty-three million feet of logs. Rafting started the twenty-fourth, and the eighty men on the drive kept the rafters busy. These men were described as "feeling a little anxious" about the West Branch drive.

The first steam mill drive cleared the Bangor boom on October 3 but it was small, only two million feet. The West Branch continued to drag along. It passed Mattawamkeag on the 11th and was then expected in, if no rain fell, about November 1st. Rains came on the 15th but not enough to aid this drive. It was strung out between Mattawamkeag and Lincoln with two hundred and sixty-five men employed in "dragging it in." Men in Bangor began to talk of wintering as much as possible in Pea Cove and behind Freeze Island in Costigan and shingling up the rest. One hundred and sixteen men were sorting and rafting at Argyle.[42]

Ayer continued his barrage of threats at the P.L.A. In letters dated October 3 and 17 he kept urging Samuel Sterns, president of the P.L.A., to hire more men.[43]

> It seems as if it would be only common decency to have a full number of men at both of these booms (Nebraska and Argyle). Not only for me, but all your company knows that the West Branch logs will soon be in the boom and the East Branch logs should all be rafted out before the West Branch logs commence to run in.

On October 21st it snowed and the temperature began to drop. Now it had obviously become a race against time. Two days later the West Branch began its sluicing at Montague. On this date the rear of the Mattawamkeag second drive, hung since June, came on, and some drivers left to shepherd these logs to Lincoln for the Katahdin Pulp and Paper Company. The West Branch cleared Montague on the 29th and began to move for Bangor. The river continued to drop steadily and men began to leave for the woods. By November 1st the crew was down to only one hundred men. The work was bitter cold. However, it was thought that if the river did not freeze solidly the logs might be made safe.

Plans were put in train to raft out as many as eight or ten million feet, and a dam was projected at Eveleth's point to aid the wintering in Pea Cove. On November 6th the last of the East Branch logs were rafted out. The West Branch was now running in very slowly with the main body still five miles from Argyle boom (two miles below Passadumkeag). By November 8th the main body was opposite Olamon stream, and on the ninth the drive reached Sugar Island. The tenth was extremely cold as the men began to put some logs in behind Freeze Island. A special meeting of the P.L.A. was called on the eleventh as the drive kept straggling in. The expense of shingling was discussed; a dam was started in order to back water up into Pea Cove, and plans were put underway to handle as many logs as possible. It was possible now to walk across the river at Pea Cove in one place the water was so low, and when the dam began to contain what little water there was the paper mill at Orono threatened an injunction so work then stopped on the dam. It made little difference. The one hundred and fifty men now working on the logs were obviously not going to move them all into safety. On the thirteenth a foot of snow fell and the rear came dragging in bringing the famous headline about the West Branch drive coming in on snowshoes.[44]

A great struggle now ensued, a struggle between men and the elements. Logs were scattered from Argyle Boom to Bangor and men attempted to get them out of the water or at least into a safe place before the river closed for good with ice. Seven million feet were in rafts on the way to Bangor, and 61,000 sticks were at Great Works when the rear of the drive appeared at Argyle. By the 19th these were clearing the dam at Bangor and by the 21st all were in the boom and rafting was progressing with great speed.

Work was being rushed up river to get as many logs as possible behind Freeze Island before the ice came. By the 22nd the river was freezing every night and the logs were barely able to move. The river had became a floating gurry of logs, ice, slabs, dirt, and sawdust. Working conditions were intolerable. On the 26th there were still 18,000 sticks of about 2,500,000 feet in the Bangor Boom, 400,000 feet in Pea Cove, 4,000 sticks frozen in the river between there and Argyle, 4,000,000 feet behind Freeze Island, and between 16 and 18,000,000 more feet in the booms waiting to be hauled out and shingled up on the banks.

On the 27th, the day before Thanksgiving, the river froze solid north of the Bangor Boom. Owners of the logs decided that the best thing to do was to break the logs from the ice and haul them to the banks as soon as the ice was safe for teams to work. The logs in the Bangor Boom area, 8,500 sticks above the dam, and 7,500 below, about 2,000,000 feet in all, were protected by an ice clause in the contract and the owners were responsible. They immediately hired Patrick Connors the son of the boom foreman to haul these logs from the ice and to shingle them up at a price of 15¢ each.[45] Wages on the boom had risen from the normal $1.75 a day to $2.50 as the weather had grown colder and the work more dangerous.

The river below the Bangor Boom was still open, and as the stevedores worked to clear the last few vessels Bangor began to batten down for the winter. The work of moving logs to the banks was slowed considerably by a storm on December 4 which dumped two feet of snow along the river. The storm was followed by 25° below zero readings on the sixth, and everyone was convinced that winter had really arrived. Maine weather however is at best unpredictable, and on the ninth, temperatures soared into the forties. A steady rain began to fall, and the ice began to break up.

The remaining logs were in great danger, and emergency crews were put on to haul them from the water. Men did this work until the twelfth at mid-day when the work simply became too dangerous. The last twelve hours work were accomplished only by ropes and boatswain chairs which prevented the men from slipping into the water. The river rose three feet and on the thirteenth the ice below the dam, along with logs lodged there, began to move slowly towards Bucksport. On the fifteenth a terrible wind and rain storm completed the work. All the ice went out before a fifty-knot wind and many logs went with it. Between five and six million feet went out, and another nine million feet was in grave danger. Tugs were called to attempt to stop the logs at Bucksport, and special trains went down from Old Town with men and batteaux to pick up what could be salvaged.

The river continued to rise eventually reaching a high point of nine feet and nine inches above the low water mark, and the ice and log mixture at Bucksport was too much for the would-be salvagers under such risky conditions. A great jam of ice, logs, dirt, slabs, and other flotsam was created at the Bucksport Narrows, although this went to sea on the nineteenth. Logs were reported all over the bay and as far south as Rockland. Only a few were salvaged. Connors offered 20¢ a stick to any picked up on the beach. One man walked the shore investigating the disaster and found some 5,300 sticks between Bangor and Winterport, most covered with ice and many battered beyond usage. The river never opened again after freezing on the 24th, and the year ended in acrimony, amidst talk of threatened lawsuits although against whom, the P.L.A., the P.L.D. or the G.N.P., it was not yet known.[46] The Bangor *News* attempted to put on a brave front but ended up clutching at some very old straws.[47]

> It is quite certain that nothing will be gained by censuring anybody for what is past. No doubt the G.N.P. Co. used up a lot of valuable time in sorting out its logs, and thus retarded the West Branch drive, but growling at the Millinocket corporation will not bring back the lost logs, or pay the expenses entailed by the flood. Whatever remedial action must be taken must be for the future.

After taking this Pollyanna view the editorialist went on to say that difficulties would no doubt increase, and to forestall them the only answer was to practice forestry and build railroads into the woods. Failing this, perhaps free lumber or free pulp might be the answer in the next tariff bill. Or, if this last was too much for the normally Republican high tariff readers of this paper, perhaps annexation of Canada would solve all the problems. The editor wound up in a blaze of imperialist glory:

> For ourselves, we believe that annexation is the easiest and most rational solution.

The paper company was also going to handle the 1902 drive, and if the last one was any indication as to the way it would be done, it was time to think of liquidation. And, although the denouement of this little play had been reached, the last act, the great spring freshet of 1902, was still to be played out before the legislature could either rectify its damages or, worse still legitimize them with passage of a West Branch bill.

The next meeting of the Penobscot Log Driving Company was lively, but the contests of 1901 were not repeated. Gilbert and A. Ledyard Smith were the main representatives of the paper company, and some acrimonious remarks were passed.[48] Logs had been lost before and drives had been late, even though not as late as this drive. It was not until spring when the ice went out that the downriver men really became unhappy with the paper company.

The river began to rise during the first week in March. The weather was unseasonably warm and the melt rapid. The river filled and the ice began to move. A giant jam of ice formed in the Bucksport Narrows south of Bangor. By the middle of March the river towns were in great danger as a week of rain coupled with the melting filled the rivers to flood pitch. By the 20th the river at Bangor was filled with running logs and ice and booms began to give way. The situation was not helped by opening North Twin Dam to relieve the strain upriver.

The great ice jam at the narrows caused another to form at High Head, the location of most of the wharves. Ice and water backed up and two bridges in Bangor were carried away. Dynamite was used on the jam but before it had much effect the whole mass, logs, ice, and so on, all went out to sea. The only logs to survive were those lodged behind Freeze Island and the few in Pea Cove. Moreover, the freshet came so quickly that it left close to a fourth of the winter's cut high and dry on the rollways in the woods. The river quieted enough so that rafting was started in early April in order to ascertain the extent of the loss. A crew was detailed to go below on the river and pick up the drift drive. Patrick Conner, who mustered this group, said that his men would live on a scow all summer, and that early estimates showed that they might salvage as much as 2,700,000 feet, or about 10,000 sticks.[49]

The crew worked hard all summer. The logs were hauled into booms at Dyer's Cove, Brewer; and Bald Hill Cove, Winterport, while those that went to sea came up in a big raft from Castine. Tugs hauled the booms upriver to Ayer's mill (the Eastern Manufacturing Company) as he had purchased the drift logs. These logs were all salvaged by September 1st, and Connor was right as to the approximate amount of logs salvaged.[50]

The 1902 drives were always in some jeopardy as a result of the early freshet, but three days of rain helped bring them on. The Mattawamkeag and the East Branch first drive cleared into the main river together and they came on to Bangor by mid-June. The men on the Penobscot boom went out on strike just as the drive came down, and as a boom broke at Montague letting down fifteen million feet the same day, their pay rise was granted. More rains came in early June and the drives were secure. Gilbert also installed a telephone between Chesuncook and the Lower Lakes which made the West Branch drive easier. On July 1, the drive began to sluice at North Twin Dam. The last of the East Branch and the West Branch drives were consolidated at Lincoln and came on together. The drive arrived on August 26th, twenty-eight million feet from the West Branch, and twenty million feet from the East Branch. The first steam mill drive arrived in the Bangor Boom on September 5th. When Argyle boom closed on October 21, one hundred and forty-two million feet had been rafted out, sixty million of it to tidewater.[51] Much of it was from the 1901 drive though as the table suggests.

The companies control the rivers

Rafting—Argyle Boom—Season of 1902

Month	Amount in Feet
April and May	22,000,000
June	21,500,000
July and August	42,000,000
September and to October 18 when closed	58,000,000
Total	142,500,000

When the river closed that fall and northern Maine settled to its usual somnolent state during the winter, everyone was aware that it would not be hibernation but only a nap until the Legislature met again. The new Legislature had one hundred and twenty-eight Republicans and twenty-three Democrats in the House and thirty Republicans and one Democrat in the Senate. Twenty-six members of the House were associated with the lumber business and four of them were representatives of paper companies. Two members of the Senate were in lumber, both of them from Penobscot County. The Great Northern Paper Company had two of its employees in this legislature. The first was George W. Sterns of Millinocket, their real estate agent and the second, and more important because he became the manager of the new bill in the legislature, was A. Ledyard Smith of Madison. Smith had been a real estate broker and insurance agent in Appleton, Wisconsin, until 1889 when he had gone with the Combined Locks Paper Company in that city. He had also worked for a Michigan firm before 1891 and 1892 when he was appointed Superintendent of the Manufacturer's Investment Corporation plant in Appleton. From 1893 to 1898 he was general manager of both the Appleton and Madison firms of the Great Northern's predecessor, and in 1899 he received the position of assistant to the president, a title which he still held in 1903. This meant in fact that he had charge of the Madison operations.[52]

Upon the convening of the Legislature the paper company's bill was presented, a bill which was almost identical to the first bill in 1901.[53] It was immediately referred to the Joint Legislative Committee on Interior Waters. This committee, which would make most of the ultimate decisions, was an interesting one. Ten men made up its personnel. From the Senate came a civil engineer, a manufacturer, and a lumberman-merchant from up the Penobscot (Patten). The

House members included four men who were intimately connected with lumber, a "law student," and two representatives who had devoted their lives to the pulp business, Ledyard Smith and Waldo Pettengill of the Oxford Paper Company.[54] Just how this committee would decide to act on the bill which would determine the future of the Penobscot River and all that that entailed was not immediately clear. There were many lumbermen, just as there had been in 1901 and would always be in a Maine legislature, but there were also manufacturers and of course Smith to represent his company interests.

The bill went to the committee on January 20, and by the end of February the committee had also received forty-two petitions in opposition to the bill. These petitions represented 3,848 names and nearly every downriver concern and corporation involved in lumbering in any form. There were in addition twelve more petitions presented to the Senate, totalling 1,147 signatures. The signers read like a Who's Who of Maine Lumbering.

The petition technique of influencing potential votes was not lost on the company however. On February 24, twenty-two petitions were presented by supporters of the bill. These petitions with 2,298 signatures also mirrored the river industries. Of the signers, 725 came from Millinocket, but conspicuous among the rest of the supporters were two groups of "log operators on the West Branch," the Bodwell Water Power Company (the firm that controlled the Penobscot at Old Town, a subsidiary of the Penobscot Chemical Fibre Corporation), the mayor and Alderman of Old Town, and many individuals from Bangor, Old Town, Orono, and other Penobscot river towns.[55] The battle of the petitions simply indicated the seriousness of the issues which were to be settled.

Just as in 1901 though, much of the discussion took place outside the legislative halls. The campaign to establish positions had begun very early. Log owners on the Penobscot had refused to pay their tolls for either 1901 or 1902 on the grounds that the driving had been done badly, losses had been incurred, and litigation might well ensue. The Log Driving Company had then initiated suit to obtain payment. On January 5th though, a special meeting was called and the members voted to non-suit the cases. James McNulty, president of the P.L.D., protested at this action but was overruled by the vote. When this occurred he took his case to the people in an open letter published in local newspapers. In the letter he claimed that the decision had been taken at the behest of the Great Northern who controlled the Board of Directors (elected in 1902). He asserted that the company wanted to join the unpaid tolls in one suit for both years in order to protect the poor driving job of 1901 and to prevent its ventilation while the new bill was before the legislators. According to McNulty both Fred Gilbert and the lawyers for the paper company had said this to him. Instead he called for a trial which would determine the causes for the failure of the 1901 drive, and he even offered to pay his 1902 tolls if the suit was continued. Furthermore he offered a bond in the amount of *all* delinquent tolls for 1902 to be paid if he were beaten in the case resulting from the 1901 drive.[56]

Newspapers in the state were hesitant in commenting. The *News* seemed to shift sides coming down partially in favor of the new West Branch bill in contradistinction to their vehement 1901 position. They extolled the virtues of water storage.[57] The *Argus* seemed to oppose the bill commenting that the International Paper Company was in opposition, had been, and would continue to be an important force in the forthcoming defeat of the new proposal.[58] The Kennebec *Journal* simply commented, "The Great Millinocket contest, so-called, is on."[59] The other Bangor newspaper, the *Commercial*, also appeared to be shifting sides as it carried more pro downriver news than any other newspaper in the state.[60]

On January 16 the first suits were filed against the paper company as a result of the 1901 drive. Suits to the amount of $96,000 were brought by eight owners of logs swept to sea. In addition it was said that six sawmills were about to file suit because they had been forced to shut down because of the erratic way in which the water was handled. In the initial complaint the downriver men said that the amount of water was exactly equal to the year 1899 when the drive had come down all right, and the fault lay in the paper company's use of the water at their Millinocket mill.[61]

At Augusta the first skirmish in the battle came on January 23rd when the Legislative Committee met to determine the date for a hearing. Smith moved a hearing in two weeks (February 5); Shaw of Bath immediately moved a hearing in four weeks (Feburary 19). Testimony was taken and it soon appeared that the company wanted the earlier date while the lumbermen wanted the later. John Ross of Bangor presented letters from several lumbering firms and prominent lumbermen asking for the later date. A heated exchange occurred. A three week compromise amendment was offered and a four to four vote with two abstentions was taken. One of the abstainees then changed his vote to defeat the measure. The two week motion was then defeated four to five with the chairman not voting and the four week motion passed by the same vote. With this the lines were drawn. Maddocks, Sweeney, Pettengill, and Smith voted for the motion, while Gardiner, Shaw, Ross, Sargent, and Putnam voted no. Burleigh, the chairman, abstained. Downriver men were exultant. They had won quite handily the first time the forces came to grips. Burleigh was the only unknown quantity now, or so they thought.[62]

Other bills were presented to the legislature now. One was a bill to charter the East Branch Improvement Company (some thought this was to divert more water from Chamberlain Lake and the St. John loggers became very upset). Bills were also offered which would allow damming of the St. John at Winding Ledges near Fort Kent. The sponsors were the same men who owned the Van Buren sawmill which caused so much difficulty in another year or so. A toll bridge at Van Buren was also advocated as part of this scheme. By the end of January bills were before the legislature which would affect, if passed, the East Branch Penobscot, West Branch Penobscot, Sebasticook, Allagash, Monson, main river Penobscot, and St. John rivers. The *Commercial* said the legislature would go down in history as the "Water Legislature".[63]

Downriver forces in opposition to the proposed West Branch bill began to amass their artillery for the big fight ahead. The Bangor Boom Company passed a resolution comdemning the act. The P.L.D. postponed their meeting date for a month until March 10 but not until a fiery preliminary board of directors meeting had taken place. A mass meeting of lumbermen, millmen, landowners, and others was called in place of the originally scheduled meeting. This meeting passed a very strong protest resolution and on the next day a bill was introduced into the legislature on behalf of this group, which would have modified the P.L.D. charter to give that corporation the right to build dams and store water. Under their bill the firm would have the right to transfer their property to the Great Northern Paper Company if a guarantee was posted which would insure water enough for driving and rafting. The new P.L.D. was constructed in such a fashion that no one company could control it. To insure their position being known, the Bangor City Council denounced the original bill, praised the new one, and instructed the mayor to go and lobby against the Great Northern bill when the hearings were held.[64]

Bills and control of the rivers seemed to be the main topics of conversation throughout the state. Everyone knew very well exactly what was at stake, and editorials were commonplace referring to the

The companies control the rivers

forthcoming legislative action as among the most important events in years.[65]

When the day set for the hearings finally arrived both sides sought postponement. New bills had been offered which tended to clarify the situation somewhat. A water storage bill was presented which called for a survey of storage and flooding conditions on all Maine rivers, to be conducted by a new and bi-partisan Water Power Commission. The original Millinocket bill was reworded, partially to protect the water rights of the downriver men and also to give more power to the supervisory water commission who would be in charge of starting the drive.[66] After this, hearings were set for February 23.

Opponents filled the air with editorial comments in the last few days. These comments went over the old ground mostly—control ought to lie in the hands of the people . . . the rights of all should be safeguarded . . . the rivers belonged to all and the legislature had no right to turn them over to a few—such were the typical comments. Only one added any heat or light to the already bright battle.[67] That one, in the Bangor *Commercial*, vented the problem of giant corporations and trusts and pointed out that the Great Northern Paper Company was in many ways a creature of these trusts. After urging postponement the editor fired his most telling piece of artillery—one which would ordinarily be kept in the rear but would invariably be brought forward at crucial moments in the Great Northern's history thereafter.

> Mr. Payne, one of the largest stockholders in the Great Northern, is also one of the principal owners in the Standard Oil Company. We do not mention this in any opposition to Mr. Payne but that the members of the legislature may realize the importance of legislation that will affect the interests of every citizen in the state.

The hearings opened the 24th of February to a standing-room-only audience in the House of Representatives chamber itself. C. F. Woodward a lawyer for the paper company opened the discussion. He said the company had spent $8 million and that these were investments not promises. He said Bangor had benefitted as the natural entrepôt to the inland waters, and the mill was not as some had said a destroyer of forests but rather a perpetuator of them. In his hour long speech he quoted from past governors, cited men in favor of conservation, and ended his peroration with statistics on lumbering and water storage. His speech was an effective presentation and was greeted by extensive applause.

The paper company officials contented themselves with calling only two witnesses, Hardy S. Ferguson their chief engineer and Fred A. Gilbert. Ferguson testified as to the plant output, the amount of timber consumed, prices of labor and stumpage, and the water flowage in the river. He also gave cost figures on driving from Chesuncook to the Penobscot Boom.

The "defense attorneys" for such they seemed to be questioned him closely but not about his previous testimony. Patrick H. Gillin, Fred Ayer's attorney and general counsel for the downriver men, was interested in preparing his ground for the pending case on the 1901 log drive. He asked Ferguson about the length of the 1900, 1901, and 1902 drives. He asked about the difficulty of moving logs across Chesuncook. Ferguson was reticent in his answers to these questions but eventually under pressure from the questioner gave "I don't know" as his answer to a series of questions on the drive and the water supply. Woodward in rebuttal asked him to testify as to the effect of the 1901 drought on paper production in that year. Ferguson seemed on surer ground here and he said that they were forced to cut back to thirty tons a day as opposed to a normal output of one hundred and sixty. After this Ferguson was excused and Fred Gilbert followed him to the stand. His testimony all concerned the 1901 drive.

Under questioning he agreed that the 1900 drive was normal, but insisted that 1901 was very dry.

Q. — Was it as dry as Augusta?
A. — About the same.

He covered the 1900 and 1901 drives in detail at the request of both Gillin and Woodward. It was obvious from the testimony that the company wanted to forget the drives of past years and focus the legislature's attention on the positive good to come in the future. The downriver men were interested in the past and in particular on the horrible consequences of having allowed the paper company control of the rivers in 1901.

Woodward recognized this in his summation. He said the company was not asking for exclusive rights to the river, only the right to improve the river at the expense of the company itself. He said the Great Northern was going to assume all debts and obligations of the P.L.D. but that the new bill did not allow them to do this. He said that the company had postponed the meeting of the P.L.D. but only because the legislature had postponed the hearings until after the meeting. He claimed that the election of 1902 was due to a misunderstanding, and said that his company was not frightened of discussing the 1901 drive. He said that they had agreed to make one drive a year and the commission was to study the water. Even if they had been at fault in 1901 it had nothing to do with the present bill which was for the benefit of all eventually. With this the paper company rested its case, and the committee adjourned until the next day. It was seven o'clock in the evening.[68]

The next day two hearings were held, one during the day devoted mostly to arguments, and an evening hearing primarily for testimony. Patrick Gillin opened for the downriver men. His speech which also was very effective said at the outset that the 1901 drive was really the crux of the matter. The paper company had made a contract and had violated it. They could not be trusted in the future. When they controlled the river the downriver men did not get their logs. The evidence was overwhelmingly on that side. His remarks were greeted with great applause when he said:

> But we are not going to have the Standard Oil Company own the Penobscot River or the great lakes which send our lumber down to our mills below.

He recalled Gilbert and took him over the ground of the 1901 drive again. He then called James F. Kimball who discussed the 1879 drive which had been similar. Kimball was a telling witness as he was then employed by the paper company as a contract logger and his criticism hit close to home.

In the evening a parade of witnesses testifying as to driving conditions came before the committee. Fred Ayer led off. He said in response to questions that the paper company could build all the dams they liked as long as the P.L.D. controlled the drive and the water. John Ross discussed his thirty-five years on the Penobscot and said that the time taken in 1901 was intolerable. In his view,

> If the company would put in a satisfactory bill, pay for the logs lost, and do what's right, there would not be so much objection to the bill.

Twelve other lumbermen spoke of their opposition to the bill. Men from the P.L.A. and the Bangor Boom Company described the difficulty of rafting and booming under adverse conditions. Some predicted two year drives if the bill was allowed to pass. The high point of the evening session was reserved however for the testimony of James McNulty, who in 1901 had supported the company but now was a member of the downriver group. The paper company obviously regarded this as apostasy.

McNulty testified that he had been told in Boston by Gilbert, Schenck, and Woodward that the P.L.D. should non-suit the original cases while the legislature was in session in order to set aside the 1901 drive as not pertinent. He said this was when he changed his mind about the merits of the bill. Woodward asked him if it were not true that he had said in Boston that if his losses were met that he would aid the paper company in passage of the bill. McNulty denied this, but Woodward implied that his memory was faulty.

At the end of Gillin's parade of witnesses Woodward called himself in rebuttal. He testified as to the accuracy of his charges against McNulty and they exchanged epithets of "liar" until the witness chamber was quieted. Fred Gilbert was also recalled and he testified that he had understood the Boston conversation as Woodward had recalled it. Again McNulty challenged his veracity and on this note the hearings adjourned to the next day when the final arguments were presented.[69]

Orville D. Baker of Augusta presented the final arguments for the downriver men. He said the P.L.D. had been doing the drive since 1846 and doing it very well indeed. The thing to do was to allow them to continue and then there would be no recurrence of the drive of 1901. At the end of his remarks he compared the company to George III in a most telling riposte.[70]

In saying that the company's desires for forestry practices and good will was nothing but a blind, he called attention to the king's defense by his ministers extolling his private virtues.

> . . . . and yet they [his domestic virtures] failed to convince our forefathers of the Revolution that there could be such a thing as a good tyrant.

J. W. Symonds also of Augusta finished for the company. He said that there was no necessary chain to the past. The Penobscot was not doomed to subserve the interests of downriver log owners in perpetuity, and if the bill passed the Penobscot would not only be a highway for logs but it would be a highway between prosperous manufacturing cities. Log driving in 1901 was not the issue, and it furthermore was not the fault of anyone, or at worst, the only fault which could be attached to the failure of the drive to come in was the weather over which no one had any control.[71]

It became immediately obvious that the committee was split.

The companies control the rivers 287

The first expectation was that two reports might emanate from the committee, one in favor and one opposed to the Millinocket bill. Both sides expressed confidence and predicted a report in their favor by a vote of six to four. No report came and it was announced that the committee would meet in executive session to take its votes. The opponents of the bill were unhappy with this pointing out that Smith, Pettengill, and Burleigh all represented interests concerned with the bill. They suggested that these three should disqualify themselves from the vote.[72]

Time passed slowly. Various log driving companies held their annual meetings. They were described as tranquil. The P.L.D. adjourned theirs on March 10th for two weeks in the hopes that the committee report would be forthcoming in the meantime. Rivers went to floodstage at about this time and some newspapers called attention to the problem saying that storage would alleviate the problem. The Kennebec had less trouble and the *News* quoted a spokesman of the Hollingsworth and Whitney corporation on that river as saying that passage of the West Branch bill would prevent a recurrence of this problem.[73]

Before this however, rumors had begun to filter out from Augusta that a compromise was in the wind. Reports from the first executive session of the Committee on Interior Waters implied the compromise would allow the paper company to build dams and store water with the P.L.D. to handle driving. The question at issue seemed to be the level of water needed below North Twin Dam.

A tremendous amount of lobbying took place at this time, and finally the committee chairman announced that the members of the committee had taken an oath of secrecy as to their deliberations and that it would do no good to issue newspaper statements. On March 9th it became known that a subcommittee had been picked to notify representatives of the opposing forces that in addition to the above provisions that the committee's bill would provide for an outside commission to determine water storage and the time of the drive, with the dividing point for the two concerns to be Shad Pond. An anonymous committee member confided to the *Commercial* that this compromise was not well liked in the committee because it did not address itself to the question of control of the rivers.[74]

At first this bill was rejected by both forces, but after another session a new and very similar bill was proposed. Under the new compromise Ledyard Smith was no longer an incorporator of the West Branch Driving and Reservoir Dam Company (which would then allow him to vote for it in the Senate); the river was split at Shad Pond, and there was to be one drive a year which had to leave the head of Chesuncook Lake by June 25th and arrive at the Bangor Boom by August 20th in 1903 and in all other years by August 5th. New driving tolls were fixed for ten years, and a water commission was set up with the duty to insure that water would flow at 2,000 cubic feet per second as long as any was stored. Both Chesuncook and North Twin Dam were to be raised. By this bill the P.L.D. was given control by its ability to control flowage and time of drive.[75]

The committee voted to accept this compromise by a vote of nine to one, with Senator Gardiner still adamant. Even he said that his opposition would end with the committee report and that the floor action would not be opposed.[76] The committee also agreed to report out a State Water Power Commission at the same time. This commission however was defeated by the legislature on March 28.[77] The *Commercial* regarded the end result as a triumph for the downriver men and said Fred Ayer was responsible for the victory.[78] The *News* simply said "Let the next legislature be as wise as its predecessor." [79]

The *Commercial* was of course premature in its exultation. The paper company had in fact won what it wanted—the right to dam the West Branch. The downriver men simply were frozen out in the long run. They moved to the East Branch for a time, but with this action by the legislature the old days of log driving in Maine were fast approaching their end. The Northern would pay for its victory, both in the cases then wending their way through the courts and later on the Kennebec, but the cost was relatively small.

The first item of business after passage of the bill was the postponed P.L.D. meeting. It was quick. The company decided to handle its own drive and the Great Northern men were voted out of office. Fred Ayer was elected president, and the board of directors became McNulty, C. W. Mullen, Harry Ross, and James Cassidy. Every man had an interest in a suit against the paper company stemming from the 1901 drive.[80]

The spring freshets though had delayed the drive. The main body did not arrive at the Penobscot Boom until September 21, and the rear of twenty-five million feet was still to come. It finally dragged in on October 3. The problem of time still plagued the downriver men. Rafting stopped on October 29 with ten million feet still in the boom. Only one hundred and thirteen million feet were rafted out and the rest had to be shingled up. The Bangor Boom work lasted until November 20th the drive was so late. They finished their work on

Headworks in Operation

NAFOH

*Grand Falls, New Brunswick Log Jam—
(Early print, 1875)*

Webster Collection of Pictorial
Canadiana, New Brunswick Museum

Ripogenus Flume   Great Northern Paper Company

Tow Boat A. B. Smith—Chesuncook Lake, before 1922

Great Northern Paper Company

Wing Jams on the St. Croix River

NAFOH, no. 180

Rafting on the River—St. John River, 1914

The Booms at Fredericton—(Early print, around 1880)

Webster Collection of Pictorial Canadiana, New Brunswick Museum

The First Leary Raft—An Artist's Conception

Webster Collection of Pictorial

The companies control the rivers

December 1st, but only by virtue of the largest raft ever seen in Bangor. Two tugs hauled a bag boom raft with 12,000 sticks in it down to the Eastern in order to clear the boom before the river froze in.[81]

It was hard for everyone and would be until the dams upriver were constructed. The Northern shut down November 24th because of low water (although the International Paper Company mills also closed this date for "overproduction"). The company did not use as many men in the woods as it had previously, slicing fifteen million feet from its projected cut. In mid-December heavy rains came and everyone was partially happy.[82]

Drives now always seemed to have trouble and this was the main reason that the paper company finally won. The sorting at both Millinocket and Lincoln was enough to delay them beyond the danger point of freezing in the fall. The drive in 1904 was no different. When the P.L.D. held its annual meeting, for the first time in years no report was issued. The lumbermen resolved themselves into executive session although it was determined later that the old officers had been reelected and that the drive was auctioned off in heavy bidding. John Ross took it at 41¢.[83]

At the beginning the driving was excellent. The new dam at 'Suncook raised the water seven feet flooding the north end of the lake all the way back to Rocky Rips, completely covering Pine Stream Falls. The logs were all in the big lake by June 1. The logs still had to be sorted, however. The East and West Branch were combined at the sorting gap in Lincoln and came down together September 17, arriving, all sixty-five million feet, about September 30. It was already too late and plans were made to shingle up thirty-million for the next spring. Rafting then commenced on the next lot. The river froze November 17 catching 12,000 pieces in the boom. The cold continued, the ice grew, and the owners finally decided to employ men to cut them from the ice. They broke a channel to boom them to the mills. Twice booms broke and logs went adrift. Part of the difficulty lay in the fact that the weather was so cold that frost in the logs inhibited the use of wedges to raft them properly. Finally a cold, miserable, unhappy group of men finished up the 1904 drive on November 27.[84] The halcyon days were over, and men who were forced to do this work were very happy for their demise. Elsewhere in that year attention had been turned to the Penobscot County courthouse and the trial of the Great Northern for its part in delaying the drive of 1901.

Several cases were pending before the Penobscot County court in regard to the 1901 drive, but when court convened that of *Herbert*

W. Marsh et al. vs. the Great Northern Paper Company was chosen as the test case. The courtroom was packed when jury selection began. The defense challenged three men of the original impanelment of fifteen; the remainder made up the jury. They came from eleven different towns in Penobscot County and seemed, upon reflection, an average group. Patrick Gillin opened for the plaintiffs using some of the same arguments used in Augusta the year before. In addition he said that he would prove that the driving season did not ordinarily extend past August 20th, and that in this year the reason for the delay was that the water had been used for manufacturing purposes. He stressed that the testimony on the dam gates opening and closing and the amount of water was important and that he would prove premeditation on the part of the company as to lower amounts of water. Forty-nine witnesses were subpoenaed by the plaintiff.[85]

In the first morning, testimony was concerned with the river surveys taken by Phillip D. Coombs, a member of the state commission, to determine starting time of the 1901 drive. Most of the rest of the day was spent in determining the amount of lumber involved in the Marsh and Ayer logs, whether the scaling was of "the full bigness", or whether there was a discount made by the scalers downstream or not.[86]

Then Herbert Marsh testified as to the time of delivery to Gilbert in Chesuncook Lake. Much of the rest of the testimony revolved about the question of where delivery took place, when it took place, the size of the driving crew, and the way the work was done. One of the points at issue was whether or not Pine Stream Falls was part of Chesuncook Lake. Another was whether Gilbert had shown due caution in handling logs in the booms as some of them had drifted into Chesuncook Cove and were brought back to the sluicing ground only by use of winches, headworks, and kedge anchors. The most damaging testimony here was offered by John Kelley a straw boss of Gilbert's on the drive He went down to Bangor during the time of the drive and described the state of the river as well. He said the deadwaters were filled with logs, and the rear was not turned over Chesuncook Dam until July 10. He said logs were sluiced on a full head, leaving a heavy rear to be dragged along.[87] Others followed who testified on the way the drive and the sluicing took place, while discussing the drive all the way to the Penobscot boom.[88]

Then Luther Gerrish, who was in charge of the water and boom at North Twin (and who had been since 1886) testified in detail as to opening and closing of the gates, and to the amount of water avail-

The companies control the rivers 291

able day by day from May 1 to November 13. Gerrish also testified that he had received a letter from P. A. Strickland asking for more water on October 21, and that he had opened another gate for a day or two to oblige an old friend even though the company had instructed him differently.[89]

After more discussion by witnesses of the drive and its methods, another witness testified as to the habits of the captains of the tow boats at Ambejejus Lake and whether the logs could have been moved sooner or not. The witness was hostile but it was fairly well developed that the logs could have come on earlier.[90] Evidence was brought forward that the paper company had installed a second sorting gap at Quakish Lake as a result of the summer's experience,[91] and then finally much more was said about dragging in the rear after leaving Lincoln. Conditions from the time of the drive arrival to the closing of the river were analyzed in detail.[92]

At this point the case was continued even though one juror was dismissed because of illness.[93] Fred Ayer was called. After discussing his expenses and the reasons for the suit he testified that he had a phone conversation with Schenck on October 6 in which Schenck agreed to order Ledyard Smith and Gilbert to give enough water to run the drive in as it was in danger of hanging and freezing.[94] After this a parade of old-time Penobscot lumbermen came to the stand to testify as to their understanding of what constituted a normal driving season.[95]

With this their case was in and the defense asked for an adjournment in order to present a motion. The judge gave them two hours and they returned asking for non-suit on the grounds that the 1901 contract was between the P.L.D. and the paper company and not these individuals, that the legislature had not imposed these obligations on the company by ratifying this contract, and that the fault of the lost booms on the Penobscot was the P.L.A., not the P.L.D.

Court adjourned at this point. The next day Orville Baker answered for the plaintiffs saying that in the case of the P.L.A. negligence was in the time the drive took, not in the strength of the booms, that the suit was on a basis of tort and not contract and therefore by failing to live up to its assumed duties the defendant had committed a tort and the paper company had assumed the role of the P.L.D. when the legislature ratified the contract. The judge then dismissed the defendant motion and, after a long discussion in the judge's chambers the jury was dismissed and the case was remanded to the law court.[96] Before it went up though the defense called two witnesses. The first was John Kelley again, who testified that he and Gilbert had had a conversation about May 1 in which Gilbert had said that he was paid by the paper company to get water enough to run the mill, and that the only way to do this was by keeping the logs back in the lakes;[97] the second witness was Charles White, who testified as to the impossibility of holding back the logs in the spring freshet of 1902.[98] The terms before the higher court were damages of $9,871.31 plus interest for the plaintiff, and non-suit for the defendant. Under the Maine code it was possible for parties in a civil suit being tried before a jury to withdraw the case from the jury and have it submitted to the higher court. The final two defense witnesses were testifying with prejudice—that is, when the court heard the case the plaintiff was resting more of the case on their testimony than on the previous testimony.

The Supreme Court of Maine heard the case on June 13 with each side presenting about 2½ hours of argument. Orville Baker appeared for the plaintiff and he determined his case primarily on the grounds of the normal driving season, the problem of misuse of waters, the Kelley-Gilbert conversation, and what he called "deliberate delay" in the lower lakes. The defense spent most of its time saying that much of the evidence was inadmissible, and the case rested on the fact that the defendant had delivered the logs and thus was not liable.[99] The court ruled for the plaintiff on July 24, 1906, accepting the argument of tort and saying that unusual climatic conditions do not excuse mistakes.[100]

The other cases pending were all settled out of court, and it is difficult to determine how much the drive cost the Great Northern Paper Company. Also, it is very difficult to determine at this juncture whether or not the paper company and Fred Gilbert were deliberately guilty or whether it was simply a very bad error of judgment. Opinions still run high on the banks of the Penobscot today concerning this problem. A close friend of Fred Gilbert said of him not long ago, "He was a very hard man." Another man said, "I never spoke to him. I met him once on 'Suncook dam. I got out of his way. He had a difficult reputation on the river." A third person remarked, "He was always faithful to his employer." [101]

The court case really marked the end of Penobscot river driving as it had been known since the 1830's. True, logs came down the river still but in smaller and smaller amounts, and after a while only from the Mattawamkeag or the East Branch. Fannie Hardy Eckstorm,

The companies control the rivers

the first historian of the West Branch drive, wrote in the year of the court case,[102] "The times had changed indeed." It was appropriate that her book, *The Penobscot Man*, was published in the year the West Branch drive died.

As the West Branch logs dwindled the Penobscot came to be a thoroughfare for logs from the other branches. The companies which did most of the lumbering here gradually took over the driving completely. The Penobscot Chemical Fibre Company formed the I. W. Buzzell Company to raft and drive from Pea Cove Boom. The Penobscot Lumbering Association purchased the Penobscot Boom Corporation in 1928.[103] They had been running it in fact since 1920 when the boom began to handle mostly four foot wood.[104] Some logs continued to come down still as the appended table shows.

Selected Years—Wood Driven

Penobscot, Maine

(in feet)

Year	Long Logs	Pulpwood	Total
1911			108,798,740 feet
1912			114,346,450
1913			113,391,470
1914			94,539,700
1915			105,137,030
1919	19,232,460	3,762,080	22,994,540
1925	8,447,870	24,715,000	33,162,870
1927	3,239,320	21,048,000	24,387,320
1928	958,030	32,508,500	33,466,530
1931		12,917,500	12,917,500
1953		10,385,465	10,385,465

The Bangor boom dwindled off spectacularly as well. The years 1911 to 1913 saw only just over 50 million feet boomed, and in 1914 the figure was only 41,000,000 of which half went to the Eastern Corporation. The average size of sticks at the two booms was 72 board feet in 1912, and this declined to 58 in 1915. This was a far cry from the 1832-1840 figure of 337 board feet per stick. Most of these logs were East Branch in origin as the West Branch drive was as small as 13 million feet in most of these years, if not smaller. The East Branch drive, some 40-45 million feet of long logs, arrived in Argyle by early August ordinarily. The remainder came from logs from the Pleasant, Piscataquis, Mattawamkeag, and the Passadumkeag as before. It was a small business though. One observer in late 1914 told the story well with his remark that, "Not in many years has the lumber trade on the Penobscot been so dull as it has been all this season." [105]

The last of the long logs appeared in 1928 when not quite a million feet came to the Jordan Lumber Company in Old Town. The last of the pulpwood came down in 1953,[106] although there were years in the thirties in which no wood at all appeared in the boom. The old river, as they say in Bangor, ran and still runs rolling to the sea. It didn't, and somehow still doesn't seem right to meet log trucks along the river road and see the river itself sparkling in the sun, empty and just a bit lonely.

NOTES

[1] Portland *Eastern Argus*, December 12, 1874.

[2] Richards, *Ninety Years On*, discusses this development *inter alia*.

[3] This involved the dumping of pulp mill waste into the Royal River. Discussion was intermittent from 1885 on and was finally brought to brook in a lawsuit filed in 1893. Eventually a referee awarded $1,650 damages to the woolen mill downriver. See *Paper Trade Journal*, December 9, 1893, April 29, 1894, and *Royal River Company vs. Forest Fibre Company*.

[4] This story is covered in Chapter Nine. A side-benefit here is that S. D. Warren has taken hourly water level readings here since 1887.

[5] Excellent discussion in Bangor *Daily Commercial*, October 26, 1878.

[6] *Lewiston Steam Mill Company v. Union Water Power Company*, SJC in Lewiston reported in *ibid.*, October 19, 1883.

[7] *Lewiston Steam Mill Company v. Richardson Lake Dam Company*, 77 Maine 337, decision handed down May 15, 1885.

[8] *SJRC*, 2022-54, 2071-2120, testimony of Walter H. Sawyer, agent for the Union Power Company in Lewiston. Also see Bangor *Daily Commercial*, December 16, 1889. An excellent description of Aziscoos Dam is A. A. Tanyane, "A Mammoth Storage Dam," *Paper Trade Journal*, February 16, 1911.

[9] See for instance the decision in *Rumford Falls Power Company v. Rumford Falls Paper Company*, 95 Maine 186, which held that the paper company was liable for payment for overuse of water even though an oral agreement had specified otherwise.

[10] *The Industrial Journal*, June 30, 1893; Bangor *Daily Commercial*, July 24, 1893.

[11] *Ibid.*, February 14, 1895.

[12] *Rumford Falls Boom Company v. Rumford Falls Paper Company*, 96 Maine 96, opinion handed down January 6, 1902. The damages amounted to $10,450.33 plus interest. On Rumford and its difficulties in regard to driving and storage see *SJRC*, Sawyer's testimony *op. cit.*, especially 2097.

The companies control the rivers 295

[13] *Weymouth v. Penobscot Log Driving Company,* 71 Maine 29, *and Weymouth v. Penobscot Log Driving Company and Trustees,* 75 Maine 41.

[14] Damages were awarded to a company whose mill was damaged by river flowage when a dam was built. Bangor *Daily Commercial,* October 2, 16, 1879; April 21, 1880.

[15] See the running battle carried on by the Commissioner of Inland Fish and Game to protect Penobscot River salmon against pulpmill pollution, *Annual Reports* of the Commissioner, 1888, 1890, 1891-2.

[16] *Charles F. Palmer v. Penobscot Lumbering Association,* SJC Penobscot County, and Bangor *Daily Commercial,* April 15-18, 1896 for case and evidence, and June 19, 1896 for decision. For related cases which went to the referee see *ibid.,* Feb. 8, 1899. Total damages were $9,104.95. The year was 1893 again. On the Mattawamkeag River similar events took place and the courts held similarly, see Bangor *Daily Commercial,* April 10-13, 1899.

[17] *C. W. Mullen v. Penobscot Log Driving Company,* SJC Penobscot County, and *The Industrial Journal,* September 24, 1897.

[18] The Mattawamkeag River cases are *Mattawamkeag Log Driving Company v. George L. Byron,* 101 Maine 181-7.

[19] *The Industrial Journal,* November 21, 1891, italics added.

[20] Bangor *Daily Commercial,* February 14, 1899. Ross was defeated for presidential re-election on the next day though.

[21] *Ibid.,* May 14, 26, 29, 30, June 13, 25, 27, 30, July 15, August 27, October 17, 18, 19, 1900. Prices were low for lumber and the sawmills didn't care at first. In fact they formed a protective association to boom prices. The GNP began to buy logs and drove the prices up, and this was when the level of the water began to bother prospective purchasers of logs.

[22] House of Representatives Document #10, *House Documents,* 1901.

[23] From the biographies in Senate and House *Journals,* The Kennebec *Journal,* January 2, 1901; and *Biographies of Members of Legislature,* 1901, Kennebec Journal Press, 1901, (a volume produced for each legislature with varying titles).

[24] These were compiled from the pages of *Legislative Record,* 70th Legislature, 1901, as they were presented at the beginning of each day's work. The remonstrances were then referred to the Committee in charge of the bill.

[25] Interview in Bangor *Daily News,* January 21, 1901.

[26] Bangor *Daily Commercial,* January 22, 23, 1901.

[27] Bangor *Daily News,* January 23, 1901. They quoted several letters, and devoted three columns on page one to a study of the question.

[28] Bangor *Daily Commercial,* January 24, 1901.

[29] Bangor *Daily News,* January 24, 1901. The headlines read, "Lumbermen Denounce 'Millinocket Bill' ". These newspapers were not published at the same time. The *News* was a morning paper; the *Commercial* an evening paper. Frequently at a time like this they are commenting on the other's story, although they may not say so.

[30] See Bangor *Daily Commercial,* January 26, 1901 for an editorial on "Water Storage" which cited the same messages used by their earlier anonymous correspondent.

[31] Bangor *Daily News,* January 28, 1901.

[32] *Ibid.*, January 29, 30, 1901, and for an answer, Bangor *Daily Commercial*, January 30, 1901, letter from Peregrine White to the editor.

[33] *Ibid.*, January 31, 1901; Bangor *Daily News*, January 31, 1901, (the best account of Prentiss' remarks); *The Industrial Journal*, February 1, 1901.

[34] Bangor *Daily Commercial*, February 2, 1901; Bangor *Daily News*, February 6, 1901.

[35] *Ibid.*, the letters were from J. P. Walker (representing mill and land owners and Bangor businessmen to the editor) and Ira Gardiner (representing the Directors of the East Branch Log Driving Company) to F. H. Strickland, treasurer of the concern, dated February 1, 1901.

[36] This meeting of the Penobscot Log Driving Company is a legendary one. It is possible to hear that men came to blows, that bribes were offered and accepted, or that events were all settled before they met that day in February, 1901. None of these things happened, as nearly as I can tell. My account is based on conversations with many men along the river as to the legends of the meeting, and most particularly the accounts in the Bangor papers, which are remarkably detailed. See Bangor *Daily Commercial*, February 12, 1901; Bangor *Daily News*, February 13, 1901 (by far the best account); *The Industrial Journal*, February 15, 1901.

[37] *House Journal*, February 20, 1901; Senate *Journal*, February 21, 1901; *Public and Special Laws*, 1901, Maine, Chapter 293; Bangor *Daily News*, February 15, 20, 1901; Bangor *Daily Commercial*, February 15 (the editorial citation), 19, 1901; Portland *Eastern Argus*, February 20, 1901; "And thus ends the much heralded fight on the Millinocket Bill." The Kennebec *Journal*, February 20, 21, 1901; *The Industrial Journal*, February 22, 1901.

[38] John Ross was interviewed after making a trip to the headwaters, and he said that all was well. He was, in fact, whistling in the wind. See Bangor *Daily Commercial*, April 13, 1901.

[39] *Ibid.*, March 14, April 4, 5, 6, 8, 9, 10, 11, 15, 17, 23, 24, May 3, 4, 13, 1901. On this drive the evidence in *Herbert W. Marsh et al. v. The Great Northern Paper Company*, 264 Penobscot County Superior Court, April, 1904, is of great importance. This trial is recorded in Vol. 136, 223, Superior Court Records, and the manuscript testimony is preserved in Bangor at the County Court House. I have used all this, and the documentary evidence, and will cite it more completely later. The newspaper citations on the drive are for the convenience of the reader. See below for a discussion of the trial which resulted from the drive.

[40] Bangor *Daily Commercial*, May 24, 31, June 4, 8, 13, 14, 21, 25, July 8, 10, 13, 15, 16, 17, 26, 30, 31, August 6, 21, 26, 27, 28, September 2, 13, 17, 19, 1901. The comments on the weather come from a daily reading of these stories. The same strictures as in note 39 apply to the trial evidence.

[41] Original Writ, September 25, 1901, cited as evidence in *Marsh et al.* . . .

[42] Bangor *Daily News*, September 21, October 3, 7, 12, 1901; Bangor *Daily Commercial*, September 23, 27, October 4, 11, 14, 15, 1901.

[43] Letters Ayer to Sterns, dates cited, in evidence, *Marsh et al.* . . .

[44] Bangor *Daily News*, October 21, 30, November 2, 14, 1901; Bangor *Daily Commercial*, October 21, 23, 24, 29, 30, November 6, 8, 9, 11, 13, 1901.

The companies control the rivers

[45] Bangor *Daily News*, November 5, 8, 11, 18, 27, 30, 1901; Bangor *Daily Commercial*, November 12, 13, 19, 21, 22, 26, 27, 1901.

[46] Bangor *Daily Commercial*, December 3, 4, 6, 9, 10, 11, 14, 16, 17, 18, 19, 21, 1901; Bangor *Daily News*, December 2, 3, 5, 9, 11, 14, 17, 18, 31, 1901. Some have forgotten just how expensive this freshet was for the paper company. There was great damage on the Kennebec. One wall of the Madison plant fell in under the flood; much machinery, including a new electric switchboard, and considerable paper was ruined. They also lost close to a million feet of logs.

[47] *Ibid.*, December 18, 1900.

[48] Annual Meeting P.L.D., reported in Bangor *Daily Commercial*, February 11, 1902.

[49] *Ibid.*, March 6, 7, 17, 19, 20, 21, 22, 31, April 2, 3, 11, 18, 21, 28, 29, 1902. This was the worst freshet since the famous one of 1846.

[50] Bangor *Daily Commercial*, May 6, 21, June 25, August 5, October 18, 1902.

[51] *Ibid.*, May 6, 21, 24, 25, 27, June 3, 6, 10, 18, 25, July 1, 11, 24, 31, August 5, 14, 26, September 5, October 18, 1902.

[52] Information on the legislation is best found in "Biographies of the 1903 Legislature", Kennebec *Journal*, January 7, 1903, and later bound separately. Some of this information is also available in *House* and *Senate Journals*, 71st Legislature, 1903, *passim*.

[53] House Documents #1, 1903, cp. House Document #10, 1901.

[54] The names of the committee members point up just how much lumbering permeates Maine history. They included a Shaw from Bath (from the old M. G. Shaw Co.), a Ross from Bangor (brother of John), a Sargent from Brewer (lumber and shipping), and Sweeny (a Democrat from Fort Kent, but in lumber nevertheless).

[55] Information on these petitions comes from House and Senate *Journals*, 1903 Legislature, on the appropriate days. Information on the makeup of the committee also comes from these sources. The two important men are Smith and Gardiner from Patten.

[56] His letter is in Bangor *Daily Commercial*, January 6, 1903; his speech before the Penobscot Log Driving Company is reprinted in Bangor *Daily News*, January 8, 1903.

[57] *Ibid.*, January 9, 10, 1903.

[58] Portland *Eastern Argus*, January 12, 1903.

[59] Kennebec *Journal*, January 15, 1903. Of course newspapers in other sections of the state were less interested in editorial positions than those in Bangor where one had to be counted on one side or another at this time.

[60] Their views are obvious every day. The best editorial is in the issue of February 17, 1903.

[61] On suits see *News*, January 17, 1903; *Commercial*, January 16, 1903.

[62] *Ibid.*, January 23, 1903; Bangor *Daily News*, January 23, 1903.

[63] Bangor *Daily Commercial*, January 24, 26, 27, 29, 30, 31, February 13 (quoting the St. John *Sun*), March 4, 6, April 2, 20, 1903. Bangor *Daily News*, January 31, February 10 (quoting the St. John *Sun*, Fredericton *Herald*, and St. John *Telegraph*, all New Brunswick), March 7, 25, 1903. Portland *Eastern Argus*, February 18, 1903. A letter was read into the legislative record from Provincial Attorney General William Pugsley to Peter C. Keegan of Van Buren.

It said in part "... this is a matter of the very gravest consequences...."
The matter at dispute really involved control of the St. John River by up-river dam and lumbering companies (mostly from Maine) and downriver sawmills (mostly from Canada). The New Brunswick legislature turned down the dam proposal, and a delegation went to Ottawa where it was eventually defeated as well. The Maine bill created the new East Branch company, but only with the express provision that it could not raise the dams, or take more water, when it was eventually passed. For the best exposition of the controversy, especially from the New Brunswick side, see St. John *Telegraph*, April 2, 1903.

[64] Bangor *Daily Commercial*, January 31, February 2, 10, 11, 12, 17, 1903. Bangor *Daily News*, January 27, February 11, 12, 1903. The new bill provided essentially that a new method of voting be used. Each operator driving a million feet of lumber ten miles had one vote, and got an additional vote for each ten miles he drove, a method designed to protect the downriver men who drove the greater distances.

[65] *e.g.*, Portland *Eastern Argus*, February 10, 1903 (they sent a reporter to Bangor to find out local feeling); Kennebec *Journal*, January 26, February 18, 1903.

[66] House Documents #88, #225, and amendment of House Document #1, presented February 11, 1903.

[67] These editorial comments come in order from Bangor *Daily Commercial*, February 17, 21, 23, 1903. The remarks about the Standard Oil connections would reappear in the hearings, and later, whenever one wanted to really attack the paper company.

[68] There is no record of these proceedings, and the reports depend on newspaper observations. The *Commercial* had a stenographer there and their reports are nearly *verbatim*, and at times they reprinted speeches *verbatim*. Bangor *Daily Commercial*, February 24, 25, 1903; Portland *Eastern Argus*, Bangor *Daily News*, Kennebec *Journal*, February 24, 25, 1903 are the references.

[69] Bangor *Daily Commercial*, February 26, 1903; Kennebec *Journal*, February 26, 1903 (Gillin's opening speech nearly complete); Bangor *Daily News*, February 26, 1903.

[70] This speech was reprinted in its entirety in Bangor *Daily Commercial*, March 3, 1903.

[71] Bangor *Daily Commercial*, February 27, 28, 1903; Bangor *Daily News*, February 27, 28, 1903; Kennebec *Journal*, February 27, 1903, (best report on Symonds' speech); Portland *Eastern Argus*, February 28, 1903.

[72] Bangor *Daily News*, February 28 (the vote predictions), 1903; Portland *Eastern Argus*, February 28, 1903 (the discussion of the propriety of these members voting); Bangor *Daily Commercial*, March 7, 1903 (on Ledyard

[73] Bangor *Daily Commercial*, March 3, 4, 1903; Bangor *Daily News*, March Smith in particular).
4, 13, 1903.

[74] *Ibid.*, March 11, 12, 1903; Kennebec *Journal*, March 10, 11, 12, 1903; Bangor *Daily Commercial*, March 7, 10, 11, 12, 1903.

[75] The bill was sent to the Governor on March 13 and he signed it into law. See *Private and Special Laws, Maine*, 1903, ch. 173, 174. Bangor *Daily Commercial*, March 12, 14, 1903.

[76] *Ibid.*

The companies control the rivers

[77] Kennebec *Journal*, March 12, 28, 1903.
[78] March 14, 1903.
[79] *News*, March 30, 1903.
[80] Bangor *Daily Commercial*, March 24, 1903.
[81] *Ibid.*, September 21, October 3, 29, 31, November 12, 27, 1903.
[82] *Ibid.*, November 24, 25, December 3, 10, 1903.
[83] Bangor *Daily Commercial*, February 9, 10, 1904, (below Shad Pond, of course).
[84] *Ibid.*, May 27, September 20, 29, 30, November 19, 21, 22, 23, 25, 27, 1904.
[85] Bangor *Daily Commercial*, January 29, 1904, and *Marsh et al. vs. Great Northern Paper Company*, 264 PCSJC, 1904, April term, manuscript testimony.
[86] Testimony, *ibid.*, 1-53 (also on amount lost at sea), 50-53.
[87] Testimony of John A. Kelley, *ibid.*, 60-125, on sluicing especially 94.
[88] Testimony of Charles A. Daisey, 125-136; James A. Dube, 136-150; Patrick O'Halloran, 151-168 (head sluicer at 'Suncook Dam); S. A. McPhee, 169-175; Charles H. Adams, (clerk of the P.L.D.), 175-176½; Louis A. Tesco, 175-194.
[89] Testimony of Luther M. Gerrish, 195-214, letter is at 212.
[90] Testimony of Purchase R. Fowler, 214-220; Joshua F. Smith, 220-224, and especially Leonard W. Vinal, 224-233. (Vinal was the timekeeper on the booming out crew at Ambejejus and in the employ of the G.N.P. Company).
[91] Testimony of Louis A. Tesco, recalled, 243-249.
[92] Testimony of William S. Porter, 249-59; Leonard B. Bryant, 259-65; Ithiel C. Blackman, 264-6; Ansel Staples, 266-8; Charles M. White (head rafter at Penobscot Boom), 269-308; Gerry B. Comstock, 308-322; Alonzo I. Mass, 322-8.
[93] Bangor *Daily Commercial*, January 30, February 1, 2, 1904.
[94] Testimony of Fred Ayer, 329-345, phone call at 336.
[95] Testimony of S. H. Prentiss, 345-50; John Ross, 350-389; Cornelius Murphy, 389-411; Obed Ireland, 411-2; Philo Strickland, 412-3.
[96] Bangor *Daily Commercial*, February 2, 3, 1904.
[97] Testimony of John A. Kelley, recalled, 414-5.
[98] Testimony of Charles M. White, recalled, 417-419, and Bangor *Daily Commercial*, February 3, 1904.
[99] *Ibid.*, June 13, 1904.
[100] *Herbert W. Marsh et al. v. Great Northern Paper Company*, 101 Maine 489-506, decision rendered July 24, 1906, SJC Penobscot County. The other cases were *Morison et al. v. GNP Co.*, 535 SJC October 1907, suit $10,000; *J. F. Smith v. GNP Co.*, 536 SJC October 1907, suit $7,300; *Ross and Appleton v. GNP Co.*, 538 SJC October 1907, suit $15,000; *Ayer and Maxfield v. GNP Co.*, 539 SJC October 1907, suit $20,000; *Charles W. Mullen v. GNP Co.*, 42 SJC October 1907, suit $10,000. A related suit, *George W. Coffin v. GNP Co.*, which contended that the plaintiff was forced to suspend his work unnecessarily (he was handling rafts from the Passadumkeag Boom), because of the lateness of the drive was also decided with judgment for the plaintiff. See Bangor *Daily Commercial*, September 15, 1904.
[101] These comments came in answer to direct questions as to his role in the 1901 drive. All were men who were too young at the time of the drive, but

who knew him well in his role as woodlands superintendent of the Great Northern Paper Company. The comments come from Louis Freedman, Old Town, Caleb Scribner of Patten, and William Hilton of Bangor, in that order. Since their comments to me, all have passed away as time moves very far from these years.

[102] F. H. Eckstorm, *The Penobscot Man*, Bangor, 1904, 190, from the chapter "Working Nights", a description of the 1901 drive and the impact it had on Bangor minds.

[103] The I. W. Buzzell Company was founded in 1916. Letter, J. F. Gould to Amor Hollingsworth (President Penobscot Development Corporation and Penobscot Chemical Fibre Corporation) December 29, 1928. The leases which they used for their control reverted to the Indians who held the right to most of the land in the mid-fifties.

[104] "Plan for handling four foot Pulpwood by the Penobscot Lumbering Association," March 2, 1920, proposed by Eastern Manufacturing Company and Penobscot Development Corporation.

[105] *Paper Trade Journal*, February 23, 1911; September 19, November 14, 1912; July 10, October 16, 23, 1913; January 22, July 23, August 13, October 1, 29, November 19, 1914; January 14, April 8, July 8, 15, November 4, 1915; *Paper*, October 28, 1914. Quote from October 29, 1914, *Paper Trade Journal*.

[106] These last paragraphs are based on a study of the concerns mentioned, the P.L.A. and the others, graciously made available to me by the Penobscot Chemical Fibre Corporation, and the last clerk of the firms, Ballard W. Keith, of Bangor. The records are all in his possession. The two basic records are File 245A, Penobscot Lumbering Association 1918-1921, and File 409A, Penobscot Lumbering Association 1926-1933. Also see the suit *Penobscot Lumbering Association v. Penobscot Development Corporation*, which was a lawyer's suit designed to settle certain legal questions after the paper company came to be the only, or main, driver on the river. This case and the correspondence surrounding it is in these files, dated August 14, 1919. The figures in the table are taken from the clerk's reports to the Penobscot Lumbering Association and certain issues of *Paper Trade Journal*. In some cases the figures are taken from cord statistics. In doing this I always have taken the standard of 500 board feet to the cord. See, for an analysis, *Royal Commission, New Brunswick Forestry*, 1927, appendix, where they deal with this problem for peeled wood, unpeeled wood, and for both spruce and fir. The technical term for peeled wood is rossed wood. It is also significant, or at least, coincidental, that the Tefft statue, *The Riverdrivers*, which graces a city park in Bangor, was dedicated in the years that the long logs were finishing up the Penobscot Drive, in 1927.

12

THE COMPANIES TAKE CONTROL OF THE RIVERS

PART TWO

> Party differences and sectional differences mean little. A new line of cleavage appears, revealing the people at large on the one hand, with most of the water power owners on the other.
>
> <div align="right">Percival Baxter (R., South Portland), commenting on the Public Utilities Commission, <i>Legislative Record, 1919</i>, 711.</div>

> Private enterprise has always developed business giants; government enterprise has always developed cunning politicians. For one hundred years we dammed the rivers and praised the builders, and now we are praising the rivers and damning the builders.
>
> <div align="right">Senator Hinckley (R., South Portland), commenting on the proposed water power amendment to the Constitution of the State of Maine (supposedly), and Percival Baxter (in fact), <i>Legislative Record, 1921</i>, 1226.</div>

On the other large rivers of Maine (the St. John, and the Kennebec) the situation of control of power was similar. The downriver men had always controlled the rivers and used the waters to drive their mills and move their logs. Increasingly their interests were challenged by upriver forces wanting to build their own mills or dams to control the water and its powers for their own interests. In the case of these two rivers the outcome was also similar to that on the Penobscot although differing times and circumstances kept the events from the court room.

There had always been a good deal of difficulty as far as the St. John River was concerned. Once the Telos Canal was dug, mill owners below Grand Falls knew in the summer that there would be periods of extreme low water, and if the logs did not get over the Grand Falls by July 1 they would probably not get to Fredericton in time to be boomed out or to St. John in time to be sawed. The New Brunswickers as a result were constantly on the lookout for anything which would deprive them of the necessary water to conduct their business. This watchfulness led to the destruction of upriver dams twice in the nineteenth century.

In the twentieth century such violence was resorted to once, but the downriver men found themselves in the position of not fearing the loss of water so much as they did the loss of time taken at a sorting gap across the river at Van Buren. This sorting gap slowed the logs arriving at Fredericton, and as the owners of the logs were Americans the dispute over free transit on the river, between upriver and downriver interests, became something of an international affair. No recourse to legislatures was possible; the long alternative was to talk and let nature take its course. This is in effect exactly what the United States and Canada did between 1909 and 1916. The two governments, after a series of diplomatic exchanges, agreed on a Joint International Commission to investigate "the obstructions" in the river St. John.[1] The commission met in session sporadically for the next seven years hearing evidence on all phases of the river traffic and then filed a final report well after the time the downriver mills were silenced.

The problem of control of the river really began when the Bangor and Aroostook Railroad made its connections with Van Buren. As the railroad moved north to the border it had tapped an ever growing hinterland of lumber and forest products which had hitherto been tributary to New Brunswick and especially the lower St. John river valley. A major reversal of trade patterns and routes had occurred between 1880 and 1902.[2] There had been some reaction before from the St. John people but nothing compared to what happened when the railroad reached Van Buren and the St. John Lumber Company built its new and highly competitive mill there.

Initially the efforts by the upriver men to control the river by building a dam at Winding Ledges was turned down. This dam, to generate power to run sawmills and pulp and paper mills, was to be erected by a company capitalized by both Maine and New Brunswick investors for $1,000,000. The men associated with the new firm included nearly every Maine capitalist of importance, several from New Brunswick, and Redfield Proctor of the Vermont Central Railroad family. The East Branch Development Company Bill which had been presented to the 1903 legislature, had been a parallel and associated attempt by the Penobscot people to gain control of part of this St. John water development. When the dam was built the flowage would come back far enough to affect the level of water in Chamberlain

Lake, Telos Lake, and eventually the East Branch of the Penobscot itself. The Maine legislature accepted the Winding Ledges bill but New Brunswick voted it down, and then the East Branch bill was passed in a compromise form which affected no change in the river at all.[3] The matter then lay in abeyance with the only exception being that the St. John Lumber Company began to control both log and water traffic on the St. John, and the downriver men grew increasingly dissatisfied.

In the next year after the Winding Ledges failure a measure was introduced into the New Brunswick Parliament to control the new Maine firm. In this proposed act logs cut on crown lands were forbidden to be exported in the round or unmanufactured state. The source of logs for the Van Buren mill would have to be confined to the Maine side of the boundary. Lengthy debate followed in the provincial parliament, and the bill was finally referred to committee. When the committee held hearings on March 31, 1904, it suggested that a better measure would be to revive the idea of an export tax on the logs, as denial of the right of export was probably in contravention of the Webster-Ashburton treaty. St. John lumbermen flocked to the witness stand to extol the new plan, and to say that it would be the main and perhaps only salvation for St. John city and its lumber mills. J. Fraser Gregory was the most effective witness as to the ultimate effect of the bill.

The Maine newspapers became very upset at the prospect of Maine money going to aid in the payment of New Brunswick debts. The *Commercial* decried these "dog-in-the-manger" attitudes and contrasted the Maine response to the log rafts of the previous century. It also offered a veiled threat of retaliation by advocating the removal of pulpwood from the free list. The committee released its report a bit later saying that the proposed legislation "was not in the public interest."[4] This first attempt at control was finished by the release of the report, but the St. John Lumber Company continued to be astride the river above the Grand Falls and, by virtue of its position, continued to maintain control of the log and lumber traffic on the river. The company was not a pleasant one to deal with, and the friction generated as a result led soon to undeclared warfare.

When the Bangor and Aroostook Railroad reached Van Buren there were about ten working mills in St. John with a yearly capacity of about 150,000,000 feet of long lumber in addition to shingles, laths, and cooperage stock.[5] The logs cut above usually came down and were sawed in St. John. The drives ordinarily came from the brooks into the corporation limits between the 18th of May (which was the limit of high water on the average) and the first of June. About one hundred and twenty-five men then took the logs over the Grand Falls. The following tables will indicate the amount of lumber handled by this corporation, the St. John River Log Driving Company. The difference in the figures can be explained by loss in driving the Falls and by the fact that logs were introduced into the river from small tributaries on the New Brunswick side which were not counted in the Madawaska Log Driving Company totals.[6]

Logs Driven to The Grand Falls By The Madawaska Log Driving Company

Year	Amount (feet)	Year	Amount (feet)
1891	71,268,534	1902	57,741,322
1892	66,053,680	1903	(combined with 1904)
1893	75,706,640	1904	173,438,000
1894	80,292,920	1905	34,794,695
1895	63,308,644	1906	138,774,138
1896	75,691,373	1907	138,132,669
1897	103,242,192	1908	84,967,000
1898	82,852,300	1909	104,500,000
1899	59,841,667	1910	91,751,000
1900	96,868,097	1911	79,200,000
1901	73,432,058		

Logs Driven Over The Grand Falls 1899-1909

Year	Amount	Cost of Drive (per M)
1899	76,454,000	15.6¢
1900	119,569,000	15.6
1901	83,017,000	14
1902	109,688,000	14
1903	54,323,000	14
1904	98,661,000	14
1905	50,258,000	18
1906	107,548,000	18
1907	118,655,000	40 (booming costs
1908	54,247,000	38 " included)
1909	97,500,000	40 "

Argyle Boom

Schooner Loading Lumber—Ritchie's Wharf, Newcastle, New Brunswick

Larson Collection, University of New Brunswick

Logs in Mill Pond and Shingled Up for Winter—St. John River, Near Woodstock, 1914

NAFOH, no. 183

Hell's Half Acre—Exchange Street Facing the River—1884   James Mundy

Suppliers to the Woodsmen James Vickery

Opening of the European and North American Railway Canadian Illustrated News, November 4, 1871

Century Magazine, December, 1885

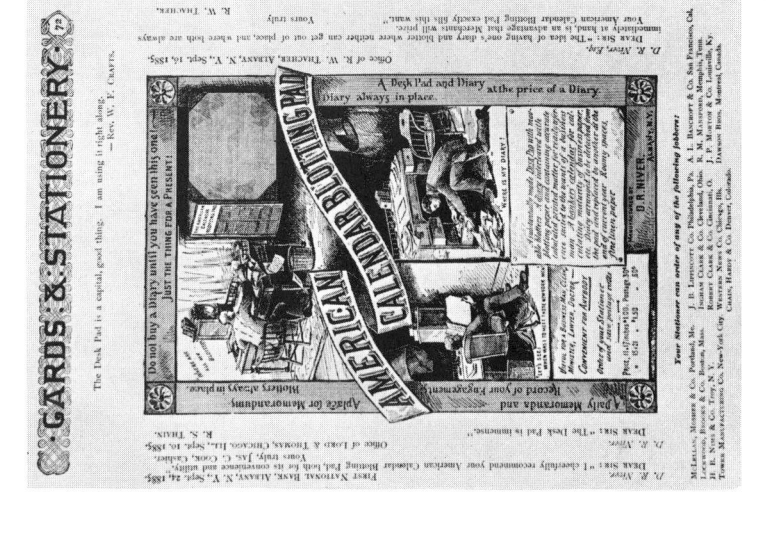

New Uses for Wood Pulp Paper

The Houses Lumber Built—Broadway in Bangor

James Vickery

The companies control the rivers

Before the time of the St. John Lumber Company there had been mills located on the upper part of the river, but they had been small and the owners usually drove their logs after the corporation drive had come down. Oftentimes farmers rafted the logs which they had felled themselves to these mills. This traffic seldom, if ever, interfered with downriver logs and water.[7]

The new mill at Van Buren which created all the trouble was started in the fall of 1903 after the necessary capital of $240,000 was raised. The money was subscribed by Portland and Augusta lumbering men and capitalists. The first mill was completed in June of 1904. A shingle mill of two stories with thirty machines, it was rated at 550 M a day. In 1904 a second mill was constructed also of two stories but with a wing alongside as well. This mill with its two bandsaws and other finishing machinery was rated at 125 M of long lumber each day. The capital of the firm was increased to $432,000 in February, 1907, and a third mill was built, somewhat smaller than the second but with the same saws and capacity. The firm had four hundred men working for it, and the monthly payroll was in the neighborhood of $20,000. The capital of the parent firm was further increased to $550,000 in February of 1909, and the firm began to purchase timberlands. By the time that the investigation was underway the firm owned several townships in northern Maine, including Big Twenty (T20R11 & 12). Its holdings amounted to about 125,000 acres. The firm was shipping about 28,000,000 feet of long lumber and 5,000,000 feet of cedar shingles each year. Eight million feet of this was cut on their own land, the rest was contracted. In order to separate their marks from the others in the drive the company hung a boom across the river and sorted their marks as the logs came down. All the logs were stopped at the boom thus delaying the drive beyond the danger point for driving the Grand Falls.[8]

The first serious difficulty occurred in 1905. The water was low and the drives were hung. The St. John Lumber Company had five and one half million feet hung at Chase's Carry on the Allagash and another seven million in the main river below Fort Kent. Several men involved went up to the dam of the East Branch Improvement Company at Chamberlain Lake to try and get some water. While the caretaker was being bemused with discussions of the open air beauties the men involved measured and estimated the amount of water available. They then stole the old man's rifle and raised the gates to let water come to the St. John.

The caretaker went down the lake to get some help and while he was gone the crew took dynamite and blew out the dam. They left, but the crew from the lake came and repaired the dam. Not enough water came to do any good; the drives hung anyway. The sawmills were forced to close by August 1st for lack of logs. Some of the men (mostly American citizens) were arrested the next spring and taken to Houlton where they furnished bond and were released. Their cases never came to trial and the bond was never repaid. Fred Ayer, who owned the East Branch Improvement Company, struck a bargain with the St. John Lumber Company, and in 1906 and 1908 when the water was again low he caused the Chamberlain Lake gates to be raised to allow the logs to come down.[9] This solved the problems of the upriver men, but it did not help the plight of the St. John city sawmills who found themselves without logs because of low water and poor sorting practices.

The 1907 drive was much delayed because of the sorting at the Van Buren booms and the 1908 drive appeared as if it would be as well. The contractor for the St. John Log Driving Company, apparently acting on orders from the downriver log owners and with the help of some of their employees, cut the St. John Lumber Company booms and drove the downriver logs through without sorting. Prospects were good for a pitched battle, but when the main body of logs cleared the sorting ground the men holding open the booms left and sorting commenced again.[10] After this episode the United States and Canada decided to hold their joint investigations.

Testimony was taken from every sort of witness and from every vantage point on the general practice of sorting and driving on this river. A general picture emerged. Sorting was irregular, depending on the water pitch and the wind. The difficulty was not in the sorting so much as it was in the jams which formed in the booms after the sorting. Even though the men worked fourteen hours a day, from 5 A.M. to 7 P.M., the logs could not be handled in the time that a good driving pitch was on the river. As a result the drive was always in danger of not arriving. Part of the trouble was caused by the fact that upriver mills threw sawdust and edgings into the river creating artificial shoals which caused, in turn, side or wing jams. The drives only came down then with the aid of horses used to twitch the logs from these shoals.[11]

Downriver men also complained of other factors. The St. John Lumber Company, by virtue of its sorting booms, was able to pull all

The companies control the rivers

the no-mark logs to saw which deprived the Fredericton Boom Corporation of income. According to the testimony they were none too careful about sawing logs which were well marked. Some said this amount came close to a million feet a year. Eventually the downriver men stationed a checker on the sorting booms and at the saws of the upriver company in an effort to minimize this carelessness.[12]

The 1908 drive was late because of the sorting and that of 1909 was also delayed. In 1910 the ice went out quickly in the spring, and the logs that had been hung because of the late drive went to sea with the freshet. Fifteen million feet went downriver to Fredericton, and some went into the Bay of Fundy. The downriver men felt that it was only fair for the upriver company to take some of the loss caused by their action. They testified that as a result of their slowness in sorting, a drift drive (that is, a collection of logs in the area past the booming grounds) had been necessary in each year after 1905 at a considerable cost to the boom concern.[13]

Drift Drive—St. John River

1906-1909

Year	Amount (in feet)	Cost (per M)
1906	5,861,389	$4.25
1907	2,964,795	6.65
1908	3,107,286	5.85
1909	714,390	9.34

The situation did not improve. The upriver company continued to sort all logs at Van Buren, although a larger crew did the work, so that the time loss did not bother the downriver men as much. The upriver mill continued to saw both marked and unmarked downriver logs, however, and this remained a bone of contention especially when it appeared that direct orders had been given to continue to sort out and saw no mark logs. The table indicates the extent of this nuisance sawing.[14]

Downriver Logs Sawed At Van Buren—1911-1912

Year	Downriver Marks	No Mark Logs
1911	173	2,202
1912	2,390	793

By the time the commission had met for two or three years it was obvious that the days of sawing were nearly finished in the lower

St. John river valley. The upriver mills had won almost by default. The commission spent the rest of its time in discussing the possibility of storage dams to control the river waters, hydroelectric dams to provide new cities in the northern wilderness, toll bridges across the river, and in general the drainage systems of the St. John and its tributaries. The first mention of what have since become fashionable plans to dam the St. John either at Winding Ledges, Rankin Rapids, or up the Allagash occurred during this last few months of rather desultory testimony.[15] J. Fraser Gregory, who was the most persistent and articulate witness, perceived the future accurately enough on one of his last appearances before the panel. He said the lumbering business in St. John was about played out, and no more mills would be built.[16] The story of the river since World War One has been one of increased growth of pulp and paper mills upriver from Fredericton.[17]

The commission finally handed in its report early in 1916. The members voted three to one (with Madigan from Van Buren in opposition) that the Telos Canal and the ultimate diversion of Chamberlain Lake was a violation of the 1842 treaty, and the Maine legislation allowing the canal was as a result *ultra vires*. They then recommended nine dams to be built, five in Maine and four in New Brunswick, at a cost of $82,565 for Maine and $99,474 for the province. The commissioners also said that both governments should pass stricter laws about mill waste, the dams should be built at joint expense, and a joint three man commission should be set up to regulate the water flow in the river. That the prize logs should be sold at auction and the proceeds be distributed equally was another recommendation. The commission also remarked that if the proposed dams were constructed there would be no need for a change in Chamberlain Lake and the Telos Canal, but if they were not the waters should be divided equitably.[18]

By the time of the commission report driving on the St. John was very nearly a thing of the past. The first World War was on and people were concerned with other things. The twenties were a time of troubles for those who had not made the shift to pulp and paper and was for them consequently a time for liquidation. There are still drives on the St. John,[19] small to be sure (only five million or so feet in 1963) but nevertheless a few logs still come over the Grand Falls, and the visitor to the valley in the summer may occasionally still see booms of logs being towed from Fredericton to St. John. At the port city itself he may see the giant Irving mill, the last of a long line. The problems of control of the river were solved, easily and basically,

The companies control the rivers

through using the human desire to talk. While men talked about the problems certain geographic and economic imperatives did their work. For those who long ago had envisioned political union in the north country as a solution to regional problems there was little comfort because, as elsewhere, the upriver mills would still have won. And today, as a result, the closeness between northern Maine and the St. John river valley is rapidly becoming a thing of the past too. The Mainers increasingly look inward; the New Brunswickers look around to Ottawa.

Although there were precedents which had been set concerning the future of downriver mills in their competition with the headwaters companies, none of these precedents had really ruled on control of the water power for hydroelectric use. In addition, no person of great prestige and position stepped forward to plead the downriver cause in the earlier cases. Only in the case of water storage on the Kennebec did such a champion come forward. His stand on behalf of the people as opposed to the companies was to have lasting significance for the state of Maine. It is for that reason that some discussion is necessary of the great water storage battles of the 1920's.

After the struggles in 1901 and 1903 little remained for the Great Northern to clean up. Routine bills were presented in the 1905 and 1907 legislatures,[20] and it was not until 1912 and 1913 that anyone even gave much thought to water control in Maine. In those years the St. Croix flowage cases were settled. The paper mill at Woodland built a dam on the Grand Falls of the St. Croix on some Indian Township land which was held for the Indians by the state. A bargain was struck which caused about $13,500 damages to be paid to the Indians, and the dam and water levels were set by agreement.[21] If further dams were to be constructed elsewhere perhaps the problems caused might not be as easy to handle. Some in the state were agitating the question of state ownership so that if dams were built the state might get the value. William T. Haines, governor in 1913, dealt with these people quite sharply in his inaugural address that year.[22] After calling for a change in state taxation to an income tax on the product rather than on the land itself, Haines said:

> Instead of opening the door to the State Treasury, as is being proposed by some, for the sale of such lands to the State to constitute a forest preserve for the growing of timber and the storage of water, which to my mind opens an avenue of great extravagance, I believe it is much better to leave all our wild lands as they are today, in the hands of private owners, with the right reserved as it now is, to everybody to go upon them for hunting and fishing, recreation and pleasure, which makes of them a great natural park.

in which all of the people have great benefits and great interests. The great asset of fish and game in these parks brings millions into our State annually, is a source of wealth, and their proper regulation and care is of very much more importance to my mind, than state ownership, and a great deal more economical for the state.

With this the issue was fairly joined. It was public power versus private, and for the next fifteen years the legislatures would wrestle with the issue.

In that very legislature a bill was offered which would preserve the fee title to water powers and water power sites for perpetuity in the hands of the state. On riparian lands the same would be true one chain from each side at high water. Leases and timber rights could be granted only by two-thirds vote of both houses of the legislature. The state itself, by this proposal, was to construct the dams and sell the power. All dams then in use or being constructed were to be purchased by the state. Needless to say the Committee on Interior Waters ruled "ought not to pass" on this legislation and it died the death usually reserved for such acts wherever they are presented.[23]

It was not until 1917 that the legislature would be forced to act on such legislation. In that year the so-called "Tracy Bill" was first presented to the legislature.[24] This measure provided that the State of Maine and its citizens were the legal owners of all water powers and rivers. It provided for the establishment of a "People's Rights and Water Power Commission" charged to make a survey, take over all unused water powers immediately, and insure that those presently in use were being used for the benefit of all. The commission was to construct dams, sell bonds, supply heat, light, and electricity to all comers (if at all possible), to use the profits to build roads, and then to distribute the remainder (if remainder there was) on a per capita basis. The commission was also given the right to control pollution and was empowered to investigate the establishment of municipal forests as well.

With this attempt at instituting public ownership before the legislature, other measures were then presented to protect the rights of citizens. All such bills were referred to the Committee on Interior Waters, made up of seven Republicans and three Democrats. Friends of public power were unhappy with the committee personnel for none of the members were advocates of the water power company position. Feelings over the makeup of the committee erupted when a bill was presented allowing the West Branch Driving and Reservoir Dam

The companies control the rivers

Company to sell its rights to the Great Northern Paper Company along with the right to sell electric power. Percival Baxter proposed referral to the Judiciary Committee after the senate had sent it to the Committee on Interior Waters. Patrick Gillin of Bangor, erstwhile lawyer of the downriver men in 1901-1903, reviewed that history after this proposal and called for fair play. Baxter answered Gillin, who by this time was representing the private power view, in a speech in which he laid the groundwork for the next few years especially in the following remarks.[25]

> The Senator from Penobscot [*Mr. Gillin*] yesterday, in the Senate, told of the long and costly fight that was made in 1903 to secure the charter of this West Branch Log Driving Reservoir and Dam Company. He said that there were millions involved in it, that private interests came down and fought throughout that session. I have been told privately, on excellent authority, and it is not mere gossip, that the lobby at that time cost those interests $150,000. Now, what were they working for? They surely were not working for the interests of the State of Maine. Those men did not spend that large sum of money having in mind the protection of the water powers and reservoirs that belong to the State of Maine; and through that large expenditure they took from The State of Maine what is, perhaps, the most valuable reservoir system in our state.

On the vote, Baxter's motion lost fifty-three to eighty-six—a geographical vote split with those opposed coming primarily from south of the Kennebec River and from Aroostook and Washington Counties. An odd few came from downriver Penobscot towns, although none from Bangor and vicinity.[26]

This discussion was just a skirmish concerning several bills, all calling for a variant of a State Water Power Commission. All the stops were pulled for this. Public power advocates asked Gifford Pinchot to come from Washington and address the legislature; he accepted but was forced to withdraw because of pneumonia.[27] He did send a strong letter advising passage of a strong bill. Baxter was floor manager for the strong bill, and after the committee reported "ought not to pass", he made an impassioned speech in favor of the minority report.

> These companies have brought upon themselves the troubles which now face them. Their arbitrary methods, their powerful lobbies, all inflame the public mind. One of their lobbyists asked me contemptuously, "Who is behind this bill?" I replied, "A good many of the people of the State of Maine are behind it, and I will do what I can to help them."

His speech did no good, the bill was killed by a vote of eighteen to ten in the senate on April 6.[28]

Other bills were presented in the legislature this year but they also all failed of passage. On the day before adjournment as the legislature was winding up its business there were several pleas for passage of some bill, and after many references to the unfavorable newspaper comment the Public Utilities Commission was suggested as a reasonable body to make an investigation. Baxter said that they had the powers and had not done anything, and the suggestion was "indefinitely postponed" as almost the last item of business in the session amidst much discussion of whether a quorum was present or not.[29]

The pressure for such laws would not be disposed of as easily in the next meeting of the solons. Among other things which passed this legislature was a fairly strong lobbying bill calling for registry and the fining of violators.[30] The legislature also accepted and tabled a report from the land agent enumerating the agricultural wild lands then in private hands, apparently in the hopes of eventual state purchase.[31] Along with this came a proposal from Baxter calling for the purchase of lands and the establishment of a forest park and preserve on Mount Katahdin. The original bill called for the elimination of timber cutting. The forest and land committee voted "ought not to pass" on this bill, but a compromise piece of legislation was voted into law which said that the state could accept such parcels of land with the provision that these lands, once accepted, could not be sold, and that they should remain in a natural state.[32]

This legislature was also concerned with water power and storage. The first item of business once the initial pomp and ceremony was dispensed with was to receive bills calling for an investigation. A number of such bills were presented. They included two calling for investigations, one to extend the subpoena power to the Public Utility Commission, one to locate and survey all the public lands near water power sites, and one to request advisory opinions from the State Supreme Court as to the extent of the state's power in regard to reservoir construction. A further bill calling for the extension of the Public Utility Commission's powers to the passing on all sales and leases was offered, as was a bill making the building of fishways compulsory in all dams.[33] The Judiciary Committee proceeded to hold hearings on the proposed legislation.

Much testimony was designed to promote some sort of compromise bill although Edward C. Jordan, who had been a member of the old Water Storage Commission, did discuss the history of large companies in regard to water power and electricity.[34] The end result was a bill calling for a Water Power Commission of ten men who

The companies control the rivers 313

would represent nearly every point of view in the state. It would have subpoena power and the right of approval on future power projects on state owned land. Its basic function was to sit as a long-range investigative force.[35] The most telling speech in favor of such legislation came from Baxter. He said that cleavage had been brought about between the people and the water power owners; he reviewed past legislation, and ended his speech with the remark, "The forces of corporate ownership and ultraconservatism are arrayed against us." The bill passed and was enacted into law.[36]

The last act in this legislature prior to the bitter battles of the 1920's concerned the fate of the questions relating to public power propounded to the courts. Baxter, who had proposed the questions, was forced to table his bill as he was called away on business, and his enemies got control of it through parliamentary maneuver. Baxter, who was helped by the Speaker of the House during this debate, rose to great heights in pleading for passage of the bill.[37] After stating that the whole future of public vs. private power was in the bill, and that the great enemies of the bill had been the Central Maine Power Company in 1917, and the Great Northern Paper Company in 1919, he then said the real battle was the people against the corporations.

> .... the gentlemen to whom I have referred before [*the lobbyists for the G.N.P. Co. et al.*] have . . . made the boast that they held the destiny of this legislation in the palm of their hands. As for myself I do not know what it is to be beaten when I am fighting a good cause, and if there ever was a good cause this water power question is one.

Baxter was right. The fight was on. This bill passed but the battle between Percival Baxter and the Great Northern Paper Company was just beginning. In this battle the stakes were in fact control of the Maine woods and although the company won, it was forced to pay more dearly than it had in the court cases decided in the first decade of the century.

The ground was well laid out when the court's answers to Baxter's questions were published.[38] Before making their answers the judges offered some facts for consideration. All ponds over ten acres in extent are "great ponds", by the Massachusetts Colonial ordinances of 1641 and 1647. The state holds these as it does tidal waters, and it cannot compel either the littoral owners or outlet mill owners to pay for the use of the natural flow. The littoral owners cannot change the level. Riparian owners own to the "thread of the stream", (*ad medium filum aequal*), and the only abridgement of this right is the public right to fish or to passage to the stream itself. There are no public rights to the power.[39] After laying this ground, which is repeated here because the rest of the discussion hangs on this interpretation in part, the judges then turned to the questions.

The state could not tax to provide public power without a constitutional amendment; once the state made a contract to give over the water powers a retroactive tax could not be levied as it too would be unconstitutional. Even a franchise tax would be only legal on limited grounds. The only way that this could be done is to repeal the ancient mill and flowage act which protects the person granted the contract. However if this act was repealed any sort of tax would be legal. Here the judges disagreed somewhat.[40] From these answers it was easy to determine that Baxter and his friends were not going to accomplish much in regard to the lands and rights already granted, but where such rights still were held by the state perhaps the companies eager for these rights might well pay and pay heavily for them.

When the 1921 legislature met the first order of business was to elect its officers. Percival Baxter was elected President of the Senate, a post of great importance and one of overriding significance when Governor Frederick Parkhurst died after a bare month in office. Baxter, as next in line, was suddenly catapulted into the position of being chief executive.

In his speech of acceptance of election to the office of Senate President, Baxter had warned the legislators of what was to come.[41]

> . . . There is danger to our institutions, however, when one group or class attempts to dictate to and control the state. Any government of the whole by a class or group would be unbearable and would mean the end of the Republic. The paid representatives of classes and groups, of special interests and special privileges are always present during Legislative Sessions. The Legislative halls are frequented by selfish plausible men who often attempt to control those of us who are here to represent all the people. We must stand firmly against these groups or class interests whoever they are and what ever they represent because for the time being we are intrusted with the perpetuation of this government, and the foundations of which are justice to all, and special privilege to none.

He then proposed an amendment to the Maine Water Power Commission law of 1919, which would have changed the makeup of the committee slightly in order to broaden the membership. In addition he offered a memorial to the Congress urging passage of an amendment to certain federal acts which would have limited the transmission of Maine power beyond the borders of the state. These

The companies control the rivers 315

two acts were passed after considerable deliberation.[42]

Baxter was a man with fixed ideas which revolved around public power, or at least public control of private power, and also around the establishment of public parks on the land once held by the state. When he was elevated to the governorship he adverted to these ideas in his inaugural address.[43]

> The forests of Maine, with our waterpowers, constitute the great natural resources of the state. The waterpowers in their nature are perpetual, while the forests may and have been, wantonly destroyed by wasteful management and by fire. A century of statehood has passed, during which period the millions of forest acres once publicly owned have passed to private hands. In the heart of the wilderness of these woodlands stands Mt. Katahdin, the greatest monument of nature east of the Mississippi River. This mountain raises its head aloft, unafraid of the passing storm, and is typical of the rugged character of the people of Maine. The purchase of this mountain will constitute a fitting memorial to the past century and in injunction to the new. . . .

With these remarks Baxter heralded his campaign. Initially it took two forms, asking the legislature for an amendment to the State Constitution in regard to water power, and asking for authorization for a state park at Mt. Katahdin. Baxter started his attack with a message to the legislature calling attention to the questions and answers on water power which the courts had given, and announcing that the commission had a state control plan for reservoirs. "The interests of the State are greater than those of any person, or corporation, within it." To accomplish this plan an amendment to the State Constitution was needed. By this method the people could decide on the future of waterpower within the state. The differences between state and federal law would also be covered with regard to taxation of power transmitted beyond state borders.[44]

> . . . . conservation, storage, control, and use of waters, and the development, improvement, transmission, utilization, electrical interconnection, control and sale of water powers . . . are declared to constitute paramount public uses. . . .

On April 5th the Judiciary Committee reported the proposed amendment "ought not to pass" by a six to four vote with the lone Democrat and three Republicans from Portland, including Brewster, in dissent. They filed a minority report. Debate on the floor of the legislature was spirited, especially after Baxter agreed to eliminate the taxation provision from the amendment.[45] Brewster, in arguing for its passage, went back to the old history of the land grants.[46]

> No discussion of the water power question is complete without a reference to timberlands. That always seems to be brought in by the heels in any consideration of this question. I do not feel that the title of the timberlands in this State needs to be questioned, nor that it is profitable for us to discuss the methods of their acquisition. The only reason that we need to refer to the nine million acres of timberland which has passed out of control of the State is that it may afford to us some lesson in preserving such rights as we may have left in the other and more valuable natural resources of this state, and that is its water power.

Debate was stormy, and those who opposed the amendment found it necessary to rise to their own defense and to clear themselves of the taint of supporting the corporation point of view.[47] When the vote was taken in the house, the opposition forces had won eighty-four to forty-four, with twenty-two absent.[48]

The senate discussed the proposition the next day. Patrick Gillin of Bangor, onetime supporter of downriver mill men, spoke at length and in an involved fashion. He rehearsed the 1901 and 1903 legislative actions, recalled his own part in the Great Northern case, and then went on to say that he approved private development. His speech was rambling and was misunderstood by other senators who thought he was going to vote for the amendment. Gillin was forced to stand again. This time he said that if the amendment had stopped at "paramount public uses" he would have voted for it, but not now.[49]

Democrats and the downstate Republican senators tended to favor submission of the legislation to the people for referendum. In discussing this possibility Senator Morrill of Cumberland County made remarks which illuminated the background of the struggle.[50]

> If there is a mystery about anything in Maine, it is how much the people ought to pay for their electric lights and power and also the reason why the power is carried from the country to the cities, while the country places are left in darkness.

The vote was eighteen to ten in favor of killing the amendment.[51]

It was a simple matter now for the legislature to kill the other part of Baxter's program, the Mt. Katahdin state park. A report came from the committee "ought not to pass". A second report was offered which eliminated the mention of money, and simply authorized the park. Both reports were accepted for debate purposes by a vote of thirteen to twelve. During the debate itself petitions and resolutions were offered by the proponents of the bill, but the most telling speech was one in opposition.[52]

The companies control the rivers

> .... the fact is, those land owners—and the Great Northern Paper Company is one of the largest of those land owners, feel aggrieved to have the state say to them: We are going to force you to sell some of your land at some kind of a price to be agreed upon by County Commissioners, or somebody else. We are going to place that right in our statutes so that we can hold you up at any time.

After calling attention to the paper company and its program in and around Moosehead Lake and the great market for farm produce at their lumber camps, Senator Sprague went on to say that he

> could not vote to antagonize such a corporation as that, that has been of so much benefit to Piscataquis County, to its farmers, and to its businessmen, as has the Great Northern Paper Company. I can say truthfully, I never took a retainer from the Great Northern Paper Company in my life. I certainly have not any at this time. I wish I had. They are good paymasters. . . .

The 1921 Legislature ended with no resolution of the water power, public and private development controversy. The 1923 Legislature would have to deal with these items. By that time Baxter had been elected governor in his own right. He was expansive in his inaugural address.[53]

> We all want development and if the State does not take it, private interests should be allowed to do so, with the State's interests fully safeguarded. In every private storage development hereafter undertaken, I would reserve to the State the right to purchase it at any time for a fair price without paying for the franchise or storage rights guaranteed by the State; the purchase price in no event to exceed the cost of the development. A clause partially covering these restitutions already has been placed in certain storage charters. . . .

He again offered the Water Storage Amendment defeated in 1921, said that the Public Utilities Commission could take over the duty of the Water Power Commission, advocated his public park and reserve in the Mt. Katahdin area, and then went on to comment on his old adversaries.

> Although some of the remnants of the old time lobby linger about the Capitol they are rapidly becoming relics of the past. Years ago lobbies exercised a considerable control over legislators and legislation. Spacious quarters at the leading hotels were maintained by these autocrats of the "Third House" and those who came and those who went paid them homage. They sought the end and were little disturbed as to the means. The old lobby leaders were able, picturesque, and powerful, but their modern successors are of smaller mental caliber.

Baxter's remarks did not represent the reality of the legislature. Almost as soon as he stopped speaking representatives presented a bill to organize and charter the Kennebec Reservoir Company. This new firm, representing manufacturing interests on the Kennebec River, wanted to take a large state-owned water power site (Long Falls) and by building a dam there control the height of water on the river and control the log drive on the river. Screams of anguish from the public power advocates were immediate. What was to be the status of electricity in the future?[54]

Emotions ran high on the bill. As one legislator remarked: "It is drawn as a log-driving proposition, but, gentlemen, it ought to be entitled a log-rolling proposition." It passed both houses of the legislature, but Baxter returned it with a stinging veto message delivered in person in which he compared the bill to the great land steal of 1868; that is, the grant to the European and North American Railroad Company.[55]

Debate over the veto message was even more heated. In the senate the vote was twenty-two to seven to pass it over the goveror's veto. The President of the Senate remarked to that body, when casting his own vote:[56]

> That men will not always agree on important matters is inevitable; that men may reach different conclusions and all be acting in the best and most honorable motives, and with honest judgment, cannot with justice be denied; that all have a right to vote according to the dictates of their own consciences, exercising their own best judgments, without being charged with a betrayal of the trust imposed on them, should not be questioned.

The house debate was also heated. When the votes were cast, the governor had lost again one hundred and twenty-one to twenty-five, with four absentees.[57]

With this action came the opening round in the battle of proclamations. Baxter issued a proclamation under the official seal of the state comparing the action in 1868 with the action in 1923, but saying that the people must now exercise their constitutional rights and call for a referendum.[58]

> Corporate interests and their lobbyists are already rejoicing over their triumph and although these lobbyists have dominated the corridors of the State Capitol they have not yet stifled the voice of the people.
>
> The men and women of Maine are yet to be heard from. Our only inheritance hangs in the balance.

The companies control the rivers 319

The governor had thrown down the gauntlet to the legislature. It was immediately picked up and a motion calling for a referendum was introduced. Argument followed in both houses, but, by voice votes, the motion for a referendum was adopted.[59]

Having achieved this initial victory Baxter then moved to consolidate his position. He appeared personally before the legislature urging them to repeal the first act, and substitute for it one of his own. This new bill provided for the land to be leased to the company for forty years, with the state having the right to take it over at not more then one half the net cost of the improvements. As evidence for public support of his bill he cited the signatures collected throughout the state favoring the referendum. He also said the bill was the product of a conference between one man who was interested in the bill and other disinterested individuals. He also thanked Owen Brewster publicly for his aid in bringing forward the new bill.[60] On the next day he sent a further message to the legislature in which he said that J. F. Wyman, treasurer of the Central Maine Power Company (the major power company involved), had assented to the new act, but that he had now refused the use of his name in a letter to the governor. Baxter went on,[61]

> When the Governor of the State of Maine enters into an arrangement it is to be expected that he will hold to it. The same properly can be expected of the others who are parties to it. As the Legislature is now considering the Kennebec and Dead River matters, I desire to give you the information that has just come to me in the letter referred to.

This message really threw everything into a frenzy.

What had happened was that after the passage of the first Kennebec Reservoir Bill the referendum threat had apparently promoted a spirit of compromise among the private power advocates on the river. They, in the persons of the Central Maine Power Company executives, had met with Governor Baxter, with Senator Brewster acting as mediator. The Baxter bill was apparently the compromise act which had been achieved at this confrontation. Baxter's remarks when presenting the bill had angered the company officials and they had withdrawn their public assent to the bill. With this an impasse had been reached. Debate was stormy. Letters from various company officials were read into the record and eventually some were discussed and read in executive sessions. The house proposed new questions for the court, but this motion was withdrawn, after it had passed.[62]

In the senate the bill met as stormy a welcome as it had in the house. Under their rules it could not be introduced without suspension of the rules, which needed a two-thirds vote. Eventually the senate president was forced to rule that all in the chamber would be counted as assenting whether they voted or not in order to allow presentation of the measure.

Senator Hinckley, long the private power advocate in the senate, rose to defend the interests which had been attacked by the governor. He said that the conference referred to by the governor had been a case of Baxter being willing to talk to the people who were maligned in his public statements. He did say that the old bill ought to be killed and the new one substituted for it, which was Baxter's point as well. Then he came to the crux of the matter. A bitter tirade was offered which implied that this whole process of meeting and compromise was designed to make Brewster succeed Baxter as governor. The peroration of the speech was surprisingly curt.[63]

> In every respect, I think it is time for this Legislature to assert its rights in the right way. I am thoroughly tired of these political plays. I am thoroughly tired of the subject of wind and water making Governors for the State of Maine.

It was now time for Brewster to answer these charges, but as he rose to deliver his rebuttal, he "seemed to be overcome" and the reading of the speech was finished by his opponent, Senator Hinckley. The speech said that on the question of motives, all motives were pure, and they should be viewed in that light, Brewster's motives as well as everyone else.

The senate then proceeded to vote the repeal of the original bill by a vote of twenty to two, and the house reciprocated the same day. Brewster later rose to a point of personal privilege and continued his remarks on motives, defending the companies and by implication himself and Baxter, saying no one should be attacked on personal grounds.[64]

The governor celebrated the triumph with another proclamation to the citizens of the state.[65]

> The referendum is vindicated. . . . The victory is won and a great principle has been established. . . . The public interests have been protected and the attempt to secure valuable water rights without paying for them has been defeated. Public sentiment has been aroused and in my opinion no legislature hereafter will GIVE away what belongs to the people. . . . The victory that has been won is a victory "OF THE PEOPLE, BY THE PEOPLE, AND FOR THE PEOPLE."

With this action the legislature was about finished. On the last day of the session an amendment was offered to the Maine Water Power Commission law which took the appointive power from the governor, but Baxter vetoed it and sent another proclamation to the people saying that the motives behind the new amendment were suspect, implying it had some connection with the people who favored private power.[66]

Baxter continued to use the proclamation method to bring his case before the people. On May 3 he issued a proclamation entitled: *The Inside History of the Kennebec-Dead River Storage Charters*. In this lengthy document (it runs eight pages as it is printed in the laws of the state) he reviewed the entire issue as it appeared to him from the Governor's Mansion.[67] According to him the lobbyists for the Kennebec project had started to work early in the session. "It was amusing to see the lobbyists race to the telephone booths to spread the news of their victory over the Governor. Their rejoicing was premature...." He went on to say that after the initial defeat the corporate interests had been then desired an early referendum because of the smaller vote total expected.

Baxter now gave his version of the famous Wyman-Baxter-Brewster negotiations. He said that they had not discussed the power because Wyman had said that "he would not know what to do with it if he had it" because it was so small and seasonal. Baxter said that this was a blind, who else could use the power once the dam for water storage was built?

The comic contretemps over introduction of the second bill occurred because Wyman, due to outside pressure, had capitulated. He said that the lobby and even the Chief Engineer of the Water Power Commission were working against the people's interests.

> Certain newspapers affiliated with the Central Maine Power Company and other corporate interests now rally to their support and the State is flooded with propaganda . . . . a deliberate policy of creating public sentiment. . . . This is all part of a general scheme of certain private interests to secure control of all the water resources of the state.

Less than three weeks later Baxter returned to the proclamation attack. In this effort, entitled *Maine Water Power, 1923-1924* [68] Baxter listed the sixteen companies of 1918 which had formed the alliance against him and the state. They were, he said, the Rumford Falls Power Company and Oxford Paper Company (Hugh J. Chisholm, president), Great Northern Paper Company (Garrett Schenck, president), Union Water Power Company, Union Electric Power Company, Androscoggin Reservoir Company (Wallace White, treasurer), St. Croix Paper Company (Arthur L. Hobson, treasurer), Central Maine Power Company (Walter S. Wyman, treasurer), Androscoggin Mills (P. Y. DeNormandie, treasurer), International Paper Company (Philip T. Dodge, president), Androscoggin Electric Company (William T. Cobb, president), Hill Manufacturing Company and Lewiston Bleaching Company (H. B. Richardson, treasurer), Pepperell Manufacturing Company (William Amory, treasurer), Bates Manufacturing Company and Edwards Manufacturing Company (H. Def. Lockwood, treasurer). Of this group he named Dodge, Lockwood, Schenck, and Wyman as concessionaires for the Kennebec Reservoir scheme. In addition he also named the presidents of Hollingsworth and Whitney, and the Shawmut Manufacturing Company on the Kennebec as being part of the heinous "concessionaire" group.

After naming his opponents he then proceeded to instruct the electorate in the realities of life. He said that there were many in the state who were offering "enlightenment" through newspapers and other means of communication. Some of these groups wanted a fact-finding board to investigate. He then proceeded to name and characterize those who wanted this board. They included James Q. Gulnoc, a representative of corporate interests and founder of the "Boom Maine" campaign; Guy Gannett, owner of woodlands, water powers, and newspapers, co-director with Wyman and a Republican national committeeman; Walter E. Elwell, President of the State Associated Industries group; Benjamin F. Cleves, from the State Manufacturer's Association; and finally Arthur R. Gould, who he described as owning practically all water power and and electrical plants in Aroostook County. He urged consideration of the aphorism, "Beware of the Greeks and their Gifts."[69]

He then went on to the crux of his argument. The problem in Maine was the lobby which had continued to grow. Probably now some form of compromise, perhaps by lease, should be found. Maine industries were not dying, and the issue was not really public against private power. It was instead whether the potential would be used by some or used by all. After citing the examples of Quebec and Keokuk he said that corporations must also share in the cost of the state's upkeep. The state should control and develop these natural resources, and the constitutional amendment advocated before was needed. He cited his own hard work in the past, and said the corporations were the "dogs-in-the-manger", not he as his opponents had said. After all, as he came to rather a plaintive end, the governor was not to blame for the recent freshets on the Kennebec, the people who stall off on the public dams were. We must as citizens get out from under this flood of propaganda and get to the reality of the situation.

After this blast the demand for compromise began to rise throughout the state. There were some possibilities. Mt. Katahdin and its hoped for state park had been defeated by a vote of twenty-three to three with five absent during the Kennebec Reservoir discussion.[70] In addition the Great Northern Paper Company had severely restricted the passage of hunters onto its lands in the area. A bill concerning the West Branch Driving and Reservoir Dam Company had sailed through the legislature but Baxter had vetoed it on the grounds that it concerned the people and that a referendum was needed. The bill had been passed over his veto, one hundred and twenty-one to nine with twenty absent in the house, and twenty-six to three in the senate, but one of the three was Owen Brewster.[71] Brewster was due to be the next governor and the old fight would probably continue. Impasse or compromise seemed to be future prospects for the water power question.

Although the legislature did not meet again during Baxter's tenure as governor, forces engendered in the 1923 Legislature and in the earlier discussions continued to operate. Pressures for compromise built up. Baxter outlined the way when he came to the legislature on January 7, 1925 to deliver a farewell address.[72]

After a series of complimentary opening remarks he launched into a governmental disquisition:

> Today when certain large, powerful and well-organized groups and business institutions exercise great influence over public affairs, "unmortgaged" officials are needed more than ever. Too many men in business and public life cater to those special interests. If the rights of the people are to be safeguarded our State needs executives and legislators who are fair-minded and free from prejudice and self-interest. Moreover, a public man who intends to control the affairs placed in his charge must be able to say "no" when occasion demands. The ability to do so often is the best of his service.

He again called attention to Mt. Katahdin and the potential for a state park. "Due to the opposition of the large timberland companies, especially the Great Northern Paper Company, no progress has been made other than to create considerable public sentiment in favor of the project. The timberland owners have repeatedly defeated the law under which the state would be empowered to condemn land after paying a fair price for it. ..." He then offered a considerable threat to the paper company.

> If, however, the orders recently promulgated by the above mentioned company restricting the use of their lands to registered persons accompanied by licensed guides are held valid, Mt. Katahdin before long will be closed to hunters, fishermen, and campers. The order referred to is the entering wedge of "regulation" that later will develop into "prohibition". The time never must come when the forest areas of Maine are made great private hunting preserves to be enjoyed only by the friends and sycophants of powerful interests. Such things savor of feudal times when the lords and barons of England claimed the sole right to the fish and game on their great estates. Before our woods are closed to us the people will be heard from.

He then offered some of his own money if the legislature would provide a group of men and some more money to attempt to buy the NW½ T3R9 (the mountain proper).

> It is interesting to remember that the Great Northern Paper Company in 1921 through its lobby intimated that it might donate to the state its undivided interest in some of the land in question. Whatever it wanted as concessions in other directions evidently was not forthcoming. It is well in the future to bear this latter thought in mind.

He finally adverted to the water power issue. He said that Maine did not have "public" control, and that the people could almost never know what was happening because of the fact that newspapers were subservient to water power interests.

Not much happened in this legislature except that the Mt. Katahdin park plan was reported out of committee "ought not to pass" and it was killed early in the session.[73] By the end of 1925 Baxter could say publicly that the paper company had assured him that the big mountain would not go into any hands but the state or local government. He also implied that they promised no logging would take place in the vicinity of Mount Katahdin. This last condition was not accepted by the company apparently.[74] The land holder experienced pressures from those who wished to utilize the mountain as the end of the Appalachian trail network, and Baxter used these pressures to his advantage. Relations between the Governor and the company deteriorated so that Schenk attempted to avoid seeing him when Baxter came to Boston for that purpose. In fact Schenck referred to Baxter "as a pretty slippery unscrupulous sort of a chap. . . ."[75]

In 1926 Baxter wanted to run for the United States Senate, and years later he said the company had fought tooth and nail against him,

The companies control the rivers 325

and the evidence for this is strong. Baxter said his loss here was worth gaining the mountain for the large state park, but the aftermath left a bad taste. The company did work hard against him, but whether this was the only thing which defeated Baxter is difficult to say. However, as Garret Schnck said to F. A. Gilbert before the election.[76]

> Use all vantage possible—. You can have some money from me. Portland should not have two Senators—strong point against Baxter if *used*.—Cobb. I believe would use his influence—Brewster would do all he could under cover Hale should not be in line to help Baxter. . . . Look out for the Katahdin crowd. They and Brewster will try and make trouble—It may come up at Augusta—and it is well to have all detail about who is there— and fires started—but I believe that we can head them off—Suppose you attend as usual to getting some friendly parties there—.

It was not until 1927 that the compromise was effected. In that year with no fuss or fanfare the legislature passed the Kennebec Reservoir Act allowing for storage and power dams in SE corner T3R4 in Somerset County. The Kennebec Reservoir Company was also given the right to acquire the Kennebec Log Driving Company. It had to provide an annual drive. The state lands were leased for fifty years, at $25,000 a year. At the end of that time the state would have to retake the lands and provide payment or release them to the dam owners. The power rights could not be used by the parent company but they could be sublet. The firm was also forbidden to sell all of its property to any one of the firms making up the parent firm.[78] In that year Baxter began his personal campaign to purchase the land around Mt. Katahdin. The land was then in turn given to the state to be used as a park, to be maintained as nearly as possible in the wilderness state. The paper company relaxed its rules on entrance into its woodlands. A law prohibiting the lumbermen from fishing while they were actually employed in "harvesting forest products" appeared on the books.[79] The paper company said in an editorial in its magazine, "It is not material to the Company's interests in any way."

The great water power battle was over. The ending was the same as on the rivers. The companies up stream had won control and quite effectively too. On the Kennebec though there were some differences. The state got a beautiful large state park. On those grounds alone Percival Baxter must rank as one of the great governors of the State of Maine.[80]

NOTES

[1] *Foreign Relations of the United States, 1909*, ix. The evidence has never been printed. The Bangor, Maine, Public Library owns the copy of the evidence presented to the American counsel, Mr. O. W. Fellows, who willed it to the library. It is in five volumes, paged consecutively, and typewritten on one side of each sheet. Errors in the transcription are not corrected, but are italicized. Most of the evidence presented in the way of documents and so on is described, but does not appear in this set of the evidence. Taken all in all those volumes present one of the most remarkable sources of evidence for river driving and lumbering practices. I shall continue to cite it as *SJRC*, witness name, and pages included. A student of mine, Thalia O. Stevens, has recently discussed the Commission and its work, "The St. John River Commission," M.A. Thesis, University of Maine, 1969.

[2] The Bangor and Aroostook was completed when the Fish River Railroad was constructed between Ashland and Fort Kent, a distance of fifty-two miles. This road, a subsidiary of the B. and A., was opened for traffic on December 15, 1902. See Bangor *Daily Commercial*, December 6 (an excellent story), and 15, 1902.

[3] This development was covered briefly in the last chapter when the 1903 legislature was discussed. For the story see Bangor *Daily News*, January 31, February 4, (the St. John board of trade protested), March 7, 25, 1903; Portland *Eastern Argus*, February 18, 1903; Bangor *Daily Commercial*, March 4 (the Telos Canal was involved and the Interior Waters Committee took evidence). They said the amount of traffic through the canal had amounted to 23,000,000 feet in 1901, 30,000,000 feet in 1902, and about 47,000,000 feet in 1903. (The increase was due to the Eagle Lake Tramway of which more later), March 6, April 2, 20, 1903; St. John *Telegraph*, April 2, 1903. The day of the vote they said that the measure elevated Maine at the expense of the province and thus was rightly defeated. Some interesting correspondence on this matter appears in the *SJRC*, at 2910. It includes letters of protest from Michael H. Herbert (Counselor British Embassy in Washington to John Hay—Secretary of State, no. 62, March 22, 1903 saying the proposed East Branch Development Company was in contravention of Article 3 of the Webster-Ashburton Treaty. Hays' answer, no. 25, March 28, 1903 said that he had communicated this protest to the Governor of Maine, and on April 3, 1903, same to same, no. 98, he communicated a copy of the Maine law which had been passed, which was not in contravention of the treaty.

[4] Bangor *Daily Commercial*, March 19, 31, April 6 (the editorial), 19, 1904 (quoting the St. John *Telegraph*). The debate in the provincial parliament was reprinted in *ibid.*, March 18, and the *Commercial* reprinted much of it on March 19, 1904. Also see on the hearing the St. John *Sun*, and St. John *Telegraph*, March 31, 1904.

[5] The provincial laws regarding river traffic are in *SJRC*, 913-1111. Mills are listed in the testimony of J. Fraser Gregory, (recalled), 744-912, and Neil McLean, 109-184.

[6] Testimony of Thomas C. Hetherington, *ibid.*, 2238-40, W. H. Cunliff, 1500, and Gregory, 758-9, 767-8, 816, 831.

The companies control the rivers 327

[7] *SJRC*, testimony of Charles E. Jones, 5-47, 52-109, who lists most of these small mills and discusses their business.

[8] The description of the mill derives from the testimony of Arthur W. Brown, *SJRC*, 299-353, taken in 1909, and James W. Parker, 354-387, 392-450. For a good description of the mill in 1913 see R. B. Miller, "Report on New Brunswick Reconnaissance Work," a typewritten manuscript from the Department of Lands and Mines, Fredericton. The firm is discussed at 78. They were then handling 45,000,000 feet of spruce each year. The booms are discussed in the cross-examination of Brown, *SJRC*, 467-539, and they are shown in an excellent map facing p. 18 of 64th Congress, 2nd session, Senate Document 724, *Uses of the St. John River*, Report of the International Commission Pertaining to Uses and Conditions of the St. John River, on the reference by the United States and Canada, February 17, 1916. The firm drove c. 52,000,000 in 1906, 28,000,000 feet in 1907, 26,000,000 feet in 1908, and 33,000,000 feet in 1909. Testimony of W. H. Cunliff, *SJRC*, 1502.

[9] Testimony of Dennis Pelkey and Nazaire St. John, *SJRC*, 2223-2232, (part of the five man crew who did the blowing); Arthur W. Brown, recalled, 2266-2302 (he entertained the old man). The firm later built a dam at the foot of Long Lake to help them with the water, *ibid.*, 2282. Also see testimony of John Sweeney, recalled, 2382-2393, 2396. He was the cook and didn't even see the dam blow, but was one of those arrested, and the testimony of John Ranney, 2353-2381 (he took the old man's rifle and supervised the placing of the dynamite). The Chamberlain Lake people apparently were nearer than was thought, and they were also more numerous.

[10] After the booms were cut a series of letters were exchanged which illustrated the bitter feelings. St. John Lumber Company to Fredericton Boom Corporation, July 21, 1908 (protesting the slowness of rafting); Andre Cushing and Co. to St. John Lumber Co., August 18, 1908 (How many of our logs have you sawed?); same to same, September 18, 1908 (no answer to our last); St. John Lumber Co. to Andre Cushing Co., September 21, 1908 (Our books are in the hands of the Chief Justice of the State of Maine; we will answer when we can.) As nearly as I can tell, they never did. From *SJRC*, 1263-6, 1446-70.

[11] *SJRC*, testimony of Charles Miller, 1373-1420; J. Fraser Gregory recalled, 1420-39; John Kilbourn 587-652; Ben Lavasseur 1631-42 (the hours worked on the boom); Jerome (Come) Cyr, 1604-29 (he was in charge of the men and teams used to card the ledges in this area). The St. John Lumber Co. burned their refuse to provide power for their mill. Another mill in the area making cedar shingles did not. They carried the sawdust right into the river by means of a carrier direct from the saw. According to the owner this had been their practice since 1881. Testimony of James Crawford, 1717-50, especially 1730, and Louis H. Bliss, 1163-1167, on the effects at Fredericton.

[12] *Ibid.*, testimony of J. Fraser Gregory, recalled, 1423-4, 870, 1698-1716; John D. Colwell, 1301-72, especially 1319; W. H. Cunliff, 1503-27; Louis R. Bliss, 1156. The percentage of prize logs was just under 1% of the total drive. 1898- .843; 1899- .220; 1900- .960; 1901- .950; 1902- 1.567; 1903- .913 1904- 1.288; 1905- 1.225.

[13] *SJRC*, testimony of Albert L. Davis, 2397-2414; Louis H. Bliss, 1112-1121, especially 1121.

14 *Ibid.*, John R. Alcorn, (log watching foreman at the upriver mills in 1911 and 1912), 2403-2433, especially 2415-7. The total of the sawing is unknown, but it must have ranged close to 300,000 feet per year, figuring ten logs to the M, which would be a conservative estimate. It stands to reason that if you are going to saw downriver marks that you will saw some of the larger logs which go through your sorting gaps.

15 *SJRC*, testimony of Arthur Noble 3045-63, on dam built at foot of Baker Lake in 1905 by Andre Cushing and Co. to control water flowage; George S. Cushing, 1283-4 on this dam; John A. Morrison recalled, 3022-3035, on dams; John Kilbourn, recalled, 3036-44, on dams; he had once attempted to dynamite the ice from the brooks in order to get the drives out earlier and discussed this at 3038. In the year 1913 with 78,000,000 feet cut on the river, 68 million was driven, but 35,000,000 stopped in Van Buren; also Hardy S. Ferguson, recalled, (Engineer of the Commission), 2510-46 on the Penobscot and Telos Canal, and on the Canal again 1937-97; Harold S. Boardman and Cyrus S. Babb, 1751-1800, a 1907 survey on the Telos canal; Boardman, 1801-25 on the Telos canal; S. Jefferson Chapleau (engineer) on water storage 2691-2732; George S. Fowler, engineer for Grand Falls Power Company, on the survey and work done at Grand Falls, 1853-1935; John Kilbourn, recalled, 2877-2908 (his dam recommendations); engineers report through June 1, 1911, 2157-2180; progress report as of June 22, 1912, 2101-2105; survey on dams of February 7, 1913, 2733-2833; statement of Charles E. Oak pursuant to International Bridge at Van Buren, May 16, 1913, 2999-3010.

16 *Ibid.*, testimony of J. Fraser Gregory, recalled, 2936. (This testimony was given on May 9, 1913). As early as 1907 the market for lumber was described as "unfavorable and crowded." *Paper Trade Journal*, October 31, 1907.

17 Primarily the Fraser Companies, Ltd. plants at Edmundston and Madawaska. See for this "Forest Heritage: The Story of Fraser Companies, Ltd." written by Mary B. Reinmuth in collaboration with Donald A. Fraser. This unpublished history is in the form of a typewritten manuscript in the hands of the Fraser Company. I am grateful to them for allowing me to read it, and to Fred Phillips and Miss Reinmuth for calling my attention to the manuscript.

18 The report is Senate Document #724, 64th Congress, 2nd Session, *Uses of the St. John River*, Report of the International Commission Pertaining to Uses and Conditions of the St. John River, on the reference by the United States and Canada, Washington, D.C., 1917 (dated February 17, 1916).

19 The last big drive by the company was in 1924. See "Early Days of Log Driving on the St. John River", Canadian *Lumberman*, September 15, 1925. This is the date cited by Lower, *Assault, op. cit.*, 41.

20 *Private and Special Laws*, Ch. 30, 1905, allowed the G.N.P. to hold stock and vote in the West Branch Reservoir Dam and Driving Company. A bill allowing them to own and operate in any county in the state and to hold stock in the Northern Maine Power Company was passed February 13, 1907; see *Journal of the Senate*, 1907, 260; they were also authorized to raise Ripogenus Dam, and the Penobscot Driving and Improvement Company was set up to handle the drive from Medway south. See Senate *Documents* 128, 129, 1907.

21 "The St. Croix Flowage Case", Third Annual *Report* Water Storage Commission, 93-96.

The companies control the rivers

[22] *Annual Address,* William T. Haines, January 2, 1913, *Public Documents, Maine,* 1913, II, 20ff; the quotation is at 23.

[23] House Document #154, 1913, *House Journal,* March 14, 1913. There was no trouble in passage of water power laws before this. See, e.g., *Private and Special Laws,* 1909, Ch. 147, chartering the Androscoggin Reservoir Company, a firm which planned to dam the Magalloway River at Aziscoos Falls to insure the correct amount of water for Rumford Falls mills. The incorporators included William P. Frye, Hugh J. Chisholm, Herbert J. Brown, Waldo Pettengill, A. N. Burbank, and Wallace H. White.

[24] House Documents #81, 1919 is the easiest way to see the bill and its purpose. It also was introduced in 1917, 1921, and 1923.

[25] Senate Document #265, 1917; *Legislative Record,* 1917, 110-116.

[26] *Legislative Record,* 1917, 116ff.

[27] House Documents #209, 211, 1917. On Pinchot, *Legislative Record,* 1917, #893. His letter is here too.

[28] House Document #710 (Baxter's bill). His speech is *Legislative Record,* 1917, 1254.

[29] House Document #10, 1917, *Legislative Record,* 1917, 726-7, 1469.

[30] Senate Documents #15, 189, House Documents #3, 1919, *Legislative Record,* 1919, 353, 466, 625, 737, 788. Eighteen petitions with 196 names were presented in favor of this legislation, 96 of them in favor of House Document #3, by far the strongest bill.

[31] House Document #1, 1919. *Legislative Record,* 1919, 135.

[32] House Document #8, "ought not to pass" March 5, 1919; House Document #499, 1919, *Legislative Record,* 1919, 1036, 1082.

[33] House Documents #4, 11, 26, 138, 216, 282, 274, 505, 1919.

[34] House Documents #134, 1919, "January 29, 1919—Public hearing of Maine Water Power Storage Bill, before Judiciary Committee, testimony of Edward C. Jordan".

[35] **House Document #436, 1919.**

[36] *Legislative Record,* 1919, 711-716 for his speech; 936 for House action, 962 for Senate action. *Public Laws,* 1919, Ch. 132.

[37] House Document #216, 1919; *Legislative Record,* 1919, 356, quote 358, debate to 361, vote and passage 366.

[38] State of Maine, *Answers of the Justices of the Supreme Judicial Court to the Water Power Questions Propounded to the Court by the House of Representatives of the Seventy-Ninth Legislature,* 1919, 42 pps., Augusta ??, 1919.

[39] *Ibid.,* 3-6.

[40] *Ibid.,* 18-20 for the answers, 21-40 for the minority opinion.

[41] Baxter's speech may be found at *Legislative Record,* 1921, 5ff.

[42] For the action on the law and amendment (discussed below), see *ibid.,* 45-6, 662, 1067, 1082, 1239, 1375, 1404; Amendment 170-5. The law is *Public Laws,* 1921, ch. 203.

[43] Inaugural Address, Percival E. Baxter, February 9, 1921, *Public Documents* 1921-2, I, 10-11.

[44] *Legislative Record,* 1921, 557-9, text at 560 where it was introduced by R. Owen Brewster, who became Baxter's floor manager.

[45] *Legislative Record,* 1921, 1062.

[46] *Ibid.,* debate 1257-1279, Brewster's speech is 1257-63.

[47] Hinckley (South Portland), *ibid.*, 1266; Bragdon (Westbrook), 1268; Maher (Augusta), 1271-5. He spoke of the possibility of solar energy in the future, and said that if the state did do the work, that Maine ought to emulate the work of the Soviet Union in this regard.

[48] *Ibid.*, 1278-9.

[49] Debate, *ibid.*, 1288-1304, Gillin and response, 1288-98.

[50] *Ibid.*, 1302-3.

[51] *Ibid.*, 1303.

[52] *Ibid.*, 624-35, Sprague's speech is at 634-5. The park was lost by a vote of nineteen to eight in opposition, *ibid.*, 635. It died in the house amidst some rancour by parliamentary ruling. See *ibid.*, 668-71.

[53] *Annual Message*, Percival O. Baxter, 1923, *Legislative Record*, 1923, 25-35.

[54] *Ibid.*, 332-346. Brewster was the most able opponent to speak. On the motion to postpone it indefinitely the bill won twenty to five support in the Senate.

[55] *Legislative Record*, 1923, 413, 430, 466, 488.

[56] *Ibid.*, 568-574.

[57] *Ibid.*, 625-637.

[58] *Laws of Maine*, 1923, 909. Proclamation, March 22, 1923.

[59] *Legislative Record*, 1923, 797-803, 897-900.

[60] His speech is at *ibid.*, 1134-6.

[61] *Ibid.*

[62] *Legislative Record*, 1923, 1240, 1258-9, 1266-7.

[63] *Ibid.*, 1153-64.

[64] *Legislative Record*, 1923, 161, 1288-89.

[65] *Laws of Maine*, 1923, 910-911.

[66] *Ibid.*, 911-3. However, see his veto message, *Legislative Record*, 1925, 9-10.

[67] *Laws of Maine*, 1923, 914-22.

[68] *Laws of Maine*, 1923, 924-38, dated May 22, 1923.

[69] A group from Portland, called the State of Maine Chamber of Commerce and Agricultural League had begun to publish pamphlets on *The Wild Lands of Maine, The European and North American Railway Grant*, etc., all designed to prove how pristine the motives of corporate interest always were. In recent books on Maine these are cited as some sort of gospel, almost as though they were official publications. Beware of the Greeks whose gifts last almost indefinitely.

[70] *Legislative Record*, 1923, 797.

[71] *Ibid.*, 578, 692, 732. Baxter vetoed thirty-four bills, and pocket-vetoed another from this session.

[72] It appears at *Legislative Record*, 1925, 14-31.

[73] *Legislative Record*, 1925, 256.

[74] Portland *Press Herald*, editorial, December 17, 1925; Garret Schenk to F. A. Gilbert, December 28, December 31, 1925. These letters are in my possession, in two files, Schenk's letters to F. A. Gilbert, 1925, 1926.

[75] Schenk to Gilbert, November 2, 1925; April 20, 1926. On the Boston visit and avoiding Baxter, Schenk to Gilbert, October 22, 1925. On Great North-

The companies control the rivers

ern feeling about the park, and towards Baxter, Schenk to Gilbert, August 31, 1925 (the quotation).

[76] Percival P. Baxter to E. A. Pritchard of the National Park Service, August 15, 1936. "Finally I concluded to take the matter into my own hands and after years of patient waiting and the exercise of some tact I acquired Katahdin, a feat considered impossible by all those familiar with the situation."; Baxter to Maud and David [Gray], April 26, 1937, "Had I not advocated Katahdin as a State Park ever since 1905, today I would be sitting in the United States Senate and would have been there since 1926, so you see I really sacrificed something for the Park, to say nothing about the money. . . . In 1926 it was understood that I was to have no opposition, but the timber barons hunted around for an opposition candidate. . . . Their money and his, accomplished their purpose and my day was lost never to return. . . ." Also on this subject, and with these letters which were engendered in an effort to block Baxter State Park from becoming a National Park, see Baxter to Senators Hale and White, April 8, 1937; Memorandum for the President (FDR) by Arno B. Cammerer, July 6, 1937, all these from Franklin D. Roosevelt Library, Office File, 6-P, Box 35. I am indebted to my student Mark L. Jacobs for calling these to my attention. This last is reprinted in Edgar B. Nixon, *Franklin D. Roosevelt and Conservation*, II, 85-7, Hyde Park, 1957. Also see Harold D. Smith To FDR, July 22, 1939, in *Ibid.*, II, 359-360.

[77] Schenk to Gilbert, September 15, 1926. Throughout these tantalizing bits of evidence is the fleeting notion that R. Owen Brewster, who succeeded Baxter as Governor, and went on to become a member of Congress, and United States Senator utilized these matters to advance himself, and although he came out looking all right, neither side ever really trusted him. As Schenk said in the last letter quoted, "Brewster would do all he could under cover." and careful readers will remember that he was "overcome" at a critical point once before.

[78]*Private and Special Laws*, Maine, 1927, ch. 113.

[79] *The Northern*, Vol. XI. No. 11, February 1927; the editorial was signed by F. A. Gilbert.

[80] It is only fair to say that the interpretation of events as herein offered was not accepted by some men of the Great Northern Paper Company recently, although others have agreed with my interpretation in private discussion. See interviews William Hilton, July, 1962. "The land was not worth a cent (T3R9). Every piece of land was a good purchase for Baxter, and a good sale for the Great Northern Paper Company." On the other hand, many others have said that the above interpretation makes good sense. On the hunting in the Maine woods, there is a persistent story that after the first order that letters threatening arson on the company lands were received in the Boston office, and this helped to create the compromise policy. Lincoln Smith, *The Power Policy of Maine*; Berkeley, 1951, covers some of this same ground, but misses the point because of his acceptance of company analysis at some points, and a too-legalistic point of view at others. In Maine it is always wise to view the effect of any and all legislation on the lumbering business before making final judgment.

13

CONSERVATION AND FORESTRY COME TO THE MAINE WOODS
1865-1930
PART ONE

> Potatoes, oats, and wheat now rarely give such crops as they did thirty or forty years ago. Fruit trees take on diseases, apples become scabbed; pears often knotty, crabbed, and extremely perverse; plum and cherry trees often forget former habits and old friendships, blight and rust and insect destroyers are everywhere. The farmer's crops are invaded from all sides. . . . Deterioration in fruits and other crops, through climatic changes in these states [*Ohio and Michigan as well as Maine*] is now clearly shown as being intimately connected with removal of these magnificent forests.
>
> <div style="text-align:right">House Document 56, <i>Legislative Documents</i>, 1869 "Memorial of the Maine Board of Agriculture to Legislature on 'Forest Trees'", with a proposed law, 14.</div>

It is surprising how early serious discussion was given to the problem of conserving the forests of Maine. Legislation was offered soon after the Civil War, and although enactment did not occur, agitation in the press continued almost without letup during the nineteenth century. Even before this much attention was given in the press to the need of conserving the forests.[1] Eventually losses from fire, insects, and natural phenomena helped to extend the publicity attendant on the forest depletion, and the State of Maine passed a law setting up a Forest Commissioner. This activity was followed by private concerns which began to apply selective cutting practices on their own lands. By 1930 applied forestry in some form was far advanced in the Maine woods. The movement was aided by many men of good will, but two stand out, Austin Cary and George F. Talbot.[2]

There were earlier discussions of forest destruction and preservation in the United States, but the original text for most conservationists in the United States during the nineteenth century was a volume produced in several editions by George Perkins Marsh, a Vermonter of strong opinions.[3] Marsh believed that indiscriminate cutting of the forest led to quicker rainfall runoff, soil erosion, and eventually to ruined land. His solution was to cut as little as possible and to reforest cut over areas.[4] These doctrines were to dominate conservationists thought throughout this time and well into our own.

Maine newspapers constantly iterated some version of Marsh's ideas.[5] Others went further as in the following quotation:[6]

> It [*The Gardiner Journal*] believes that it would pay to let a large portion of the waste land in this state grow up to lumber. It would help the fertility of the rest of the land, benefit our waterpowers, add to the beauties of our scenery, and increase the healthfulness of our climate.

Some said that the forests were in real danger of exhaustion within five years, or perhaps even sooner.[7] In an effort to control this waste the use of the sawdust was advocated. "The sawmills of this state are reaping sawdust enough annually to freight a navy—a very small part of which is utilized in packing ice. It might and ought to be all saved and manufactured into planks, floorboards, and moulded into vessels, furniture, and hundreds of forms. . . . When mixed with insoluble glues and pressed, as it may be, there is no calculating the profit that is yet to be utilized in manufacturing. . . ."[8] The best liked alternative to the destruction was to plant a tree for each that was felled,[9] or at least to plant some anew.[10] The amount used for fencing, railroad ties, and fuel alone used up the annual growth, to say nothing of the wood for construction purposes. ". . . Some scale of tree planting on an immense scale ought at once to be adopted."[11]

Many felt that lowering the tariff would help. Then Canadian lumber would enter freely and Maine and New Hampshire trees could be saved. With luck this proposal could be combined with a proposal to keep the cheap Canadian labor at home to cut their trees as well, and the Yankee could cut conservatively with Yankee labor.[12] Of course not all journals believed that indiscriminate cutting was dangerous. Those that held that cutting was healthy constantly brought forth the champions of the industry to say that the growth was in excess of the cut, and as a result the forests were not in danger.[13] Still most northeastern newspapers were on the side of Marsh and his theories. The following editorial of 1883 was an example of this mature thought.[14]

SAVE THE FORESTS

> The terrible floods in the valleys of the Ohio and its tributaries are the result of the destruction of the forests of Western New York, Western Pennsylvania, and Western Virginia, where the headwaters of the Ohio are fed; and New York journals improve the occasion to warn New Yorkers not to bring similar disasters upon the Valley of their noble river by destruction of the Adirondack forests now going on. Forests retain snow and rain, moderating alike freshets and drought. When they are destroyed, the waters pour unobstructed down the valleys, creating devastating floods, unheard of before; and in the dry season leaving channels almost bare of water. This has been the effect in France; it has become the effect on the Mississippi and the Ohio. It will be the same on the Hudson, and the rivers of Maine, if the forests are destroyed. There is a determined effort in the State of New York to avert the danger; but in Maine we are actually courting it. Our Congressmen are demanding, to the point of party revolt, that a premium shall be paid for felling our forests [*the current tariff battle*], and our customs regulations generously facilitate the coming in of our provincial neighbors with their teams to help in the destruction. Weeks ago our correspondent at Calais told of scores of teams admitted there to our territory, in bond, to be used with crews, also from the Provinces, in getting timber during the winter, in the Spring to return without paying the slightest tax for the, to them, valuable privilege, but which is expensive to our owners of teams and to the lumbermen. The Aroostook *North Star* makes a similar report. Thus is natural greed stimulated to hasten the destruction of all our forests and bring upon the valleys of the Kennebec and Penobscot the devasting floods that are sure to follow such destruction. If it were wise to live today, as if there were no tomorrow, this might be excusable; but with the existing general impression that the country will remain after we are gone, and that our children will need, and will be thankful for the heritage unimpaired, it seems like the folly of madness, not the wisdom of wise men, to pursue the course we are now pursuing.

By this time reasons for and belief in conservation were well established. But the demands on the forests created by the pulp and paper industry, along with an increased demand as construction began to pick up after the worst years of the depression, caused a flurry of restatement in the period from 1885 to 1895.

Newspapers printed letters from correspondents attacking the methods of forestry then in vogue.[15]

> Be assured that we have not a single tree to spare for the sole purpose of giving men work. While the best minds of other and older countries are devising means for the preservation of their forests we are calmly contemplating the wanton destruction of our own.

Editorialists advocated selective cutting,[16] or planting of forests and the growing of tree crops.[17] Old lumbermen said in answer that some were cutting in a scientific way already and the agitation had paid off.[18] Still many were fearful. After discussing the pulp revolution one newspaper remarked, "A well-known lumberman estimates that if the manufacture of wood pulp increases for the next ten years as it has recently it will require all the spruce production in the State to supply the demand." [19]

Demands for strict laws and closer observation of good cutting methods caused lumbermen to defend their interests. One lumberman down from the woods for Christmas consented to be interviewed. He maintained that the spruce gum pickers were the villains in the forest as they set the fires (a theory which ranked with one later to the effect that loss of the forests in northern Maine was due to the charcoal burners), but his remarks otherwise were important.[20] He said,

> This forestry agitation is, for a large part, based on wrong premises. I believe in taking care of the forests, but when it is said that the forests of Maine are disappearing, such statements are dead to the wrong. Why, two cords of wood are growing in Maine, today, where one was growing ten years ago. I never saw sapling-pine growing faster than it is today. What Maine wants is not so much a protection of the forests as more railroads to bring them to the markets. On wild lands the growth of our hard woods is wonderful. I agree with the forestry folks about fires in the woods. . . .

Most Maine people apparently felt this way. This same newspaper, a leader in forestry agitation, said less than four months later that nothing could really be done except random seeding by those who went to the woods.[21] Agitation continued however.[22]

Indiscriminate cutting received the most attacks, including even the sale of Christmas trees, firs which taken all in all were considered in that time as never much more than a weed.[23] Most of the arguments centered on the desirability of controlling the cut,[24] but few went further than did William A. Russell of Bellows Falls, Vermont, president of the Fall Mountain Paper Company, in his presidential address to the American Paper Makers Association in 1892.[25] Russell came directly to the point:

> We have gone on since our last meeting denuding the forests, building up huge machine shops, and vying with each other to furnish to the consumer paper at the smallest margin of profit to the manufacturer. I suppose we shall ever do this. Certainly we shall keep on

denuding the forests, for as we are turning almost wholly to wood as a fibre, not almost wholly, but largely, we are drawing on the forests rapidly. I hope that some wiser way of cutting our timber in this country will be devised so that we shall not at so early a day see the end of our spruce forests. We must either of our own volition, or by some government control, prevent the destruction of the forests in the way that is now going on. There is no necessity for it. The cutting of timber, if you remember, is usually carried on in this way. A lot of land is purchased by a lumberman and it is cleaned of all its trees of any size. In Germany—and I hope later in this country we shall adopt the plan—they only cut the maturer trees and leave the younger trees to grow. In that way we may have a perpetual supply of fibre. That is being adopted by some parties now in connection with paper industry.

No matter how many denudiac speeches were made however the operators continued to insist that the supply was inexhaustible.[26] The same papers which printed the speeches also printed news stories which ended like this one[27] discussing the operations of Charles Palmer in the winters of 1892 to 1894 in the Molunkus and Smyrna areas:

In order to assure a superior quality of pulp wood, free from knots, Mr. Palmer saves only the butts of the logs, the tops being thrown away.

Not all the dangers were those of the ax and the pulp grinder. Natural hazards always lurked in the background of forest ownership. Great gales blew down swaths of trees. The worst in this period were the so-called Saxby gale of 1869 and the hurricane of 1883. Both caused heavy damage.[28]

The great loss to this country from the Saxby gale will be to the woods. We have had some of our men up exploring and they say they can walk ten miles at a time on the trees that are down without stepping on the ground. In some places for half a mile about every tree is down. The bridges and buildings can be put back, but the woods all down will soon get on fire and burn all over the down district.

Men harvested blown down timber from the hurricane of 1883 as late as 1885, and in the winter of 1884-5 much of the Penobscot cut consisted of blowdowns from this great wind. One man ran a portable mill near Katahdin Iron Works to saw shingles, clapboards, and box boards for the Philadelphia and New York trade from the blown down trees.[29]

Even a worse hazard were forest fires. Nearly every year saw some, with the fires of 1864[30] being perhaps the worst until the dry years near the end of the eighties. In those times reports of fires filled the papers. In 1885 a disastrous fire burned a tract forty miles long and twenty miles wide in Franklin and Somerset counties. The van of this fire was in the blowdown territory of 1883 so the damage was not as great as it might have been.[31] Another bad year for fires was 1889.[32] Frequently newspapers used the fire threat to preach the conservation gospel[33] as in the following screed:

> The destruction to property in Maine during the past fortnight by forest fires, which fortunately have been extinguished or in a large measure checked by the rains of the past two days, brings conspicuously to the front that that has never received, in this section at least, the attention its great importance demands. . . . Of what use is it to talk of the preservation and propagation of forests, and the waste of timber in cutting, as long as whole townships are devastated by the fiery element every season? Clearly, while such conditions exist, the owner of timber land has very little enducement to cull this growth instead of stripping it, or to leave young timber standing until it shall have acquired greater value by reason of increased age. The risk is too great. Rather will he be inclined to strip his tract of whatever may be of value upon it, while yet happily the fires have left it untouched.

The editorialist went on to advocate passage of a law similar to a law of New Brunswick which caused campers from May 1 to December 1 to clear a five foot radius around camp fires and to extinguish the fire when finished with it. Those in charge of the parties in the woods were directed to read the law to their clients.[34]

Interestingly enough there was already a fairly strict law on the Maine books in regard to fires, but failure to enforce it had in effect rendered it null. This law said:[35]

1. No fires can be set on another person's land without permission.
2. Such a fire, if set, will cost the instigator a fine of from $10 to $500.
3. If the fire is malicious the fine rises to from $20 to $1,000, or a jail sentence of from three months to three years.
4. The fire kindler shall be responsible for the prevention of his fire spreading.
5. While hunting or fishing everyone is subject to the provisions of 1, 2, and 3.
6. Log Drivers may have fires, but must use caution with them.

In the early 1880's a further difficulty was the depredations caused by insects. C. H. Fernald, a professor at the University of Maine, was busy in attempting to isolate what seems to have been the spruce budworm. He was active in writing letters to newspapers and lobbying for the appointment of a state entomologist. His letters only served to swamp him with specimens.

Conservation and forestry come to Maine 339

By 1883 the trees were dying, "dying at a fearful rate," in northern Aroostook County. Franklin B. Hough, of the United States Bureau of Forestry, came to investigate with an entomologist, and, accompanied by Fernald, they traveled to the north county. The controversy over the cause of the deaths prompted letters from aged residents who said that a similar destruction, also thought to be the fault of worms, had occurred in the period 1808-1813 and that this depredation had led to the Miramichi fire of the early 1820's. Lumbermen thought the worms only came in trees harmed by a gale in 1872 and the hurricane in 1883. They also said that the damage was worse on the north side of hills. One newspaper said that after the trees had disappeared settlement would follow thus the damage was not an unmixed blessing. Fernald said that most of published statements were false and he and the newspapers called for further research. The entomologist's reports were apparently not clear and although the destruction was probably caused by a worm, and probably the budworm, the scare soon died out.[36] At the end of the century a sawfly attacked the hackmatack in eastern Maine and nearly wiped it out. For years there were almost none. It is only within the past twenty years that this beautiful tree has begun to flourish once more in this area.[37]

Whatever depradations natural hazards may have caused, man and his works were still the primary spoilers. As the pulp demand grew so did the production of the chemical mills. These mills dumped their refuse into the rivers, and spawning fish in particular began to die. The Penobscot sea salmon suffered extensive damage from the refuse of the mills on the Penobscot, especially the Penobscot Chemical Fibre Corporation. In addition to having to buck pollution salmon found it impossible to get up the river over the dams built in their way. E. M. Stilwell, Commissioner of Fish and Game for Maine in the late 1880's, carried on a long and lonely fight in this quarter until he retired from the office. With his retirement little was heard of the salmon until recently when those interested in attracting visitors to Maine began to call for fishways, pollution controls, and fish planting. Needless to say a salmon swimming up the Penobscot today is about as rare as a great auk flying along State Street.[38]

The earliest efforts for a state program of practical forestry date back to the immediate post-Civil War period, with the first formal appeal being penned by the Secretary to the Maine Board of Agriculture in 1869. As he said in his annual report, while proposing a bill for the legislature:[39]

> The people need to be agitated and a wholesome public sentiment created. . . . While the state has manifested a laudable ambition in developing its resources, while it has wisely provided guardians for the Fisheries, and a Commission on Waterpower, it has not yet recognized the more important public concern that underlies both those, and all other public interests. We believe this to be an important public matter that does not lie outside of legitimate legislation. Shall the legislative voice continue silent on the matter of *forests*, till the last tree shall be cut, thus assuring dry channels to the *rivers* and the consequent death of the *fishes*?

The proposed legislation provided that if a person would plant trees (2,000 to the acre) within three years time, the land would be tax-exempt for twenty years if the trees remained in a thriving condition. Trees should be planted as windscreens on highways, and those who destroyed such marginal trees should be given fairly heavy fines. This bill had an all-encompassing title which included the phrase, ". . . to suggest the expediency inaugurating a State policy, encouraging the preservation and production of forest trees; . . ." The memorial document which accompanied the bill covered the evils of deforestation in detail and pointed a significant moral with the lines:[40]

> No country possessed by a civilized people, has ever been seen to preserve for any considerable time a proper proportion of its surface in forest growth. It is but the work of an hour to destroy a tree, that has been reared by the patient work of centuries.

Appeals of this sort were but little listened to at this time.[41] It did however mark the beginning of what might be called governmental attention to forestry in the northeast. Nearly all later action would flow from this source.

Some attempts at legislation continued to be put forward, but in the main most people felt that such agitation was of little moment.[42]

> The time is far distant when the situation will require any interference of any kind by the government with the management of timberlands owned by individuals, or when the conditions will be such that the interference of the government will be at all useful. . . . The suggestion is sometimes made that government should establish forestry schools or have a corps of professors of forestry to give instruction, not orally to attending pupils, but by public lectures or by the preparation of essays to be printed and circulated at public expense. . . . Until conditions materially change, we can safely leave it to the self-interest of the owners of timberlands to determine when the time shall have arrived to save timber instead of using it.

Fifteen months later this same periodical was advocating selective cutting, and two months after that it could even regard a congress of such professors with sympathy.[43]

> The American Forestry Congress has been in session at Boston during the past week. The association includes in its membership a number of first-class cranks, whose sayings have a tendency to make the whole business ridiculous; but it also comprises a large number of intelligent businessmen and scientists, who have given the subject of forest preservation much study, and many of the papers read and addressed delivered are of exceeding value.

Much of the reason for this change in attitude in Maine in particular, and to some extent throughout the whole northeast, came from the work of George F. Talbot, then of Portland. Talbot was born in East Machias, Maine, on January 15, 1819 into the famous Talbot lumbering family. He graduated from Bowdoin in the class of 1837, read law and taught before becoming active in the anti-slavery campaign. He was a delegate to the 1860 Republican Free Soil convention in Chicago. After the campaign he was United States attorney for Maine till 1870, and he served as Solicitor of the United States Treasury from 1875 to 1877. He was interested in such matters as history and theology as hobbies while he continued his law practice. In addition to this he expressed his views on the tariff, child labor, ten hour days and other matters of social significance.[44] His most important contribution however came in the field of forestry where through his efforts the Maine forestry commission was created. His propaganda activities in this area make him one of the significant pioneers in the fields of conservation and forestry.

Talbot's activity consisted basically in sending letters to newspapers calling for a state forestry convention to discuss conservation practices elsewhere. He also suggested a series of laws in these early letters, laws which would provide stringent penalties for the setting of forest fires; laws which would set up mutual companies to insure woodlands against fire; laws calling for state promotion of tree planting on state owned lands; laws which would exempt those planting trees from taxation; a law which would prohibit the cutting of trees below eight inches diameter breast high; and finally a law which would provide funds for a survey to be taken to determine what potential timberlands there were in the state.[45]

These letters elicited response from others praising Talbot for his views and throwing their support behind his efforts.[46] About all that came of this effort though was the observance of the first Arbor Day in Maine on May 10, 1887. The State Pomological Society took that occasion to warn of the reckless cutting being brought on by the pulp craze, and it exhorted farmers to join in the campaign to preserve the forest.[47] As they said,

> Look well after the woodlands, and instead of sending all your hard-earned dollars to western investors, expend some of it at least in the improvement of your woodlands. The investment will be safer than Kansas farm mortgages, because it will continually enhance the value of your farms.

The agitation began to pay dividends. At the State Grange at Skowhegan in the fall of 1888 a call was issued for a State Forestry Convention to be held at Bangor in mid-December in conjunction with another state Grange meeting. The call which was signed by four ex-governors, a former vice-president of the United States, and nearly every lumberman of consequence in the state, as well as people like Talbot, was an impressive document.[48]

> The defoliation of the forests is becoming a serious matter. Timber is becoming scarce. In many of the states hardwood timber is becoming scarce and manufacturers are looking about for the occupancy of new sections of hard wood growth. The forests of Maine are exceedingly valuable in these woods. Their value should be appreciated and the care and preservation of our forests and timber trees be made the protection of our State. More stringent laws than we now possess to prevent and punish the destruction of these woods, by the careless setting of fires, are needed; and the planting of trees, as well as the better care of those already set, are also subjects about which public information should be disseminated.

On the surface this document dealt only with hardwoods, but Talbot and those fostering the movement needed the signatures of the prominent men concerned. When the meeting occurred in Bangor those attending could then turn their attention to the real matter at hand, proposing a Forest Commissioner for the State of Maine.

Attendance at the Bangor gathering was small. Hannibal Hamlin chaired the meetings, which ran like clockwork. First, Talbot proposed a lobby in Augusta to promote conservation desires. Then a three man committee, including Talbot, was appointed to make an agenda for the meetings. While the committee met the meeting was given over to a paper on colonial methods of preservation of the forest. A committee was then chosen to draft a bill for the legislature consisting of Fernald (the University of Maine entomologist), John T. Adams, (a Portland newspaperman), and Henry F. Prentiss (a Bangor lumberman and author of the bill presented in 1869). Immediately thereafter the committee returned with a bill for the legislature.

Conservation and forestry come to Maine

On the second day John E. Hobbs of South Berwick, who had made the remarks on the colonial experience, and Talbot were appointed a committee to memorialize the President of the United States to the effect that he should withdraw timberlands from the public domain and give over the lands to the public use. At this meeting letters from dignitaries, including B. E. Fernow, were read. Professor F. L. Harvey of the University of Maine read a paper entitled "Importance of Forestry to Maine—Forestry Problems of the State" which was well received. The Maine Forestry Association then adjourned, happy with their accomplishments, even though attendance had been sparse.[49]

The State Pomological Society continued its attack on the problem, and when the legislature convened in 1889 one of the first bills introduced was the Forest Commissioner's bill. In conjunction with the lobbying for the bill Talbot and others began a series of public lectures throughout the state.[50] Such individuals as J. B. Harrison, corresponding secretary of the American Forestry Congress, appeared as support for Talbot. He and others advocated cutting the Maine woods so as to make a crop. Talbot accompanied and introduced these speakers and took most of them on to Augusta to testify before the state legislature as to the merits of the bill.[51] Unfortunately the bill was declared to be "inexpedient", and it was only saved by a parliamentary maneuver in which Talbot was instrumental. The house reconsidered the bill and proposed that it be put over to the next legislature for action. With this move it only remained for the campaign of education to continue.[52]

The new technique involved more praise for (and fewer attacks on) the large land owners and the farmer on his woodlots. Talbot said that the bill was "not destructive of free enterprise," and he himself began to stress the fire prevention and protection aspects of the bill.[53]

> A proper policy for our state has in view two ends, the preservation of the forest, the restoration of the forest. . . . We shall have a system either while our forests survive or *after they are destroyed*. It seems wiser to have it while we have woods to preserve.

In 1890 Talbot went as the delegate of the Maine Forestry Association to the American Forestry Congress. The Congress welcomed him and urged the State of Maine to adopt his proposals. Edwin C. Burleigh in his annual address as Governor took that occasion to praise Talbot and urge passage of his bill. With such strong support the bill's success was assured.[54] It passed the legislature with almost no opposition. The new law,[55] which went into effect in the summer of 1891, provided that the state land agent should take on the duties of state forest commissioner. He was charged with the duty of collecting and classifying statistics relating to forestry. He was to ascertain how the forests of Maine were being destroyed by fires, wasteful cutting, and what effect this was having on the state's watershed. His report on his findings was made mandatory. Fire wardens (usually the selectmen) were appointed and many minor laws were enacted. They provided for fines for not extinguishing fires; for the erection of notices; for the use of non-combustible wads in firearms; for railroads to cut, burn, and remove their slash; for the railroads to use spark arrestors; and for making the railroads responsible for their employees. The state forest commissioner was charged with awakening interest in conservation in schools and in the state university, and he was supposed to prepare and circulate tracts on the subject.

A major battle had been won with the establishment of the commission, but the war was not over. The first report of the state commission kept up the attack.[56] Articles dealing with the subject were the feature of the publication. Professor F. L. Harvey of the University of Maine wrote on "Preservation of Our Forests". George T. Godfrey dealt with "The Relation and Importance of our Forests to Summer Tourists, Sportsmen, etc.". Wilson Crosby of Bangor wrote on the "Economical Cutting of Our Forests" where he advocated less skidding, narrower roads, lower stumps, and taking the trees higher up into the limbs. George F. Talbot returned to another love of his, "Forest Planting and Municipal Ownership of Forest Lands," in which he advocated using tax-delinquent lands for municipal forests. John E. Hobbs of South Berwick sounded a denudiac note with his article "The Depreciation of Our Forest Growth and its Effects Upon Our Forest Industries" in which he predicted that without some sort of stringent controls of cutting Maine would be barren of trees by 1941.

Talbot continued his attack. A bill was submitted to the legislature calling for support of municipal forests, but it was in advance of its time. Only newspapers could really afford to advance that sort of thing.[57]

The inital work was over. The state was actively involved. The precedents had been set; the laws were on the books. Now it would depend on the activities of men who were involved in the woods work as to how much would be accomplished. The most important man here was Austin F. Cary, also of East Machias and also a graduate of Bow-

Conservation and forestry come to Maine

doin. His class was 1887, just fifty years after Talbot, and he was in Talbot's tradition.[58]

Cary's life falls into three chronological periods. The first of these was the time when he was a student and (later) preparing himself to be a forester, laying the foundation for scientific forestry in the Maine woods. The second period was the time of his employment by the Berlin Mills Company and his tenure at Harvard. The third period of Cary's work was his life in the national service, spent mostly in the south and lying outside the scope of this work.

Cary is usually described as a "down-to-earth" fellow, "a real Maine man," "a prophet," a "man who knew what he was talking about," "never that much of a book man" or in similar ways. His correspondence tends to bear these statements out. He had a tendency to be blunt and he might have been described as a man of single purpose. He usually saw the New England picture more clearly than he did that of the entire United States. Throughout his early life he was always conscious of his role as a pioneer and his task as a teacher and instructor in the necessity of forestry. For all of these reasons he left a considerable mark on the Maine woods.[59]

In the early days of the Maine Forestry Commission the commissioner was usually a harried man frequently filling a political void, and the work of the commission was carried on by part-time employees who had the Maine woods as their sole interest. Cary was the most important of these men. It was due to his early work in the nineties that the commission became a respected part of the Maine woods. Long after he had gone to work for the Berlin Mills Company he was still worrying about the commission and its chances for success. For this reason conservation and forestry in Maine depend on Austin Cary.

Cary's first contribution to scientific Maine forestry occurs in the second report of the Maine Forest Commissioner in 1894, "On the Growth of Spruce." [60] This report analyzed growth patterns and cutting techniques useful in spruce forests. Partially as a result of this work he did a fair amount of surveying for the Bureau of Forestry under Fernow.[61] Fernow was more willing than some to put up with Cary's rather explicit commentary, in part because Fernow was using some of Cary's filed work for his own publications.[62] Forestry was not an easy business in those days. Cary was willing to work for $75.00 a month and expenses but occasionally he didn't get that, he had to wait nearly six months for the department to pay him for six weeks work in Washington. His letters got increasingly bitter; ". . . how tired I am of hand-to-mouth dealings. . . ." he once protested to

Fernow.[63] Still the prospect of his beloved state adopting some of his principles was frequently enough to sustain him.[64]

> It is work that will tell with the hard-headed lumbermen and land owners and I think with some other materials I have for Mr. Clark (the State Forestry Commissioner), will make the fortune of forestry in this state. We expect to strike for an appropriation this winter that will keep one or two men at work, and I see the possibility of a forestry commission being a power in this state.

The difficulty, as he saw it, was to get an accurate estimate of the condition and the producing power of the cutover spruce lands.[65] The summer of 1895 was spent in doing just such extensive cruising and, although when this was over no work was in prospect, he said in a confident letter,[66] "It is my hope too that this season's work will be the cause of having forestry put on a permanent footing in this state."

Cary missed few opportunities to get his message into public print. After the busy summer of 1895 he submitted to several interviews in all of which he proposed better cutting.[67]

> There is no doubt that the lands drained by the Kennebec are being overcut. Neither is there any doubt that the cut will have to shrink largely within a few years. A great many towns have been greatly damaged by cutting so far as their producing capacity in the future is concerned. The cutting has been done carelessly.

There is no question but that Cary preferred to give his message in this form rather than submit formal writing. As he once said querously,[68] "I am done cruising for this season and it only remains to figure results and drag the work into literary shape. That however is a task which I approach with dread." In fact Cary's problem was that he was too direct for his readers who wanted the usual dull language of official reports. Cary also continued to have his troubles in getting both new work and compensation for that work which he had already done.[69]

Whether it was the slow pace of the work or the general lack of results it is difficult to say, but Cary felt it necessary to discuss what governments could do to promote scientific forests. Forestry practices are the most important thing,[70] he said. "Approximate results will do now if they are to the point. Symmetrical monographs may well be left to the time when forestry commissions have plenty of leisure and are abundantly provided." Intensive work was as important as extensive.

> In other words, this work ought to be confined to a limited field, directed toward some clearly-defined, practical and attainable result, the attainment of which would recommend the work, its methods, and

Conservation and forestry come to Maine

purposes, to the country. . . . If forestry is going to give any better or surer information, it cannot be done off-hand, but only after patient and well-directed investigation. . . . State forestry reports can be so remodeled as to contain something besides pictures and prattle. . . . purpose needs to be defined, and our work to be organized in accordance with it . . .

Others were as concerned with Cary's position as he, although they were not as blunt as he in expressing their position.[71]

> Instead of teaching the people how to grow timber as a profitable crop on cheap lands, our speakers, writers and forest officials have been prone to make doleful predictions about the speedy destruction of our forests, our close approach to a wood and timber famine, the decrease of rainfall, the decrease of farm crops and water power, the increasing severity of our climate, till a follower of theirs might well regret that Father Miller's prophecy of the world's end in 1843 had not been fulfilled.
>
> You can grow timber in Maine as a crop as well as you can grow grass or potatoes or presidential timber and in this line also make true your motto, "Dirigo".

It was in this year that Cary decided that if he were to make a serious contribution to his state he should observe forestry practices abroad, especially in Germany.[72]

> It doesn't seem to me that as a help in doing what I see to do in this country, a long course of study in the forest schools is worth the time and expense it would require. For much that can be so learnt one can depend on literature and observation.

The Berlin Mills Company was apparently willing to foot part of the bill of Cary's expenses and so he went for three months to Germany, Italy, and Switzerland. He viewed the trip as a necessity for his home state.[73] "Maine is ready for forestry. I hope in this trip to be helpful in providing her with it in form adopted to her conditions." He thought the trip valuable, and it apparently paved the way for his employment as Forester by the Berlin Mills Company in 1898.[74] Upon his return, his written missionary work went on unabated. He gave addresses before the Maine State Board of Trade,[75] and his articles appeared in such important places as the Maine Forest Commission *Reports* and in *Paper Trade Journal*.[76]

Cary thought the prospects for good forestry practices increasing all the time. As he said, some lumbermen were beginning to treat the forest like a field, not a mine. The Berlin Mills Company for instance was using saws in the woods, taking the trees down to very low stumps, running the logs well up into the limbs, and getting more wood per tree. Nothing was cut under twelve inches at the stump. The men were unhappy and the new plans were difficult to enforce, but success in the first winter (1893) on the demonstration town (T5R3 WBKP—Parkertown) had caused the promulgation of the plan over all their cuttings.[77]

Still, even with all this evidence of the impact of his and other's teaching, Cary felt that the scientific forestry cause needed reinforcement. A full-scale survey of the spruce situation was needed, operating from this beginning, and designed for aiding those "ready to apply genuine forestry principles in their cutting". He asked Fernow to write to the Maine Forest Commission extolling the accomplished work and calling for its continuation. This, he felt, would assure the success of the $1500 appropriation needed.[78]

The work never was completed. Cary went to work for the Berlin Mills Company and his new position took most of his time. His campaigns did not stop however. He, Fernow, and Pinchot made addresses in 1898 at a meeting of American Pulp and Papermakers Association telling them what should be done in the future.[79] Cary spoke on woods and woods work. Pinchot talked about the concept of "sustained yield" and the work of W. Seward Webb who had implored the department to provide him a working plan, and Fernow spoke generally about forestry and wood pulp supplies. The meetings were successful but the work still went slowly. Gradually Cary had come to the conclusion that the future lay in private employment. And although he was to return to governmental employ eventually and hope for it at other times, forestry work went on in the northeast largely through support from private sources.

The first stage of Cary's life was over. In attempting an evaluation one is struck with the fact that Cary always regarded himself as a pioneer and always comported himself as a teacher. His early training had been as an entomologist and botanist, and he had taught at Bowdoin prior to spending full-time in the woods. A private income had allowed him to indulge himself as to his life work, and he chose forestry because it was ". . . . natural, desirable, inevitable even . . . ," and his motive was to introduce ". . . any ideas found in it that might be applicable to the old established interests and thereby securing plain substantial benefit for my state and its people." [80]

His work for the Maine Forest Commission had sharpened his thought somewhat from this point, but in a real sense the only difference was in the methodology to be used. Cary no longer thought as he once had that learned reports would do more than pave the way.

Conservation and forestry come to Maine

Now the work had to be carried forward by men who were trained foresters, but unlike many in his generation Cary believed the road to success lay in private hands. As he once said in an article, "You can't stop the economic forces that are cutting down our forests, but you can guide them if you go at it right." [81]

Throughout his life he was to follow this philosophy. Simple business interests would dictate conservative management,[82] and good business calculation had to go hand in hand with good forestry.[83] He believed that by showing private companies and land owners that long term profits would be substantially more than in the short run that they could be made to adopt progressive measures. For that reason he did not care much for book forestry although he was able to hold forth with the best of them.[84] He suggested that people should work in the woods as common laborers to understand the work. After achieving this goal one could then go to the companies with long-term growth plans providing for scientific cutting, regeneration of the forest cover by re-seeding and re-planting, and better use of waste materials. Such long term plans were a must, but their acceptability depended, for Cary, on the men who put them forward.[85]

His mature views were exposed to public criticism in a paper delivered to the Society of American Foresters, and published in the *Journal of Forestry*. This paper was a strong call for the role of private interests, and it helped pave the way for the marriage of private and public forces so common today in forestry.[86] Cary was an individualist in a time when many were advocating public control and ownership. For that reason he had difficulties with Gifford Pinchot who seemed always to avoid the New England woods in favor of the west.[87] In a sense this may account for the thrust of New England forestry today which seems so far removed from public control. Cary was the driving force in private forestry, and Pinchot, who expressed the other side during the formative period, allowed events to follow Cary's lead. For it was in the second stage of Cary's life that private forestry got its start in the northern New England woods.

NOTES

[1] As some examples of the early interest, see *Sandy River Yeoman*, October 26, 1831, "We are glad to find that the public mind has begun to be somewhat directed to the planting of trees." *Maine Farmer*, November 28, 1834; Bangor *Whig and Courier*, February 16, 1837, discusses forest depletion and overcutting; *Yankee Farmer*, January 4, 25, November 21, 1840 (on management of woodlots); Franklin *Register*, May 10, 1841 "Report of the Trustees

of the Franklin County Agricultural Society" which discussed planting and reforestation; Oxford *Democrat*, March 3, 1854, editorial, "The Cultivation of Forest Trees,"; *Drew's Rural Intelligencer*, January 13, 20, advocating legislation forcing planting, February 24, September 29, 1855 (time to reforest); Oxford *Democrat*, February 16, 1855, calls for conservation in cutting and refers explicitly to European difficulties from overcuttting.

[2] A recent M.A. Thesis, Charles G. Roundy, "Changing Attitudes toward the Maine Wilderness," University of Maine August, 1970 has cast doubt on my analysis. See especially 81-128. Unfortunately his facts lie in a procrustean bed created by his theories. One is still advised to look for the economic realities behind social and political "facts" especially in the Maine woods.

[3] David Lowenthal, *George Perkins Marsh, Versatile Vermonter*, Columbia University Press, New York, 1958.

[4] This doctrine is best stated in his *Man and Nature*, New York 1867. The last edition was entitled *The Earth as Modified by Human Action*, New York, 1898. For an earlier view see Frederick Starr, "American Forests: Their Destruction and Preservation," *Annual Report*, Department of Agriculture, 1865, Washington, 218-9.

[5] Portland *Eastern Argus*, March 2, 1871, quoting Lewiston *Journal*.

[6] Bangor *Daily Commercial*, January 20, 1872, quoting Gardiner *Journal*.

[7] Bangor *Daily Commercial*, January 29, 1872 (Brunswick and Topsham area, "soon to be exhausted"); April 13, 1872 (within five years); July 30, 1872 (also within five years); November 19, 1873 (within fifteen years on the Kennebec); Portland *Eastern Argus*, August 13, 1872 (five years); November 22, 1877 (soon and it may have already happened); Oxford *Democrat*, December 31, 1872 ("Sad, that this part of the country is being so extensively detreed."

[8] Bangor *Daily Commercial*, August 6, 1872 (editorial—"Economizing Wood").

[9] *Ibid.*, February 4, 1873.

[10] *Paper Trade Journal*, June 1, 1874 (editorial—"Preservation of Trees").

[11] Bangor *Daily Commercial*, January 3, 1876 (editorial—"Preserve Our Forests").

[12] *Ibid.*, January 5, 1883, quoting the Belfast Age; Portland *Eastern Argus*, March 1, 1883 (editorial—"Protecting Such Labor").

[13] *The Industrial Journal*, February 15, 1884 (editorial) quoting the wood purchaser for the Maine Central Railroad, May 2, 1884, (editorial) quoting Alexander Gibson, the New Brunswick lumberman.

[14] Portland *Eastern Argus*, February 17, 1883 (lead editorial).

[15] Bangor *Daily Commercial*, February 6, 1886, letter from "A Correspondent."

[16] *Ibid.*, March 18, 1886, quoting from editorial published in *Manufacturer's Gazette*.

[17] Bangor *Daily Commercial*, May 5, 1888.

[18] *The Industrial Journal*, December 2, 1887.

[19] Bangor *Daily Commercial*, May 3, 1888.

[20] *Ibid.*, December 28, 1888, interview with an anonymous Bangor lumberman.

[21] *Ibid.*, March 20, 1889.

22 *Ibid.*, November 23, 1889; Portland *Eastern Argus*, June 19, 1890, editorial "Destruction of Forests" quoting Philadelphia *Record;* The *Industrial Journal*, September 12, 1890, editorial.

23 *Ibid.*, July 31, 1891, quoting the Boston *Commercial Bulletin;* Bangor *Daily Commercial*, January 4, 1892, quoting the Boston *Advertiser* on the Christmas tree depredations, and then refuting their argument.

24 See, for instance, *Paper Trade Journal*, November 28, 1896, and *The Industrial Journal*, October 21, 1892, quoting the *Manufacturer's Gazette* on the Glen Manufacturing Company, Berlin Falls, New Hampshire, and how they should not only try to control lands, but control the cut on their lands as well.

25 From the address quoted in the report of the annual meeting at Saratoga Spring, New York, of the APPA, *Paper Trade Journal*, July 30, 1892.

26 *The Industrial Journal*, September 23, 1892. After Senator Eugene Hale made a denudiac speech at Skowhegan, the Auburn *Gazette* interviewed prominent lumbermen who all agreed that he was dead wrong.

27 *The Industrial Journal*, January 19, 1894.

28 On the Saxby gale, Fisher, *Story of a Downeast Plantation, op. cit.*, 155-7, and Samuel E. Boardman to Professor Spencer E. Baird of the Smithsonian Institute, October 14, 1869, quoted in *Naturalist of the St. Croix, op. cit.*, 56-8.

29 Bangor *Daily Commercial*, March 8, 1884; March 18, April 3, 1885; Portland *Eastern Argus*, November 28, 1887.

30 *Land Agent's Report*, 1864, 7-8, "Some portions of the timberlands lying within the easterly ranges were visited by fire during the summer of the present year . . . . the inhabitants residing in the town of Washburn and upon Township B, Range 1, called "Alva" have suffered great losses. Mills, dwelling places, crops and fences have been swept away." Rain came at an opportune time.

31 Bangor *Daily Commercial*, July 2, 1886. Another bad year was 1880. A train went from Bangor to Kingman with firefighters, a pumper loaded on a flat car, and hose. See *ibid.*, July 14, 15, 16, 1880.

32 *Ibid.*, July 22, September 10, 1889.

33 *The Industrial Journal*, May 13, 1892, editorial entitled "A Burning Question".

34 Also see Bangor *Daily Commercial*, July 16, 1886 for comment on this law. Another bad year which brought forth editorials of a similar nature was 1881 when fires ranged over T14, 15R10, and 13R12. See *Ibid.*, September 12, 1881.

35 *Public Laws*, 1855, Ch. 132, p. 134-5. I have paraphrased the provisions of the law.

36 Bangor *Daily Commercial*, October 29, November 8, 1881, letters from C. H. Fernald; Portland *Eastern Argus*, May 3, July 9, 1883; July 11, 1883, letter from "S" on the earlier attacks; on the lumberman's views see Bangor *Daily Commercial*, July 15, 1883; for the pollyanna view see *The Industrial Journal*, July 27, 1883; C. H. Fernald to Lewiston *Journal* cited in Bangor *Daily Commercial*, August 1, 1883 refuting all and calling for research; editorials to this point appeared in Portland *Eastern Argus*, July 9, 1883, and *The Industrial Journal*, August 3, 1883. For Fernald's research and observation see his "Spruce Tortrix" in *American Naturalist*, January, 1881.

[37] On the beginning of this blight see Bangor *Daily Commercial*, January 27, 1899; also interviews Harold A. Noble, Topsfield, Maine, July, 1962, and personal observation.

[38] See *Annual Report*, Commissioner of Fish and Game, 1888, *Public Documents*, Maine I, 1889. "The admission of pulp mills upon our streams, more especially upon the Penobscot . . . . was and is a most grievous error. . . . No manufacturer of any kind whatever should be allowed to throw its waste into a river any more than into our highways. No argument or demonstration is necessary here. . ." This statement was at page eight. For the difficulty of no fishways and the resultant failure to breed see *ibid.*, 13; 1889-1890, quoting a letter Marshall MacDonald to E. M. Stilwell, January 25, 1890. MacDonald was speaking for the Federal Commissioner of Fish and Fisheries. "It seems to me that in view of the present condition of affairs, the attempt to establish good fishing at Bangor and above is hopeless." Also *Annual Reports*, 1891-2, *Public Documents*, VI, 6f, "Of course, the fish are decimated, but what does that matter if it does not affect the price of pulp or leather? Has it not created a new industry in the sale of spring water for the drinking of our population?"

[39] *Fourteenth Annual Report of the Secretary of the Maine Board of Agriculture, for the year 1869*, Augusta, 1870, 82-3 for quote, 85 for the bill itself, his italics.

[40] *Legislative Documents*, 1869, House Document 56, 4 for the quotation, and 14 for the bill.

[41] See, for instance, the impact of Franklin B. Hough, "On the Duty of Governments in the Preservation of Forests," *Proceedings*, American Association for the Advancement of Science, August, 1873.

[42] *The Industrial Journal*, April 25, 1884 quoting and commenting on an editorial in *The Banker's Magazine*.

[43] *The Industrial Journal*, July 17, 1885 quoting and commenting on articles and editorials in the Farmington *Journal* and the *Country Gentleman*. The quotation here is from *The Industrial Journal*, September 25, 1885.

[44] Talbot's papers seemed to have disappeared. Information in this paragraph comes from George W. Drisko, *Narrative of the Town of Machias*, Machias *Republican* press, 1904, 561-3; George F. Talbot to Wilmot B. Mitchell, February 4, 1901, a letter in the Maine Historical Society about his Bowdoin class, and from his letters to editors. See, e.g., G. F. Talbot to editor, Portland *Eastern Argus*, March 25, 1890 on the desirability of free trade.

[45] G. F. Talbot to editor, Portland *Eastern Argus*, January 25, 1887.

[46] G. Morgan (Fryeburg, Maine) to Portland *Eastern Argus*, February 17, 1887.

[47] *Annual Report*, 1887, *Public Documents*, II, 1888, 60 ff. "Arbor Day and its Observance" by D. A. Knowlton.

[48] Bangor *Daily Commercial*, November 23, 1888. B. E. Fernow was extended an invitation to appear, but apparently could not. It was announced that papers would be given at the Bangor meeting.

[49] Bangor *Daily Commercial*, December 14, 18, 19, 1888.

[50] *Annual Report*, State Pomological Society, 1888, *Public Documents*, II,

Conservation and forestry come to Maine

1889 "The Forestry Question", 131-2; Bangor *Daily Commercial*, January 1, 24, 1889.

[51] Portland *Eastern Argus*, January 26, 1889.

[52] *Ibid.*, February 9, 1889.

[53] See G. F. Talbot to Portland *Eastern Argus*, February 16, 1889 (his italics).

[54] Annual Message, *Public Documents*, I, 1891 (January 8, 1891), 24.

[55] *Laws of Maine*, 1891, chapter 100. The first Commissioner was Cyrus A. Packard.

[56] *First Annual Forest Commissioner's Report*, 1892, *Public Documents*, II.

[57] *The Industrial Journal*, December 9, 1892, reprinted his proposed bill, and commented favorably on it. See also Bangor *Daily Commercial*, February 28, 1893, which reprinted an article from the New York *Tribune* advocating such work. They called it a life insurance policy for the children of the country.

[58] A good contemporary sketch of his life is in Bangor *Daily Commercial*, October 3, 1896.

[59] There is a good picture of Cary on the cover of *Forest History*, Vol. 5, No. 1 (spring, 1961). The comments here come from interviews conducted with men who knew Cary well; Louis Freedman, one of Cary's students and long-time friend and Woodlands Supervisor of the Penobscot Chemical Fibre Corporation for many years, and William Hilton, Vice President in Charge of Woodlands for the Great Northern Paper Company for many years. The Maine and New Hampshire period is covered in the correspondence of the Chief Division of Forestry, 1894-1905, Records Group 95, National Archives. Also see Roy A. White, "Austin Cary, The Father of Southern Forestry", *Forest History*, Vol. 5, No. 1 (spring, 1961); Professor White's dissertation at the University of Florida, "Austin Cary and Forestry in the South", and Austin Cary, "Forty Years of Forest Use in Maine," *Journal of Forestry*, XXXIII, No. 4, (April, 1935), as well as the notes below. I shall cite the correspondence simply by person and date. It all comes from RG 95, NA, unless otherwise specified. Professor Ring makes a great deal of Cary's odd and peculiar personality. He was withdrawn, and without doubt had fixed ideas about forestry, but his personality was not much different than many woods personalities of that era. He was constantly aware of his position as pioneer, and was not above using his reputation as an idiosyncratic personality to advance his views. See my sketch of his life forthcoming in Henry Clepper, ed., *Pioneers of American Forestry*, National Resources Council, 1971.

[60] Austin Cary, "On the Growth of Spruce", *Second Forest Commissioner Report*, Maine, 1894, *Public Documents*, 1895, I 28ff. See also on the work Cary to B. E. Fernow, January 9, 20, October 3, 1894. This same report reprinted from the 1885 New Hampshire *Report*, "Forest Management and Reforesting", one of the earliest clear calls for this sort of work.

[61] In Wisconsin and Michigan, see Cary to Fernow, December 25, 1894; February 19, March 3, 9, 1895.

[62] Cary to Fernow, July 10, 31, 1895, on his disagreements with Fernow's work in the 1894 New Hampshire Forest Commission *Report*. Also Cary to

Fernow, August 31, 1895, and to . . . Sudworth, July 10, 1895 on cedar reseeding.

[63] Cary to Fernow, July 31, 1895 in particular.

[64] Cary to Fernow, April 16, 1895.

[65] Cary to Fernow, June 30, 1895.

[66] Cary to Fernow, October 20, 1895.

[67] *The Industrial Journal*, October 11, 1895, interview. Also see editorial in the same issue reprinted from *Manufacturer's Gazette*, to the same point.

[68] Cary to Fernow, November 18, 1895. The man of the woods continues to echo these remarks.

[69] Cary to Fernow, December 18, 27, 1895, January 15, 18, February 10, 11, 1896.

[70] Cary to Fernow, November 20, 1895.

[71] John D. Lyman (Exeter, N. H.) to editor *Commercial*, February 20, 1896, extolling Cary's work.

[72] Cary to Fernow, February 23, 1896.

[73] Cary to Fernow, March 2, 29, 1896. Fernow gave him advice as to who to see as well as letters of recommendation, as apparently did Gifford Pinchot.

[74] Cary to Fernow, September 18, October 18, December 1, 1896. Throughout these letters which discuss his work he alludes to the prospect and fairly soon of private employment. Apparently the Berlin Mills Company used him as sort of a spare time employee for a while.

[75] This important address was reprinted in *Paper Trade Journal*; see "The Forests of Maine", by Austin Cary, in *ibid.*, April 25, May 9, 16, 1896. He made the point here that the investment in pulp and paper mills was so large that self-interest would provide conservative cutting and forestry practices.

[76] Austin Cary, "How to Apply Forestry to Spruce Lands", *Paper Trade Journal*, February 19, 1898. In his article, which is one of his more important contributions, he calls for applied forestry, citing *in extenso* his own experiences for the Berlin Mills Company. He called for the hiring of college graduates for two or three years of woods work of all kinds, then an extensive two year scholarly tour of study in Europe, and finally to give them some time to experiment. He counseled the paper companies not to be impatient. See on this same theme *Paper Trade Journal*, August 14, 1897; "Need of Spruce Farming". Among other early articles by Cary which are still important are "Spruce on the Kennebec River", "Spruce on the Androscoggin"; "Exploration on the Magalloway", "Our Forests and the Future". All of these are in the *Third Forest Commissioner's Report*, 1896, *Public Documents*, 1897, II, generically under the title "Report of Austin Cary to the Forest Commissioner", 15-203 plus appendix. The best review and criticism of the work is in B. E. Fernow, *Garden and Forest*, 1897.

[77] Cary, "Spruce on the Androscoggin", 113-6.

[78] Cary to Fernow, February 1, 1897; Cary to Fernow, February 22, 1897.

[79] On the addresses to the APPA, see Cary to Fernow, January 2, 1898; Fernow to Cary, January 5, 1898. The articles appear in *Paper Trade Journal*, February 19, 1898. Gifford Pinchot, "The Sustained Yield of Spruce Lands",

B. E. Fernow, "Forestry and Wood Pulp Supplies". Cary's article is discussed in note 74.

[80] Austin Cary, "Common Sense in Conservation", *Journal of Forestry*, XXXIV, No. 3 (March, 1936), 235.

[81] "Report of the Third Annual Meeting of the Canadian Forestry Association, 1902," *Forestry Quarterly*, I, No. 2 (January, 1903), 68.

[82] Austin Cary, "Unprofessional Forestry," *Forestry Quarterly*, IV, No. 3 (September, 1906), 187.

[83] Cary to Pinchot, December 14, 1898.

[84] "News and Notes", *Forestry Quarterly*, XIII, No. 2 (June, 1915), 284; W. B. Greeley, "Austin Cary as I Knew Him", *American Forests*, LXI, No. 5 (May, 1955), 30.

[85] "News and Notes", *Forestry Quarterly*, XIII, No. 2 (June, 1915), 187; Austin Cary, "Reflections", *Journal of Forestry*, XVIII, No. 5 (May, 1920), 474.

[86] Austin Cary, "How Lumbermen in Serving Their Own Interests Have Served the Public", *Journal of Forestry*, XV, No. 3 (March, 1917), 271-89.

[87] Samuel P. Hays, *Conservation and the Gospel of Efficiency*, Cambridge, 1959, 28-35, 47-8 especially note, on Pinchot's feeling about the east. Also see M. Nelson McGeary, *Gifford Pinchot: Forester-Politician*, Princeton, 1960, Chapters 1-6, and Gifford Pinchot, *Breaking New Ground*, New York, 1947, *passim*.

14

CONSERVATION AND FORESTRY COME TO THE MAINE WOODS

1865-1930

PART TWO

> The Berlin Mills Company is to be congratulated in securing Mr. Cary's valuable service, and it is noticeable in this matter private enterprise goes ahead of and it outstrips public and official authority. The State of Maine should have secured Mr. Cary's services for the benefit of its great forestry interests, and not allowed a private corporation to have controlled them.
>
> > Bangor *Daily Commercial*,
> > March 26, 1898, editorial.

> I have no patience with these theorists who are continually talking about the destruction of the forests and urging the preservation of the timberlands for future generations. In the first place, they are, as a rule, highly impractical and their arguments are nothing but theories, and in the second place, the argument for saving the trees for future ages, is most absurd.
>
> As long as Americans remain as they are at present, they'll not be so considerate for the welfare of future inhabitants, as to cease money making and close down big industries. They'll run the thing as long as there is any money in it. . . .
>
> > Bangor *Daily Commercial*,
> > February 18, 1899, interview with A. H. Carter, Berlin Mills Company.

The Berlin Mills Company, Cary's new employer, was often in the van in woods work. When it hired Cary the firm apparently gave him a fair amount of leeway in what he would do. Primarily his job was to act as missionary for the gospel of scientific logging on their lands, to map their lands and make recommendations as to long range cutting practices, and to train young men in the field of forestry for the future. Apparently he was free to pursue some other work on his own as long as it did not detract from his time with the company. Once his recommendations were made it was, of course, up to him to see that they were followed out.

The epigraph quotation from one of his employers is an indication that not all in the firm favored his work. For this reason, and the fact that he was a pioneer, Cary's new position was not a sinecure, as he was aware. "Of course, it isn't the same as a forester's position in Europe, but I am pleased with the prospects on all accounts." [1] Cary had apparently viewed his position as being temporary eastern representative of the Bureau of Forestry which under Fernow and Graves had been quite closely associated with eastern ventures. However soon after Cary went to his new job Gifford Pinchot became Chief of the Bureau in Washington. Although Cary congratulated him on the new position one gets the feeling that he thought the future was somewhat less bright than it had been before.[2]

> Your aim I am sure will be to bring things actually to pass, not merely in the public lands of the west, but here in the east as well and elsewhere where the methods will be indirect and the forces to be used are private. This latter is my field as well, and few of us there are in it. I can hardly think that any can be beyond the point where comparison of ideas and experience will be of service.

Cary did receive an appointment as unpaid agent of the division in the Northeast which entitled him to the publications of the division free and gave him something of a forum for his views on forestry.[3] In the long run though, the appointment of a forester by the International Paper Company was of more direct use to him as it gave him more private support.[4]

The work went slowly. "It is a very imperfect business. We cut too hard, the men are dull and indifferent, the bosses think they know it all and are jealous of their rights. However, this is the only way I see to commence." And, ". . . . closer organization will pay." [5] His work then consisted mostly of making maps of the company lands. Some outside work was done on the lands of his old *alma mater*, Bowdoin, but this was a sideline, emanating from his Brunswick home in spare time. The company maps done in fifty foot contour levels was his first concern and must have given his employers fairly close knowledge of their holdings. The first year he was employed by them he mapped 40,000 acres.

In the spring of 1899 he continued the missionary aspects of his work by offering a short lecture course to his logging bosses when they came from the woods. The lecture was illustrated with magic lantern slides some of which he had prepared and some which he borrowed from the Bureau in Washington. At about this same time his men began to experiment with using one horse alone in the yarding of logs

Conservation and forestry come to Maine

in order to preserve as much reproduction as possible. Pinchot apparently wanted the company to organize a working plan similar to one taken up by W. Seward Webb, but Cary was unable to accomplish this. Instead he attempted to get Pinchot and Graves to come and see his work in the Maine woods.[6]

The Berlin Mills Company had increasingly adopted principles leading toward scientific logging such as cutting low on the stump, introducing the saw rather than the ax, and cutting only larger trees, but Cary's work was still not the same thing as supervising a government forest. "You understand that good business calculation has got to be hand in hand with good forestry," he once warned Pinchot. At another time when evaluating the work of the International Paper Company's forester and himself, he said, "Griffiths and I have been a year now in our present employ and no doubt we have each learned something, not only about forestry in its application to our spruce lands, but as to the way in which businessmen view it, what they are in a way to adopt and what to reject, etc." [7]

Cary was happiest when he was teaching and doing his missionary work, his life's vocation in his eyes. In 1899 Herman Van Schrenk, a bureau entomologist, came to Maine and Cary was able to display his accomplishments. Upon leaving, the scientist wrote a long letter extolling Cary's work.[8]

> I found Mr. Cary a most delightful man and thorough and enthusiastic in a position which few men would fill with such an interested devotion. I was much struck with the difference in the lumbering methods of the Androscoggin and Penobscot regions, very much in favor of the former. The waste which I saw north of Moosehead in the way of top timber was astounding, due no doubt, as Mr. Cary says, to the use of the *board* measure in scaling lumber in these regions where the cutting is done by stumpage. It would be very desirable to have the great mills use the top lumber likewise, which would most assuredly be the case were they to adopt the scaling by caliper.

Cary was always in demand as a public speaker. One address before the Boston Society of Civil Engineers was published as a pamphlet and was regarded as an important work in the ongoing fight for scientific forestry.[9] In addition he kept up part time work for the State Forestry Commission.[10]

By this time Cary was supervising all the lumbering done by his company, visiting each jobber for two or three days every three or four weeks all winter long. Under his guidance detailed analyses of the cuts made in each camp were compiled. By introducing his methods he was able to reduce the wastage to 1.47 per cent of the total cut on seven camps—250,000 feet in 15,500,000 cut.[11] He continued to advocate mapping, and urged the introduction of papiermache maps of the contour levels. By use of such maps, he thought that contracts could be drawn with jobbers practically eliminating waste. He was instrumental in starting small nurseries for replanting purposes especially in burned-over sites.[12] At least once he worked against his own employer in order to put the government onto a good thing as to land to be purchased for experimental purposes.[13] After five years of this work though he was ready to go on to other work which he hoped to be with the Bureau but which was instead in academic life.[14]

He felt a necessity to keep up his proselytizing work in Maine, which never moved as rapidly as he liked. As he said to Pinchot:[15]

> Much depends on the impression forestry makes in Maine in the next ten years. Our professor at Orono needs help, otherwise the next legislature may let the public instruction side of it at least drop.

As long as he remained with the company the bright young Yale forestry boys went to him for advice and aid. They learned logging, cutting, organization—that is, the serious side of forestry.[16] The department always valued his advice too on which young men would make the best foresters for them.[17]

By and large one gets the feeling that Cary was happy and relatively satisfied with his accomplishments. As he described them,[18]

> While I was with the Berlin Mills Co. (I am no longer in their employ) some 200,000 acres of land were got under a thoroughgoing map system, and a large section of their cutting (the business handled some 70,000,000 feet per year annually) was got under a careful control and when possible handled conservatively.

In later years the Berlin Mills Company was ready to take a great deal of credit as to the beginning of forestry in the northeast.[19] They hired Cary it is true, but one gets the feeling from reading the papers of the time, and Cary's letters in particular, that the work went forward in spite of, not because of, the Berlin Mills Company. Cary was an effective preacher of this new scientific gospel, and because he was a woods man, not a book man, he could be and was accepted as a man who had the welfare of the woods, not himself or the company, at heart.

In the few remaining years that Cary was in the north he continued his practice of teaching lessons about the value of forestry wherever he went. Occasionally he was very blunt with those who were less receptive to his message, as here.[20]

It is bad to mince matters in dealing with things of this kind. There are wild-land men who talk as if they thought they ought to be pensioned for the public service they render in holding and administering their properties. Personally, I could never see any reason for that. The value of their property is mainly the creation of the community, and its safety and the income derived from it is dependent on the community no less. . . .

This was about Cary's last blow in the service of Maine forestry. Soon he would leave teaching and join the Forest Service in the south, still preaching his gospel wherever he went.[21] Summers he was back in his beloved Maine working on the Bowdoin forests and offering his experience to those who came to his Brunswick home. He was always happy about his role as the pioneer private forester in New England and once said that he was "staggered" by the results.[22] Well he might have been. His philosophy had been clear-cut and New England practices derive directly from them. His strategy had been to proselytize and to demonstrate by actual work. His tactics were those of selective cutting, regeneration, sustained yield, and demonstrable profits. Few prophets have been as successful as was Austin Cary. In his time Maine forestry had come to the realization that private forestry and public aid went best hand in hand. In New England he thought the private sources should be superior, and so they were, although never so superior as to completely dominate.

The most interesting early example of this dual work was the Great Northern Paper Company's use of Bureau of Forestry men to make a working plan for their woodlands. This was the sort of work that Pinchot wanted to do; the influence of the Division was enhanced, and the results were widely spread.[23] Others were less sanguine about the prospects of really scientific forestry practices emerging from such studies. Cary insinuated as much in a letter to Pinchot after the first year of the two year survey.[24] "I should be glad to know if you still feel satisfied with your relation to the Great Northern Paper Company. You remember perhaps my feeling as to their sincerity and the likelihood of their actually doing the work on forestry principles." Pinchot did not answer the letter, but his assistant Overton W. Price did reply in a stiffish sort of paragraph.[25] "Our work upon the tract of the Great Northern Paper Company has gone steadily on during the summer and we hope to have the working plan in final shape by spring. The Great Northern Paper Company has given us so far no reason to doubt their sincerity in the matter."

Not all were skeptical,[26] and some listened with great interest to the results of the first year's work. Henry Grinnell, who was in charge during that summer (1901), addressed a group of fifty lumbermen and landowners. He suggested the introduction of selective cutting under the guidance of a trained forester.[27] "This necessary expenditure will be more than repaid by the increased production of timber on your lands."

By the terms of the contract between the paper company and the forestry division, a complete survey of the company lands was to be made "to determine the amount of timber standing on the company lands, and the method by which it should be timbered to maintain the timber supply and improve the condition of the forest."[28] The company contributed $5,000 to pay for the survey, and the foresters used company men, horses, and tools although they paid for the usage.

During the first summer the men made a general survey of the land, and the plan was worked out successfully the second year. William C. Hodge, who was then in the process of working his way up in the division, had charge during the second year. Grinnell, who did the first survey, briefed him before he left for the work. A base line was drawn through each township and half mile stations were marked on it. From these stations the towns were analyzed in both directions with extensive topographic and silvicultural notes being taken in addition to the measurements. The measured strips were the usual sixty-six feet wide through each township. More than 7,400 trees were measured for volume, rate of growth, height, and length. The trees were measured by the usual local methods (that is, by the Bangor rule), and also with the scalage taken far up into the branches. On the felled trees diameters inside and outside the bark were taken. Scalage and diameters were taken every ten feet on the logs. More than 2,600 such surveys were made in all. There were no roads from the main camp, and Hodge sent his men in on camping trips of three or four days duration. The men were split into three groups of five men each, while the sixteenth man acted as doctor, and did stem analysis as well as some height computations when he was not needed elsewhere. All told, 2.5 per cent of the total area was completely covered, which provided for a very accurate survey.[29]

The work did not proceed quite as easily as the description suggests. It cost more than the $5,000 and the bureau had to make up the difference. Some of the Great Northern employees regarded the young foresters with a good deal of indifference if not actual enmity.[30]

Conservation and forestry come to Maine

> I have not received but sent Farley off with the copy [*an order on the company*] to try and get horses etc. on that. I hope he will succeed though Jim Devine who has charge of the horses is a man of considerable force of character and may require bulldozing.
>
> I received a letter the other day from Gilbert, manager of the Great Northern, asking for a memorandum of the supplies I have obtained from their depot camps—the thunderbolt has not yet fallen in the shape of a bill but I expect it daily.
>
> . . . the men's wages are scandalously high and many other irregularities.

These were comments tossed off by Hodge in his regular letters reporting to Washington.

The final report from the work was thorough. It covered eighteen townships, 365,787 acres in detail. The projected figures for yield were given depending on whether the trees were to be cut to eight, ten, twelve, or fourteen inches of diameter at breast height. Hodge recommended twelve inches on the slope, twelve on the flat, twelve on the hardwood, and eight in the swamps of Somerset County. In Piscataquis County the corresponding figures were ten, ten, ten, and eight. As he said:

> Being cut to these limits, the forest will yield but little less than if lower limits were used, and will be left in far better condition.—The fir does not need to be cut to diameter limits, but adoption of the spruce limits for fir is earnestly recommended.

Hodge estimated that if the company cut an annual acreage of 6,338 acres the average annual yield would be 13,437,603 feet, 86 per cent of which would be spruce. On this basis the average complete rotation on the lands would be fifty-three years. He recommended that the cutting should always be supervised by a "local man with exact knowledge of local conditions and the ability to make necessary changes with discretion."

He made all the correct recommendations in terms of the conservationist. They included:[31] All trees should be marked; the loggers were allowed no discretion, and spruce poles were not to be used for road making. Lodged trees must be removed; stumps could not be higher than eighteen inches, and merchantable stuff had to be run up into the branches to the four inch diameter line. Branches should be lopped off so as to lie flat on the ground in the slash area. The company should employ foresters who would

1. decide the areas to be cut.
2. supervise the marking.

3. inspect and criticize the logging contracts.
4. inspect areas during and after the cutting to check violations.
5. supervise protection of the tract generally.

To a very great extent the paper company adopted his recommendations, or at least, began to adopt them.

By 1907[32] the company was employing foresters who marked all trees to be felled and who visited the camps each week to enforce the rules. Rules, similar to these printed here, were posted in a conspicuous place in the camps for all to see.

1. Roads to be swamped just wide enough and no more.
2. Roads, skids, etc., to be made of unmerchantable timber.
3. Scattering trees are to be cut as they are met.
4. A sound stick eighteen feet long and six inches in diameter at the small end is a merchantable log.
5. The cuts are to be twelve inches breast high for spruce, eight inches for fir.
6. Four foot wood is to be cut seven inches breast high, and each stick must make at least three four foot lengths.
7. Nothing smaller than this is to be cut.
8. Merchantable stuff in swamping cuts is to be hauled at once.
9. Dry spruce is a merchantable cut.
10. Trees are to be felled at the swell of the roots.
11. Snow is to be shoveled away to achieve this.
12. The saw is to be used if at all possible.
13. Felling must harm as little reproduction as possible.
14. Butting off with an ax is to be discouraged.
15. Limbing is to take place as soon as felling.
16. Trees must be run up to six inch diameters in tops.
17. No lodged trees are to be left.
18. Four foot wood must be full length, sound and of correct diameter.
19. Short logs are to be cut if undersized stuff is needed for long logs.
20. All merchantable stuff is to be hauled.

Although Cary was the important pioneer in woods practice, and the Great Northern Paper Company took the first important tentative steps in management of its own woods, others had gone on attempting to have laws passed which would have made such practices the rule rather than the exception in the woods. The work of George F. Talbot in setting up the Forest Commission had to be extended, and there were many who undertook this work while Cary and the paper company had gone their way.

Many advocated laws or regulations which would set diameter limits on trees to be felled,[33] while others contented themselves with issuing gloomy *ex cathedra* views of the future.[34] In an effort to con-

Conservation and forestry come to Maine

trol traffic into the woods some advocated the licensing of guides.[35] Nearly everywhere though the future was regarded as being dark. Fourteen years only were left for logging, floods must inevitably come, and the doctrine of state ownership was the only answer.[36] One editorialist remarked:[37]

> Nothing is more true than that notwithstanding the extent and value of our forest lands, what with the danger from fire and the constant demands upon them for modern industrial uses, they cannot be too carefully guarded or too strictly economized.

As the century ended, the picture seemed to grow even blacker. When others called for compulsory planting and rotational cutting, Maine adherents to the conservation gospel simply relied on the printing of their remarks to take the message to the unconverted.[38]

It was not until the fall of 1900 that any substantial gains were made in this quarter. In that year the University of Maine, located at Orono, began to offer a course in forestry. By 1903 Samuel N. Spring, the first forestry professor at the university, a graduate of the Yale School of Forestry in 1898 and a veteran of the bureau's work on the white pine, was even offering a course entitled "Lumbering" during which one week was spent in an actual lumber camp. Students earned one credit.[39]

Not much was really accomplished because of the necessity of fighting rear-guard actions against those interested in gaining as much as possible from the state through hunting and fishing revenues. Under these pressures the state extended hunting privileges into September, the worst month of the year for fires, and for a time the conservation adherents and the big landowners joined in a campaign to roll back this tide. Forest fires became a major topic of conversation. It was prophesied freely that those who came to kill the deer would instead kill the trees with their indiscriminate camp fires and matches. The watchword was "Let the September law be repealed!! " Finally in 1901 after lengthy debate the obnoxious law was repealed.[40] That same legislature attempted to extend the tax on wild lands in order to gain additional revenue. The newspapers attacked this venture as strongly on the grounds that the trees were less likely to be cut if the tax burden was lighter. They charged that Oak, forestry commissioner, was a tool of special interests, and they said, rightly enough, that the big companies were the ones in the van of conservation and the only ones who practiced a twelve inch cutting regulation. Instead of this indiscriminate taxation it was time, according to these supporters, for the state to move into the field of conservation and provide protection against strip cutting and usage especially in the crucial area of forest fires. One writer threatened court action to determine the constitutionality of the game laws unless the state moved quickly to appoint good men with scientific training.[41] If these laws were to be maintained and the taxes increased, the editorialist believed that some protection should then be demanded in return. Under such stiff attacks the tax increases failed as had the extension of the game law.

Others were as convinced that action in the area of state regulation was imperative. Among them was Francis Wiggin of Portland who carried this message to the State Board of Trade. In his widely-reprinted speech,[42] he called for the establishment of forest reservations by the right of eminent domain. He thought they ought to be located in the Rangeley Lakes area, around Moosehead Lake, on the West Branch of the Penobscot River, near Mt. Katahdin, and along the Allagash River. In making this call for action, he said that the study of forestry at the university had to be advanced. To him no work was more important.

The chorus increased. The cut was too heavy and the pulp mills and the paper makers were the cause. Something must be done to control the "big, ungainly, ill-smelling, and profitable pulp mill." [43]

The attack was of such proportions that it could not be ignored. Hugh Chisholm, still president of the International Paper Company, was the first to spring to the defense of paper companies. He declared that the paper industry was not to blame for deforestation nor for the reduction of the water supply. According to his view it was the huntsman, the charcoal burner, and the fuel cutter who were really at fault. The paper companies cut only spruce, nothing else, therefore it could not be their fault. Besides, there was not really less water, it only seemed so.[44]

This rather extraordinary statement was followed by the publication of the Forest Commissioner's *Report* for 1902.[45] Edgar E. Ring, the forest commissioner, implied strongly that the view of the "denudiacs" was wrong, and he lent the prestige of his office to the side of Chisholm:

> From my knowledge of the conditions existing in the timber producing regions of Maine, supplemented by the competent judgment of many men versed in woodcraft, whose services I have been able to command, I am satisfied that the supply of available material to carry on our present pulp and lumber manufacturing establishments,

and such others as may be built as time goes on, is sufficient and ample; and I firmly believe that there is no immediate danger of a timber famine in this State, ...

A few seized on Ring's analysis immediately as evidence that all was well in the Maine woods,[46] but these words were like red flags to the old bulls of conservation. George F. Talbot leader of the earliest fight roused himself to lead this new one.[47] He doubted Ring, called his estimate "a very extravagant one," and cited the work of Austin Cary as evidence of the time it took a tree to grow to maturity. He also cited his own experiences as evidence for the case against Ring. In the 1840's when he was a law clerk in Gardiner great pine logs two feet at the butt were commonplace at the mills, but now only small stuff was at the saw and none of it pine. On his own land in 1856 the logs ran three and four to the M; now it was eighteen on the spruce and eleven on the hemlock. Who could believe the commissioner in the face of such evidence? He said that the state needed better fire protection than the law of 1889, selective cutting and replanting, and finally the repurchase of lands from private hands to place them into state forests.

> In that way we may ultimately retrieve the most fatal mistake in our state history, of fooling away for trivial purposes and nearly without compensation a magnificent fund on the income of which our State might have been permanently maintained without taxation.

Little was heard from the Ring forces after this blast from the past. Chisholm continued to defend his position on the grounds of self-interest, and he quoted Ring as saying some years later (October, 1906), "I know of no reason why I should change my mind in reference to the estimates as to the stand of spruce in Maine and the annual growth and consumption in my report for 1902. I think the figures on the stand of spruce at the time were very conservative. . . I believe that the average annual growth of spruce throughout the state will equal three per cent, providing careful cutting is carried on and forest fires are kept out." Ring added, "I am glad to say the owners of wild lands are, as a general rule, cutting their tracts with more care, but there still is a great deal of wasteful cutting." Chisholm was more interested in "over-production" and the "disturbing" tendency for workmen to "follow unwise leaders." [48]

In the same publication however that Chisholm used to spread his gospel the results of a special census of the pulp and paper companies in the state brought forth the comment that "the most serious problem in this industry is the question of a wood supply for the pulp mills." They even cited Ring's figures to prove their point of deforestation.[49] The American Pulp and Paper Manufacturer's Association apparently agreed as they proposed the adoption of conservation, better forest fire protection, and the use of state and federal aid in water storage systems.[50] Some old-time lumbermen had another solution: ". . . manufacture should be made subservient to other branches. Pulpwood should be a by-product of the mills, as any refuse, even sawdust, would make pulp." [51]

As this controversy was raging, it became, in fact, moot. The big companies won their battle for control of the rivers, and inevitably thereafter as the landowners and lumbermen capitulated their land fell into the hands of these same big companies. Once this became true the doctrine of self-interest did begin to take hold. And, although it was oftentimes exercised by the simple procedure of purchasing someone else's lumber and letting your own grow, it was in the final analysis conservation and good forestry.[52]

More important was the on-going campaign to cut down the fire menace. Landowners, big companies, conservationists—all could join in this movement. No more respectable position could easily be found in Maine in the first decade of this century.

In the struggle against forest fires 1903 was the key year. As a result of the controversy concerning the taxation of wild lands a forest fire patrol had been started in the Maine woods. The legislature appropriated $10,000 for use in obtaining fire wardens. Spring came early, with little rain, and the woods were very dry. From April 8 to June 9 only 1.8 inches of rain fell in the north woods. During this time there were twenty days in a stretch without any rain followed by heavy winds beginning June 3. Local fires which had been burning were fanned by this wind. These fires, most extensive in T3, 4, 5R9, 2R8, and Long A towns, burned until a heavy rain on June 9 put them out. All told in the state that year 269,451 acres of land, of which 220,232 acres were wild land, were burned over. The damage to this land was estimated to run at least $750,000.[53] Such staggering losses caused most people to urge even greater efforts by the state.

Through the rest of 1903 and 1904 newspapers and others clamored for more money, more wardens, and more patrols. When the North American Fish and Game Protective Association held its annual meeting in Portland the delegates were greeted by Charles E. Oak, a former forest commissioner, who said that it was in the best interest of sportsmen and others to join the landowners in calling

Conservation and forestry come to Maine

for more state aid here. Some said that the forest commissioner ought to have the right to close the woods to persons during drought times in an effort to cut down the fire danger.[54]

In 1905 three lookout stations were in operation. The landowners and the lumbermen paid for the construction of the stations and for the telephone lines while the men were paid by the state. These first stations, at Squaw Mountain, Attean Mountain, and Mount Bigelow, all on the Kennebec, were the first in the United States. The Squaw Mountain lookout, William Hilton, spotted twenty-five fires between June 10 and September 12. The other two lookouts spotted twenty-three between them.[55]

The system worked so well that the state began to expand the program. Charts, telescopes, and range finders were given to the lookouts and a regular schedule of observations set up. Three more stations were constructed, at Skinnertown, Spencer Mountain, and on Whitecap Mountain.[56] Another dry year was 1908 with 142,000 acres burned over in unincorporated towns. By 1910 however, twenty-four stations were in operation and a schedule of cooperation was set up between Maine and New Hampshire. 1911 was also a bad year with nearly 100,000 acres under the fire attack, but this was the last such year before World War One.[57] To a very great extent the fire damage was beaten, even though Maine still lives in a fear of the terrible days of forest fires similar in extent to the great fires of 1947.

Other natural disasters were not so easily kept in check. However their effect was mitigated somewhat by a society which was willing to aid sufferers. The Internal Revenue Service allowed the large landowners to prorate and deduct the losses caused by the insect plagues, and when great winds came the government stepped in and purchased most of the downed timber.[58] Not all losses were thus subsumed by a gracious government but the effects of the losses were lessened.

Both the state and local governments were active in the new world of the twentieth century. Money was increased for instruction purposes at the state forestry school; by 1906 the thirty-seven students (seventeen of them forestry majors) could take sixty-one hours of instruction in this field. The courses included:[59]

General Forestry	18 hours
School Course in Forestry	3
Silviculture	5
Field Work in Silvi-	

culture	5
Forest Measurements	6
Field and Office Work in Forest Measurements	6
Lumbering	6
Forest Management	6
Thesis	6

The state had turned over Indian Township (TS), which was owned by the state although the proceeds went to the Indians at the reservations at Peter Dana Point, to the university for demonstration purposes in forestry. By this time the larger companies were beginning to use University of Maine graduates in their work. As practical men they oftentimes were given better jobs than some of the Yale graduates at the same time.[60]

The forest commissioner also used the services of the Federal Bureau of Forestry to aid him in his work. In 1902 Ralph S. Hosmer and ten other young foresters came to Maine in July. They stayed two and one half months in the Squaw Mountain township area (T2R6 BKP, WKR) then owned by the M. G. Shaw Company of Greenville and Bath. The township was completely surveyed in every way. The report and recommendations were not much different from the one done for the Great Northern Paper Company except that this one was made public.[61]

Much of the forestry work both public and private would in the next few years consist of publications. Official documents urged the cultivation of the farm woodlot and the growing of blueberries on burned land.[62] By 1907 there were 150,000 acres of blueberry barrens under cultivation in Washington County. The state obtained some stumpage income from berries grown on the public lots.[63] Articles also appeared which called for better forestry practices, especially during the controversy over the establishment of the White Mountain National Forest.[64] Others deduced figures which proved that there was more forest land in that period than there had been before when abandoned farms reverted to woodland.[65] At the end of World War I though, a special report to the Federal Bureau of Forestry said that under the then obtaining conditions the forest resources of Maine were rapidly diminishing in quantity and deteriorating in quality. "These facts mean that Maine is on the high road to a state of complete timber exhaustion." [66]

Many held the opinion that some state action was necessary to preserve the forest wealth, and as a result throughout the period to

1930 there were attempts in the Maine legislature to pass laws which would make conservation and forestry practices mandatory or at least profitable.

The most serious attempt of this kind began in 1907. It would have regulated the cutting of timber below certain diameter limits. The initial plan, which would have prohibited the cutting of spruce or pine below twelve inches in diameter, was presented in various ways. Two bills came up in the first legislature to consider them. The first called for a ten inch law with a $1.00 fine for each violation, and the second called on the state to acquire and reforest cutover lands. The legislature, which had lumbermen in most of the key posts, created a special committee "On Forests and Water Supply" to handle these bills. All members of the committee were either farmers or small businessmen. They voted eight to one with one abstention that a bill combining these provisions "ought to pass." The original sponsor from York County, heaviest hurt of the cutover areas, moved acceptance of the report in the legislature. "Many landowners", he said in his remarks, "both large and small have advised and heartily approved of this measure, though coming late, as being needed to stop the present most wasteful and destructive (to use no stronger term) method of handling our timber trade."

Two days later he was answered in a speech which characterized the bill as, ". . . a piece of freak and vicious legislation . . . . the bill is unscientific; it is improperly and loosely drawn; it is freak legislation. I claim that it is outrageous to take from the small property owners of Maine their individual rights in this matter. Shall the State of Maine say to me that I shall not cut a pine or hemlock tree on my land that is under ten inches in size at the butt? . . ." An eight inch amendment was defeated and the bill was then indefinitely postponed. Later one of the members of the committee tabled the bill and asked for and received permission to ask the court whether the proposed law would be constitutional.[67] On March 10, 1908, the court held the legislation to be "perfectly constitutional". The new governor, Bert M. Fernald, commended the courts on their holding in his inaugural address in 1909 and proposed the passage of the law, in conjunction with the creation of a Water Storage Commission.[68]

> In the past two years history has been making along these lines, and the various Congresses for conservation and irrigation and the gatherings of Governors of States which have been in 1908, have developed public thought along the lines of the welcome opinion of our Maine Supreme Court. I am persuaded that the waste of public equity,

notably in our forests, makes it imperative that the present legislature, should at least make a beginning in the direction of State Forest reserves and State water storage. From year to year, the Nation has gone on hit or miss, our rivers have shrunk, and our forests have been invaded, so that, from one end of the country to the other, there is reasonable alarm. Exhaustion of natural resources, of water power and forests would be a grave disaster to corporations as well as to the people. Floods waste and destroy in the spring, and in midsummer, autumn, and winter we go dry. Unless we preserve the forests on our mountains and hillsides the soil will be carried by the heavy rains into the valleys and the mountain slopes will become forever desolate. Thousands of acres in other countries now desert were made by such waste and pillage of wild lands. To preserve and cultivate the forests at our river headwaters, seems to me to be our immediate duty.

Even with such highly placed appeals being addressed to them the legislature was in no mood for this kind of law. Lumbermen, lumber dealers, paper manufacturers, and lumber manufacturers all had strong voices in Augusta. A new select committee was set up to deal with the controversial proposals. Of the ten members four were lumbermen, one was a Rumford merchant, one a contractor, and one a paper manufacturer. The committee received the bills and proposed that they be put off until the next legislature. The action was taken posthaste. In 1911 the legislation was not called up, and in 1913 when it was, the solons ruled that "the legislation was inexpedient." [69]

In 1915 the proponents tried again. An attempt was made to create "auxiliary forests" with a twelve inch cutting law. These privately owned forests would come under state control as to their operation. The inevitable special committee was set up, and the bill was put off until the next meeting of the legislators.[70]

In 1917 a new and more exact bill was offered. It would apply only to pine. The forests would be exempt from taxation if the twelve inch law (at a one foot height) was observed. An excise tax of one half per cent per year would be levied until five per cent was achieved; the funds were to be used for seeding and reforestation costs. The bill died in committee.

Conservation supporters would not quit. In the seventy-ninth legislature which met in 1919, they produced proposed legislation which met most of the older objections to the bill.[72] The recommended law would have set up state forests under control of the land agent. The agent was instructed to recommend lands for purchase by the state. Owners of lands could denominate "auxiliary forests," but such forests came under the control of the state for cutting purposes. A

Conservation and forestry come to Maine

twelve inch law was appended for pine. There was provision for the introduction of long range selective management plans by the owners which would supersede any state plans. There was to be no taxation on the trees except that the excise tax on the cutting would still apply to be used for reforestation purposes.

A special joint committee on "State Lands and Forest Preservation" was set up to deal with this bill. The committee voted seven to three "ought not to pass." The minority however produced their own bill with a favorable recommendation. This bill prohibited the cutting of pine ten inches in diameter four feet high. All pine less than sixteen inches, and all hemlock and spruce less than twelve inches four feet high were to be free from taxation. The auxiliary lands could be withdrawn from the state control if it became more valuable for other purposes and if the full tax was paid. After considerable discussion the majority report was accepted, and a special seven man committee was created to investigate and report to the next legislature. The committee consisted of the forest commissioner and six others, one of whom was a lumberman and one other prominently placed as a devotee of the Passamaquoddy tidal bore project.[73]

In 1921, without any fanfare a measure was adopted called "An Act for the Preservation, Perpetuation, and Increase of the Forests of the State of Maine." [74] This legislation authorized the establishment of state forests, and called for further recommendations to be made. It also allowed "auxiliary forests" to be set up. Cutting on such forests would be allowed by the state, and all white pine ten inches in diameter at four feet high were to be left along with "three seed trees per acre." All pine sixteen inches in diameter at four feet high was to be tax-exempt, and the land itself could only be assessed at a maximum valuation of $5.00 per acre. The excise tax provisions only applied to land in municipalities, and the money reverted to them. Counties could benefit under special programs. Returns had to be made annually, and the forests could be withdrawn from state control on payment of the full tax plus expenses.

In 1923 the law was amended to include trees above the stated diameter limits as well. The forests were made more difficult to withdraw by making the owner liable to all taxes from the date of the listing at the full value at the time of withdrawal plus expenses. Few people took advantage of this law and in 1929 it was amended further. This time the provisions were made easier but they were interpreted to apply only within organized towns. In 1933 the law was repealed.[75] Just how many people took advantage of the law is unknown. The only discussion of it is in Coolidge, and it is marred by bad reading of the law and a rather naive assumption that the law was desired by all.[76] As a matter of fact, the law was not wanted, and it was worded in such a way as to make it unreasonable that the big landowners would involve themselves. Like many, if not most conservation methods and techniques in Maine, it was applied when it was to the self-interest of the big landowners who by this time were the big paper companies.

These companies had by this time adopted scientific logging, forestry, conservation, and were generally cultivating the woods as though they were a farm. The development of these attitudes came as the big firms began their program of large purchases. The following tables show the amounts of land purchased each year from 1899-1929 by the largest of the firms, the Great Northern Paper Company.[77]

On the eve of the great depression the Great Northern Paper Company owned well over one and one third million acres of land. By this time the firm was purchasing land for investment purposes. It had decided to remain in Maine and that meant preserving the forest heritage. An investment of this sort was worth preserving[78] and for that reason scientific forestry came to the Maine woods.[79]

The long battle and struggle over control of the Maine woods was over in effect. The paper companies, especially the Great Northern, were to add to their holdings very extensively in the thirties and forties but the pattern was well set long before. The upshot of the battle though had been to preserve the woods, and today Maine looks more as it did in 1660 than it did in 1860 when we began this story.

Conservation and forestry come to Maine

Purchases and Sales—Penobscot Lands

Great Northern Paper Company

1899-1929

Year	Acres	Price	Average cost per acre	Pulpwood (cords)
1899	252,060	$1,042,575	$4.136	2,004,103
1900	1,293	5,353	4.11	12,255
1901	144,019.65	500,457		500,247
1902				
1903				
1904				
1905				
1906	99,762	488,818	4.541	570,388
1907	14,070	74,220		74,357
1908				
1909	9,603	83,500	8.695	48,594
1910	19,993	111,973		152,743
1911	23,089	108,493		151,323
1911 cont.	19,552	82,609		128,523
1912	581	14,367	24.728	842
1913 2 mo.	8,961	53,284	6.946	80,059
1913 10 mo.	64,107	392,194		134,838
1914	1,839	8,211	4.466	2,771
1915	70,571	675,917		476,885
1916	142,507	829,592		750,038
1917	987	3,000	3.039	1,326
1918				
1919				
1920	78,411	1,758,837		395,883
1921	164,645	3,060,750		649,016
1922	31,029	515,007		176,157
1923	6,921	77,545		18,324
1924	14,519	146,643		25,777
1925	1,027	34,000		5,348
1926	18,448	336,513		94,589
1927	103,336	674,597		216,168
1928	84,816	1,237,000		407,892
1929	19,945	212,370		41,401

Total   1,127,916 acres***
***Sold 1899-1929—24,065.75 acres

Purchases and Sales—Kennebec Lands

Great Northern Paper Company

1899-1929

Year	Acres	Price	Average cost per acre	Pulpwood (cords)
1899	86,851	259,948		592,514
1900	15,899	39,075	2.457	95,544
1901	4,160	6,080		2,331
1905	20,800	67,000	3.221	135,288
1906	876			2,901
1909	14,216	115,000	8.082	54,794
1912	5,852	50,750	8.55	34,898
1921	300	8,114	27.05	990
1923	23,520	467,340		74,752
1924	10,076	108,676		1,446
1925	611	28,328		
1926	2,656	33,489		7,516
1927	40,228	374,613		106,269
1928	15	1,500	100.00	
Total	226,597 acres***			

*** Sold 63.31 acres
Grand Total 1,354,513 acres

NOTES

[1] Cary to Roth, April 8, 1898.
[2] Cary to Pinchot, September 18, 1898.
[3] Cary to Pinchot, November 29, December 21, 1898.
[4] Cary to Pinchot, September 18, 1898.
[5] Cary to Pinchot, November 29, 1898.
[6] Cary to Pinchot, December 14, 21, 1898; January 15, 27, April 9, 1899.
[7] Cary to Pinchot, December 14, 1898; March 30, 1899.
[8] Letter appears, along with a glowing editorial, in Bangor *Daily Commercial*, September 19, 1899. One of Cary's interests was in promoting the cubic foot method of scaling lumber.
[9] *Forest Management in Maine*, Boston, 1899. The favorable reception, "Such work could be in no more capable hands than those of Mr. Cary." Bangor *Daily Commercial*, November 18, 1899.
[10] Austin Cary, "Management of Pulpwood Forests—System of Forestry Practiced by Berlin Mills Company", Forest Commissioner's *Report*, 1902, *Public Documents*, 1903, II, 125-52.
[11] From *ibid.*, 134; also see "The Management of Pulpwood Forests" by Cary in *Journal* of the Canadian Forestry Association, 1902.

Conservation and forestry come to Maine

[12] Cary to Pinchot, May 29, 1903. Apparently they were started that spring.

[13] See, on a township then owned by Dartmouth College, but which was going on the market, Cary to Pinchot, October 6, November 9, 1902; Overton Price to Cary, October 13, November 14, December 2, 1902; Cary to Price, November 19, 1902.

[14] Cary to Pinchot, July 22, 1903 (from Austria), and September 3, 1903.

[15] Cary to Pinchot, October 31, 1903.

[16] Same to same, December 14, 1903.

[17] Overton Price to Cary, April 20, 1905.

[18] Cary to Pinchot, May 9, 1904.

[19] In particular in a book published in 1958, William Robinson Brown, *Our Forest Heritage: A History of Forestry and Recreation in New Hampshire*, New Hampshire Historical Society, Concord. According to White, "Austin Cary . . .", *op. cit.*, 28, notes 44-5, Cary resigned as a failure, but was asked to return in seven years to reinstitute his methods. He cites as evidence, Austin Cary, "The Forester Comes Back", an address delivered February 24, 1926 to the American Pulp and Paper Association in New York. The letters do not bear this interpretation out.

[20] See his article on German forestry in *Forestry and Irrigation*, November, 1903, and the one from which the quotation was taken, "Maine Forests: Their Preservation, Taxation, and Value," a paper read before the State Board of Trade, September 26, 1906, and reprinted in *Annual Report*, Department of Industrial and Labor Statistics, 1906, 180-193. In 1907 he made lecture tours of Maine towns using stereopticon slides to carry his message on growth utilization, disease, soils, cutting methods, and scientific forestry. *Paper Trade Journal*, October 31, 1907. He summarized his Maine career in a Forest Service Circular No. 131, in late 1907. *Paper Trade Journal* reprinted it at January 9, 16, 1908. Of course they also published his famous *Manual for Northern Woodsmen*, New York, 1910. In this year Cary was doing his missionary work as Superintendent of State Forests in New York.

[21] This period is well covered in Professor White's dissertation, aforecited. Also see Frank Heyward, "Austin Cary, Yankee Peddler in Forestry", *American Forests*, LXI, No. 5 (June, 1955), Part I, 29, Part II, No. 6, 28.

[22] Austin Cary, "Letter from Austin Cary to Mr. Earle Kaufman, January 22, 1935", *Journal of Forestry*, XXXIII, No. 6 (June, 1935), 643.

[23] See the remarks of Pinchot as reported in Bangor *Daily Commercial*, January 9, 1904. He was speaking specifically of the Great Northern Paper Company work, which he described, and badly.

[24] Cary to Pinchot, October 6, 1902.

[25] Price to Cary, October 13, 1902.

[26] Bangor *Daily Commercial*, March 28, 1901; an editorial commended the company.

[27] Bangor *Daily News*, November 11, 1901. "Kill off the fir and give the young spruce a chance," was his prescription.

[28] William C. Hodge, Jr., "A Working Plan for the Penobscot Timberlands of the Great Northern Paper Company: Somerset and Piscataquis Counties, Maine." This typewritten manuscript is in the files of the Great Northern Paper Company and I was able to use it because of the generosity of the late William Hilton, and the present Vice President, Woodlands, John T. Maines. Mr. Hilton

was very proud of the document and said that his company still followed it. "Would you like to look at our Bible?" was his comment to me.

[29] This paragraph is derived from a reading of *ibid.*, whose 71 pages cover it in detail; from Pinchot's comments in Bangor *Daily Commercial*, January 9, 1904, and from a study of the correspondence in the files of the Bureau, especially Pinchot to Hodge, June 12, 14, 19, 21, July 2, 1902: Hodge to Pinchot June 13, 1902; Hodge to Price, August 12, 1902; Price to Hodge, July 10, 1902. All letters RG 95, National Archives.

[30] Hodge to Price, June 24, July 3, August 30, October 13, 26, 1902; Price to Hodge, June 28, September 3, 1902; Olmstead to Hodge, September 20, 1902. One of the reasons the cost was higher was the inclusion of Pittston Town (T2R4 NBKP) after the work was underway. It had had a bad burn in 1894 and it needed the work to determine the extent of the reproduction.

[31] Hodge, *op. cit.* His recommendations with his tabular proof run generally from 54-71 in the document. The towns included in the report were 1R9, 2R10, 3R11, 4R12, 4R13, 4R14 in Piscataquis County, and 4R17, 4R18, 5R17, 5R18, 5R19, 5R20, 6R17, 6R18, Pittston, Dole, and Prentiss.

[32] From rules preserved in George T. Carlisle and Frank Shatney, "Report on a Logging Operation in Northern Maine", unpublished manuscript, University of Maine, Orono, 4-6. There is a good deal of evidence that the rules were honored as much in the breach as in the observance. Other firms active in the State, especially in plantation experiments were International Paper Company and Pejepscot Paper Company. See *Paper Trade Journal*, May 26, 1910; February 7, 1918, Julian Rothery, "Forest Planting and Protection".

[33] *Paper Trade Journal*, December 28, 1895 (eight inches twenty feet up); Bangor *Daily Commercial*, December 2, 1895 (C. A. Packard, a one-time land agent, and J. F. Sprague (Monson), also eight inches twenty feet up.)

[34] *Ibid.*, January 7, 1895, the land agent Charles Oak, ". . . it seems the wise thing to take count of stock."

[35] *Ibid.*, a meeting of landowners was held to advocate this in 1893, and Oak was recalling the meeting to the new solons.

[36] *Ibid.*, February 26, 1897, editorial "Floods and Forests", July 9, 1898, "Forest Protection", January 9, 1899, "Uses of Wood Pulp", February 2, 1899, "A Warning".

[37] *Ibid.*, September 7, 1895.

[38] *Ibid.*, May 27, 1899, citing a paper by W. F. Ganong in New Brunswick, June 27, July 6 (New Hampshire advocates), March 24, 1900, an editorial book review of *North American Forests and Forestry*, written by Earnest Bruncken, State Forestry Commission of Wisconsin.

[39] Bangor *Daily Commercial*, July 26, 1900, editorial "A Course in Forestry", and June 6, 1903; *Catalogue*, University of Maine, 1902-3, 75-7, 83-5. By 1912 the course had 46 students and 35 graduates were working in professional forestry. Arthur Briscoe, Professor of Forestry, called for an increase in funds. *Paper Trade Journal*, October 31, 1912.

[40] Bangor *Daily Commercial*, October 23, 25, 31 (from Philadelphia *Record*), 1899; January 4, 1900, editorial, "Forest Fires and $3,000". It called on the ghost of the 1825 Miramichi fire; January 1, February 20, 22, March 12, 18 (full page), 19, 20, 1901.

Conservation and forestry come to Maine 379

[41] Bangor *Daily Commercial*, February 20, 21, 22, especially 25, "Before the Taxation Committee", 26, 27, 1901. A petition was presented to the legislature with the names of two hundred and fifty lumbermen on it in opposition to the tax increase.

[42] Francis Wiggin, "The Preservation of Maine Forests", an address delivered before the Board of Trade at Rockland, Maine, October 15, 1901, *Industrial and Labor Statistics*, 1901, 103-118; also Bangor *Daily Commercial*, October 17, 1901. The paper said, "the preservation . . . is imperative."

[43] *Industrial and Labor Statistics*, 1899, 60; *The Industrial Journal*, February 16, 1900, a special illustrated pulp and paper edition. The quotation is from Kennebec *Journal*, January 12, 1901.

[44] Bangor *Daily Commercial*, April 12, 1902; the best account of the interview which appeared in most Maine papers.

[45] Forest Commissioner's *Report*, 1902, *Public Documents*, 1903, II; his statements appear at 1-10 especially.

[46] Portland *Eastern Argus*, January 10, 1902.

[47] G. F. Talbot to Editor, *ibid.*, January 13, 1903.

[48] Chisholm, *op. cit.*, 165-6 for Ring's quotes, 167, 168-9, for production, and labor difficulties.

[49] *Industrial and Labor Statistics*, 1906, 141.

[50] *The Industrial Journal*, April, 1907; *Paper Trade Journal*, all April, 1907.

[51] J. Fraser Gregory in a Forestry Convention in New Brunswick, 1906, reported in *New Brunswick Yearbook, 1907*, 96. See Portland *Eastern Argus*, March 19, 1903, quoting Worcester *Post* for a somewhat similar view. "It will make some difference to Maine and to New England when these spruce forests and the pulp mills they support have disappeared."

[52] William Hilton said to me in an interview July, 1962, "We always did cut a lot of wood to accommodate other people." One suspects that it also accommodated the Great Northern Paper Company as well.

[53] Bangor *Daily Commercial*, April 13, 27 (letter from W. M. Munson, a professor at the University on the Minnesota forestry system), June 9, 13, 17 (and all spring actually), 18 (on the Minnesota system again), 1903. *Report*, Forest *Commissioner*, 1904, *Public Documents*, II, 1905, 7-26, "Forest Fires of 1903."

[54] Bangor *Daily Commercial*, July 21, 1903, July 31, 1903, October 7, 1904; the report of the meeting is at January 21, 1904, see especially the editorial "Forest Protection and Water Storage". Also *Argus*, January 22, 1904.

[55] Sixth Forest Commissioner *Report, Public Documents*, II, 32-37, Hilton's diary is at 33-6. See the story in *Forest History*, V, No. 2. Hilton told me in an interview that the driving force here was the M. G. Shaw Company of Bath, and especially Billy Larkin, master driver on the Kennebec River. Larkin was experimenting with yarding logs down Squaw Mountain with a cable and Hilton stayed there nights and went up to the mountain days. The big trouble was the Maine Central Railroad which threw out sparks into the slash.

[56] Forest Commissioner's *Report*, 1906, 37. It cost $750 to build and equip each one of these three new ones.

[57] *Seventh Report*, Maine Forest Commissioner, *Public Documents*, 1909, II, 11-19. "Forest Fires of 1908", Department of Forestry *Report*, 1910-11;

Ninth Report, Forest Commissioner, *Public Documents*, 1913, II, "Forest Fires 1911", *Tenth Report*, Forest Commissioner, 1914; *Eleventh Report*, Forest Commissioner, 1916; Charles B. Fobes, "Lightning Fires in the Forests of Northern Maine 1926-1940", *Journal of Forestry*, Vol. 42, (April, 1944), *Idem.*, "History of Fires", *Economic Geography*, Vol. 24, (October, 1948); Samuel T. Dana, "Forest Fires 1916-1925", Maine Forest Service Bulletin #6, 1930; Royal Shaw Kellogg, "Cooperation in Forest Protection", *Journal of Forestry*, (October, 1921). In 1909 the program of forest protection became the responsibility of the landowners with the chartering of the "Maine Forestry District", to supervise the wardens, and the lookout towers. This service is achieved by a tax on the owners based on the dollar valuation. See *Thirty-Third Report*, Maine Forest Commissioner, 1959-60, "Fiftieth Anniversary of the Maine Forestry District, 1909-1959", 126-134, with an important chronology 133-4. One of these events of importance here was the use of railroad patrols, started about 1913-14 by Louis J. Freedman, later to be Director of Woodlands, Penobscot Chemical Fibre Company. On this see "A Friend Recalls A Young Bangor and Aroostook," *Mainline*, Vol. 14, No. 6, [Nov. - Dec. 1966], 15-17. He discussed at some length with me the early days in Maine conservation and fire fighting practice. The woods were closed by administrative fiat for the first time in 1911. The courts upheld the constitutionality of the Maine Forestry District in 1912 and the wardens were put under federal employment for a while through the Weeks Law. Later this became cooperative through the Clarke-McNary Law (1924). A good discussion of the District on the eve of World War One occurs in T. G. Loggis (Deputy Minister Lands and Mines, New Brunswick) to George J. Clarke (Premier and Minister Lands and Mines, New Brunswick), February 22, 1915, reprinted in Joseph Herbert Sewall, "A Comparative Study of the Development of Forest Policy in Maine and New Brunswick", unpublished M.A. thesis at University of Maine, Orono, 1957, in appendix 195-200. "I consider the mode of forest protection in the State of Maine ahead of any state in the Union so far as they are applicable to us. . . ." At this time there were fifty-five forest lookouts in use at one time or another.

[58] On the budworm damage, William Hilton to Commissioner of Internal Revenue, November 9, 1928; November 5, 1930 for the G.N.P. Company. I am indebted to Hilton for calling my attention to these letters. The biggest hurricane occurred in 1938. See Historical Records Survey, *The New England Hurricane*, Boston, 1939 for a photo-history. The trees which came down were boomed and sawed in mills in Saco and Norway. The job was not finished until 1942 and much of the lumber was used in barracks construction. The five million feet in the Saco boom was finished in 1939. The Norway operation produced 4,750,000 feet. The government paid $665,000 for the down timber. In this last of the pine operations of any size, the largest log, sixteen feet long, scaled out 772 feet of boards. See Portland *Sunday Telegram*, September 28, 1941 for a nice story.

[59] Bangor *Daily Commercial*, April 13, 1903; Forest Commissioner, 1906, *Public Documents*, II, 1907, 56-8, "Forestry Course at the University of Maine", and for a history of the cutting on Indian Township, see *ibid.*, 102ff.

[60] This last few lines is based on several interviews and discussions with men active in the woods then, especially William Hilton, Philip Coolidge, and Louis Freedman. Their comments lead one to the conclusion that no matter

where these people came from, they had to be retrained, and it took patience. Another conclusion might be that not all men in the companies, in those days, were blessed with that virtue. The University of New Brunswick forestry program was also active. Maine began its nursery programs in 1913, added new professors and courses, and began its famous summer courses. *Paper Trade Journal*, February 20; April 24, June 5, July 17, 21; August 7, 14, 21, 28, September 4, 1913.

[61] *Fourth Report*, Maine Forest Commissioner, 1903, *Public Documents*, 1904, II, 59-71. Ralph S. Hosmer, "A Study of the Maine Spruce", and "A History of Squaw Mountain Township", in *ibid*. For Pinchot's comments see Bangor *Daily Commercial*, January 9, 1904.

[62] *Fifth Annual Report*, Forest Commissioner, *Public Documents*, 1905, II, 46-8, "The Woodlot"; also see James W. Toumey, "The Woodlot, A Problem for New England Farmers", *Scientific Monthly*, V, (September, 1917).

[63] On blueberries see the files of the Forest Commissioner, and stumpage contracts located therein.

[64] The New England Governors and Gifford Pinchot, "Deferred Forestry", *New England Magazine*, XXX, (December, 1908); Philip W. Ayres, "Is New England's Wealth in Danger?", *ibid.*, XXX (1908); "White Mountain National Forest—Will It Pay?", *The Northern*, Vol. 2, No. 1 (April, 1922). This last was published at the time of their controversy with Baxter. Their answer was predictable. According to them it didn't, at least from 1915-1921, and profitability was their criterion for public forests.

[65] See among others, P. L. Butterick, "Forest Growth on Abandoned Agricultural Land", *The Scientific Monthly*, V, (July, 1917); Roland M. Harper, "Changes in Forest Area in Three Centuries", *Journal of Forestry*, XVI, (April, 1918); Ben Ames Williams, "The Return of the Forest", *American Forests*, (1927), and *Literary Digest*, XCIV, (September 10, 1927).

[66] "The Timber Situation in Maine", May 31, 1919, a typewritten manuscript in the files of the Bureau of Forestry, RG 95, NA, Washington, D.C. Not all felt this way, or were willing to allow it to occur. The S. D. Warren Company was planting white pine in special forestry plantations as early as 1921. They were pruned, thinned, and watched very carefully, and grew about 1.1 cords per acre each year. Warren's *Standard*, III, No. 11 (January, 1954).

[67] There is a general discussion of this in Forest Commissioner, 1909, *Public Documents*, II, 30-2; *Legislative Record*, 743-4 (Perkins' remarks); 816, (the rebuttal); 890-2. My compilation of the economic structure of the legislature depends here, as is usual, on the published biographies of the members in the Kennebec *Journal*, and my general knowledge of lumbermen at this time and place. All votes were voice votes.

[68] Inaugural Address of Bert M. Fernald, January 7, 1909; *Public Documents*, 1909, II,—the sections on "Conservation of Our Natural Resources" occurs at 6-8. The decision was also hailed by *Paper Trade Journal*, in its June 25, 1908 editorial. It called Maine "a pioneer in forest legislation and management."

[69] *Legislative Record*, 1909, 619-740; *House Journal*, 370, 647; *Legislative Record*, 1913, 1149; Senate *Document* 1, 1909.

[70] Senate *Document* 99, 1915; *Legislative Record*, 1915, 891.

[71] Senate *Document* 37, 1917; *Legislative Record*, 1917, 442.

[72] House *Document* 20, 1919.

[73] House *Documents* 20, 514, 1919; *Legislative Record*, 1919, 975, 1125, 1257. There was in addition House *Document* 147 which called for mandatory work of this sort on timberlands in unincorporated towns. This legislation was, needless to say, declared to be "inexpedient". *Legislative Record*, 1919, 470.

[74] *Public Laws*, Maine 1921, Chapter 78.

[75] *Public Laws*, 1923, Ch. 138; *Public Laws*, 1929, Ch. 306; *Public Laws*, 1933, Ch. 139.

[76] Philip T. Coolidge, *A History of the Maine Woods*, Bangor, 1963, 504-7. Austin Cary was somewhat more analytical of all paper companies in a letter to Morris Westfeld, April 8, 1929. This letter brought an agreeable response from George S. Kephart, April 12, 1929. Both agreed that economic self-interest was the key. Letters in my possession.

[77] These figures are abstracted from a volume entitled "Millinocket—Purchases and Sales." The next table is from "Madison—Purchases and Sales". I am indebted to William Hilton, former Vice President Woodlands, now deceased, of the G.N.P. Co., for allowing me to use these materials, and explaining them to me. For other sales of lands and mills see *Paper Trade Journal*, September 17, 1908, T7R9 NWP for $180,000; July 31, August 21, 1913, 84,000 acres and three steam sawmills from William Engels estate for $525,000.00.

[78] Just what the G.N.P. Co. investment was worth, it is hard to say. On March 1, 1913, when an evaluation had to be struck for income tax depletion purposes the company valued its 534,817.61 acres of Millinocket (Penobscot) lands at .415¢ an acre, or $221,949.31. The pulpwood, 3,034,519 cords, was rated at $1.86 a cord, or $5,644,944.08. The Millinocket total was $5,866,893.30. The Madison, or Kennebec, lands, 148,655.31 acres were valued at $61,691.95. The pulpwood on that land, 575,954 cords, was valued at $1,071,414.66, for a total of $1,133,106.81 and a grand total of $7,000,000.00 book valuation. This, of course, does not include any of their other holdings—pulp mills, ships, railroads, and so on. It was hardlly a fly-by-night company. This data comes from File #194, dated April 23, 1929. I am again in the debt of William Hilton.

[79] See, on this point, at that time, Ernest F. Jones, "Forestry Conceptions Among Timberland Owners in the Northeastern Spruce Region", *Journal of Forestry*, Vol. 39, 175-80 (1931). Jones was a Great Northern Paper Company forester.

15

THE MAINE WOODS IN THE TWENTIETH CENTURY

> Driving on the West Branch is much different than it was then [1860]. At that time the men were about all Yankees and Indians. A man had to have a recommendation to get on the drive at all. Now almost anyone can get a job. The Yankees have almost all dropped out and so have the Indians and their places are filled by men from the Provinces and other foreign places. The great cause of this is the decrease in pay.
>
> In old times many of the men got $5 a day for driving and now if they get $2, they think they are going pretty well.
>
> Charles Hathorn, reminiscing to a reporter, Bangor Daily Commercial, October 1, 1895.

In 1860 a traveler from the 1820's would not have noticed much difference in the life of the Maine woods. Indeed, a traveler from the 1860's would not have found that much difference in the 1890's. However a traveler from the nineties, put down in the Maine woods in 1930, would have recognized the trees but not much else. A revolution had been consummated which involved methods of work as well as the off-duty hours of the men.

With the paper companies in full control of the woods and rivers the cuts were much less. During the time of controversy, 1901-1904, conservative estimates of the amount of soft wood cut in the Maine woods ranged from 682,000,000 feet in 1901 to 762,000,000 feet in 1904.[1] By 1909 the cut was well over a billion feet, although this figure was not often reached again as both pulpwood and sawlogs continued to be felled. After this time sawlogging was much diminished and the cut ranged between 800 million and one billion feet until 1917 when it began to tail off even more. By 1920 the cut was less than 500 million; in the mid-twenties it had dwindled to less than 400 million, and by 1930 it was only about 300 million feet, and some thought that it had gone lower. The cut was never to go over the 300 million foot mark again until World War II.[2]

The big totals still came from the Penobscot river basin, then the Kennebec, thirdly the Androscoggin, with the St. John-Aroostook area cut fourth. The rest of the cut came from downeast and occasional small operations on the Saco and other similar rivers. By 1900 half the cut was pulp and the percentage of pulp increased each year until by 1930 the cut was nearly all destined for the pulp mill. It probably ranged as high as 90 per cent of the total across the state.

On the Androscoggin nearly all trees cut went for pulp. In 1895, of the 200,000,000 foot total at least 125,000,000 feet went to the pulp mills, and this was the last big year for logs. The Berlin Mills Company and the International Paper Company were purchasing land and controlling production.[3] Their logs were also beginning to move more by railroad. On the Kennebec the story was similar.[4] Downeast saw mills still took an occasional four or five million feet, but a more ordinary year was that experienced on the Narraguagus in 1903. Fourteen operations used one hundred and forty-six men and twenty-seven horses for the small amount harvested (perhaps 3,500,000 feet). The St. Croix cut was 32 million feet in 1903, but half of it was pulp and soon it would all be. The Todds who had been big operators on that river, began to act as middlemen purchasing pulpwood on commission. Even in the north country the impact of the pulp grinder began to be felt.[5]

Sometimes the totals were still fairly big; 128 million feet above the Nashwaak on the St. John in 1896, 135 million in 1897 but in 1898 the cut total fell back to 80 million feet. This was due mostly to the big mill at Ashland which did not live up to its expectations, sawing only twelve to fifteen million feet a year rather than the thirty million it was rated for.[6] Big operations were made by people like William Cunliffe who usually worked three hundred men and seventy-five horses in eight to ten camps,[7] but those days too were numbered. On the Aroostook the cut ran between fifteen and twenty-five million feet a year, but more than half of it stayed upriver to go for shingles. In 1904 the St. John cut including the Van Buren mill was only 40,000,000 feet with 22,000,000 more cut on the Aroostook. This was more or less due to a hung drive in 1903, but even in 1905 the state only produced about 70,000,000 feet in the north country.[8]

By 1913 the cost of getting logs cut and to the boom was about $10.15 M which made the profit margin very narrow. As one man said when asked about railroad ties,[9]

> I only got out about 6,000 last year as the hauling gave out and left some on the yards. It is a hard business getting ties and costs all you can get for them. . . . The cost will depend on the chance and the

length of the haul. If you can get .50¢ a piece for ties, get a contract; if you cannot, go easy.

Occasional deal shipments to England continued but the profit was disappearing. Times were bad enough in the lumbering industry so that by the end of this period a Royal Commission investigated. They found mostly loss and suggested lower stumpage rates, longer leases, better rules for scalage, and a shift to pulpwood operations.[10] All of these things came about, and the greatest of the changes was in pulpwood.[11]

The West Branch Penobscot had become a Great Northern stream to all intents and purposes by 1900, thus it only remains to tell briefly the last days on the other branches. For a short time the East Branch Penobscot did very extensive lumbering. From the twenty-one million feet felled in 1895, all cut around Chamberlain Lake, (the branch did a business of 31,000,000 feet all told, but ten million was out for Sherman, Staceyville, and Island Falls mills), the cut was to rise fairly high. In 1897, the year of the organization of the East Branch Driving Company, the cut was 18,000,000 feet. By 1901 however it was 41,600,000 with twelve million feet going to the Katahdin Pulp and Paper Company at Lincoln. In 1902 forty-five million was felled, in 1903 sixty-six million feet. Fifteen million feet was hung and did not come down in the drive, and when the 1904 cut went to 59,500,000 feet the drive was 75,500,000 feet. This was the banner year though. In 1905 the cut slackened to 22,500,000 feet, and it continued to diminish.[12]

East Branch Penobscot Log Cut—1895-1905

Year	Total (board feet)
1895	31,000,000
1896	c. 20,000,000
1897	18,000,000
1898	24,000,000
1899	not known
1900	not known
1901	41,600,000
1902	45,000,000
1903	66,000,000
1904	59,500,000
1905	22,500,000

A similar story was true on the Piscataquis. The river's record was fifteen million feet in 1901; thirty million feet in 1902; and back to twelve million feet including the hung logs of 1903. The Passadumkeag was also about played out, but during these years it produced a few logs as the table indicates.[13]

Passadumkeag Cut—1901-1904

Year	Amount (board feet)	Remarks
1901	16,000,000	
1902	12,000,000	13 operations
1903	11,000,000	3 million hung
1904	17,000,000	nearly all hemlock

The 1904 drive with the hung logs from 1903 was the largest drive in years on that branch.

Big loads were still dragged, a team of 3,000 pounds hauling 4,500 feet of spruce on T2R3, WBKP, in 1895; and big trees were still cut. T. A. McPherson cut a pine scaling 4,200 feet the same season.[14]

Basically though the story all over the state was pulpwood and most importantly the pulpwood cut by the big companies.[15] The following table shows the amount of the Great Northern Paper cut on their own lands. It does not, of course, include the pulpwood purchased from others, which was considerable.[16]

Pulpwood Cut—Great Northern Paper Company Lands
1899-1929

	Millinocket Block (Penobscot Lands) **			Madison Block (Kennebec Lands)	
Year	Cut (cords)	Budworm Damage	Year	Cut (cords)	Budworm Damage
1899	—		1899	—	
1900	19,558		1900	20,641	
1901	36,962		1901	24,310	
1902	59,401		1902	21,632	
1903	53,986		1903	14,977	
1904	69,102		1904	22,894	
1905	52,744		1905	26,544	
1906	56,924		1906	19,482	
1907	64,239		1907	21,002	
1908	139,032		1908	29,255	
1909	111,376		1909	43,615	

Maine woods in the twentieth century 387

1910	77,714		1910	27,653	
1911	65,334		1911	25,604	
1912	38,104		1912	44,707	
1913	34,390		1913	9,899	
1914	47,783		1914	16,024	
1915	66,641		1915	2,950	
1916	96,954		1916	29,067	
1917	96,420		1917	123	
1918	172,255		1918	10,790	
1919	175,165		1919	44,903	
1920	95,809	259,395 (fir)	1920		33,971
1921	222,168	259,395	1921		33,971
1922	100,179		1922	3	
1923	229,436		1923	11,224	
1924	98,265		1924	34,595	
1925	35,449		1925	53	
1926		184,835 (spruce)	1926	6,024	22,300
1927	73,504	184,835	1927		22,911
1928	49,326	184,835	1928	14	12,715
1929	18,165	184,835	1929	11,505	12,715

**For comparison purposes a cord equals 500 board feet rossed and approximately 400 feet in the rough state.

Life in the woods was also changing rapidly.[17] Maine lumbermen were still in demand at times of national emergency. When the Spanish-American War came a U.S. Army recruiter hung a placard in Pickering Square in Bangor.[18]

AXEMEN WANTED

THE LUMBERMEN OF MAINE ARE NEEDED NOW
Axemen and Riverdrivers

WANTED
for the 1st U.S. V.
Engineers

Within a week eighteen enlisted and left for Boston.[19] In 1917 a similar situation occurred. Forrest Colby, then the State Forest Commissioner, sent out a call for enlistments:[20]

> You are doubtless aware of the immense amount of lumber needed for war purposes. Old England is willing to give her groves, parks, and orchards, and even her shade trees, to provide this timber; but it is in great need of skilled lumbermen and portable mill outfits to furnish the men and the material. Now is the opportunity for New England to help Old England.

Some went and served with Greeley in France, but most Maine lumbermen were not called except as draftees. The paper needed to fight the war made the Maine forests as much of a premium as the spruce forests of the west producing their lumber for aircraft.[21]

The big changes came when men went to the woods,[22] much earlier than previously, and in the men who went. The traditional crews had been made up of French Canadians, "Bangor Irishmen," and State of Mainers in about equal proportions. By 1896 however, many of the crews consisted of drunks and bums who were shanghaied into woods work. Local saloon operators in Bangor often received $1.00 for each person they provided and what better way to get them than through the use of liquor? By the end of the century agents for these saloon keepers were scouring the Boston waterfronts for woods help. One of these saloon keepers said in self-defense, "Many of these men being foreigners and of a low character, it is not strange that at Old Town and Greenville there are disturbances when these men come out in the spring." He was attempting to play down the drinking and shanghaiing, but it certainly existed. At least once there was a major riot and brawl in the woods among woodsmen, over religious differences. These men resorted to axes and handspikes to settle their quarrel. None were killed, but severe injuries were inflicted before the riot could be stemmed.

Later the Great Northern Paper Company employed a full-time agent in Boston to hire their crews. A Bangor agency, Largay's, usually supplied the men for other operators. By 1912 crews were made up of French, Poles, Finns, Russians, Swedes, Irish, and "Boston men," whom everyone agreed would not stay long and were of little use. These were the people who were responsible for the standard remark, "We had three crews, one in the woods, one coming from Bangor, and one leaving for Bangor." There was some discussion in 1907 of a law which would have punished men for taking "French leave", but some felt that this would create an ill-disguised peonage. The discussion petered out in an argument over who was responsible in such cases—the men or the employment agencies.

By the mid-twenties the crews were made up, in good part, of what Maine people call "Polacks," a generic name for east and north Europeans, except Finns or Swedes. These people were not good river drivers, and their woods work was not as pretty as the Finns who had no equal, but "they were the best all-round men on four-foot stuff that goes to the woods today."[23]

Maine woods in the twentieth century

Methods used in the woods varied a good deal in the new century. In addition to mechanical methods of moving logs which will be discussed later, the saw began to replace the ax almost entirely. The years between 1890 and 1910 were the transition years, and generally the saw became predominant in a west to east pattern across the area. By 1899 a newspaper could remark, "The saw is making rapid inroads in felling over the axe." By 1902 standard Penobscot stumpage permits had as contract provisions the use of the saw, the cutting of trees to the swell of the roots, the prohibition of wagon sleds in the cut, and finally diameter limits to the cutting. St John permits included these provisions a year later. The diameter limits varied from as high as fourteen inches on the stump for spruce and cedar to as low as six inches in diameter four and one half feet from the ground.[24] Occasional contracts called for culling the fir or for the removal of all tops as well.[25] Sometimes contractors pressed for lower limits on the cutting but the stated limits were ordinarily observed where it was possible.[26]

Techniques of forestry practice had made their way quite well in the woods by the new century's first decade.[27] It was a much different world than before. Men refused to let their wood go to the first bidder. "It would be a waste to have it cut for a number of years yet." "Do not have it cut too small." "We want the timber scaled so that the State will receive pay for each 1,000 feet and the purchaser shall also receive 1,000 feet." These were typical comments used in transactions over one piece of land in the period 1900-1915.[28]

Food and living conditions also changed rather markedly in this time. The old days of beans, biscuits and the occasional piece of venison were gone. In the middle nineties camps were beginning to add sugar cake, fruit cake, molasses cookies, apple sauce, and even some vegetables to their menu. Conditions improved rapidly even beyond this point. The cook often became almost a caterer, purchasing the supplies with his employer's money but being paid primarily for his ability to serve good food. Cooks could and did command $50.00 to $60.00 or even occasionally $65.00 a month. At one camp in Greenwood in 1908 a reporter said that "a large beef creature is eaten every five days", and the bill of fare in another camp included such items as coffee, mincemeat, raisins, turnips, tripe, lemon pie filling, and cabbage. In a third a barrel of pickles was provided each month; ham, tomatoes, corn, salmon and pumpkin were also served. One old timer reminisced about clam chowder served in a camp he worked in about 1911. That must have been the height of luxury to a downeaster.

The Great Northern provided to its camps just before World War One both fresh pork and ham as regular items. The hams and pork came from a piggery which the company ran outside Millinocket. Visitors to the town on the railroad well remember its location.[29]

The woods farms which had been used to supply food for the men, and hay for the horses, continued to be used by the giant paper companies after they moved into an area. The Grant Farm in 1897 produced one hundred tons of hay, five hundred heads of cabbage, thirty-five bushels of beets, fifty of turnips, and seven hundred and fifty of potatoes. They had twelve acres down in oats. Pittston Farm, on the west shore of Moosehead Lake, produced five thousand bushels of potatoes and kept three hundred pigs which were allowed to run loose in the woods. By the end of the twenties hay and potatoes were the chief crops raised with 2,500 bushels of potatoes at the Grant Farm in 1927 and 4,500 bushels at the Pittston Farm. The Northern cut 225 tons of hay on its farms as well, 135 at the Grant Farm and ninety at Lily Bay. As the truck replaced the horse there was less reason to maintain the farms and they gradually dwindled into non-usage, but in their time they were a sight to see, great broad expansive areas cut from the woods, with modern machinery intermixed with stumps and wild animals. It was not unusual for bears, moose, and other animals to become quite tame in the vicinity of one of these farms.[30]

The woodsmen began to be treated more and more like human beings in other ways. Ministers of the gospel began to preach fairly regularly at camps in the woods. Organizations provided reading materials. Three Catholic priests toured the Kennebec camps regularly, and a Protestant, G. W. Berie, toured the camps each winter from 1895 to 1912.[31] At this time some camps treated holidays as festive occasions. Such a one was Bradeen and Edgerly's operation at Lily Bay in 1903. They served five kinds of meat, six kinds of vegetables, five varieties of pastry, four types of sauce, and four kinds of pie. The menu astonished one observer.[32]

> Jam on the lumberman's table! That's enough to make Father Shaw stir in his grave. The traditions of almost a century have been destroyed. Soon the reports from the primeval forest will tell of gout and bad stomachs due to high living, and possibly the woodsman will become mixed up in politics.

In that same season, during the holidays in camps of Lawrence, Newhall, and Page on the Dead River the men were treated to target shooting, wrestling matches, a revival meeting, and a series of prize fights culminating with the champion of the Dead River being named.[33]

The health of the woodsman was also better protected. In 1903 a smallpox scare caused the Secretary of the State Board of Health to issue an order causing all woodsmen to be vaccinated. The state law at the time demanded a vaccination within three years as a condition of hiring. Young, author of the notice, offered to supply men to do the work cheaply and threatened to enforce the law. The big operators complied immediately, and the state built a pest house at Jackman to isolate suspects.[34] By the beginning of World War One the Woodsman's YMCA at Greenville had installed a hospital of sorts where from six to ten men were treated each day for cuts, frozen toes, and similar complaints. This fifty room hotel was heated by steam, had showers, tubs, and featured bowling alleys, billiard tables, a reading room and a restaurant for the lumbermen out of the woods.[35]

The ultimate in treatment came from the Social Service Department of the Great Northern Paper Company. This department which was under the general control of Alfred Geer Hempstead, minister of the gospel and historian of the West Branch, provided films, libraries, lectures, and a monthly news magazine for the woods operators. These features were prevalent throughout the twenties, and it was not until the companies began to pay their men by the piece work methods that such treatment was discontinued.[36] This company also pioneered in making a film which was described in a publicity release as:

> ... a tribute to King Spruce, for King Spruce is democratic. He prefers service to sovereignty through his conversion of pulp and paper, and in our achievements from the president to the office boy....

The film, a twelve reeler, was entitled "Jack Spruce—or—Life in the Northern Woods" and it showed the entire process of the Great Northern Paper Company from trees to paper. For some time the film was much requested in the camps of the company itself.[37]

The Social Service Department was busy. For instance, from September 1, 1926 to June 30, 1927 it put on two hundred and sixty-four moving picture shows to 12,996 spectators. In the same period Hempstead gave illustrated lectures on religious themes, and the company provided several newspapers in addition to the regular library of fiction. During the twenties the company also held regular field days on the Fourth of July and other holidays. It was altogether a much different sort of life than one had seen seventy-five years before.[38]

Perhaps the most interesting changes which occurred in these times was in the development of mechanical methods of moving logs once they were cut. Big loads were still being hauled. Two teams hauled 205,000 feet in one week on the Androscoggin in 1899, with the average load being 3,410 feet. Working on a four turn road with teams averaging 2,700 pounds the men moved loads of 5,060, 4,920, and 4,860 feet during this time.[39] Still, this was not quick enough and men were experimenting with methods of eliminating the horse in the woods completely. Experiments were successful, and the log hauler and tractor replaced horses by 1930 for most purposes just as the horse had replaced the oxen by 1890.[40]

Experiments in improved transportation started as early as 1889 although they were apparently unsuccessful.[41] The first one which received any sort of a trial appeared in the fall of 1895. Ira Peavey, son of the inventor of the Peavey cant-dog, had been working since 1891 on a steam log-hauler. His machine was propelled by a toothed iron wheel driven by steam, and it would haul a four sled train. The main sled was in fact a flat-bottomed scow rounded at each end, with a small cabin built on it. Inside the cabin was a five horsepower boiler and a three horsepower engine used to drive the wheel which was actually a hollow cast iron drum five feet in diameter. The wheel was moved by chain drive and a sprocket gear which absorbed through a cam shaft the irregularities of the road. Steam flowed through a pipe into the drum so that snow would not clog the teeth. The entire affair was steered by a wire rope to a single load runner. The main sled was twenty feet long, six feet wide, and ten feet high, and it cost him $2,500 to build. The log-hauler weighed four tons, and when Peavey exhibited it he said that the scow could be used as a slow moving boat if necessary.[42] Peavey's log hauler was also unsuccessful. (I have never been able to find anyone who has ever heard of it, much less who might have seen it.)

It is Alvin O. Lombard to whom most authorities give the credit for the first workable log hauler. Lombard, who was from Waterville, invented the caterpillar or lag tread in 1899.[43] The first machine which was used for more than experimentation was on an operation of Lawrence, Newhall and Page in 1900. Lombard patented his haulers in the early summer of 1901 after the winter's work had convinced him that it was feasible. His first machine had a seventy-five horsepower boiler feeding a seventy-five horsepower engine, with the whole affair on a sled. It weighed fourteen tons and could haul as much as sixteen horses at a speed of three miles an hour. This machine was sent in to haul on Alder Stream Town (T2R5 WBKP). It ran about a week, then it broke down, and was not used again.[44] Lombard changed his mind about the feasibility and experimented during the rest of the winter with a log hauler similar to a trolley. The electricity was to be supplied by an overhead line run into the woods. An experimental tractor of this type failed that spring and was not used again.[45]

The next fall Lombard went back to making steam log haulers, and his firm manufactured about one hundred of them by 1917. As early as 1908 though Lombard was experimenting with gasoline driven log haulers. The steam log hauler by this time took an engineer, fireman, and driver to operate them. They were mostly about thirty-two feet long and weighed twenty tons. They moved at about five miles per hour and could haul one hundred and twenty-five tons (30,000 feet spruce.) The machines were expensive, $5,000 each F. O. B. Waterville in 1908, and took a good deal of maintenance. Before they operated well the roads had to be leveled and swamped out much better than for ordinary team log hauling.[46]

Eventually the big Lombards drove out even the Eagle Lake Tramway, but only in conjunction with the continued development of the iced road. With a sprinkler sled holding 13,000 gallons it was possible to keep the roads at glare ice. In that area Lombards (on operations of the Eastern Manufacturing Company on T9R14) hauled from seven to ten sleds that were loaded with a cable and pulleys. There were four of them and a dispatcher was used to control the traffic. A ten telephone line was set up and each time one of the big machines passed, word was sent to the main dispatcher. They operated from 1907 to 1913 in that location.[47]

The Great Northern purchased its first Lombards in 1918 and 1919, and they were used mostly for plowing and toting. The first hauling on a large scale did not occur until 1920-21 at Caucomgomoc. They moved out 10,000 cords that winter from very rough country. On the Cuxabexis operations near the Telos watershed the next year four Holts and twenty-one sleds moved 8,864 cords in thirty-three working days. The paper company which had had as many as 1,700 horses found that eight hundred were enough by 1922. As an observer said, "The tractor and truck are replacing them somewhat." While horses were idle they ate two quarts of oats, one quart of shorts, and thirty pounds of hay each day. The tractor ate nothing except while it was working. By April of 1926 the Northern had twenty-one log haulers, four 10-ton Lombards, ten 10-ton Holts, and various other odd kinds, including the first of their famous Twin tractors.[48]

The original of these was a tank type tractor made by Lombard. It had a duplex engine, with four cylinders on a side, two crankshafts, two clutches, and two transmissions. The tracks were independent of each other, and the affair was steered by speeding or slowing the engines.[49] The first one worked rather well, but it was discarded in 1927 and a new one manufactured to the specifications of the company. It had two motors hung on special mounts, each one developing sixty-three horsepower at eight hundred revolutions per minute. The brakes, transmissions, rear ends, and gearing were all hung separately as well. The steering wheel controlled the throttle of both engines and by turning it one changed the revolutions and thus steered the machine in the direction desired. This machine would do eight miles an hour in high gear, so the company claimed.[50] It mostly was used for big toting jobs as the necessary highways were just too difficult to maintain to keep it in ordinary use.[51] The paper company had however built 150 miles of good highway in the woods for such usage by 1925 as they were in the woods to stay.[52]

The fact that they were to stay allowed them to plan their operations for a long period of time, and thus they were always able to expand large amounts of money to improve transportation. Two such developments at the end of this period were the improvements at Grindstone in 1924 and 1925,[53] and the great operation at Cooper Brook in the late twenties. The Grindstone operation was simply an effort by the use of haulers, roads, trestles, and electric loaders to avoid driving and hand piling the large amounts of logs needed. With the use of trucks they were able to haul in all weather and without snow if necessary.

Cooper Brook was a difficult affair, and in some ways it was one of the more interesting operations ever undertaken in the Maine woods. Here the paper company through mechanical ingenuity changed the normal course of rivers by simply ignoring them. Cooper Brook feeds the West Branch Penobscot and eventually flows into the Lower Jo Merry Lake. The lumber which came down it in this operation though came from the East Branch Pleasant River, above the Gulf, across the watershed, and ordinarily would have been driven down the Piscataquis coming out at Howland below Millinocket.

In the winter of 1926 and 1927 the first road was built for log haulers. It had a natural grade of seven and eight per cent, though, and Lombards did not operate well at more than two percent grade. The next summer was spent in reducing the grade with a steam shovel and twenty horses. The men made a cut at the height of land, twenty-seven feet deep, which was equal to 330,000 cubic yards of gravel. A trestle 1,250 feet long and twenty-five feet high was then built to

Maine woods in the twentieth century

the cut. The whole thing was completed just in time for the 1927-8 season. Seven camps with three hundred men went in to cut pulp. The depot camp alone had eighty men, and plans called for a production of 19,000 cords the first winter. The crews consumed about two and one half tons of fresh beef a week, and fifty bushels of potatoes every eight days. One hundred and fifty tons of hay and nine hundred tons of coal were toted in from Greenville for use by horse and machine. The first year's work was actually much more than was expected as the log haulers moved the sleds (an average of seven with each load) the nearly thirteen miles over the trestle to the road and finally to Cooper Brook. The largest load of the winter had seventeen sleds and it held one hundred and eighty-five cords of spruce. In the one hundred working days for the log haulers and their crews, they moved 29,494 cords of wood to Jo Merry Lake ready to go to the pulp grinders at Millinocket. The twentieth century had truly arrived in the Maine woods.[54]

Log haulers were not the only mechanical marvels in the woods. Of equal and perhaps more interest are two other attempts to move logs mechanically from difficult locations—the Bradstreet Conveyer and the Eagle Lake Tramway. The Bradstreet Conveyer was an attempt to tap Penobscot trees for Kennebec waters, and it was reasonably successful. Four thousand, seven hundred feet of iron cable, with dogs every eight feet, carried logs across Northwest Carry at Moosehead Lake. Four engines and four boilers ran the Conveyer, and the device carried close to a million feet every twenty-four hours. An electric light plant was built to provide light throughout the night during the driving season, and two dams were built to ensure a water supply for the boilers. A house was erected for the headquarters crew right at the conveyer itself. This conveyer, which was really just a mechanical sluice, was operated for several seasons.[55] When the Great Northern Paper Company moved into this area it bought and dismantled it.

Other attempts to ease the work were also tried. In the season of 1902-3, for instance, N. N. Jones, operating at Wassataquoik Mountain, used a cable and a donkey engine to yard his lumber out on the steep hills.[56]

The Eagle Lake Tramway was the most ambitious transportation innovation. Marsh and Ayer (of the famous suit against the Great Northern) obtained control of the Eagle Lake Country (T8-R13) in 1901 and 1902. The old dams in the area had gone into disrepair, and that meant that if the logs cut here were moved at all they would have to go to St. John. Marsh and Ayer moved in and replaced some dams on Wassataquoik Stream and above Sourdnahunk. The firm also leased the Chamberlain Farm for a headquarters, but even so more than one thousand tons of supplies had to be toted in from Patten in the first year. The machinery, consisting of two boilers each weighing eight tons and a wire rope one and one half miles long weighing thirteen tons, went to Greenville from Bangor on the train in the fall of 1901. It was ferried across the big lake to Northwest Carry and then began the long trek (forty-two miles) to Eagle Lake. It was a terribly difficult job, and it looked finally as though the cable could not be moved. Eventually a man from Bangor agreed that he could splice the cable if it were cut and the transport of it was eased considerably. On March 1, 1902, the foundations for the new tramway were started after a new towboat was built on Chamberlain Lake to tow the logs once the tramway had moved them.

The cable, an inch and one half in diameter, ran about 3,000 feet with the logs and returned. A steel saddle with dogs was clamped on to the cable every ten feet to hold the logs while they moved. Logs moved at about two hundred and fifty feet per minute. On the first run the bolts on the dogs and clamps had not been threaded deeply enough, and they all (4,800) stripped their threads and had to be regrooved. Later during that run, Herbert Marsh who was supervising the loading of the logs became insistent that the men put a log in each clamp. The weight was too much and the cable kept stopping. O. A. Harkness the ostensible straw boss came from the Chamberlain Lake side to see what was causing the delay. Marsh asked him as he approached if there was anything he could do. According to the testimony of people who saw the exchange, Harkness answered, "Yes, go back to Bangor and let us handle the damn logs as they ought to be handled." Marsh in any event went back to Bangor and the logs began to move. For six years the tramway moved 500,000 feet every working day, about fifteen or sixteen million feet a year. Harkness later estimated that 100,000,000 feet went over the tramway and down to Bangor mills, wood which would have otherwise gone to Fredricton by the St. John River. Yankee ingenuity was able here to extend the lumbering era somewhat longer even after the press of economics had forced the coming of the pulp paper revolution elsewhere.[57]

In more settled sections of the state this revolution was forestalled by the use of the railroad, and it is to that story that we now turn.

The railroads have a double impact in the twentieth century. The first is simply the extension of roads into the woods to aid in the harvesting, and the second is the moving of more and more forest products away from their place of manufacture by rail. This second influence is simply the continuation and extension of the tendencies noticed above in the chapter on transportation. Elaboration of a few statistics will serve to illustrate this point. The first impact though demands a bit more analysis. There are two railroads of some importance to the Maine economy in this category, the first the Rumford Falls and Rangeley Lakes Railroad, and the second the Chesuncook and Chamberlain Lake Railroad with its interconnection, the Eagle Lake and West Branch Railroad. Neither lasted very long but both were important in their time.

The first road, the Rumford Falls and Rangeley Lake, came about apparently because of trials made by the Berlin Mills Company on the amount of pulpwood in a cord of wood. Their experiments proved that wood peeled in the woods in the summer made about four hundred and fifty pounds more pulp per cord, which was about $2.20 per cord in terms of the 1895 prices. They also decided that this wood could not be driven without some loss and as the manager of the firm said, "They must be railed. . . .time will come when all mills on the Androscoggin will be supplied with their stock by rail." As a result of these findings a railroad was built from Mooselookmeguntic Lake in the Rangeleys to connect with Rumford Falls, and then to Berlin, New Hampshire.[58] What an impact the road had. Within a year prominent lumbermen were predicting the end of driving. In 1896 the Androscoggin Water Power Company moved all of its cut by railroad car, and an Oxford County observer said:[59]

> This is only an indicator, but it shows which way the wind is blowing. The days of the drive in Maine are numbered and as the railroads push farther into the woods, fewer and fewer logs will go out into the stream. The modern rush demands that no time be lost in getting the trees from the woods to the mills and the only sure way to do this is to car the logs. In five years I do not think there will be a drive of any consequence on the rivers of Maine.

When the shipping of lumber was started over the road it was at first slow with only thirty-five to forty thousand feet moving each day. A month later though the men were shipping eighty thousand feet a day and were only limited by a shortage of cars. During the week of Washington's Birthday in 1896, 1,300,000 feet went over the road, and on March 5, 1896, one train of fifty-three cars carried 175,000 feet. Nearly twenty-five million feet came out the first winter.

During the year 1900, 22,500,000 feet were carried by the road, and as late as 1910, 13,539,000 feet moved to market over the road.[60] Certainly this short railroad had revolutionized lumbering in the northwestern part of Maine.

In northern Maine, in the Great Northern's woods, another railroad was built at a later time, and it was also thought to be quite revolutionary. This railroad started at Eagle Lake near the old tramway ran across T8R13, through 7 and 6R13 to the east shore of Umbazooksus Lake and was designed to tap the Chamberlain Lake area, which was still as in the time of the tramway or the earlier Telos Canal, in the St. John water shed. In June of 1925 Fred Gilbert supervised and a crew surveyed the road, and by February, 1926, the sixteen miles was ready to be grubbed out. The northern part, eleven miles, which was known as the Eagle Lake and West Branch, was started in April, 1926, and finished by August, 1927. Edward La Croix, sometime contractor for the paper company and part owner of the Madawaska Land Company, built this section. He sent in men for three camps and they toted all their supplies from Lac Frontiere, Quebec. The most difficult section was the trestle over Allagash Brook in T8R13. This trestle was 1,800 feet long and ten feet above the water when it was completed. C. M. Hilton, brother of a later vice-president of the paper company, was the engineer on this road. The Great Northern end, the Chesuncook and Chamberlain Lake, was completed at about the same time and the two were connected. When the routine of operation was established three trains of twelve cars each moved across the road each day averaging about 6,500 cords of pulpwood each week. Two 225 foot conveyers were constructed to raise the wood and load the cars. It was possible to raise a cord every ninety seconds with this device. As the cars were tilted, and so was the trestle, dumping them was a fairly simple matter using the hinged sides developed by the Maine Central Railroad to be used in delivering pulpwood to downriver mills. It was necessary however to develop a sort of dredge to clear the bark away from the unloading trestle which extended six hundred feet into Umbazookus Lake. The builders thought that the road would last until 1947, or twenty years, and although it didn't, the engines are still something of a wonder to those who penetrate that far into the north country.[61]

In addition to this sort of experimentation the railroads also wanted very badly to supplant the rivers for the movement of logs to downriver mills. After the victories of the paper companies in the first decade of the century they were sure that the inevitable result would be for this trade to fall into their hands. When this didn't occur quite as rapidly as they thought it should, fairly close investigations were made of the costs of transporting logs both by rail and water. The results of the investigation showed that water transport was still cheaper but not by much, and wherever there were any sort of obstructions, probably the railroads would win out on the basis of the guaranteed transport. The average cost per ton-mile by river was .0082 cents, and by railroad .0125 cents. However on long distances the differences were more marked, .0034 cents for the river drive to .0125 cents for the railroads.[62]

After the time of this investigation, which dealt with the competition only up to 1910, the rivers were more and more given over to the companies for waterpower generation. Long log sawing was done for. Pulpwood in four foot sticks, from tiltable cars, could be moved more cheaply by railroad than on the rivers. Wood by rail could move all year round; it came and went when the mills wanted it, it took less labor, and by and large it cost less. Today it is usual to see railroad cars filled with pulpwood and most unusual, except on the Kennebec, to see pulpwood in the rivers.

Just as had been predicted earlier the railroads also completely supplanted vessel traffic in this time. Starting with a rate decrease on the Bangor and Aroostook in 1899 (about 40%), the business began to boom.[63] The table shows the total amounts carried in the eight month period July 1—March 1 for the next six years.

Bangor and Aroostook Tonnage, July 1 - March 1

1899-1904

Year	Amount (in pounds)
1899	512,477,421
1900	687,802,957
1901	957,642,450
1902	1,090,643,088
1903	1,032,343,544
1904	1,352,068,312

These figures are for the total of goods carried in that time period. In 1904 when the report was issued, the results for the fiscal year 1904 were 1,046,693,232 pounds of forest products, 452,510,523 pounds of agricultural products, and 531,332,733 pounds of other products for a total of 2,030, 336, 488 pounds moved over the road. It was no wonder. The road said that in Aroostook County alone mills produced each day on an average 1,207,000 feet of long lumber, 1,275,000 shingles, and 621,000 laths. The big bonanza predicted for the northern railroad had actually occurred. The next table breaks down traffic in 1903 and 1904 and drives the point of success home even more securely.[64]

Bangor and Aroostook Railroad Shipments
July 1—March 1, 1903-1904**

Item	1903	1904
Bark	11,724,200	20,600,580
Cordwood—Slabs	76,814,649	69,122,915
Excelsior	7,934,427	9,450,100
Kindling	1,324,500	2,962,800
Last Blocks	5,022,545	1,927,900
Laths	18,141,700	11,607,205
Logs, poles, posts	60,356,967	62,979,195
Long Lumber	151,462,759	155,432,093
Paper	104,093,122	95,413,172
Pulpwood	69,748,416	124,782,200
Shingles	22,182,480	22,127,103
Ships Knees	54,000	73,480
Ship Timbers	2,694,170	4,533,120
Shooks	10,036,035	14,605,471
Spool Stuff	24,868,470	44,539,180
Ties	34,649,530	16,294,110
Wood Pulp	9,511,350	35,166,990

**all figures in pounds

Nearly as revealing is the following table, abstracted from a series of *Annual Reports* of the Maine Central Railroad. They are the figures of the tonnage carried of wood pulp, pulpwood, and paper, in the years 1900-1915.

Truly the railroad had made its way. The promise of the short route to Europe may never have materialized, nor did Portland or Bangor ever really tap the west and St. John remained a back water port, but the railroads did do something their promoters had not envisioned. They completely dominated the transport of the leading product of the northeastern spruce region—wood and wood products.

As far as driving and rafting were concerned relatively few changes occurred until World War One and after, with the exception of the marked diminution in the number and amount of logs moving down the rivers. Certain minor technical improvements were made— The Katahdin Pulp and Paper Company strung a telephone up the

Maine Central Shipments

Wood Pulp, Pulpwood, and Paper

1900-1915**

Year	Pulpwood	Woodpulp	Paper
1900		220,566	141,881
1901		254,428	186,678
1902		305,129	247,380
1903		255,010	258,489
1904		271,085	267,902
1905		267,916	317,144
1906		271,010	324,823
1907	357,919	330,571	395,905
1908	705,601	395,029	494,695
1909	631,418	315,855	499,487
1910	699,409	349,281	573,523
1911	629,463	365,988	500,368
1912	737,503	410,953	477,447
1913	819,177	398,861	530,065
1914	855,053	354,656	496,578
1915	781,347	428,639	477,672

**All figures in tons

Wassataquoik to aid the men, and others were put in on the East Branch proper.[65] There was a good deal of experimentation with the poplar being used for pulp. Some was driven with the bark still on. Others drove the sticks in long logs and still others attempted to drive their logs in the first rush of water in the hopes of moving them without loss.[66]

By and large though, poplar had to be peeled before it could be driven and this meant, for ease of handling, moving it in four foot sticks.[67] Otherwise things were much the same at least on the surface. For a time excursions still went to North Twin Dam to see the sluicing—in 1896 a special train with eighteen cars filled with nine hundred people went to watch. The West Branch drive fed them baked beans, corned beef, hot biscuit, molasses cookies, doughnuts and tea at the wangan.[68] River drivers from Maine still went elsewhere to demonstrate the skills inherent in being a Bangor Tiger—as late as 1900 two hundred of them were on the Connecticut and some went out even later.[69] The men also acted about the same when they came out of the woods, but then conditions had not changed that radically for them.[70] River driving was still rugged, difficult, and dangerous work.

As the West Branch Penobscot came under the control of the paper companies the sawmill men had moved to the East Branch, and this led them to form a driving company to protect themselves, the East Branch Log Driving Company. The drives were ordinarily auctioned off to the lowest bidder. Three drives were often held, the first on the spring waste water of those logs which were able to be gotten into the river, the second with water from Grand Lake Dam, and the third drive on the end of this water and whatever rain had fallen. The table indicates the prices for the years in which they are known.[71]

East Branch Log Driving Company

Prices—1898-1904—Cents Per M

Year	1st Drive	2nd Drive	3rd Drive	Assessments	Contractor
1898	47				Con Murphy
1899	61	80			Con Murphy
1900	65	79½	72	3-5¢ (location differed)	Gardner and Kellogg
1901	72	68	75		Murphy, Fleming and McNulty
1902	85				Oscar Thomas
1903	73	81½	77¾		Con Murphy
1904	86½				Fred Ayer

Although we have discussed the end of the Penobscot boom work with the coming of the pulpwood, it is necessary to look at the operations of the boom, and its parent company the Penobscot Lumbering Association in the interim years while logs were being sawed. The boom was still a big business; in 1895 it did a business of $77,558.41 and had assets of over $10,000 to start the new season. That year was a successful year rafting with some 63,787,280 feet of logs moving through the Bangor boom alone to the tidewater mills. The table gives some indication of the sawing done in the tidewater mills on the Penobscot at this late date.[72] In addition another 12,000,000 feet was shingled up for spring sawing. The 75,000,000 total compared favorably with the sixty million feet sawed in 1895.

In the fall of 1896 booms filled with pulp at Lincoln, Enfield, and Howland were torn away, and the 35,000,000 feet of pulp that came down so filled thePenobscot Boom that no rafting could be done.

Penobscot Tidewater Sawmill Production

1896

Firm	Production (feet)
C. G. Sterns & Co.	11,128,750
Hodgkins & Hall	12,665,960
F. W. Ayer	20,121,350
Horse & Co.	6,158,920
Hastings & Strickland	7,857,420
D. Sargent's Sons	5,854,880
Total	63,787,280

The logs were hauled out and shipped to Kennebec paper makers in order to allow the logs to be hauled. This event cast a premonitory shadow for many. The boom had handled 22,000 sticks in one day of the flood and only made 5,000,000 feet.[73] The days of the boom were in fact coming to a close.

It hung on well though. The Bangor Boom handled sixty-one million in 1899, and there were so many prize logs that a rebate of 3¢ M was granted to all who had logs rafted.[74] By this time it was costing 28¢ M just for the surveying alone, 1¢ to the Surveyor General, 10¢ for the deputy surveyor, and 17¢ for the four overhaulers, so the rebate was undoubtedly welcome.[75] The last really big year for the boom was 1904. Eighty-five men were employed, fifty-five on the boom sorting, and thirty to take down the steam mill drives. In that year the men handled nearly eighty-five million feet, moving 3,500,000 feet through the boom each week.[76] By this time though the cost of driving and rafting had risen to the point where weaker forces were beginning to think in terms of closing their mills in face of the lateness of the drives and the difficulty of moving the logs under extremely adverse conditions. The following table illustrates the rising costs.[77]

As the improvements on the Penobscot are discussed in detail in Hempstead[78] it serves little purpose to recapitulate them here except to fill in details which he missed. There was almost constant work going forward on the river. On the West Branch from Chesuncook Lake to Medway $38,817.09 was expended for dams, $24,747.29 for steamers, $2,631.98 for piers and booms and $25.00 for blasting from 1898 to 1907. This was a total of $66,221.36 for relatively minor improvements in a decade's time. From 1898 to 1907 the company expended $23,325 for improvements from Medway to the Penobscot Boom limits at Argyle. During these same years

Penobscot Lumbering Association—Rafting Prices

Penobscot Boom

Year	Price (in cents per M)	Contractor	Remarks
1896	37 7/8	C. M. White	
1897	34 7/8	C. M. White	F. A. Gilbert
1898	37 5/8	C. M. White	spirited bidder
1899	37 15/16	C. M. White	
1900	50 15/16	C. M. White	
1901	37 7/8	C. M. White	
1902	50 13/16	C. M. White	
1903	55 15/16	C. M. White	
1904	56 15/16	C. M. White	Spirited bidding

$9,579.37 was expended on the East Branch from Grand Lake Dam to Medway.[79]

These improvements were not enough for the controlling firm, the Great Northern Paper Company, and beginning in 1912 a new series of dams was built and rebuilt culminating in the construction of the massive Ripogenus Dam, fourth largest in the United States at the time of its construction. The original rebuilding was of the dam at Seboomook Falls in 1912 and 1913, with its eight gates, three spillways, and two sluices. This dam was 808 feet long with a head of twenty-eight feet.[80]

Ripogenus was a much more difficult matter. To begin with it was almost impossible to get to so in 1914 a road had to be built from Lily Bay on Moosehead Lake to the dam site. The cement for the dam (13,000 tons) was ferried to Lily Bay by scow, thence by road to the Grant Farm, and finally to a location near the old Chesuncook dam. The road carriage was accomplished in seven and one half ton trucks. These trucks had little power, were fed with gravity feed fuel lines and thus they had to be snubbed over Sias Hill (the height of land between Kennebec and Penobscot waters near the Grant Farm, a hill which no longer exists). The snub line was an 1,100 foot cable. The dam when it was finally constructed was 860 feet long, ninety-two feet high, and sixty-four feet thick at the bottom. On top of the dam was built a sixteen foot roadway.

Before the dam was started eleven acres of land was cleared and a saw mill, four big cranes, and a camp for the men, lighted by electricity, were set up. Rock and gravel from the area was brought in endless chain conveyers to a rock crusher. The mixing and measuring of the cement ingredients was practically automatic, and when the mixing was finished two large buckets operating on cables dumped the cement into place. A brook nearby was diverted to wash the gravel used in the cement. In addition to the dam the ledges nearby were all bored and filled with grout and cement to make the entire area absolutely solid. From the crest of the dam to the lower part of the gorge was a drop of 261 feet. The lake created by the dam was forty-two miles square, and it held about 21,500,000,000 cubic feet of water. This was one of the great engineering feats of its day.[81]

With the construction of the big dam little remained except to move the logs at the will of the dam owner, the paper company. In 1922 the company ran twenty-two short brook drivers to make its total, although this seems to have been a rather large amount.[82] Improvements on the river continued but most of them consisted simply of rebuilding earlier dams such as Seboomook Falls in 1926-7, and Canada Falls in 1922 and later in 1927.[83]

Even a company as large as the Great Northern though could not always control the weather. In the fall of 1927 strong freshets broke up the booms upriver on the Kennebec and 12,500 cords of pulp went down stream. Hollingsworth and Whitney and S. D. Warren responded to a plea for help by stringing their booms across the river to contain the logs, but the flood was too strong and they broke as well. New booms were stretched across the river between Hallowell and Gardiner to stop the lost logs, and it was December 10 before the drift drive cleared up. Logs were scattered all over the river with 4,000 cords at Skowhegan, 5,000 at Waterville, 1,300 at Five Mile Island (The Augusta Boom); 1,600 at Brown Island (The Gardiner-Hallowell Boom); 300 at Richmond and Bath, and 300 more lost at sea.[84] By and large though, the paper companies began to control the rivers through their great dam constructions during this period.

One other upriver innovation deserves some notice. That was the growth of the steamboat towing which had been pioneered by the *John Ross*. After the paper companies gained control of the rivers, new boats were built. The first of these boats of importance was not built by them but by Marsh and Ayer to tow from the Eagle Lake Tramway. This boat, the *H. W. Marsh,* ran from 1903 to 1909 when its bow was crushed in the ice. Another boat used in this area was the *George A. Dugan* built in 1902-3 at Chamberlain Farm. This boat, seventy-one feet long with a twenty foot beam and six and one half foot draft and operated by two single engines from two vertical boilers, was used to tow logs on Chamberlain Lake until 1913.[85]

On Chesuncook Lake, which was still the main towing thoroughfare, the *John Ross* was replaced in 1902 by the *A. B. Smith*. The new vessel used the same engine and boilers as had the *John Ross*. It was scow-rigged with a false prow, and the engine was too heavy for the boat so that it hung by chains from stanchions in order to move the logs in the booms. It ran through the 1926 season when it was replaced with the *West Branch* #2. The latter vessel was built in a shipyard at Chesuncook Dam, utilizing among other things a steam box to bend the oak used in the frames. The keel was laid in May, 1926, and the boat then built the old-fashioned way from the shoe through the keelsons and up the planking. The planking was three inches thick, with the strakes five inches thick and edge-belted every three feet. The vessel was fitted with an engine house, dining room, galley, and staterooms for crew and guests. The forecastle berthed six men. The vessel was propelled by a 360 horsepower diesel engine, which came to Greenville by train, then to Lily Bay by scow, and finally to the dam by trailer. It had five one-thousand gallon tanks for fuel, enough for ten or twelve days hauling, even for round-the-clock work. Another small diesel was abroad to generate electricity for lights, searchlights, and an electric windless to aid in making the work smoother.[86] The new craft was an instant success as the following comparison will readily show.[87]

Towing Comparisons

A. B. Smith (1926) and West Branch #2 (1927)

	A. B. Smith (1926)	West Branch #2 (1927)
Started Towing	May 19	May 21
Finished	September 4	July 27
Time Spent	3 months, 15 days	2 months, 6 days
Crew	10 men	7 men
Time Taking Fuel	100 hours	13½ hours
Av. Fuel Used Daily	10 tons coal	301 gal. crude oil
Fuel Cost Per Day	$250	$50
No. of Booms Towed	62	51

As far as work on the lakes is concerned one other development in which the paper company pioneered is worth mention. That is the so-called "boom jumper". These were dories or motor boats with three keels and skegs to protect the propeller, bottom, and wheel from floating logs and for jumping over boom sticks. The first of these was built in 1912, and by 1928 thirty-two of them were in operation. They rapidly replaced the older batteaux for work around the logs.[88]

Downriver the story was as one might expect, whether it was sawmills or pulp mills. Gradually the big upriver mills simply forced them out of business, and the last years for all except the giant mills were a constant struggle to stay alive. As early as 1895 the steam sawmills on the Penobscot shut down to try and force prices higher. Some mills were down for over a month before they began to saw again. This was apparently the work of another organization, the Northeastern Lumberman's Association, formed in Boston to keep a floor under prices. They were supposed to cooperate on sales and distribution but nothing came of it.[89]

The first year the Great Northern bought logs, 1899, was a good price year, but this artificial boom did not last long and in the next year prices sagged off badly. A small panic occurred on Exchange Street in Bangor which was followed by a meeting of sawmill owners and operators attempting to keep prices up. A new organization, the Eastern Lumber Manufacturer's Associates, was formed for this purpose. There should have been no amazement over the sluggishness of prices though as the amount of lumber available for sale, counting the new drive, was very high.[90] The table indicates the amounts.

Lumber To Be Sold—1900—In Northern Maine

River Basin	Old Logs	Lumber	New Drive
		(million board feet)	
Kennebec	5	1	40
St. John	6½	4	86
Penobscot	19½	4	55
Ashland (by water)			15
Totals	31	9	196

These figures only included the spruce, no pine, hemlock or pulp was listed, and furthermore this was only the lumber which would have been shipped by water transport. Prices continued to decline and the protective associations were of little value in retarding this downward drift.[91]

The next year the companies began to face demands from the local unions for a cut in hours. The new union backed by the American Federation of Labor issued an ultimatum after their demands were ignored. In response to this the noon hour was lengthened, but by this time union recognition was at stake. On Labor Day a giant parade was held to demonstrate solidarity with twenty-three separate union organizations passing the reviewing stand. The companies then cut the length of the working day but would not recognize the union. Most men on the river went out on sympathy strike, and although one Bangor mill worked with a scab crew of schoolboys, the majority of labor in the Bangor area was out on strike. They finally were granted all their time demands but at the price of no union recognition. Even this was considerable from people in as much difficulty as downriver sawmills at this time in their history.[92] If sawmilling was to survive it would do so upriver.[93]

Much the same story is true for the smaller pulp mills. Occasionally there were reports of new pulp mills to be built, both upriver and down, but with the exception of the construction of the new plant at East Millinocket by the Great Northern Paper Company, few if any of them came to anything.[94] The older companies remained, and some of them began to provide quite decent working conditions.[95] Those that did remain either grew larger or declined precipitously. The Northern for instance grew. A new storehouse was built at Stockton Springs, housing paper to be sold and to be used to dispatch it either by railroad or by vessel.[96] The paper company did get most of its coal by vessel, and it ran a fleet of vessels until after World War One.[97] The Northern had by this time expanded into the inland transportation business as well, purchasing the Georges Valley Railroad in 1918. This road, which ran eight miles from Union to Warren, was built in 1894. The paper company wanted it for the lime taken from a quarry owned by the road.[98] And, as the period came to an end the paper company was constructing a new hydroelectric plant in Anson to be used to supply the Madison mill with power.[99]

Smaller mills like the Forest Fiber Company gradually gave up the ghost after remodeling their boiler rooms in an effort to get more power and eventually shifting to electricity.[100] It was all in vain; by 1923 the mill was so uneconomical that it was discontinued.

The main S. D. Warren Company mills survived, just as they had in the difficult times of an earlier era. Rag fibers went by the board, and until 1906 the company purchased their sulphite pulp. In that year they discussed the problems entailed in shifting to sulphite in their own mills, but the fact that they "were not too well placed to obtain low-cost spruce" dissuaded them. In 1917 the shortage created by the war caused them to build a sulphite plant, and when

Maine woods in the twentieth century

prices rose the new plant paid for itself easily. The firm ran it until 1929.[101]

In 1910 the company was studied thoroughly by efficiency engineers, and fairly detailed accounts and records have been kept since then. The last of the bonus payments, by this time confined to such places as the coating rooms, were paid in 1909.[102] The mill was a money-maker even though costs of production continued to rise.[103] Wages rose from just under $2.00 a day in 1913 to as high as $6.50 by 1920. Women were getting $4.18 a day by this time.[104] Production increased fairly steadily from 3,200 tons of paper in February, 1909 to 5,500 tons by December, 1916.[105] The war of course was a great boon. During that period mill production went to 200 tons of paper a day or 6,000 tons per month.[106] Costs continued to increase during this time but the mill still made money. The following table will indicate how the transition period looked in terms of costs.[107]

Raw Material Costs Per Ton of Paper

S. D. Warren Company, 1909-1920

Year	Raw Material Costs	Labor	Total Cost	Price Received
1909	37.18	16.05	75.40	81.90
1910	38.15	17.45	76.40	81.60
1911	36.93	17.70	74.48	80.20
1912	37.46	17.09	74.11	79.80
1913	38.36	16.79	76.33	79.80
1914	37.49	16.10	74.23	77.20
1915	36.72	16.38	74.76	
1916	41.90	17.26	80.11	
1917	45.54	23.99	101.12	
1918	52.76	31.60	118.85	
1919	59.82	37.49	132.16	
1920	66.53	42.74	149.26	

Profitability of course was the crucial point and, in fact, has been the central point throughout this work. Survival depended on the extent of profits. But there was no aid for the unprogressive or the slow to learn. Transition periods were short and inevitably quite cruel. Those who were intelligent or lucky enough to catch these tides at their flood were very well off indeed; such were the S. D. Warren Company, and the Great Northern Paper Company. Those who did not, as the Dennison firms in the 1880's or nearly all sawmills in any era, were less fortunate and their passing was rapid and mostly unmourned.

Throughout the period however, even though it sometimes appears to be only a rather desolate account of the advance of the pulp grinder, it is really an account of the tremendous resurgent ability of the forest. Unlike Wisconsin, Minnesota, New York, and Pennsylvania, Maine had solved the great problem of the North American continent. By trial and error, with the aid of good men such as Austin Cary and George Talbot, and through the sufferance of such men as Fred Gilbert, the Maine woods had indeed survived. Common sense, self-interest, love of nature, the educational process, or as some might have it just plain blind Yankee luck, whatever the reason, Maine people did have the last word. They had their woods and they cut them too. Modern man is thankful.

NOTES

[1] Bangor *Daily News*, November 18, 1901; *The Industrial Journal*, April, 1902; Bangor *Daily Commercial*, February 16, 1904.

[2] Henry Bake Steer, *Lumber Production in the United States, 1799-1946*, USDA Miscellaneous Publication #669, Washington, 1948, 11. It is my impression that these figures will run a bit high, perhaps about 5%. Comments indicate the extent of the decline. In 1907 logging conditions were "unusually unfavorable," and by 1914 with the smallest cut in 30 years, the market was described as "demoralized". *Paper Trade Journal*, January 10, 1907; October 29, 1914 and throughout the period. The cut in 1968 was just over 2 million feet of all kinds. *Forestry Facts*, June, 1970.

[3] Forest Commissioner, Second *Report*, 1892, 151; *Paper Trade Journal*, November 14, 1896 (on land purchasing); *The Industrial Journal*, May 1912.

[4] Bangor *Daily Commercial*, November 15, 1904; Hollingsworth and Whitney purchases.

[5] Bangor *Daily News*, January 31, 1901; November 18, 1901; Kennebec *Journal*, January 3, 1903 (the Narraguagus); Bangor *Daily Commercial*, November 28, 1903 (the St. Croix); *Todds of the St. Croix Valley*, 23, (on the pulpwood commission work); interview Harold Noble, Topsfield, Maine, July, 1962.

[6] Bangor *Daily Commercial*, April 2, October 3, 1896; January 31, 1898.

[7] *Ibid.*, March 31, 1896; camps at 12R10, 12-13R11, 13-14R12, 13-14R13.

[8] Bangor *Daily Commercial*, March 28, 1899; January 14, November 15, 1904 (the last quoting the St. John *Sun*); Bangor *Daily News*, November 8, 1901.

[9] A. M. Thurrett to Miller, August 20, 1913 for the quote, and for a similar statement, F. S. Murchie to Miller, August 2, 1913, quoted in Appendix, to Miller, *op. cit.*, *New Brunswick Reconnaissance*.

[10] *New Brunswick Yearbook, 1907*, 81 (on the deal trade); *Report* of the

Royal Commission in Respect to Lumber Industry, 1927, Fredericton, N. B., 1927, 6-15.

[11] See on this "Forest Heritage", *loc. cit.*, a history of the Fraser Companies, Ltd.

[12] Bangor *Daily Commercial*, February 18, April 25, 1895; April 15, 16, 1897; December 3, 1901; May 7, 1902; January 22, 1903; January 22, 1904; February 16, September 16, 1904; Bangor *Daily News*, March 9, 1901; March 11, 1903.

[13] Bangor *Daily News*, March 9, 1901; Bangor *Daily Commercial*, May 7, 1902; February 16, 1904.

[14] *Ibid.*, April 1, 8, 1895. The pine was on Wade Plantation, T13R4.

[15] For newspaper accounts of the Great Northern operations, see *The Industrial Journal*, November 9, 1900; April, 1907; April, 1908; May, 1909. (In this year they had 2,700 men working on the Penobscot, 1,300 on the Kennebec, and used 1,200 horses. They expected to cut 110,000,000 feet plus purchasing another 20,000,000). Also April, 1918, (they made no summer cut because of the impact of the war on their woods help.)

[16] From the Great Northern Paper Company files. These are abstracts of their depreciation lists filed for the International Revenue Service.

[17] For a look at two lumber towns, see Bangor *Daily Commercial*, July 15, 1899 (an article on Exchange Street, "Lumber Row", in Bangor), and April 12, 1902 (a long story on Patten and its place as a center for toting of supplies.)

[18] An advertisement appeared in Bangor *Daily Commercial*, July 7, 1898.

[19] Among those going in this first group were such famous loggers as "Roaring" Jack McCarthy, and "Bummer" Morgan. See *ibid.*, July 12, 1898.

[20] Quoted in *The Industrial Journal*, May, 1917, in editorial.

[21] This story is best told in Vernon Jensen, *Lumber and Labor*, New York, 1945. Interview, Louis Freedman, July, 1962 (Old Town, Maine), who served with Greeley's Engineers in France.

[22] The earlier moves to the woods created comment; see *The Industrial Journal*, November 9, 1900; Bangor *Daily Commercial*, July 3, 11, 1902.

[23] *The Industrial Journal*, January 4, 1895; Bangor *Daily Commercial*, January 11, 14, 1896, interviews and comments from Barney Kelley, a notorious saloon keeper and "labor agent" of this period; as one verse of the "Boys of the Island" was sometimes sung:

> Brade Kelley will poison a man with bad whiskey,
> For pasttime they will banish their lager and ale;
> Then on the corner when he does get frisky,
> They will call for Tim Carey to take him to jail.

In *Minstrelsy of Maine*, 119, n. 2, F. H. Eckstorm apparently mistakes him for someone else; also see Bangor *Daily Commercial*, December 1, 1899; July 17, 1903 (on the Great Northern Paper Company employment agent in Boston); Portland *Eastern Argus*, January 20, 1903, "Life in the Lumber Camps", Bangor *Daily News*, March 23, 1903, a story to the effect that shanghaiing doesn't really occur very often; on the fight among the religions see a long article in *The Madison Bulletin* (Madison, Maine), December 27, 1900, quoted in *Northeast Folklore*, Vol. 2, no. 1 (Spring, 1959), 17-8, (the battle occurred at the Forks on Christmas Eve). One of Larry Gorman's songs, "The Hoboes of

Maine", *Minstrelsy of Maine*, 140-4 is an excellent description of the men at the turn of the century, also see E. D. Ives, *Larry Gorman, op. cit.*, 118-22. These paragraphs are also based on Interviews, Harold Noble, July, 1962; William Hilton, July, 1962, who told of his father going to Canada on the railroad to get "a box of Frenchmen" (a box car); and John P. Flanagan, "Labor in the Maine Woods", First *Report*, Department of Labor and Industry, 1911-12, Waterville, 1913, 206-27, especially 220, on the men, 223-6, on the shanghaiing, and agencies. Also see *The Northern*, May, 1925, "The Polack". On Great Northern hiring costs in 1916-1918 see *Pittston Farm Weekly*, September 10, 1964. The figures were 1916—$44,141.83, 1917—$168,974.98 and 1918—$64,968.70. Other comments on workers are in F. A. Gilbert to T. S. Ramsey, April 19, 1920 quoted in *Pittston Farm Weekly*, July 9, 1964, and P. E. Whalen to all clerks, December 19, 1913 in February 11, 1965. Also see *Paper Trade Journal*, October 28, 1909; July 27, 1911; May 8, 1913 (an attempt to use Italian labor in the woods); October 9, 1913; December 7, 1916.

[24] The newspaper was the Bangor *Daily Commercial*, December 1, 1899. This paragraph is based on study of all stumpage contracts in the Forest Commissioner's Files, from 1899 to 1930. These files are located in Augusta. C. W. Robbins, Old Town, printed the standard permits, and those bearing his stamp 1/1/03 have the provisions mentioned. (They may have come earlier, but not much). See e.g., Owners to Thomas Phair, Presque Isle, November 1, 1905; (T13R4); Owners to Thomas Phair, Presque Isle, T13R4, November 8, 1902; Owners to Neil MacLean, October 17, 1903 (T17R10). The first two were twelve inches at the stump, the last fourteen; also State to Birch River Company, cedar fourteen inches, no tops in the woods, on T16R7, September 10, 1909. All these contracts and many more are in Box 209. Interviews, Harold Noble, July, 1962; William Hilton, July, 1962; Caleb Scribner, August, 1963; I. H. Bragg, August, 1963.

[25] State to J. C. Horsman (W½ Indian Township TS) cull the fir entirely, July 26, 1915; in unmarked box, Forest Commissioner Files; State to John Morrison (T4R1 NBPP), six inches diameter four and one half feet high, November 1, 1923, in Box—"Stumpage Contracts #1".

[26] S. F. Peaslee to E. E. Ring (Forest Commissioner) December 11, 1905. He wanted to change the fourteen inch diameter provision in his contract to read twelve inches. Box 209.

[27] For a fairly good specific picture of lumbering techniques in the camps the reader is referred to several unpublished theses for the bachelor's degree in the University of Maine library. These theses were required for the major in Forestry at that time. Some of them were illustrated with snapshots which have been abstracted, enlarged, and are now in a box marked "Lumbering Operations—Photographs", also in the Library. They are: David Nathan Rogers, "Lumbering in Northern Maine", 1906, 55 pps., (T5R9); R. J. Smith and S. B. Locke, "A Study of the Lumbering Industry in Northern Maine", 1908, 22 pps., (T1R13)—The Boynton Blocks); George T. Carlisle and T. Frank Shatney, "Reports on a Lumbering Operation in Northern Maine", 1909, 16 pps., (G.N.P. Company in area above Millinocket on the West Branch Penobscot); Bernard A. Chandler, "Lumbering in Northern Maine as Illustrated by an Operation of the Emerson Lumber Company", 1909, 29 pps., (T6R6 and SW¼ of More Plantation, T6R5); Lewis F. Pike and Jonathan M. Jewett, "A Report on a

Lumbering Operation on Township No. 29—Washington County", 1908, no pages, (T29 M.D.—Devereaux Town); W. J. H. Miller, J. P. Poole, and H. H. Sweeter, "A Lumbering Report of Work on Squaw Mountain Township—Winter of 1911-1912", 1912, 28 pps., plus appendex, (T2R6 BKP EKR); Glenn C. Prescott and Raymond E. Kendall, "Lumbering in the Dead River Region—Somerset County Maine", 1916, 68 pps., plus appendix, (T3R6 BKP WKR)—Upper Enchanted Town, T2R6 BKP WKR—Johnson Mountain Town, T3R5 BKP WKR—Salmon Stream Town).

[28] From a box marked "Grand Lake Stream" in Forest Commissioner's Files. The quotations come from Assessor's Report, 1902; Assessors to Viles, March 3, 1914; Commissioner to J. W. Edgerly, Princeton, November 28, 1914. Grand Lake Stream is the modern name of Hinckley Town where the big tannery was located (T3R1 TS). See Minnie Lee Atkinson, *Grand Lake Stream, op. cit., passim*.

[29] *The Industrial Journal*, April 3, 1896; Bangor *Daily Commercial*, May 29, 1902; *Studies of the Food of Maine Lumbermen*, C. D. Woods and E. R. Manfield, Bulletin #149, U.S.D.A., 1904; *The Industrial Journal*, January, 1908; Rogers, "Lumbering in Northern Maine", *op. cit.*, 18-23, (There is a complete list of every item used in food preparation, and a detailed menu for a week in these pages); Prescott and Rendall, "Lumbering in the Dead River Region", (They kept their own pigs for swill); Flanagan, "Labor in the Maine Woods", *op. cit.*, 209-213, the quotation is from 210; Interviews, Harold Noble, I. H. Bragg, William Hilton, Caleb Scribner. Noble worked as a cookee in his youth. Bragg was the cook on the Eagle Lake Tramway operation. See also *Paper Trade Journal*, February 24, 1910; May 10, 1917 on Great Northern feeding practices.

[30] *The Industrial Journal*, November 19, 1897; Flanagan, *op. cit.*, 211; The Northern, Vol. VII, No. 8, A. G. Hempstead, "A Visit to Chamberlain Farm"; Vol. III, No. 8 and 9 and Vol. III, No. 6 on production.

[31] Bangor *Daily Commercial*, March 15, 1901; Bangor *Daily News*, March 16, 1901, commenting on visits of W. H. Jackson, of the Advent Christian Church in Bangor; Flanagan, *op. cit.*, 227, on religious services in 1912.

[32] Bangor *Daily News*, January 3, 1903.

[33] *Ibid.*, January 6, 1903.

[43] Bangor *Daily Commercial*, October 29, 1903, Letter, A. G. Young to All Operators; October 30, November 7, 9, 1903. The last an editorial.

[35] Flanagan, *op. cit.*, 221-3. *Paper Trade Journal*, May 11, 1911; November 19, 1914 (on a program like G. N. P. Company's); February 18, 1914, Charles R. Towsen, "Bringing Out the Best in the Workers," which surveys the YMCA's all over New England woods. Occasionally there were problems as in 1924 when the Klan and the I.W.W. clashed in Greenville. See Portland *Press Herald*, February 5, 1924 and Rita Breton, M.A. thesis on Maine 1917-1925, June, 1972.

[36] *The Northern*, Vol. 1, no. 7; Vol. 5, no. 2, "Social Service at Musquacook", Vol. 8, no. 6, letter F. A. Gilbert to A. G. Hempstead, on the purposes of *The Northern*, 1. Dissemination of information to employees; 2. "To bring out certain principles beneficial to employer and employees"; 3. Publicity. The magazine cost, in 1927, without adding Hempstead's salary, $7,508.25, which may have been a factor in its discontinuance.

[37] *Ibid.*, Vol. 1, no. 7; Vol. 2, no. 9. I have not seen the film, nor do I know anyone who has.

[38] *Ibid.*, Vol. 7, no. 4; "Social Service Department—1926-27 Season". It is sometimes difficult to realize just how much life had changed, but in 1927 the State Forestry Department hired a plane and pilot (90 hp, 8 cylinder Waco bi-plane) to act as fire warden. It flew from July to September 5, 1927 when it wrecked on Moosehead. Its replacement cracked up on Caucomgomac with the loss of the pilot. See *ibid.*, Vol. 7, nos. 4 and 7. Even these marked changes were a long way from the "Family Camps" of today with schools, roads, houses and so on in the woods themselves. See Warren *Standard*, Vol. II, No. 1, March 1952, "Family Camps and Mechanized Logging".

[39] Bangor *Daily Commercial*, March 16, 1899.

[40] The horse lingered strongly. Kimball and Nutter, a firm in Woodsville, New Hampshire had a lumbering business, but as a major sideline it provided horses for other firms. See Kimball and Nutter collection, 7 ledger books, Cornell University, especially woods ledger 1910-1915 and horses 1903-1904.

[41] *The Industrial Journal*, December 20, 1889, contains a description of what sounds like a steam log-hauler in the Canadian woods, but I have been unable to trace it further.

[42] *The Industrial Journal*, November 22, 1895; Bangor *Daily Commercial*, December 19, 1895. The latter reference is an account of the trials, and it includes sketches of the machine.

[43] Prescott L. Howard, "The Era of the Lombard Log Hauler," *Forest History*, Vol. 6, Nos. 1 and 2 (Spring and Summer, 1962); also see *Northeastern Logger*, March, 1962, Walter M. MacDougall, "Alvin O. Lombard"; both of these articles leave a good deal to be desired. Another similar device was invented and patented in 1901 by Alfred Kilby in the Dennysville area. It was used with limited success. Recently the old veteran was located and will be preserved in the Smithsonian. *Paper Trade Journal*, March 28, 1912 (which says it is still in operation); Bangor *Daily News*, April 7, 1970 with photo.

[44] *The Industrial Journal*, January 11, 1901 (the only contemporary account which I have found).

[45] Kennebec *Journal*, February 18, 1901. Also see, for a good account, Portland *Sunday Telegram*, May 28, 1939. This is a short history of the company.

[46] *The Industrial Journal*, April, 1908. This is an excellent account with pictures illustrating the early years of the company, and discussing its status in 1908. The only account that I know of on the economics of using a log hauler, both steam and gasoline powered, is in Prescott and Rendall, *op. cit.*, 25, 27-8, 42-50. They made comparisons with the cost of using horses as well. Those who would write about log haulers would do well to read this thesis.

[47] O. A. Harkness, "Log-Haulers 20 Years Ago", *The Northern*, December, 1927.

[48] *Ibid.*, Vol. II, No. 7 (on horses); Vol. VI, No. 1, Hugh Desmond, "Steel Horses for Long Hauls". F. W. Ayer hauled three four-horse sleds loaded with pulpwood with an automobile in the winter of 1912. *Paper Trade Journal*, April 3, 1913. Great Northern used trucks to tote. Teamsters were sent to Boston to

learn driving and maintainence. In these same years 1911-1915 roads began to be built. *Paper Trade Journal*, May 20, 1915.

[49] *The Northern*, Vol. II, No. 3.

[50] *The Northern*, Vol. 7, No. 1, F. V. N. Schenck, "The New Twin Tractor."

[51] *Ibid.*, Vol. 2, No. 11, on toting by tractor in 1923.

[52] *Ibid.*, Vol. 5, No. 5, No. 6, "Highways in the Realm of Old King Spruce".

[53] *Ibid.*, Vol. 5, No. 11, "Recent Developments at Grindstone"; Vol. 6, No. 9, No. 11, "Grindstone Bridge".

[54] *The Northern*, Vol. 7, No. 9, A. G. Grover, "Cooper Brook Log Hauler Road", Vol. 7, No. 12, "Chopping Down Newspapers", reprinted from the Boston *Evening Transcript*; Vol. 8, No. 2, "Cooper Brook Operation 1927-1928". The amount cut and hauled amounted to about 14,500,000 feet board measure which was an excellent winter's work.

[55] Bangor *Daily Commercial*, July 22, August 10, October 30, 1893.

[56] *Ibid.*, January 6, 1903.

[57] Bangor *Daily Commercial*, January 1, 23, 1902; June 16, 1903; *The Northern*, Vol. 7, No. 8, O. A. Harkness, "The Eagle Lake Tramway"; interviews, I. H. Bragg, who cooked for the builders and operators of the tramway; and Caleb Scribner, both of Patten, Maine, in August, 1963, and letters I. H. Bragg to author, March 21, April 4, 9, 12, June 21, 1963 in which he described most vividly the building of this innovation. See below on the towboat, the *H. W. Marsh*.

[58] Letter, Crawford (Mgr., Berlin Mills Company) to Austin Cary, (no date, early 1895?), quoted in Cary, "Spruce on the Androscoggin", *op. cit.*, 201-2. *Paper Trade Journal*, March 7, 1912 has a discussion of the building.

[59] Oxford *Democrat*, October 28, 1896; *The Industrial Journal*, November 6, 1896.

[60] *Ibid.*, December 6, 1895; January 24, March 6, 13, April 3, 1896; April 20, 1900; Third Water Power Commission *Report, Public Documents*, Maine, 1913, IV, 42-60, *passim.*

[61] *The Northern*, Vol. 6, No. 8, "Another Advance Step in Woods Transportation"; Vol. 7, No. 8, "The Pulp Wood Express".

[62] Third Water Power Commission *Report, op. cit.*, 1-62 *passim*, especially 44-52, 59, 61-2.

[63] Bangor *Daily Commercial*, January 31, 1899. The rate was cut from $65 to $38 per car on long lumber, from 72¢ M to 37¢ M on shingles, and from 21¢ to 11¢ each on railroad ties of cedar.

[64] *Ibid.*, March 16, November 15, 1904.

[66] Kennebec *Journal*, March 14, April 4, 1903; *The Industrial Journal*, April, 1906.

[67] Four foot wood became the norm on Great Northern waters in 1912-1913. Both were driven and the costs compared. The four-foot wood was driven for $4.994 per cord, while long logs cost $6.425. See *Pittston Farm Weekly*, February 6, November 5, 1964 for the cost analysis.

[68] Bangor *Daily Commercial*, July 20, 1896.

[69] *Ibid.*, April 16, 1900; Robert E. Pike, *Spiked Boots, op. cit.*, and his "65,000,000 feet of logs" in *Saga*, January, 1964. His book is about the last years of the Connecticut drive, the article about the 1915 drive, the last one. There were a lot of Mainers that year.

⁷⁰ Kennebec *Journal*, March 14, 1901. "Coming Out of The Woods".

⁷¹ From the reports of annual meetings; see Bangor *Daily Commercial*, March 8, 1899; March 7, 1900; March 6, 1901; April 5, 1902; March 4, 1903; March 2, 1904. Also see Bangor *Daily News*, March 7, 1901; March 5, 1903.

⁷² Bangor *Daily Commercial*, March 3 (annual meeting); November 21, 1896 (final report). The total rafted at the Penobscot boom in 1896 was 157,-994,450 feet. *Ibid.*, March 2, 1897 (annual meeting).

⁷³ Bangor *Daily Commercial*, April 7, 1897.

⁷⁴ *Ibid.*, October 14, 1899; March 6, 1900.

⁷⁵ Bangor *Daily Commercial*, November 3, 1900 (the men worked a twelve hour day, and handled about 100,000 feet of sawed lumber, for a wage of about $4.25 each). The survey was conducted by state law.

⁷⁶ *Ibid.*, September 24, 1904.

⁷⁷ From reports of the annual meetings of the Penobscot Lumbering Association; see *ibid.*, March 3, 1896, March 2, 1897; March 7, 1899; March 6, 1900; March 5, 1901; March 4, 1902; March 3, 1903; March 1, 1904. As above, although the boom was active until 1915 or so, the amounts were 50 million, 40 million or less. In 1915 only 80 men were employed, and half of them were boys.

⁷⁸ Hempstead, *The Penobscot Boom, op. cit.*

⁷⁹ These figures come from tables in Third Water Supply Commission's *Reports, op. cit.*, 53ff. By 1910 the G. N. P. Co. had dammed enough so their storage capacity was 32 billion cubic feet, through the North Twin and Chesuncook systems. An excellent description of the system before Ripogenus is in *SJRC, op. cit.*, testimony of Hardy S. Ferguson (the engineer for the GNP Co.) given March 11, 1910 in Bangor, 1937-97; also see *The Northern*, Vol. 6, No. 11, F. A. Gilbert, "Early Improvements on the South Branch of the Penobscot River"; Vol. 6, No. 10, "Improvements on the South Branch (Canada Falls)"; Vol. 6, No. 11, "Dams at Seboomook Falls"; Vol. 2, No. 6, F. S. Davenport, "Some Pioneers of Moosehead—Chesuncook and Millinocket" and Portland *Sunday Telegram*, November 13, 1938.

⁸⁰ *The Northern*, Vol. 6, No. 11, "Dams at Seboomook Falls".

⁸¹ *The Northern*, Vol. 1, No. 8 (by far the best account); *Pulp and Paper Magazine of Canada*, April 1, 1915, "Great Northern Paper Company Builds a New Dam", and Hempstead, *op. cit.*, 118-9.

⁸² *The Northern*, Vol. 1, No. 2, No. 3, "1922 Drives"; Vol. 6, No. 3, "1926 Drives", Vol. 6, No. 9. In 1926-7 the company obtained between 75 and 100,-000 cords from farmers. It was peeled and then hauled in by railroad.

⁸³ *The Northern*, Vol. 6, No. 11 "Dams at Seboomook Falls"; Vol. 6, No. 10, "Improvements on South Branch (Canada Falls)".

⁸⁴ *Ibid.*, Vol. 7, No. 10, "Kennebec River Drive of 1927-28".

⁸⁵ *Ibid.*, Vol. 7, No. 8, A. G Hempstead, "A Visit to Chamberlain Farm—Old Towboats".

⁸⁶ *Ibid.*, Vol. 6, No. 9, "Towing Boats on Chesuncook Lake".

⁸⁷ *Ibid.*, Vol. 8, No. 6, "Towing".

⁸⁸ *The Northern*, Vol. 8, No. 6. For towing on Moosehead Lake, by this time, mostly of tourists, see Bangor *Daily Commercial*, March 14, 1896, "Steamboats on Moosehead".

[89] *Ibid.*, February 9, 22, March 20, April 22, July 2, August 6-15 (especially the 6th), September 1, November 30, 1895.

[90] Bangor *Daily Commercial*, April 15, 17, May 5, 1899; April 21, 23, 24, 25, 1900.

[91] *Ibid.*, April 3, 1901; Bangor *Daily News*, January 3, 1901, for prices. The same story was true for all rivers, not just the Penobscot. In an account of the Machias Lumber Company operations in 1901 a newspaper reporter remarked that the 14,000,000 feet sawed on that river was the smallest amount in sixty years. During the same year the Hodgkin and Hall mill in East Hampden sold at auction for $17,000, which was certainly low for a property which had 27,000 feet of frontage on the river, at a place where there was fourteen feet of draft even at low tide. See *ibid.*, December 6, 1901 for Machias, and October 3, 1901 for an account of the auction.

[92] Bangor *Daily Commercial*, August 20, 23, 26, 27, 28, 29, September 1, 2, 6, 10, 11, 1902.

[93] Bangor *Daily News*, January 29, 1903 for an account of mills in Aroostook County owned by Charles A. Milliken of Augusta; Portland *Eastern Argus*, March 3, 1903, for one at Madrid, owned by the Berlin Mills Company. This one ran until the eve of World War One. The James Walker mill in Orono closed down in 1912 ending continuous operation since 1838. At one time this was the largest mill in the world. *Paper Trade Journal*, October 10, 1912. Small portable mills continued. For descriptions of two see *Mars Hill View*, December 30, 1909.

[94] Bangor *Daily News*, February 23, 1901, projected mill at Vanceboro; October 3, 1901, projected mill at Biddeford or Bonny Eagle; March 2, 1901, mills at Mattaceunk on the Penobscot, and at Lincoln. The East Millinocket growth would bear investigation, if the records could be made available.

[95] *The Industrial Journal*, May 24, 1901. Hollingsworth and Whitney were constructing a clubhouse, gymnasium, library, reading room, bowling alley, and billiard room for their employees.

[96] *Ibid.*, May, 1907.

[97] Kennebec *Journal*, March 10, 1903. *Paper Trade Journal*, July 7, 1910; November 14, 1912. Also see Rexford Sherman, "The Bangor and Aroostook Railroad and the Opening of the Port of Searsport," M.A. Thesis, University of Maine, 1967.

[98] *The Northern*, Vol. 2, No. 8; interview William Hilton July, 1962; he is the source for the comment about the lime.

[99] *The Northern*, Vol. 3, No. 10, "New Hydroelectric Plant at Anson."

[100] On this mill, generally, see Edward C. Plummer, *Reminiscences of a Yarmouth Schoolboy*, Marks Printing House, Portland, 1926, 106-8. In the S. D. Warren files are a sheaf of letters which tell this story well. "Blueprint, March, 1911, Coal Required Expressed in Live Steam Uses and Loss at Boilers", "Report," H. L. Strong and F. L. Wellcome, July 8, 1914; F. L. Wellcome to S. D. Warren Co., June 25, 1914; S. D. Warren Co. to Brandeis, Dunbar and Nutter, Boston, Mass., June 30, 1914; Undated Report c. 1914 F. L. Wellcome, "On Use of Waterwheels", F. L. Wellcome to S. D. Warren Co., April 28, 1920; "Estimate of New #3 Boiler—Forest Paper Company", August 11, 1920; "Revised Estimate", August 30, 1920; F. L. Wellcome, "Electrification of Forest Paper Company", no date, but about 1920-1; Welcome ?? to F. E. Core,

no date, but probably about 1920-1; "Report on Power (50 Ton Basis)—Forest Paper Company". In addition, one should consult *A History of the S. D. Warren Company, op. cit.* This was predicted as early as 1914 after a shutdown. See *Paper Trade Journal*, June 25, 1914.

[101] Warren's *Standard*, III, No. 12, February, 1954. Labor problems plagued them briefly during the war and a short strike occurred in 1916. See *Paper Trade Journal*, September 21, 28, October 5, 1916; April 19, 1917. Others thrived as well. *Paper Trade Journal*, July 26, August 9, December 20, 1917; February 14, March 14, 1918. G. N. P.'s earnings were the largest in their history—12% both years, even with a 25% increase in capital. See *Paper Trade Journal*, July 5, 12, 19, 26, August 2, 1917 for a detailed look at the Brewer and Lincoln paper mills, "Development of Scientific Methods of Management in a Manufacturing Plant."

[102] *Warren's Standard*, II, No. 6, "Record of Bonus Payments—Coating Room 1908-1909".

[103] Four notebooks, "Cost Statements January to July 1910, A and B, and 1911 B"; "Deliveries to Mill in July 1910"; "July 1910". These are apparently the residue of efficiency engineers report.

[104] *Timebooks*, 1913, 1919, 1920.

[105] "Production" from *Balance Sheets Ledger*.

[106] Warren *Standard*, I, No. 8 (October, 1951).

[107] "Raw Material Cost per Ton" 1909-1914, from "Balance Sheets—Tonnage Analysis", a huge ledger running from 1909 with all sorts of this kind of information in it up to 1920. For a general study of the paper industry with more on Maine mills see my *History of The Paper Industry, 1691-1969*, New York, 1971.

16

FACTORS IN THE MODERN MAINE WOODS INDUSTRY
1930-1960

As one comes closer to the present the historical documentation is much less, and the story becomes more fragmentary. Newspapers, a bountiful source for the historian in the nineteenth century, also are worth much less as they cater to different interests. As a result categorization of the industry in recent times is a difficult and probably somewhat inaccurate task. One of the factors which is present is the increasing interest and in fact interference of government in private enterprise. Although grumbling takes place most persons welcome it, especially as it involves such items as the Civilian Conservation Corps, help in obtaining woods crews, expenditures for forest fires, or entomological control. Other factors which change are the shift to contract logging for the largest companies, and in the past years the massive mechanization of all aspects of woods operations. The dominant firm in the state remained the Great Northern Paper Company although others still remained, and small operations, both for portable sawmills and specialty factories and mills, were also a part of the Maine picture; however, they seemed to be dwindling in importance.

The thirties brought national economic disaster and with it, severe economic and social problems to Maine. Population had been dwindling for some time, but now that decline stopped as there was nowhere to go and nothing to do. The problems of unemployment in the state were great and were on the increase when Franklin Roosevelt took office in March of 1933. Maine had not voted for him, but it awaited his administration with curiosity. It soon became apparent that one of the New Deal's new efforts to deal with the problems of economic recovery and unemployment would have an impact on rural and forested Maine. This was the Civilian Conservation Corps.[1] This agency, founded on April 5, 1933, would employ young men from 18-25 years on basically conservation projects in which the departments of War, Interior, and Labor all were involved.

Maine had great hopes for the CCC. However that depended on allowing these men to work on private lands for, with the exception of the White Mountain National Forest, little if any forested land was held by the federal or state governments. The Governor, the Forest Commissioner, and others were able to prevail by citing eleven specific Maine projects dealing with fire control and eradication of disease and insects.[2] It was well that this occurred. So many young men had arrived in Bangor to enlist in the CCC that housing facilities were overtaxed and some men had to be housed in temporary quarters while awaiting their disposition.[3] Fort Williams in Portland Harbor was eventually designated as a receiving station where the enrollees received about ten weeks of indoctrination and basic training before reporting to their project camps all over the state.[4]

Throughout the CCC period an average of a dozen camps were in operation in Maine with most of them coming under the supervision of the Maine Forest Service. Altogether over 16,000 Maine youths participated in the program in some 28 camps. In fact the number of potential enrollees was always greater than the billets that were available, a factor which created distress in some quarters. The last Maine CCC camp at Bar Harbor closed in June of 1942.

The major areas in which the CCC had an impact on the state and the state forest industries were in fire protection, road building, insect eradication, and disease control. In addition a good deal of beautification was undertaken as well. Fire control was the first duty in a state so heavily forested. The Maine Forest District, founded for the purpose of fire control on private lands, had reached by 1933 a good state of preparedness. Commissioner Neil Violette saw the CCC as a means of strengthening its position. The CCC campers strung telephone lines, removed slash, constructed campgrounds, and built roads for access to the woods. At the end of the period, although fires were still a possibility, the probability of containing them was better than at any time in the state's history.

Altogether some 400 miles of roads were constructed in the Maine woods by CCC workers. Many of them were short loops for fire fighting purposes, but several of them were major woods highways and are still travelled by many today. In fact in some areas "old-timers" still mention going for a drive "on the CC road," as the Corps was invariably called in the woods. The most significant roads constructed were the Wilson's Mills road in Rangeley, the Shirley Mills road south of Greenville, the beginning of the Kokadjo road, the Shin Pond-Grand Lake Matagammon road on the East Branch Penobscot, the Ambejejus road to Togue Pond at Mount Katahdin, the Beddington-Nicatous Lake road on the Machias River waters, and

Factors in the modern woods industry 421

the network of roads from Myra north to Springfield and east to Deer Lake. Other important roads were the Tomah Stream road and the various roads in and around Indian Township. Altogether 279 miles of gravelled truck trail were constructed, along with the clearing of 596 miles of roadside slash; 592 miles of foot trail were cleared, and 440 miles of telephone wire strung. The footpath mileage included much of the area known as the Appalachian Trail.[5] After the 1938 hurricane emergency, CCC camps cleared many of the storm-blown trees from the trail as part of their last efforts in the state.

The other major factor in which the CCC modified the state's woodlands was through the program of insect and disease control. Henry B. Pierson, state entomologist from 1921-1956, was chiefly responsible for this work which involved mostly control of the gypsy moth which had spread into southern Maine in some numbers by World War I and which by 1933 was a major pest. A quarantine zone had been established to control import or export of suspected woods products infested with the larvae of the pest. The chief method of control was to destroy the eggs masses with creosote as they lay dormant in the winter months. Hundreds of boys combed thousands of acres in southern Maine to accomplish this work. The brown tail moth was also an object of their work. Perhaps more dangerous was the spruce saw-fly which also ruined many trees of considerable economic worth. The pest spread south from the Maritimes and had caused great damage in the 1920's. In 1937 it arrived again in Maine in large numbers. The method of control this time was by introducing a wasp which acted as a parasite and eliminated the sawfly.

Although the insect control was of major significance, perhaps of equal importance was the work on plant disease and especially the white pine blister rust. The disease, first brought to the United States in 1909, was caused through a peculiar life cycle which involved the gooseberry or currant bush. Elimination of these bushes would cut the disease substantially and save many trees. More than 400,000 acres had been treated and the offending bushes destroyed by the CCC when it ended its work in 1942; nearly all this acreage was in southern Maine where the danger was especially great.

Other items which should be mentioned include picnic ground construction, the building of an arboretum in Capital Park, Augusta, a small amount of field planting, and nursery work (a major item in southern CCC camps), development of the land of the Moosehorn Wildlife Refuge in Washington County, construction of trails and campsites in Acadia National Park (in fact the present park remains very much a CCC memorial), and highway beautification of the road to the park from Ellsworth. Other cleanup projects were involved in several historical sites in the state, especially in the Camden Hills area. Some major work was also put in operation during the period in Baxter State Park. Emergency firefighting, and aid to persons effected by the two major natural disasters of the 1930's, the great floods of 1936, and the hurricane of 1938 also figured largely in the work of the CCC. McGuire summarized the work of the Corps well when he said,[6]

> ..., it is apparent that the program to a degree helped prepare hundreds of Maine youths for the demands of adult life.
> The effects of the CCC upon the land have had a measure of permanancy. Although postwar changes in protection techniques came, particularly in insect control, and such disasters as the fires of 1947 at Bar Harbor and in southern Maine, detracted from that long-range significance of CCC work, other accomplishments remain. In the Forestry District, many connecting roads have become major routes for further road construction. Along the Appalachian Trail CCC shelters are maintained. In the Moosehorn Wildlife Refuge CCC trails, roads, and picnic sites still serve the areas. In southern Maine, the Massabessic Forest has continued to develop, as has the Sebago Lake Park, and the White Mountain National Forest. In Augusta the entomological laboratory is still in operation. The Camden Hills State Park, begun by the CCC, is a major recreation area. These surviving accomplishments of the CCC in Maine, along with the undetermined social value of the agency demonstrate the fact that in many ways, the CCC was more than a crutch—it was a foundation, a facet of the Permanent New Deal.

As far as the woods themselves and woods operations were concerned, although they were certainly modified and changed by the New Deal, much went on as it had with the exception that more and more modern technology began to be involved. River driving effectively disappeared except for pulp wood and much of that was upstream. The men in the woods were still mainly French, with a smattering of Yankees, but increasingly they began to go home weekends in their own cars and occasionally even at night. Some moved their family into the camp sites, and the so-called family camp, or woods village, began to be discussed by the 1940's and implemented in the 1950's. Much of the lumber in the state was cut by farmers and others to supplement their income. It was possible for a high school student to cut, limb, and yard enough pulpwood Saturdays and holidays to support himself, and it was not unusual for this to happen.

We have an excellent account of how woods operations were conducted in the period just before World War II. A summary of those

Factors in the modern woods industry

findings will set the stage for the wartime changes. This involved a detailed study of the Great Northern operations in northwestern Maine from 1935-1940.[7] Basically this gives a picture of woods operations on the verge of the major mechanizations of today. The thirties were still a transition period as horses were still being used, and the saw and the ax were the major tools for the men themselves. On the other hand trucks and tractors were coming into increasing use.

By this time nearly all operations were pulpwood, and chiefly unpeeled pulpwood at this time. Most operations included from three to five camps, with each camp producing 6,000 cords and the entire operation 30,000 cords each year. This would be approximately 3 million board feet for each camp or 15,000,000 board feet for the whole operation.

Preliminary work remained much the same. The operation had to be laid out, and this was still done according to whether it was "a good chance" or not. Dams were constructed, roads built, camps built, and so on. The men then proceeded to the cutting, hoping to get their stint before the deep snows when hauling would begin prior to the spring drive.

The chief difference between the operations of the 1930's and those conducted in the 1860's was the scientific nature of the operators' approach. Now detailed maps were ordinarily available, and the camps were located so that the maximum walking distance for the workers was two miles. A dry yard with good drainage, excellent water, and a large cleared space were imperative for a camp usually planned for three years use. The depot camp with its headquarters for supplies, camps for clerks, walking bosses, storage for surplus equipment would ordinarily have been near a main highway or the railroad. The camps were different however. By this time they were usually of square timbered frame work with roofing paper on the roofs. Sky lights and windows gave light and ventilation at all times to the occupants. Sanitation was a problem which was no longer solved in the hit or miss fashion of earlier times.

Beds, not bunks, were the rule for the men although the deacon seat remained. Sinks for washing, and towel racks were further amenities. Nearly half the camps had electric light, while the rest were also lit, usually with gasoline lamps. The blacksmith shop was often larger as it involved a garage for the small tractors increasingly used on the operation. Hilton estimated camp costs at about $3,000 each in this period.

The roads into the camps were much better built than before as it was felt that additional fire protection and long use repaid many times the greater initial costs. Of course the bulldozer, tractor shovels, and heavy graders made road building a much easier operation than it had been previously. By the end of this period toting was being done with great frequency with heavy tractors as the horse was increasingly replaced. The men ate well in these camps as did the animals. Hilton estimated the cost of 20.68 tons per 1,000 cords for food for the men, and not quite seven tons of oats and hay for animals. This would be about 55 tons per million board feet.

At the time that Hilton wrote, the Great Northern Paper Company was in a period of flux in regard to cutting. It had operated through contractors in its early history, and then from 1912 or so through the middle thirties it had operated its own woods camps. This was the heyday of the social service department, and a period when the company manufactured its own axe handles, harness, and many of the items used in the woods. A major company subdivision was located at Greenville to manufacture these items and all others which might be called for. By the late 1930's the firm was going back to contract cutting, and although this would be delayed somewhat by the advent of World War II with its labor shortages, contract cutting has been the chief mode of operation in recent years.

Under the contract system the men were paid by the cord which usually involved felling, limbing, topping, and then twitching the stick to the yard or roadside. The yarding crew would saw the trees into four foot sticks, and pile them in cords. On these operations the average man was producing 1.608 cords, while the average three man crew produced 8.846 per day. Average horse production was about 4.309 cords per day. Tractors on these operations, in their infancy of use, were yarding 10.72 cords, and observers thought 12-14 cords would be a better average with experience.[8] Again this was a transition period for hauling to landings after the cut was in. Horses, small tractors, medium and large sized tractors, Lombards, and trucks were all utilized on these operations. On the basis of these experiments it was felt that tractors were probably the least expensive method, but that depended on so many variables that it was obvious that many of these other methods would continue to be used. After the war of course, medium tractor hauling became the norm with trucks hauling mostly to the mill. Horses and Lombards as well as the large sized tractors were not much used.

Even with four foot wood the streams to be driven demanded a great deal of work. Smaller rainfall, greater runoff, and the fact that these operations were deeper in the bush were all factors in these costs. Hundreds of small splash dams with their fancy crib work and ornate gates were built and often can still be seen in the woods. One of the interesting factors in Hilton's discussion is the routine use of dynamite to move the drive along, otherwise the techniques of driving, sluicing, and booming were not much different than they had been for close to a hundred years. The work was still wet, dirty, and dangerous, and the men received more pay for it than other work.

What seemed to be the case with all this work, from building the camp to piling the wood in the mill yard, was that the work was now done more scientifically and, where possible, with more labor saving devices. Costs had risen and the profit margin was in the willingness to experiment and use these new methods. Much more control of the men and their techniques, even to the place of using time-study men, was a part of all operations. In fact Hilton's book was a detailed study for five years in an effort to cut costs and make the woods work much more rational as this company looked to the future.

The future involved fighting a massive, long-term, arduous, and expensive global war.[9] For the Maine woods this meant continued production of logs for pulp and paper manufacture. As the labor drain for the military and the war plants grew, this caused a great problem in the woods, the problem of labor supply. The problem was eventually solved through the use of farmers, a greater introduction of Canadian labor, and finally in 1944 and 1945 with the rather extensive use of prisoner of war labor. These demands helped expedite the thrust toward mechanization already in evidence before the war.

The Great Northern and other Maine companies had traditionally used French labor from Canada for much of its pulpwood operation. However in the years since 1931 this wood had been cut with other labor, and since 1934 no Frenchmen had come across the line to cut pulpwood. In 1941 and 1942 small quotas were allowed. By 1943 great pressures were beginning to mount to allow more and more Canadians to come to Maine to cut pulpwood. Coincidental with this was the "farmer pulpwood" campaign with advertisements in newspapers and over the radio pleading with farmers to cut pulpwood for the war effort. County communities were set up, often in cooperation with the Extension Service, and prizes were offered to the men who produced the most pulpwood in a given time. This campaign continued throughout 1946 and 1947 as postwar paper demands were greater than the ordinary labor sources could provide. The war offered for many pulpwood cutters a chance to make more money in more salubrious work, and when the war was over they simply did not return to the woods.

In 1943 attempts to increase the Canadian imports also were mounted. Maine's senior senator, Ralph O. Brewster, was on Harry Truman's Senate Investigating Committee and he made a useful contact for the companies in Washington. More and more Brewster was asked to use his political leverage with the State Department bureaucrats.

Other methods were used. The Governor of the state issued a proclamation urging farmers to cut wood, and he mentioned the Canadian labor supply in his message. Negotiations opened between the two governments, and the American negotiator was replaced as he was "not big enough to deal with the men up there." A congressional committee conducted investigations of the problem in the House. In the fall 3,500 Quebeçois were allowed to come to Maine and later in the winter another 1,000 were permitted. This was still not enough.

The *Boston Herald* sent a reporter to the border towns and wrote a series of articles telling of unemployment by woodcutters who could not come to Maine. This set off rebuttal statements from Halifax to Vancouver as the French-English split in Canada fueled the flame. The next year the pressure continued to be built up until the State Department told the industry very frankly that further efforts would be useless. The Canadians had resisted some U. S. pressure, and to some extent feelings were damaged between the two countries.

The real solution to this problem was the utilization of prisoners-of-war. This labor source was discussed quite early, but many felt the Maine woods were no place for enemies with axes. The pressures continued and in the summer of 1943 as the shortage intensified the War Manpower Commission investigated the use of P. O. W.'s and by October these troops were being trained for work in the woods. Early in 1944 2,500 came to Maine, and the Bangor *Daily News* welcomed them by saying, "However, the labor situation is so serious that even a NAZI is welcome, provided he sticks to his wood cutting."

Production of P. O. W.'s picked up steadily until they were averaging about two thirds of a cord a day or roughly about one half normal labor. It was this labor which solved the manpower problem in the United States pulpwoods in the spring of 1945. In fact these prisoners were so important that another campaign was mounted to allow them to remain through 1946, and P. O. W.'s did cut pulpwood

Factors in the modern woods industry

in the United States until that June, although apparently all were gone from Maine by the early spring of 1946. At the height of usage about 1,000 P. O. W.'s were cutting wood in Maine, based in camps in Princeton, near Moosehead, and in Aroostook County.[10]

The other area in which the war made great impact was in hastening the coming mechanization ranging from large crawler tractors, hoisting and loading devices to small compact portable saws. In June, 1945 one of the major factors in this was a three day conference on mechanization held at Gorham, New Hampshire. The work of the Brown Company was featured at this conference attended by woods people from all over the United States and especially the northeast. At that time the items of difficulty included the fact that some machines were not too well engineered as yet. Firms present were not ready to move to the chain saws, citing cost of maintainence as one reason, but it was only a matter of five or six years until these problems were overcome and the chain saw, the loader, the clam shovel, the backhoe, and fellers were beginning to be commonplace in the woods. Other wartime innovations were the use of helicopters, Bailey Bridges, radio communication between crews, and in fact a general air of experimentation in woods work which was not so noticed before.[11]

Utilization of the woods had by the end of the war fallen into a pattern. The huge companies controlled nearly all the cut, and the cut was to their specifications. Farmers still tended their farm wood lots, some hardwood was cut for dimension lumber, and occasionally portable sawmills still worked in the woods. Mostly though the woods were pulp woods and increasingly the cut went to the Millinocket and East Millinocket mills. The other mills in the state purchased much of their wood although S. D. Warren purchased land, planted trees, and experimented with cutting in southern Maine. Increasingly some mills in the state were becoming obsolete in face of southern competition, and it was the larger and more progressive firms with money or credit that survived best.

Great Northern found itself in a good place with the war over. Its 1946 production was a record. The firm generated all of its power and its machines were nearly all new and quite good shape. By 1953 with a massive expenditure the mills were in excellent shape as the three new paper machines at East Millinocket were rated at 325 tons a day. The firm borrowed $32 million to install these machines and its new chemi-hardwood method which would open up new acres for the mill to use. In fact the utilization of hardwood seems to be the major factor in the postwar period. New products such as corrugating mediums were developed in the GNP mills for the new pulp production. In fact as the sixties opened the firm began its expansion into the south which led to major development there and its final merger recently with Nekoosa-Edwards. This firm like so many others was national in scope. The industry in Maine was no longer Maine owned or controlled. Scott bought Hollingsworth-Whitney; Georgia Pacific the Woodland facility, and S. D. Warren was absorbed by a larger firm. The moves begun with the formation of the International Paper Company in the 1890's were clearly part of the twentieth century woods picture.[12]

Some statistical account of the Great Northern Paper Company's cut may indicate the immense size of their operations.

St. John Watershed	1934-1956	592,742 cords
Main River	1934-1963	436,670
Caucomgomac	1934-1963	49,038
North Branch	1934-1963	858,561
South Branch	1934-1963	605,025

In addition the firm purchased and drove itself in this area 185,896 cords. Although this meant that the river drive in this thirty years amounted 2,727,932 cords, or by the standards of an earlier day about 1,500,000,000 feet. The Penobscot valley was still a major woods producer.[13]

Another way to look at this period is to indicate the amounts cut by the individual contractors. These men were responsible for the great cuts with their crews of men from the states, the provinces, and even in later days from such exotic locations as Tibet. The table summarizes their work.[14]

Willie Caouette	1948-1964		192,387 cords
Alfred Nadeau	1947-1964		195,217
Lucien Gosselin	1952-1964		121,339
Henry McMahon	1934-1943		144,183
Adelard Gilbert	1956-1965	(Scott Brook)	88,870
Leo Dumas	1934-1965		294,965
Adelard Gilbert	1951-1965	(other ops)	191,145

Most of the rest of the story of the recent past can be told fairly quickly. Continued control of the two main enemies of man in the woods, tree disease and fires, was effected. Occasionally the spruce bud worm caused problems, and aerial spraying was instituted with success except apparently for the fauna elsewhere. The fire danger was always present, and several years were bad, but 1947 was the worst with massive damage in York, Oxford, and Hancock counties. Christmas trees continued to be a major export although increasingly after 1960 they are being replaced by metal and other artificial trees.

The university continued to be an active force in the state's woodlands and in planning for their usage. In 1950 a Pulp and Paper Foundation was formed to insure quality education for graduates in this area, and the forestry school grew especially in recent years with the efforts of Harold Young, who began to pioneer usage of "the complete tree." In other ways the university was involved as well with a substantial record in manuscript collection, subsidization of research, and some publication. Increasingly such aspects as the literary nature of the woods, a photographic and oral history record, as well as many others were being studied. The woods continued to provide employment for most of the state's citizens as it always had in one way or another.[15]

As always the history of the state was bound up in its woodlands. In 1969 for instance, the cut was more than many of the estimates made in 1850 of the total. Hardwood amounted 197,000,000 feet; softwood not for pulp was 511,000,000, and the pulpwood cut was 2,668,000 cords, or about 1,334,000,000 board feet. Altogether the cut then amounted to 2,042,000,000 board feet.[16] One of the things that the history of the Maine woods seemed to show was that it was possible to have your woods and cut them too, but an ancillary lesson was that this was only done through close attention to the workings of the political process and continued analysis and publicity for events as they occurred. Pressure on the woods continued to build as vacationers could now travel to "Those distant shores of 'Suncook", and in the future, as in the past, usage would continue to be the central problem of life for the citizens of the Pine Tree State.

NOTES

[1] I am following here H. Paul McGuire, "The Civilian Conservation Corps in Maine, 1933-1942," M.A. thesis, University of Maine, August, 1966, and John H. Salmond, *The Civilian Conservation Corps*, Duke University Press, 1968. On the depression in the woods see Bangor *Daily News*, September 18, October 11, 1933.

[2] *Kennebec Journal*, April 4, 8, 1933.

[3] Bangor *Daily News*, April 13, 14, 1933.

[4] *Kennebec Journal*, April 24, 25, 27, June 5, 1933; Bangor *Daily News*, April 8, 1933.

[5] "Accomplishments of . . . Camps in Maine, 1933-1939," cited in McGuire.

[6] McGuire, "The CCC in Maine," 142-3.

[7] C. Max Hilton, *Rough Pulpwood Operating In Northwestern Maine 1935-1940*, University of Maine Studies, Vol. XLV, No. 1, Orono, 1942. For a look at the State in this period see, George Otis Smith, "Maine, The Outpost State" *National Geographic*, LXVII, No. 5 [May 1935], 533-592.

[8] Hilton, 21-32.

[9] This discussion of wartime is based on my *History of the United States Paper Industry, 1690-1970*, New York, 1971, chapter 14, and sources cited there. The discussion itself was based on the papers of the West Virginia Pulp and Paper Company, Cornell University Regional History Collection, and Great Northern Paper Company, loaned to me by John McLeod, especially file 518 "Canadian Labor"; 518-B "Canadian Labor Articles,"; 533-A, "Wage and Hour Laws"; 518-A "Canadian Labor"; Boston *Herald*, February 8, 11, 26, March 5, 1944; file 518-D 'Great Northern Paper Company".

[10] Lt. Col. George G. Lewis and Capt. John Mewha, *History of Prisoner of War Utilization by the United States Army, 1776-1945*, Washington, 1955; *Paper Trade Journal*, throughout, but especially September 16, October 14, November 4, 1943; May 11, 1944; August 2, 1945. Bangor *Daily News*, August 31, 1944. *Brand Names and Newsprint, Hearings*, Part 3, *Newsprint*, 78th Congress, first Session, Washington, 1944; *Final Report*, 79th Congress, 1st Session, Washington, 1945.

[11] Again, this is based upon my *Paper Industry*, and in particular West Virginia Paper Company, Box 78, letter M. H. C. to W. J. Bailey on the New England conference with a detailed report of all aspects of the conference. Another similar was held in Cooperation, New York, the same year.

[12] *Paper Trade Journal*, April 10, 1947; January 30, August 21, 1953; October 15, 22, 1954; October 29, 1956; February 21, 1955; March 26, April 16, October 28, November 12, 1962; May 4, 1964; May 31, August 23, October 4, 25, 1965; August 1, October 24, 1966.

[13] *Pittston Farm Weekly*, April 18, 25, May 23, 30, June 6, 1963.

[14] *Pittston Farm Weekly*, July 24, 30, August 13, September 10, 24, October 22, 1964; February 4, 11, 1965; February 3, 1966.

[15] *Paper Trade Journal*, October 19, 1950, Howard A. Keyo, "The University of Maine," October 16, 1953, "University of Maine Celebrates,"; April 30, 1956, Harold E. Young, "Where's The Wood Coming From?" David C. Smith, "Forest History Research and Writing at the University of Maine." *Forest History*, Vol. 12, no. 2, [July, 1968], 27-31 summarizes the progress till that date. Since then both Richard Sprague and Edward Ives have continued their work along with the writer.

[16] *Forestry Facts*, June, 1970 University of Maine Extension Service.

APPENDICES

APPENDIX I

The Log Drive of the Kennebec—1866-1912

Year	Board Feet	Year	Board Feet
1866	47,035,278	1890	168,882,451
1867	75,635,602	1891	147,460,585
1868	52,044,483	1892	130,125,015
1869	91,436,205	1893	188,446,509
1870	80,881,519	1894	129,716,614
1871	81,701,944	1895	119,984,514
1872	119,578,190	1896	118,201,902
1873	128,695,162	1897	142,511,888
1874	97,427,499	1898	101,444,363
1875	105,978,538	1899	107,452,801
1876	113,646,782	1900	147,424,579
1877	50,698,148	1901	136,063,291
1878	93,500,719	1902	133,772,610
1879	82,667,270	1903	146,413,732
1880	77,044,898	1904	163,894,303
1881	144,379,343	1905	132,025,401
1882	127,332,195	1906	148,726,278
1883	124,247,085	1907	127,955,309
1884	106,187,334	1908	128,472,904
1885	105,702,925	1909	107,985,561
1886	103,489,535	1910	117,007,177
1887	160,975,657	1911	115,626,169
1888	137,375,736	1912	95,665,550
1889	134,644,947		

Source: J. F. Defebaugh, *Lumber Industry of America*, II, 76; Third Annual *Report*, Maine Water Storage Commission, 57-8.

APPENDIX II

Penobscot River Navigation Dates, 1818-1904

Year	Open to Navigation	Closed to Navigation
1818	May 1	December 10
1819	April 19	December 8
1820	April 18	November 18
1821	April 15	December 4
1822	April 10	December 5
1823	April 19	December 9
1824	April 10	December 12
1825	April 11	December 14
1826	April 5	December 16
1827	April 2	December 6
1828	April 1	December 18
1829	April 15	December 16
1830	April 9	December 8 and January 9, 1831
1831	April 9	December 3
1832	April 19	December 4
1833	April 9	December 1
1834	April 8	December 9
1835	April 17	November 27
1836	April 12	November 27
1837	April 15	November 27
1838	April 21	December 24
1839	April 17	December 1
1840	April 1	December 1
1841	April 17	December 19
1842	March 21	November 19
1843	April 21	November 30
1844	April 12	November 27
1845	April 21	December 7
1846	March 29	December 10
1847	April 23	December 21
1848	April 12	December 21

Appendix, Penobscot navigation dates

Navigation Dates—Penobscot River
(continued)

Year	Open to Navigation	Closed to Navigation
1849	April 1	December 7
1850	April 12	December 8
1851	April 8	December 30
1852	April 21	December 15
1853	April 5	December 8
1854	April 27	December 5
1855	April 15	December 7
1856	April 16	December 1
1857	April 6	December 10
1858	April 11	December 12
1859	March 30	December 9
1860	April 16	December 7
1861	April 11	December 20
1862	April 18	December 3
1863	April 19	December 2
1864	April 8	December 12
1865	March 31	December 8
1866	April 1	December 13
1867	April 18	December 4
1868	April 18	December 10
1869	April 11	December 9
1870	April 8	December 21
1871	March 13	November 30
1872	April 19	December 10
1873	April 19	December 1
1874	April 16	December 12
1875	April 16	November 29
1876	April 18	December 10
1877	March 29	December 30
1878	April 2	December 19
1879	April 24	December 19
1880	April 6	November 26
1881	March 18	December 12 and January 2, 1882
1882	April 10	December 8

Navigation Dates—Penobscot River
(continued)

Year	Open to Navigation	Closed to Navigation
1883	April 13	December 16
1884	March 25	December 16
1885	April 18	December 17
1886	April 16	December 5
1887	April 15	December 23
1888	April 15	December 13
1889	March 30	December 14 and January 9, 1890
1890	April 6	December 2
1891	March 31	December 16
1892	January 2, 1892 January 19, 1892 April 2, 1892	January 16, 1892 January 21 December 12
1893	April 14	December 13
1894	March 23	December 22
1895	April 5	December 23
1896	April 12	December 11
1897	April 7	December 22
1898	March 26	December 14
1899	April 10	December 27
1900	April 3	December 12
1901	April 2	December 9, open December 16, closed 17
1902	March 22	December 8
1903	March 12	December 18
1904	April 4	December 6

Dates compiled from list in Bangor *Daily Commercial*, April 17, 1875, and thereafter from the daily "Commercial Marine Register" in *ibid.*, and *Reports of the Harbormaster*, yearly 1845-1927, in *Annual Reports, City of Bangor*, Bangor, various publishers, 1845-1960; Also Bangor *Daily Commercial*, December 8, 1904; Bangor *Daily News*, December 7, 10, 17, 18, 1901.

APPENDIX III

Pulpwood Consumption—United States—Maine Selected Years 1899-1930

Year	United States	Maine	No. of Mills Maine
1899	1,986,310	*	
1905	3,192,123		
1906	3,661,176		
1907	3,962,660	942,437	38
1908	3,346,953	717,813	35
1909	4,001,607	903,962	37
1910	4,094,306	917,029	37
1911	4,328,052	955,768	38
1917	5,480,075	1,309,239	33
1918	5,250,794	1,234,969	33
1919	5,477,832	1,279,852	34
1920	6,114,072	1,389,495	35
1921	4,557,179	1,005,158	32
1922	5,548,842	1,238,910	32
1926	6,766,007	1,298,357	26
1927	6,750,935	1,273,368	26
1928	7,160,100	1,309,988	25
1929	7,645,011	1,311,577	24
1930	7,195,524	1,203,377	25

* Maine not reported separately until 1907. Quantity in cords. Based on serial publications of Department of Commerce and Labor: Bureau of the Census. *Forest Products*, various subtitles and various dates.

APPENDIX
IV

ICE-FREE DATES—St. *John River at Fredericton Boom 1875-1915*
taken from table in Fredericton *Daily Gleaner*, April 21, 1967

Year	Open	Close
1875	May 2	November 17
1876	April 24	November 29
1877	April 15	November 23
1878	April 20	December 18
1879	April 9	November 21
1880	April 24	November 22
1881	May 2	November 21
1882	May 1	November 26
1883	April 18	November 16
1884	April 16	November 19
1885	April 23	November 27
1886	April 24	November 23
1887	April 27	December 1
1888	April 26	November 21
1889	April 16	December 1
1890	April 21	November 24
1891	April 14	November 20
1892	April 15	December 5
1893	April 14	November 14
1894	April 16	November 20
1895	April 10	December 6
1896	April 18	November 22
1897	April 22	December 2
1898	April 16	December 7
1899	April 21	November 13
1900	April 20	November 28
1901	April 8	November 22
1902	March 23	December 3
1903	April 1	November 20

Appendix, St. John ice-free dates

1904	April 21	November 13
1905	April 11	November 20
1906	April 19	November 28
1907	April 24	December 3
1908	April 26	November 19
1909	April 15	December 12
1910	April 4	December 4
1911	April 19	November 18
1912	April 18	December 2
1913	April 1	December 7
1914	April 27	November 19
1915	April 12	December 12

APPENDIX

V

ICE-FREE DATES—Moosehead Lake   1860-1900
from *Pittston Farm Weekly*, February 25, 1965

Year	Open	Close
1860	May 11	December 12
1861	May 12	December 18
1862	May 18	December 9
1863	May 18	December 7
1864	May 6	December 14
1865	May 4	December 11
1866	May 11	December 11
1867	May 19	December 6
1868	May 18	December 4
1869	May 10	December 4
1870	May 4	December 20
1871	May 13	November 29
1872	May 11	December 12
1873	May 16	December 1
1874	May 25	December 13
1875	May 24	December 11
1876	May 23	December 11
1877	May 6	December 4
1878	May 29	December 4
1879	May 14	December 14
1880	May 6	November 27
1881	May 9	December 12
1882	May 18	December 10
1883	May 13	December 12
1884	May 8	November 29
1885	May 16	December 15
1886	May 2	December 7
1887	May 13	December 21

Appendix, Moosehead Lake ice-free dates

1888	May 22	November 27
1889	April 30	December 7
1890	May 9	December 2
1891	May 14	December 7
1892	May 4	December 7
1893	May 18	December 3
1894	May 1	December 1
1895	May 6	December 5
1896	May 8	December 2
1897	May 8	December 3
1898	May 4	December 14
1899	May 7	December 8
1900	May 11	December 7

APPENDIX

VI

Raw Materials Cost—Papermaking—1872-1879
(in cents per pound)

Year	Bleach	Soda Ash	Alum	Cotton Waste	Colored Rags	White Rags	Wood Pulp	Common Newsprint
1872	4.8	4.12	3.37	3.75			5.	12.5
1873	3.31	3.19	3.37	3.75			4.87½	12
1874	2.71	2.75	3.37	2.50	3.	6.	4.5	10.75
1875	2.56	2.37	3.37	2.75	3.	6.	4.	9.62
1876	1.75	2.37	3.31¼	2.62½	2.78	5.25	4.	9.
1877	1.65	2.12	3.25	2.62½	2.53	4.5	3.5	8.5
1878	1.25	1.62	3.03	2.	2.12½	3.69	3.25	6.87
1879	1.25	1.6	2.62	2.	2.	3.41	3.	6.06

Source: *Paper Trade Journal*, February 14, 1880.

BIBLIOGRAPHY

Manuscript Sources and Collections of Papers

Bangor Harbormaster: These papers consist of various logs, both yearly and daily, of vessels entering and leaving Bangor. They are quite complete after 1917, and are embellished with newspaper clippings and photographs of important vessels. An annual report was published. The earlier records may exist, but apparently are in private hands. Office of the City Clerk, Bangor, Maine.

Austin Cary: Correspondence with Bureau of Forestry, 1894-1905. An excellent source on Cary's work both with the Maine Forestry Commision and the Berlin Mills Company. Located in Record Group 95, National Archives. (Microfilm).

Chandler Papers—University of Maine—Portable Sawmills in 1880's.

Forestry Commission: *Maine*: Useful letters, correspondence, papers, contracts, reports, surveys, on Maine forestry in general, and on public lots in particular, from 1889 to 1930. Forest Commissioner's Office, Augusta, Maine.

Great Northern Paper Company: I used various ledgers and files relating to annual cut, land purchases, and land sales from 1899 to 1930. In addition I used their copy of W. C. Hodge's cutting plan for their woods. Also many detailed files 1930-1947 from their Boston lawyers.

Hinckley and Egery: Various ledgers, account books, letters, and maps in Bangor Historical Society, Bangor Public Library, and Baker Library, Boston. This firm ran a foundry, and manufactured machinery, especially sawmill equipment, from 1828 to 1895. For this work the materials on land holdings were valuable.

W. C. Hodge: Correspondence 1901-3. This correspondence relates to the development of the Great Northern Paper Company's working plan in their woods. Record Group 95, National Archives, (Microfilm).

Land Agent: *Maine*: Letterbooks, account books, deed books, bills paid, survey books, maps, survey notes, and the other paraphernalia of a land office, 1820-1889. Forest Commissioner's Office, Augusta, Maine.

Major Lord: Six small account books, two or three letters, a few contracts, and a yearly balance sheet of a small Portland lumber merchant in the 1867-1875 period. Maine Historical Society, Portland, Maine.

Herbert W. Marsh et al vs. Great Northern Paper Company: Manuscript testimony, exhibits, and items introduced in testimony of the great trial resulting from the 1901 log drive on the Penobscot River. Penobscot County Court House, Bangor, Maine.

Mattawamkeag Lake Dam Company: Records, ledgers, account books, some letters relating to the operation of this small dam in northern Maine, 1860-1910. In custody of Ballard F. Keith, Bangor, Maine.

Mattawamkeag Log Driving Company: Some letters, record books, mostly accounts of annual meetings 1875-1915. In custody of Ballard F. Keith, Bangor, Maine.

Miscellaneous Papers: Letters, invoices, receipts, way-bills, a few ledgers, and a few stumpage contracts of different firms. These are the property of the Bangor Historical Society, although some of them are found in the Bangor Public Library.

New Brunswick Department of Land and Mines: Various type-written reports of surveys in New Brunswick woods, all cited in text or below.

Penobscot Lumbering Association: Fairly complete record of their transactions, and some letters as well, 1917-1953. In custody of Ballard F. Keith, Bangor Maine.

Joseph W. Porter: Although Porter was a big lumberman on the Penobscot and Passadumkeag from 1855 to 1900 these papers are of relatively little value as those records have disappeared. Consists now mostly of his correspondence as State Penal Institution Inspector. In Bangor Historical Society and Bangor Public Library.

Saint John River Commission: Testimony taken before this international group, 1909-1916. I used a lawyer's copy of this testimony now bound in five volumes and paged consecutively. Some of the exhibits are only described. An excellent and valuable source. Bangor Public Library.

Wakefield Papers: The family of O. H. Wakefield ran a small sawmill in Burlington, Maine from 1870 or so to 1910. These records, along with miscellaneous records of a store, dam companies, logging companies, sales of pulp wood, and so on, from 1869 to 1940 were in the author's possession, and are now in Fogler Library. Records are not cited in the text but were very valuable in understanding the change from sawlogging to pulp wood.

S. D. Warren Company Papers: These papers consist of all the company records which could be located in the period 1860-1930. They improve in quality after 1885. However, the total of materials was very valuable consisting of time books, letters, account books, ledgers, photographs, albums of clippings, files of letters and reports, as well as their working library of bound volumes. There is probably more material available somewhere, but it would not seriously change the account in the text.

Newspapers and Periodicals

Bangor (Maine) *Daily Commercial* 1872-1906.

Bangor (Maine) *Daily News* 1900-1905.

Kennebec *Daily Journal* (Augusta, Maine) 1895-1905.

Bibliography

Oxford *Democrat* (Paris, Maine) 1850-1883
This weekly covers Oxford County, Maine, and northeastern New Hampshire. Located in South Paris, Maine Public Library. Their run starts in the late 1830's and comes forward to the current issue.

Paper Trade Journal 1872-1971
The outstanding trade journal published weekly in New York City.

Portland (Maine) *Daily Eastern Argus* 1860-1892, 1900-1905.

Records and Proceedings, Cumberland Institute for Paper Perfection 1912-1914. Monthly reports of meetings to discuss the paper business. Mostly of technical interest, although occasional discussions of pulp wood procurement and mill operation shed light on the history of the S. D. Warren Company.

Superior Facts 1927-1932
A monthly magazine, published in Easton, Pennsylvania, devoted almost entirely to paper manufacturing history.

The Industrial Journal 1880-1918
This weekly which started as *Maine Mining Journal,* January 2, 1880, changed its name to *The Mining and Industrial Journal,* September 22, 1882, and *The Industrial Journal,* January 2, 1885. It became a monthly under the later title, June, 1901. It is an outstanding source for economic developments throughout its period. One file exists in a nearly complete form (99%) in the Bangor Public Library. I have cited it throughout in the final form of its name.

Katahdin Kalendar 1880-1881

The Northern 1921-1928
Monthly published by the Great Northern Paper Company, invaluable for the twenties and the coming of mechanization in the woods. Some useful historical information as well.

Mars Hill View 1909-1911

Presque Isle Sunrise 1867-1869

Warren's Standard 1951-1963
A monthly published by the S. D. Warren Company. Historical articles are helpful.

Farmington Chronicle 1870-1876.

Phillips Phonograph 1887-1888.

Wilton Record 1886-1888.

Official Publications—State of Maine (Serials)

Bureau of Industrial and Labor Statistics, *Annual Report,* 1887-1910, Augusta, Maine.

Commissioner of Forestry, *Annual Report,* 1892-1960, Augusta, Maine.

Commissioner of Inland Fisheries and Game, *Annual Report,* 1888, 1889, 1890, 1892, Augusta, Maine.

Department of Labor and Industry, *Biennial Report,* 1912-1930, Augusta, Maine.

Inspector of Factories, Workshops, Mines, and Quarries, *Annual Report,* 1893-1910, Augusta, Maine.

Land Agent, *Annual Report,* 1828-1889, Augusta, Maine.

Secretary of the Maine Board of Agriculture, *Annual Report,* 1860-1890, Augusta, Maine.

Trustees of the Maine State College of Agriculture and the Mechanic Arts, *Annual Report,* 1872, 1900-1910, Augusta and Orono, Maine.

Legislature, *House Documents,* 1860-1930, Augusta, Maine.

Legislature, *Senate Documents,* 1860-1930, Augusta, Maine.

(The above all found serially in the publication *Public Documents, Maine,* 1860-1930, Augusta, Maine.)

The following are bound separately:

Legislature, *House Journal,* 1860-1910, Augusta, Maine.

Legislature, *Senate Journal,* 1860-1910, Augusta, Maine.

Legislative Record 1917-1930.

Maine: Present Condition of the State—Its Agricultural, Financial, Commercial and Manufacturing Development, Advantages of the State as a Summer Resort, published under direction of the Secretary of State (Oramandel Smith), Kennebec Journal Print, Augusta, 1885.

Statistics of Industries and Finances of Maine for the Year 1883—Second Report—Secretary of State (Joseph O. Smith), Sprague and Sons, Augusta, Maine, 1883.

The Wealth and Industry of Maine for the Year 1873—First Annual Report—prepared by William E. S. Whitman, Sprague, Owen and Nash, Augusta, Maine, 1873.

Books

Frederick S. Allis Jr., editor, *William Bingham's Maine Lands 1790-1820,* Vol. XXXVI-XXXVII of the *Publications,* Colonial Society of Massachusetts, Boston, 1954.

Minnie Lee Atkinson, *Hinckley Township or Grand Lake Stream Plantation: A Sketch,* privately printed at the Newburyport *Herald* Press, Newburyport, Massachusetts, 1920.

Bibliography

J. W. Bailey, *The St. John River in Maine, Quebec, and New Brunswick*, The Riverside Press, Cambridge, Massachusetts, 1894.

H. K. Baring and O. O. Babb, *Water Resources of the Penobscot River Basin, Maine*, prepared in cooperation with the Maine State Survey Commission by the United States Geological Survey, Department of the Interior, Water Supply Paper 279, Washington, D. C., 1912.

Captain F. C. Barker, *Lakes and Forests as I Have Known Them*, Lee and Shepard Company, Boston, Massachusetts, 1903.

Stanley Foss Bartlett, *Beyond the Sowdyhunk*, Falmouth Book House, Portland, Maine, 1937.

Bruce W. Belmore, *Early Princeton, Maine*, n. p., Princeton, Maine, 1945.

P. Vidal De La Blache et L. Gallois, edit., *Amerique Septentrionale*, Part I, *Generalities—Canada*, Part II, *Etats-Unis*, Vol. XIII of *Geographic Universelle*, Librairie Armand Colin, Paris, France, 1935, 1936.

W. F. Blanding, compiler, *Bangor, Maine—Industries and Resources*, Bangor Board of Trade, (*Journal* Publishing Company), Bangor, Maine, 1888. (Subtitled—The City of Bangor—Queen City of the East—Manufacturing Advantages, Commercial Relations, Transportation Facilities, Business Resources and Social Features—Products of Her Industries for 1887.)

Edward Mitchell Blanding, compiler and publisher, *The City of Bangor—The The Industries, Resources, Attractions and Business Life of Bangor and Its Environs—Manufacturing Advantages, Commercial Relations, Transportation Facilities, Business Resources, Educational Opportunities, and Social Features of the Metropolis of the Northeast*, The Industrial Journal Press, Bangor, Maine, 1899.

Samuel Lane Boardman, M. S., *The Naturalist of the Saint Croix—Memoir of George A. Boardman*, (A Selection from his Correspondence and Published Writings, Notices of Friends and Contemporaries with his list of the Birds of Maine and New Brunswick), privately printed, Charles H. Glass and Company, Bangor, Maine, 1903.

W. J. Brennan et al. *Survey and Report of River and Stream Conditions in the State of Maine*—1930, n. p. (Maine Water Power Commission?), Augusta, Maine, 1931.

William Robinson Brown, *Our Forest Heritage: A History of Forestry and Recreation in New Hampshire*, New Hampshire Historical Society, Concord, New Hampshire, 1958.

Ralph C. Bryant, *Logging: The Principles and Common Methods of Operation in the United States*, McGraw-Hill Company, New York, 1923.

Laura J. C. Rullard, *Now-A-Days!* T. L. Magagnos and Company, New York, B. B. Mussey and Company, Boston, 1854.

Ava L. Chadbourne, *Maine Place Names and the Peopling of Its Towns*, Bond-Wheelwright Company, Portland, Maine, 1955.

Edward E. Chase, *Maine Railroads: A History of the Development of the Maine Railroad System*, Southworth Press, Portland Maine, 1926.

J. P. Cilley, *Bowdoin Boys in Labrador*, Rockland Publishing Company, Rockland, Maine, c. 1900.

City of Bangor, *Annual Reports*, 1861-1930, various publishers.

John H. Clapham, *An Economic History of Modern Britain*, 3 volumes, Cambridge University Press, Cambridge, England, 1926-1937.

Victor S. Clark, *History of Manufactures in the United States*, 3 volumes, published for the Carnegie Institution of Washington by the McGraw-Hill Company, New York, 1929.

——— Colby, *Colby's Atlas of the State of Maine Including Statistics and Descriptions of Its History, Educational System, Geology, Railroads, Natural Resources, Summer Resorts and Manufacturing Interests*, 4th Edition, Colby and Stuart, Houlton, Maine, 1888.

Forest H. Colby, *Forest Protection and Conservation in Maine*, (Maine Forestry Commission?, Augusta, Maine?), 1919.

Robert P. Tristram Coffin, *Kennebec: Cradle of Americans*. (The Rivers of America Series edited by Constance Lindsay Skinner), Farrar and Rinehart, New York, 1937.

D. C. Coleman, *The British Paper Industry 1495-1860: A Study in Industry and Growth*, Oxford University Press, New York, 1958.

S. W. Collins Company, *Lumbering Then and Now: S. W. Collins Co. 1844-1959*, S. W. Collins Company, Caribou, Maine (?), 1960 (?).

Edwin T. Coman Jr. and Helen M. Gibbs, *Time Tide and Timber: A Century of Pope and Talbot*, Stanford University Press, Stanford, California, 1949.

Philip T. Coolidge, *History of the Maine Woods*, privately printed by Furbush-Roberts Printing Company, Inc., Bangor, Maine, 1963.

Samuel T. Dana, *Forest Fires in Maine 1916-1925*, Bulletin No. 6, Maine Forest Service, Augusta, Maine, 1927.

Samuel T. Dana, *Timber Growing and Logging Practice in the Northeast: Measures Necessary to Keep Forest Land Productive and To Produce Full Timber Crops*, United States Department of Agriculture, Technical Bulletin No. 166, Washington, D. C., 1930.

A. H. Davis, *History of Ellsworth, Maine*, Lewiston *Journal* Press, Lewiston, Maine, 1927.

Harold A. Davis, *An International Community on the St. Croix 1604-1930*, University of Maine Studies, Second Series, No. 64, Orono, 1950.

Bibliography

William T. Davis, editor, *The New England States: Their Constitutional, Judicial, Educational, Commercial, Professional and Industrial History*, 5 volumes, D. H. Hurd and Co., Boston, c. 1897.

Clarence A. Day, *Farming in Maine 1860-1940*, University of Maine Studies, Second Series, No. 78, Orono, 1963.

Clarence A. Day, *A History of Maine Agriculture 1604-1860*, University of Maine Studies, Second Series, No. 68, Orono, 1954.

Holman Day, *King Spruce: A Novel*, A. L. Burt and Co., New York, 1908.

Holman Day, *The Rider of the King Log*, A.L. Burt and Co., New York, 1919.

James Elliot Defebaugh, *History of the Lumber Industry of America*, 2 volumes, The American Lumberman, Chicago, 1907.

Lew Deitz, *The Allagash*, Holt-Rinehart, Winston, 1968.

Department of the Interior, Canada, *The Province of New Brunswick, Canada—Its Development and Opportunities*, Ottawa, Canada, c. 1924.

Dominion Bureau of Statistics, *The Maritime Provinces 1867-1934, A Statistical Study of Their Social and Economic Condition Since Confederation*, (Published by the Authority of the Hon. H. H. Stevens, M. P., Minister of Trade and Commerce), Ottawa?, 1934.

George W. Drisko, *Narrative of the Town of Machias: The Old and the New, The Early and the Late*, Press of the Machias *Republican*, Machias, Maine, 1904.

Josiah H. Drummond, *The Maine Central Railroad System—An Uncompleted Historical Sketch*, n. p. (The Maine Central Railroad?), (Portland, Maine?), 1902.

Lewis Woodbury Eaton, *Pork, Molasses and Timber*, The Exposition Press, New York, 1954.

Fannie H. Eckstorm, *David Libbey: Penobscot Woodsman and River-Driver*, Vol. IV of True American Types, American Unitarian Association, Boston, Massachusetts, 1907.

Fannie H. Eckstorm, *Old John Neptune and Other Maine Indian Shamans*, The Southworth-Anthoenson Press, Portland, Maine, 1945.

Fannie Hardy Eckstorm, *The Penobscot Man*, new edition, Jordan-Frost Printing Company, Bangor, Maine, 1924.

Fannie Hardy Eckstorm and Mary Winslow Smyth, *Minstrelsy of Maine: Folksongs and Ballads of the Woods and the Coast*, Houghton-Mifflin Company, Boston, 1927.

L. Ethan Ellis, *Newsprint: Producers, Publishers, Political Pressures*, includes the text of *Print Paper Pendulum: Group Pressures and the Price of Newsprint*, Rutgers University Press, New Brunswick, New Jersey, 1960.

Edward H. Elwell, *Portland and Vicinity*, Loring Short, Harmon and Jones, Portland, Maine, 1876.

Federal Writers Project of the Works Progress Administration, *Maine: A Guide 'Downeast'*, American Guide Series, Houghton-Mifflin Company, Boston 1937.

Lewis B. Fisher, *The Story of a Downeast Plantation—Facts and Fancies About the Pine Tree State—By One of the Tribe of Fishers*, University of Chicago Press, Chicago, 1914.

Edwin M. Fitch, *Agricultural Tariffs—The Tariff on Lumber*, series edited by J. P. Commons, et. al. Tariff Research Committee, Madison, Wisconsin, 1936.

Robert J. Fries, *Empire in Pine: The Story of Lumbering in Wisconsin 1830-1900*, State Historical Society of Wisconsin, Madison, Wisconsin, 1951.

E. H. Frothingham, *White Pine Under Forest Management*, United States Department of Agriculture, Technical Bulletin No. 13, Washington, D. C., 1914.

Charles Glaster, *The West Brancher*, Vantage, 1970.

Roland Palmer Gray, *Songs and Ballads of the Maine Lumberjacks With Other Songs of Maine*, Harvard University Press, Cambridge, Massachusetts, 1924.

W. B. Greeley, *Forests and Men*, Doubleday and Company, New York, 1951.

W. B. Greeley, *Forest Policy*, (The American Forestry Series), McGraw-Hill Company, New York, 1953.

William E. Greening, *Paper Makers in Canada: A History of the Paper Makers Union in Canada*, International Brotherhood of Paper Makers, Cornwall, Ontario, 1952.

John A. Guthrie, *The Newsprint Paper Industry—An Economic Analysis*, Harvard Economic Studies Vol. LXVIII, Harvard University Press, Cambridge, Massachusetts, 1941.

Helen Hamlin, *Nine Mile Bridge: Three Years in the Maine Woods*, W. W. Norton Company, New York, 1945.

————, *Pine, Potatoes, and People*, Norton, 1947.

George W. Hammond (?), *A Mill-Built Dwelling House—Yarmouth, Maine*, Yarmouthville, Maine (?), n. p., 1892 (?).

James Hannay, D. C. L., *History of New Brunswick*, 2 volumes, John A. Rowles, St. John, New Brunswick, 1909.

James Hannay, *Saint John and Its Business*, n. p., St. John, New Brunswick, 1875.

Bibliography

Seymour E. Harris, *The Economics of New England: A Case Study of an Older Area*, Harvard University Press, Cambridge, Massachusetts, 1952.

W. E. Haskell, *Newsprint*, The International Paper Company, New York, 1921.

William E. Haskell, *The International Paper Company 1898-1924*, The International Paper Company, New York, 1924.

Keith Havey and Robert Davis, *Water Resources of Washington County*, Rural Areas Development Committee, Washington County, Maine (Office of the County Agent, Machias, Maine), 1962?

Samuel P. Hays, *Conservation and the Gospel of Efficiency: The Progressive Conservation Movement 1890-1920*, Harvard Historical Monographs, Vol. XL, Harvard University Press, Cambridge, Massachusetts, 1959.

Alfred Geer Hempstead, B. D. M. A., *The Penobscot Boom and the Development of the West Branch of the Penobscot River for Log Driving*, University of Maine Studies, Second Series, No. 18, Orono, 1931.

C. Max Hilton, F. E., *Rough Pulpwood Operating in Northeastern Maine 1935-1940*, University of Maine Studies, Second Series, No. 57, Orono, 1942.

Stewart H. Holbrook, *Yankee Loggers: A Recollection of Woodsmen, Cooks, and River Drivers*, The International Paper Company, New York, 1961.

Stewart H. Holbroook, *The American Lumberjack*, enlarged edition, of *Holy Old Mackinaw*, Collier Books, New York, 1962.

Homer Hoyt, *The Effect of the War Upon the World Lumber Situation*, United States War Trade Board—Bureau of Research, Misc. Reports, Vol. 1, No. 3, Washington, D. C., 1918.

Lucius L. Hubbard, *Hubbard's Guide to Moosehead Lake and Northern Maine, being the third edition, revised and enlarged of "Summer Vacations at Moosehead Lake and Vicinity," Describing Routes for the Canoe-man over the Principal Waters of Northern Maine, with Hints to Campers and Estimates of Expense for Tours*, A. Williams and Company, Boston, 1882.

Lucius L. Hubbard, *Woods and Lakes of Maine—A Trip From Moosehead Lake to New Brunswick in a Birch-Bark Canoe To Which Are Added Some Indian Place Names and Their Meanings New Published*, second edition, Tickner and Company, Boston, 1883.

John G. B. Hutchins, *The American Maritime Industries and Public Policy, 1789-1914: An Economic History*, Harvard Economic Studies, Vol. LXXI, University Press, Cambridge, Massachusetts, 1941.

Multiple Use of the Forests: A Progress Report From the International Paper Company, International Paper Company, New York, 1962?

Edward D. Ives, *Larry Gorman: The Man Who Made The Songs*, Indiana University Press, Bloomington, Indiana, 1964.

Edward D. Ives and David C. Smith, editors, *Fleetwood Pride 1864-1960: The Autobiography of a Maine Woodsman*, Vol. IX of *Northeast Folklore*, 1967 [Orono, Maine, 1968].

Vernon H. Jenson, *Lumber and Labor*, (Labor in the Twentieth Century, edited by Henry F. David et al), Farrar and Rinehart, New York, 1945.

J. F. W. Johnston, *Report on the Agricultural Capabilities of the Province of New Brunswick*, n. p., (Province of New Brunswick?), Fredericton, New Brunswick, 1850.

Herbert C. Jones, *Sebago Lake Land*, The Bowker Press, Portland, Maine, 1949.

Royal Shaw Kellogg, *Pulpwood and Wood Pulp in North America*, McGraw-Hill Company, New York, 1923.

Thomas Starr King, *The White Hills: Their Legends, Landscape and Poetry*, Crosby and Nichols, Boston, 1864 (first published 1859?).

Charles Lanman, *A Tour to the River Saguenay in Lower Canada*, Carey and Hart, Philadelphia, 1848.

L. N. Lamm, *Tariff History of the Paper Industry of the United States 1789-1922*, American Pulp and Paper Association, New York, 1927.

Wiliam Berry Lapham, *History of Norway, Oxford County, Maine*, Brown-Thurston Company, Portland, Maine, 1886.

Wm. B. Lapham and Silas P. Maxim, *History of Paris, Maine From Its Settlement to 1880 With a History of the Grants of 1736 and 1771 Together With Personal Sketches, a copious genealogical register and an appendix*, Paris, Maine, n. p., (The Authors), 1884.

Agnes M. Larson, *History of the White Pine Industry in Minnesota*, University of Minnesota Press, Minneapolis, Minnesota, 1949.

Leading Businessmen of Bangor, Rockland, and Vicinity, Embracing Ellsworth, Bucksport, Belfast, Camden, Rockport, Thomaston, Old Town, Orono, Brewer, Mercantile Publishing Company, Boston, 1888.

Richard Gordon Lillard, *The Great Forest*, A. A. Knopf, New York, 1947.

Lockwood Trade Journal, *1690-1940—250 Years of Papermaking in America*, Lockwood Trade Journal Company, New York, 1940.

Lockwood Trade Journal, *The Progress of Paper With Particular Emphasis on the Remarkable Industrial Development in the Past 75 Years and the Part that Paper Trade Journal Has Been Privileged to Share in That Development*, Lockwood Trade Journal Company, Inc., New York, 1947.

David Lowenthal, *George Perkins Marsh: Versatile Vermonter*, Columbia University Press, New York, 1958.

A. R. M. Lower, W. A. Carrothers, and S. A. Saunders, *The North American Assault on the Canadian Forest: A History of the Lumber Trade Between*

Bibliography 451

Canada and the United States, part of the *Relations of Canada and the United States,* published for the Carnegie Endowment for International Peace, by the Ryerson Press, Toronto, Canada, 1938.

M. Nelson McGeary, *Gifford Pinchot: Forester-Politician,* Princeton University Press, Princeton, New Jersey, 1960.

Lilian M. Beckwith Maxwell, *An Outline of the History of Central New Brunswick to the Time of Confederation,* York-Sunbury (N. B.) Historical Society, Inc., The Tribune Press, Sackville, New Brunswick, 1937.

Annual Report, Maine Central Railroad Company, 1890-1920, Tucker Printing Company, Portland, Maine (to 1911), after 1911, no publisher (by the company?).

Maine Chamber of Commerce and Agricultural League, *History of the Public or School Lots,* Maine Chamber of Commerce and Agricultural League, Portland, Maine, 1923.

Maine Forest Service, revised by H. B. Pierson, *Forest Trees of Maine,* revised edition, Maine Forest Service, Augusta, Maine, 1951.

State Forestry Department (Lillian S. Tschamler), *Report on Public Reserved Lots,* State of Maine, Augusta, Maine, 1963.

State of Maine, *Answers of the Justices of the Supreme Judicial Court to the Water Power Questions Propounded to the Court by the House of Representatives of the Seventy-Ninth Legislature,* State of Maine, Augusta, Maine, 1919.

George P. Marsh, *Man and Nature or Physical Geography as Modified by Human Action,* Charles Scribner's and Company, New York, 1867.

George P. Marsh, *The Earth as Modified by Human Action, A Last Revision of "Man and Nature",* Charles Scribner's and Company, New York, 1898.

Millinocket, Maine—50th Anniversary 1901-51, n. p., no place, no date listed (Millinocket, Maine, 1951?).

A. C. Morton, Civil Engineer, *Reort on the Survey of the European and N. American Railway: Made Under the Authority of the State of Maine,* Harmon and Williams, Portland, Maine, 1851.

Joel Munsell, *Chronology of the Origin and Progress of Paper and Paper-Making,* 5th edition, with additions, n. p., Albany, New York, 1876.

Guy Murchie, *Saint Croix: The Sentinel River,* Duell-Sloan, and Pearce, New York, 1947.

Government of New Brunswick, *The New Brunswick Official Yearbook—First Year of Issue,* Fredericton, N. B., 1907.

Northern Maine—The Sunrise Land, introductory number of the New Northeast, June, 1894, Bangor, Maine.

The New Northeast, Vol. 1, No. 1, July 1894, Bangor, Maine (all published?).

David Norton, Esq., *Sketches of the Town of Old Town, Penobscot County, Maine From Its Earliest Settlement, to 1879; with Biographical Sketches,* S. G. Robinson, Printer, Bangor, Maine, 1881.

H. W. Owen, *History of Bath, Maine,* The Bath *Times* Company, Bath, Maine, 1936.

Robert E. Pike, *Spiked Boots: Sketches of the North Country,* Cowles Press, Saint Johnsbury, Vermont, 1961.

———, *Tall Trees, Tough Men,* W. W. Norton, 1967.

Gifford Pinchot, *Breaking New Ground,* Harcourt, Brace and Co., New York, 1947.

Edward C. Plummer, *Reminiscences of a Yarmouth Schoolboy,* Marks Printing House, Portland, Maine, 1926.

Dorothy N. Prescott, *Cornelia Warren and the Story of Cedar Hill,* privately published, Belmont, Massachusetts, 1958.

Proceedings of the Thirty-First Legislature of the State of Maine in Relation to the European and North American Railway, Foster and Garrish, Portland, Maine, 1852.

A. Proteaux, with Additions by T. S. Le Normand, *Practical Guide for the Manufacture of Paper and Boards,* translated from the French with notes by Horatio Paine, A. B., M. D., to which is added a chapter on the manufacture of paper from wood in the United States by Henry T. Brown of the "American Artisan", Henry Carey Baird, Industrial Publisher, Philadelphia, 1866.

J. A. Purinton, Immigration Agent, (European and North American Company?), *Situation, Character, and Value of the Settling Lands in the State of Maine Published for the Information of Immigrants,* Samuel S. Smith and Son, Printers, Bangor, Maine, 1871.

John Rattray and Hugh Robert Hall, editors, *Forestry and Forest Products— Prize Essays of the Edinburgh Forestry Exhibition, 1884,* David Douglas, Edinburgh, 1885.

Rev. William O. Raymond LLD., F.R.S.C., *The River St. John: Its Physical Features, Legends and History From 1604 to 1784,* edited, by Dr. J. C. Webster, C. M. G., The *Tribune* Press, Sackville, New Brunswick, 1st edition, 1910, reprinted 1943, 2nd edition, 1950.

Parker McC. Reed, *History of Bath, Maine,* n.p., Portland, Maine, 1894.

George W. Rice, *The Shipping Days of Old Boothbay,* Boothbay Harbor Press, Portland, 1945.

Henry Richards, *Ninety Years On 1848-1940,* privately printed at the Kennebec *Journal* Press, Augusta, Maine, 1940.

Bibliography

Laura E. Richards, *Stepping Westward*, D. Appleton and Co., New York, 1932.

C. P. Roberts, editor, *Knowles Bangor Business Almanac for 1875 With Historical Sketches of Bangor and Its Business Enterprises*, O. F. Knowles and Co., Bangor, Maine, 1875.

Ernest R. Rowe *et al*, edited by Marian B. Rowe, *Highlights of Westbrook History*, Westbrook Women's Club, Westbrook, Maine, 1952.

William Hutchinson Rowe, *The Maritime History of Maine: Three Centuries of Shipbuilding and Seafaring*, W. W. Norton and Co., New York, 1948.

Report of Royal Commission in Respect to Lumber Industry, 1927, Fredericton, New Brunswick, 1927.

S. A. Saunders, *The Economic Welfare of the Maritime Provinces*, Acadia University Press, Wolfville, Nova Scotia, 1932.

Scenic Gems of Maine, Geo. W. Morris, Portland, Maine, 1898.

C. A. Schenck, *Logging and Lumbering or Forest Utilization: A Textbook for Forest Schools*, L. C. Wittich, Darmstadt, 1913-14.

Upton Sinclair, *The Autobiography of Upton Sinclair*, Harcourt-Brace-World, New York, 1962.

David C. Smith, *History of the Paper Industry of the United States 1690-1970*, Lockwood's, New York, 1971.

Lincoln Smith, *The Power Policy of Maine*, The University of California Press, Berkeley, California, 1951.

John S. Springer, *Forest Life and Forest Trees: Comprising Winter Camplife Among the Loggers, and Wild-wood Adventure, With Descriptions of Lumbering Operations on the Various Rivers of Maine and New Brunswick*, Harper and Brothers, New York, 1851.

State Chamber of Commerce and Agricultural League, *History of the Land Grant to the European and North American Railway*, Portland, 1923?

State Chamber of Commerce and Agricultural League, *History of the Wild Lands of Maine*, Portland, Maine, c. 1923.

Thomas Sedgewick Steele, *Canoe and Camera or Two Hundred Miles Through the Maine Forests*, Orange Judd Company, New York, 1880.

Thomas Sedgewick Steele, *Paddle and Portage From Moosehead Lake to the Aroostook River, Me.*, Estes and Lauriat, Boston, 1882.

Henry Bake Steer, *Lumber Production in the United States 1799-1946*, United States Department of Agriculture Miscellaneous Publication No. 669, Washington, D. C., 1948.

Isaac Stephenson, *Recollections of a Long Life: 1829-1915*, privately printed by R. R. Donnelley and Sons Company, Chicago, 1915.

James Stacey Stevens, *Meteorological Conditions at Orono, Maine*, University of Maine Studies (Old Series), No. 7, Orono, Maine, 1907.

Louis Tillotson Stevenson, D. C. S., *The Background and Economics of American Papermaking*, Harper and Brothers, New York, 1940.

Edwin Sutermeister, *The Story of Papermaking*, S. D. Warren Company, Boston, 1954.

The Loggers or Six Months in the Forest of Maine, (The Dirigo Series), Horace B. Fuller, publisher, Boston, 1870?

L. C. Thomas, *The Province of New Brunswick, Canada: Its Natural Resources and Development*, Department of the Interior, Ottawa, 1930.

Henry D. Thoreau, *The Maine Woods*, (Riverside Edition), Houghton-Mifflin Company, Boston, 1906.

William Todd, *Todds of the St. Croix Valley*, privately printed at Mount Carmel, Connecticut, 1943.

United States, 60th Congress, 2nd Session (1908-1909), House Documents Vol. 132-6 (Document No. 1502), *Pulp and Paper Investigation Hearings*, Washington, 1909.

United States, 64th Congress, 2nd Session Senate Document No. 724, *Uses of the St. John River—Report of the International Commission Pertaining to Uses and Conditions of the St. John River, on the reference by the United States and Canada*, Washington, D. C., 1917 (dated February 17, 1916).

United States Congress, Senate, Committee on Labor and Public Welfare, *Importation of Canadian Bonded Labor, Hearing Before Committee on S. Res. No. 98*, 84th Congress, 1st Session, Washington, D. C., 1954.

United States Department of Agriculture, Forest Service, *Forests and National Prosperity: A Reappraisal of the Forest Situation in the United States*, Miscellaneous Publication No. 668, Washington, D. C., 1948.

United States Deartment of Commerce and Labor, Bureau of the Census, *Forest Products of the United States*, 1907, 1908, 1909; *Pulpwood Consumption* 1910, 1911; *Pulpwood Consumption and Woodpulp Production*, (Forest Products 1921, 1922); *Paper and Paper Board: Production and Papermaking—Equipment in Use—Pulpwood Consumption and Wood Pulp Production* (Forest Products 1928, 1930, 1932), Washington, D. C., 1909, 1911, 1912, 1913, 1923, 1924, 1929, 1932, 1934. (The same publication, but the title varies as indicated).

United States Department of State, Bureau of Statistics, *Special Consular Reports—American Lumber in Foreign Markets*, Vol. XI, 53rd Congress, 3rd Session, Miscellaneous Document No. 92, Washington, D. C., 1896.

Charles R. Van Hise, *The Conservation of Natural Resources in the United States*, The Macmillan Company, New York, 1910.

Bibliography

Alfred J. Van Tassel with the assistance of David Bluestone, *Work Mechanization in the Lumber Industry*, Work Projects Administration—National Research Project, David Weintraub director, Report no. M-5, Philadelphia, 1940.

Frederick William Wallace, *In the Wake of the Wind Ships*, George Sully and Co., New York, 1927.

Frederick W. Wallace, *Record of Canadian Shipping 1786-1920*, Hodder and Stoughton, London, England, 1929.

Cornelia Warren, *A Memorial of My Mother*, privately printed at the Merrymount Press, Boston, Massachusetts, 1908.

S. D. Warren Company, *A History of the S. D. Warren Company, 1854-1954*, Anthoensen Press, Westbrook, Maine, 1954.

Samuel Dennis Warren—September 13, 1817—May 11, 1888: A Tribute from the People of Cumberland Mills, The Riverside Press, Cambridge, Massachusetts, 1888.

George Wasson and Lincoln Colcord, *Sailing Days on the Penobscot: The River and Bay as They Were in the Old Days*, Marine Research Society, Salem, Massachusetts, 1932.

Charles E. Waterman, *The Oxford Hills and Other Papers*, Merrill and Webber Company, Auburn, Maine, c. 1930.

Lyman Horace Weeks, *A History of Paper-Manufacturing in the United States 1690-1916*, The Lockwood Trade Journal Company, New York, 1916.

Walter Wells, *The Water Power of Maine*, Sprague, Owen, and Nash, Augusta, Maine, 1864.

Westbrook *American*, May 19, 1954, *S. D. Warren Centennial Edition*, Westbrook, Maine, 1954.

West Virginia Pulp and Paper Company, *Fifty Years of Papermaking*, West Virginia Pulp and Paper Company, New York, 1937.

William Bond Wheelwright, *From Paper Mill to Pressroom*, George Banta Publishing Company, Kenosha, Wisconsin, 1920.

Charles F. Whitman, *A History of Norway, Maine From the Earliest Settlements to the Close of the Year 1922*, The Lewiston *Journal* Printshop and and Bindery, Lewiston, Maine, 1924.

Austin H. Wilkins, *The Forests of Maine—Their Extent, Character, Ownership, and Products*, Bulletin No. 8, Maine Forest Service, Augusta, Maine, 1932.

Charles E. Williams, *The Life of Abner Coburn: A Review of the Political and Private Career of the Late Ex-Governor of Maine*, Press of Thomas W. Burr, Bangor, Maine, 1885.

Harold Fisher Wilson, *The Hill Country of Northern New England: Its Social and Economic History 1790-1930*, Columbia University Studies in the History of American Agriculture, Vol. III, edited by Harry J. Carman and Rexford G. Tugwell, Columbia University Press, New York, 1936.

Theodore Winthrop, *Life in the Open Air and Other Papers*, Henry Holt and Company, New York, 1876.

C. D. Woods and E. R. Mansfield, *Studies of the Food of Maine Lumbermen*, United States Department of Agriculture, Office of Experiment Stations, Bulletin No. 149, Washington, D. C., 1904.

Frederick James Wood, *The Turnpikes of New England*, Marshall-Jones Company, Boston, 1919.

Richard G. Wood, *A History of Lumbering in Maine 1820-1861*, University of Maine Studies, Second Series, No. 33, Orono, Maine, 1935, new edition with introduction by David C. Smith, Orono, 1972.

Articles

Philip W. Ayres, "Is New England's Wealth in Danger?", *New England Magazine*, XXXVIII, (May, 1908).

————, "The Outlook for Forestry in New Hampshire," *Granite Monthly*, LIX, (January, 1927).

"Austin Cary Retires", *Journal of Forestry*, XXXIII, No. 9, (September, 1935), 820-1.

"Austin Cary Speaks Out", *Journal of Forestry*, XXXIII, No. 11, (November, 1935), 916-22.

Alfred G. Bailey, "Railroads and the Confederation Issue in New Brunswick, 1863-1965", *Canadian Historical Review*, XXI No. 4, (December, 1940), 367-383.

————, "The Basis and Persistence of Opposition to Confederation in New Brunswick," *Canadian Historical Review*, XXIII, No. 4, (December, 1942), 374-397.

Bangor *Daily News*, November 8, 1956, "The Telos War."

Bangor *Daily News*, April 15, 1957, "Land Agents Letter".

Liberty Dennett, "Maine Wild Lands and Wild Landers", *Pine Tree Magazine*, Vol. VI, No. 6; Vol. VII, No. 1-6; Vol. III, No. 1, (January-August 1907).

C. A. Brautlecht, "Pulp and Paper in Maine", *Paper Mill*, Vol. 57, No. 35, (September 1, 1934).

Charles F. Brooks, "New England Snowfall", *Geographical Review*, Vol. 3 (1917), 220-240).

Bibliography

P. L. Buttrick, "Forest Growth on Abandoned Agricultural Land", *The Scientific Monthly*, V, (July, 1917).

Austin Cary, "A Defense of Private Forest Ownership", *Journal of Forestry*, XXXIII, No. 12 (December, 1935), 964-7.

Austin Cary, "Forest People", *American Forests and Forest Life*, Vol. XXI, No. 383 (November, 1925), 664-6.

Austin Cary, "Forty Years of Forest Use in Maine", *Journal of Forestry*, XXXIII, No. 4 (April, 1935), 366-72.

Austin Cary, "How Lumbermen in Following Their Own Interests Have Served the Public: An Address Delivered Before the Society of American Foresters, Feb. 10, 1916", *Journal of Forestry*, XV, No. 3 (March, 1917), 271-89.

Austin Cary, "How to Apply Forestry to Spruce Lands", *Paper Trade Journal*, February 19, 1898.

Austin Cary, "Letter From Austin Cary to Mr. Earle Kaufman, January 22, 1935", *Journal of Forestry*, XXXIII, No. 6, (June, 1935).

Austin Cary, "Maine Forests, Their Preservation, Taxation, and Value: Paper Read Before State Board of Trade, Bangor, Sept. 26, 1906", *Annual Report*, Commissioner of Industrial and Labor Statistics, Maine, 1906, 180-193.

Austin Cary, "Management of Pulp Wood Forests: System of Forestry Practiced by Berlin Mills Company", Fourth Annual *Report* of the Forest Commissioner, Maine, 1902, 125-52.

Austin Cary, "Notes on Relative Frost Hardiness", *Forestry Quarterly*, II, No. 1 (November, 1903), 22-3.

Austin Cary, "On the Growth of Spruce", Second Annual *Report* of the Forest Commissioner, Maine, 1894.

Austin Cary, "Report of Austin Cary to the Forest Commissioner", Third Annual *Report* of the Forest Commissioner, Maine, 1897, 15-203 plus appendix, consists of "Spruce of the Kennebec", "Spruce on the Androscoggin", "Explorations on the Magalloway", and "Our Forests and the Future".

Austin Cary, "The Forests of Maine", *Paper Trade Journal*, April 25, May 9, May 16, 1896.

Austin Cary, "The Future of New England Forests", *Journal of Forestry*, XXI, No. 1 (January, 1923), 15-24.

Austin Cary, "Unprofessional Forestry", *Forestry Quarterly*, IV, No. 3, (September, 1906), 183-7.

Gilbert Chinard, "The American Philosophical Society and the Early History of Forestry in America", *Proceedings* of the American Philosophical Society, Vol. 89, No. 2 (July, 1945), 444-88.

Chief Engineer, "Law of Waters", Third *Report* of the Water Storage Commission, Maine, 1912, 63-90.

Hugh J. Chisholm, "History of Papermaking in Maine, the Future of the Industry", Twentieth Annual *Report* of the Bureau of Industrial and Labor Statistics for the State of Maine, 1906, 161-9.

"Early Days of Log Driving on the St. John River", *Candian Lumberman*, September 15, 1925.

A. C. Eckler, "A Measure of the Severity of Depressions 1873-1932", *Review of Economic Statistics*, XV, (May, 1933).

Rendig Fels, "American Business Cycles 1865-1879", *American Economic Review*, XLI, (June, 1951), 325-48.

B. E. Fernow, "Forestry and Wood Pulp Supplies", *Paper Trade Journal*, February 19, 1898.

John P. Flanagan, "Labor in the Maine Woods", First Biennial *Report* of the Department of Labor and Industry, Maine, 1912, 206-227.

Charles B. Fobes, "Lightning Fires in the Forests of Northern Maine, 1926-1940", *Journal of Forestry*, XLII, (April, 1944), 291-3.

"Forest Management and Reforesting", Second Annual *Report* of the Forest Commissioner, Maine, 1894, 60.

"Great Northern Paper Company Builds a New Dam", *Pulp and Paper Magazine of Canada*, April 1, 1915, 211-2.

W. B. Greeley, "Austin Cary As I Knew Him", *American Forests*, LXI, No. 5 (June, 1955), 30.

Roland M. Harper, "Changes in Forest Area of New England in Three Centuries", *Journal of Forestry*, XVI, (April, 1918).

A. B. Hastings, "Appreciation of Austin Cary", *Journal of Forestry*, XXXVII, No. 4 (April, 1939), 287.

Frank Heyward, "Austin Cary, Yankee Peddler in Forestry", *American Forests*, LXI, No. 5-6, (June, July, 1955).

Ray R. Hirt, "Fifty Years of White Pine Blister Rust in the Northeast", *Journal of Forestry*, Vol. LVI, No. 7 (July, 1956), 433-8.

Franklin B. Hough, "On the Duty of Governments in the Preservation of Forests", *Proceedings,* American Association for the Advancement of of Science, (Augusta, 1873).

Ernest F. Jones, "Forestry Conceptions Among Timberland Owners in the Northeast Spruce Region", *Journal of Forestry,* Vol. XXXIX, (1931), 175-80 (typescript in my possession.)

E. O. Merchant, "The Government and the News-Print Paper Manufacturers", *The Quarterly Journal of Economics,* Vol. XXXII, (1918), 238-56.

Ralph H. McKee, "The Training of Men for Positions in Pulp and Paper Mills", Second Biennial *Report,* Department of Industry and Labor, Maine, 1914, 149-159.

J. C. Nellis, "Wood-Using Industries of Maine", Ninth Annual *Report of Forest* Commissioner, Maine, 1913, 83-185.

Max Phillips, "Benjamin Chew Tilghman, and the Origin of the Sulphite Process for Delignification of Wood", *Journal of Chemical Education,* Vol. XX, (September, 1943), 444-7.

Gifford Pinchot, et al, "Deferred Forestry", *New England Magazine,* Vol. XXX (December, 1908).

Gifford Pinchot, "The Sustained Yield of Spruce Lands", *Paper Trade Journal,* February 19, 1898.

"Report of the Third Annual Meeting of the Canadian Forestry Association, 1902", *Forestry Quarterly,* I, No. 2, (January, 1903), 67-9.

"Report on the Drainage Basins in Maine", Maine State Planning Board, under direction of Arthur C. Cuney, and Alfred Milliken, Augusta, Maine, 1937 (mimeographed).

Samuel Rezneck, "Distress, Relief, and Discontent in the United States During the Depression of 1873-1878", *Journal of Political Economy,* Vol. 58 (1950), 494-512.

Ralph S. Sawyer, "Katahdin Iron Works", *Sun-up Magazine,* September, 1927.

David C. Smith, "Wood Pulp and Newspapers, 1867-1900", *Business History Review,* Vol. XXXVIII, No. 3 (Autumn, 1964), 328-345.

―――――, "Virgin Timber: The Maine Woods as A Locale for Juvenile Fiction," in Richard Sprague, ed., *A Handful of Spice,* University of Maine Studies No. 88, Orono, 1969, 186-205

―――――, "Maine and Its Public Domain: Land Disposal on the Northeastern Frontier," in David M. Ellis, ed., *The Frontier in American Development: Essays in Honor of Paul Wallace Gates,* Cornell University Press, Ithaca, New York, 1969, 113-137.

―――――, "Bangor—The Shipping and Lumber Trade," in James Vickery, ed., *A History of Bangor,* Forbush-Roberts, Bangor, 1969, 23-37.

―――――, "Toward A Theory of Maine History—Maine's Resources and the State," in Arthur Johnson, ed., *Explorations in Maine History, Miscellaneous Papers,* Orono, 1970, 45-64

―――――, "Forest History Research and Writing at The University of Maine," *Forest History,* Vol. 12, no. 2 [July, 1968], 27-31

―――, "A Look at The United States Paper Industry in its 19th-Century Growing Years," *Paper Trade Journal*, Vol. 152, no. 37, [September 9, 1968], 60-63.

―――, "Paper Mill Problems in Middle of Last Century Resemble Ours," *Paper Trade Journal*, Vol. 153, no. 2 [January 13, 1969], 32-6

―――, "Wood Pulp Paper Comes to The Northeast, 1865-1900," *Forest History*, Vol. 10, no. 1, [April, 1966], 12-25.

George Smith, "King Axe", *St. Croix Observer*, Vol. 38, No. 7, (January-February, 1963).

Herbert A. Smith, "Forest Education Before 1898", *Journal of Forestry*, Vol. XXXII, (October, 1934), 684-9.

Herbert A. Smith, "The Early Forestry Movement in the United States", *Agricultural History*, Vol. 12, No. 4 (October, 1938), 326-46.

Frederick Starr, "American Forests: Their Destruction and Preservation", *Annual Report*, United States Department of Agriculture, 1865, 218-9.

"The Wood Lot", Fifth Annual *Report*, Forest Commissioner, Maine, 1904.

James W. Toumey, "The Woodlot: A Problem for New England Farmers", *Scientific Monthly*, V, (September, 1917), 193.

Orville Wells, "The Depression of 1873-1879", *Agricultural History*, XI, (July, 1937), 237-49.

"When Did Newspapers Begin to Use Wood Pulp Stock?", *Bulletin*, New York Public Library, Astor, Lenox and Tilden Foundations, Vol. XXXIII, No. 10, (October, 1929), 743-9.

Byron Weston, "History of Paper Making in Berkshire County, Mass." *Collections*, Berkshire Historical and Scientific Society, II, 1895.

Marinus Westveld, et al, "Natural Forest Vegetation Zones of New England", *Journal of Forestry*, Vol. LIV, No. 5 (May, 1956), 332-8.

Francis Wiggin, "The Preservation of Maine Forests: An Address Delivered Before the Board of Trade at Rockland, Maine, Oct. 15, 1901", Annual *Report*, Commissioner of Industrial and Labor Statistics, Maine, 1901, 103-118.

Ben Ames Williams, "The Return of the Forest", *American Forests*, (1927); *Literary Digest*, XCIV, (Sept. 10, 1927).

Roy N. White, "Austin Cary, the Father of Southern Forestry", *Forest History*, Vol. 5, No. 1, (Spring, 1961).

Unpublished Manuscripts and Theses

George T. Carlisle and T. Frank Shatney, "Report on a Logging Operation in Northern Maine", unpublished thesis, University of Maine, Orono, Maine, 1909, 16 pps.

Bibliography

Bernard A. Chandler, "Lumbering in Northern Maine as Illustrated by an Operation of the Emerson Lumber Company", unpublished thesis, University of Maine, Orono, Maine, 1909, 29 pps.

Elda Garrison, "The Short Route to Europe: A History of the European and North American Railroad", M.A. thesis, University of Maine, Orono, Maine, 1950.

Robert Donald Goode, "The Economic Growth of the Pulp and Paper Industry in Maine", M.A. thesis, University of Maine, Orono, Maine, 1934.

Wm. C. Hodge, Jr., Field Assistant, Bureau of Forestry, "A Working Plan for the Penobscot Timberlands of the Great Northern Paper Company: Somerset and Piscataquis Counties, Maine", type-written manuscript, c. 1903, 71 pps. files of the Great Northern Paper Company.

R. B. Miller, "Report on New Brunswick Reconnaissance Work", type-written manuscript in Department of Lands and Mines, Fredericton, New Brunswick, files, c. 1913, 106 pps. plus several appendices.

W. J. H. Miller, J. P. Poole, and H. H. Sweetser, "A Lumbering Report of Work on Squaw Mountain Township: Winter of 1911-1912", unpublished thesis, University of Maine, Orono, Maine, 1912, 28 pps. plus appendix.

Lewis F. Pike and Jon N. Jewett, "A Report on a Lumbering Operation on Township No. 29—Washington County, Maine", unpublished thesis, University of Maine, Orono, Maine, 1908, no pps.

Glen C. Prescott and Raymond E. Rendall, "Lumbering in the Dead River Region—Somerset County, Maine", unpublished thesis, University of Maine, Orono, Maine, 1916, 68 pps. plus appendix.

Mary B. Reinmuth in collaboration with Donald A. Fraser, "Forest Heritage: The Story of Fraser Companies, Ltd.", typewritten manuscript in files of Fraser Company, Ltd., Edmundston, New Brunswick.

David Nathan Rogers, "Lumbering in Northern Maine", A Report Presented to the Department of Forestry of the University of Maine, January, 1906, University of Maine, Orono, Maine, 1906, 55 pps.

Joseph Herbert Sewall, "A Comparative Study of the Development of Forest Policy in Maine and New Brunswick", M.S. thesis, University of Maine, Orono, Maine, 1957, 200 pps., plus appendix.

R. J. Smith and S. B. Locke, "A Study of the Lumbering Industry in Northern Maine—Winter 1907-1908", unpublished thesis, University of Maine, Orono, Maine, 22 pps.

"The Evidence Before the Committee on Interior Waters, on petition of Wm. H. Smith, Daniel M. Howard, Warren Brown, and Theodore H. Dillingham, for leave to build Sluiceway From Lake Telos to Webster Pond", 1928, and 1941, 91 pps., (mimeographed).
reported by Isreal Washburn, Jr., Maine State Library, Augusta, Maine,

"The Timber Situation in Maine, 1919", dated May 31, 1919 in files of Forest Service, Record Group 95, National Archives (microfilm).

Harris E. Videto, "The Development of the Forest and Forest Industries of New Brunswick Resources Development Board, Fredericton, New Brunswick, 1951, 84 pps., (mimeographed).

Roy Ring White, "Austin Cary and Forestry in the South", Ph.D. dissertation, University of Florida, 1960.

ORAL INTERVIEWS

I. H. Bragg, Patten, Maine, August, 1963.
Louis Freedman, Old Town, Maine, July, 1962.
William Hilton, Bangor, Maine, July, August, 1962.
Harold Noble, Topsfield, Maine, several, 1961, 1962, 1963, 1964.
Caleb Scribner, Patten, Maine, August, 1963.

INDEX

Amounts of cut: 1870-1890, 38, 39, 40; 1890-1900, 40
Androscoggin area: Twentieth century cut, 384
Androscoggin drives: 70-2; long drive of 1880, 71; pulpwood drives, 71-2; no organization, 72
Androscoggin River, 2, 41-3
Androscoggin valley: small operations described, 42-3, 58n
American Wood Paper Company, 234
Arbor Day, 341-2
Aroostook County: as alternative to West, 187; bond issues for Bangor and Aroostook Railroad, 223, 229-30n; needs of pioneers, 205n
Ashland Lumber Company: described, 223-4; shipments, 230n
Auxiliary forests, 372-3
Ayer, Fred W.: log corner, 129-30; 1901 drive, 273, 274; sawdust balers, 130; testifies, 1903, 285; testifies in Marsh case, 298
Ayer, F. W.; mills, description and production, 123
Axes, types and uses, 17

Band saw, 123, defined, 134n
Bangor: dependency on lumbering, 37; imports, 149; related exports, 149; sawmills, 118-119; shipping, amounts and destination, 147-8
Bangor and Aroostook Railroad: 222-4, construction, 223; impact on area, 223-4; tonnage carried, 399-400
Bangor and Piscataquis Railroad, 44, 222-3
Bangor Boom: 89-90; last days, 402-3; log sizes, 293
Bangor Navigation Company, 144
Bangor Tigers, 100n
Baring, great fire of 1875, 126
Basin Mills: description and production, 120-1; closes, 417n; production, 123
Bath: sawmills, 117; shipping, 147, 166n
Baxter, Percival W.: becomes Governor, 315; before legislature on Kennebec reservoir bill, 319; calls for Katahdin park, 1923, 317-8; comments on park provenance, 331n; comments on Central Maine Power Company, 313; comments on Kennebec Reservoir Co. referendum, 318-9; comments on his enemies, 322; comments on lobbyists, 317-8; comments on public power, 313; comments on Public Utilities Commission, 312; describes compromise, 324; farewell address, 323-4; floor manager for public power bill, 311; inaugural address (1923), 315, 317; issues proclamation, 321-2; Mount Katahdin State Park, 312, 315; Offers the Mount Katahdin compromise, 324; President of Senate, 314; proclamation on Kennebec Reservoir Co., 320; sets tone for public/ private power battle, 315-6; speech as senate president, 314; United States Senate campaign, 1926, 324-5
Baxter State Park, 325
Berlin Mills Company: comments on Cary's Work, 357; driving, 72
Bethel Steam Mill Company, 115
Bingham purchases, 5
Blueberries, 370
Blunt and Hinman: description and production, 119-120
Booms: described, 85-6
Boom jumpers, 406-7
Box boards, 122
Box shooks, 141
Bradstreet Conveyer, 395
Brewster, Ralph Owen: Baxter floor manager, 315; "overcome", 320; World War II labor shortage, 426
Bureau of Forestry: works on Great Northern Paper Company woods, 361-3

Canadian: food and provisions, 20
Canadian Pacific Railroad: construction and route, 221, 222, 229n
Canadians: arrival in Maine, 19; discontent in Maine, 19-20; hiring, 20-1; visit to logging camp 20; workers, 19
Cant dog business, 96n
Canton Steam Mill Company, 115-116
Camp construction: 14-15; in the 1930's, 423
Camp interiors, 15, 16
Cary, Austin F.: biography, 344-5; analyzes his impact, 361; Berlin Mills Forester, 347-8, 359-60; comments on early work, 345-6; on Great Northern Paper Company,

361, on landowners, 361, to Pinchot, 358; on University of Maine, 360; early reports, 345; European study, 347; further assessment, 353n; gives lectures, 358-9; teaching techniques, 377n; views assessed, 349; on what is to be done, 346-7
Carratunk Falls, 73
Chamberlain Farm: described, 23, production, 23
Chesuncook and Chamberlain Lake Railroad, 398
Chisholm, Hugh J.: Biography, 259n; offers advice to pulp and paper workers, 256; responds to attacks, 366; makes announcement, 249
Christmas trees, 159
Civilian Conservation Corps: 419-422; assessment, 422; camps, 420; insect control, 421; minor work, 421-2; roads, 420-1
Climate and weather, 23-4
Coats, J. & P.: spoolwood, 142
Coburn farm: production, 22
Coburn Land Company, 195
Coburn operations, 43
Connecticut River log drives: 77-9; 1876 drive, 78
Connors, Patrick: 1901 drive, 276
Connors, William: master rafter, 89
Cooks: attributes, 15; methods, 15
Cooper Brook operations, 394-5
Contract Cutting, 424
Cost of Living, 231n
Costs: factors in, 225-6
Cruising, 14
Cunliffe, William, operations, 384
Cunliffe and Stevens, 54, 55
Cutting Laws: defeated, 371-2, 373; passed, 373; amended, 373; repealed, 373
Cutting legislation, 371-4
Cutting records, 25

Dale, Samuel H., land sales, 196, 197
Dams, 14, 64
Deacon seat, 16
Deal: defined, 132n; European destinations, 149-150
Dennison, A. C.: mills, 235; Mechanic Falls operations, 236; other ventures, 236-7; Canton mill, 236-7; labor difficulties, 237
Dennysville, Production in 1871, 135n
Depletion, impact of natural disasters on, 337-8
Depot Camp, 423
bering, 211
Depression of 1873: impact on lum-

Depression of 1930's: impact on lumbering, 419
Dingle, 15
Downeast operations, 45-6
Downeast river driving, 75-6
Driving techniques, 65-7
Dry Town, 2
Dynamite, 14

Early Fire Control Laws, 338
East Branch, Penobscot River, cut statistics, 1895-1905, 385
East Branch Log Driving Company, 1898-1904 prices, 402
Eagle Lake Tramway: 395-6; Building, 396; impact, 396
Eckstorm, F. H.: comments on 1901 drive; 292; writes *Penobscot Man*, 292-3
Ellsworth: lumber production, 1877, 124
Ellsworth sawmills: impact of Bar Harbor, 124-5
European and North American Railway: Aroostook lands and trespass, 182; disposal of land, 184, 185; failure, 179; further investigations, 184; land grant, 177-8; location, 203n; money raising, 178-9; New Brunswick section, 178; publicity, 181; purpose, 177; relations with Maine Central, 183-4; sale of land, 180-1; trusteeships, 179-180
European Trade: from Canada, 153-4
Exploration, 14

Fairfield: sawmills, 117; Lawrence Phillips and Co., 117
Fernald, Bert: calls for cutting law, 371-2
Fernald, C. H., entomological studies, 338-9
Fires in Camp, 24
First settlers, 5
Flies, 23
Flooding: and driving, 68
Forest Commission Act: passes, 344-5, early responses, 344
Forest depletion, efforts to ameliorate, 336
Forest Fires: of 1903, 368; of 1947, 428-9
Forest Fibre Company, 238, 239-40
Fredericton Boom: described, 95; holdings, 104-105n
Freight Costs, 139, 152; Bangor-Boston, 1876-1894, 152
Freshets: 1887, 69; 1902, 277-8; 1927, 405

Index

Gang Saws: defined, 133n
Gibson, Alexander "Boss", 45
Gilbert, Fred H.: comments on 1901 drive, 272, 273; described by acquaintances, 292; master driver in 1901, 271; testifies in 1903, 284, 285, 286
Gillin, Patrick H.: attorney in 1903 cases, 284, 285, 286; speech of, 311; speech on public power, 316
Grand Lake Stream Plantation, 218
Granger Turnpike Company, 176-7
Grant Farm, production, 390
Greenock, Scotland, 142
Great Northern Paper Company: cut, 1899-1929, 386-7; cut, 1934-1963, 428; contract operators and cut, 428; early history obscure, 251; early growth, 255; earnings, 418n; expansion, 458; land purchases, 374, 375-6; minor legal matters, 309; modern development, 251; New York men replace Bangor men, 253; recent years, 419; since World War II, 427-8; Social Service Department, 391; woods operations, 1935-1940, 423-5
Great Ponds Act, court rulings on, 313-4
Gulf Hagas: description, 50; lumbering difficulties, 50

Haines, William T.: comments on public versus private power, 309-310
Hall Brothers; described and production, 124
Hauling records, 25-6
Hayford and Stetson (Stetson Cutler and Co.) description and operations, 126-7
Headworks, 66
Hempstead, Alfred Geer, 391
Hinckley and Egery: land holdings, 196-7
Hobson, Joseph, operations described, 132n
Houlton, as railroad town, 223
Hodge, William C., and Great Northern Paper Company working plan, 362-3
Hurricanes, 24

Ice and shipping, 139
Indians, 21
International Commission on Saint John river, report, 308
International Paper Company: 247-251; birth and first year, 249; early discussions of, 248; Maine mills described, 249-250; pulpwood importations, 160
Investment and speculation, pre-Civil War, 7
Immigrants in Tanneries, 218
Impact of railroad on land sales, 199-200

Johnson and Phair, 127
Joint Committee on Interior Waters: 1901 legislature, 268; 1903 legislature, 279-280
Jordan Lumber Company, 294
Jewett, E. D., land sale, 198-199

Kennebec River: location, 2; log driving 72-5; log driving and jams, 73-4; operations in area, 43-5; organized for log driving, 73; pulp wood drives, 74-5; area residents' relationship with Canadians, 44; specialty mills, 118; strikes in saw mills, 111
Kennebec Log Driving Company: dates, 1900-1912, 98n; amounts, 1866-1912, 431
Kennebec Reservoir Dam Company Bill: compromise, 319-320; passes, 325; referendum, 318-9
Kilnwood, 168n
Kineo House, 32n
Knights of Labor, 111

Labor difficulties, 407-8
Land Disposal, 4; amounts, 1860-1872, 176
Land Grants: 182-3
Land Sales: auctions, 191-3; general history, 200; amounts 1873-1881, 189-190; political item, 193-4; Revolutionary War claims, 193
Large Lumbermen, fill out holding, 190-1
Latin American trade: 145, 146; freight costs, 146
Legislature of 1903: described, 279
Lewiston Steam Mill Company: 42, 115
Location of paper mills, proposed, 243n
Log Haulers: Lombard Co. history, 393-4; Lombard perfects, 392-3; Peavey invents, 392; Twin Tractors, 393-4
Log sluicing: excursions to see, 84, 401
"Loggers Boast," lyrics, 29n
Logging chance, 13-14

Lookout stations, 369
Lord, Major: Portland shipper, 140
Lumber business: Bangor employment, 91
Lumber camps: diversions, 17; Sundays, 18
Lumber Cut, twentieth century, 383-4
Lumber shipments: Provinces to U.S., 160
Lumber surveyed: Bangor's greatest year, 27n
Lumber statistics: 11-12, Bangor, 12-13
Lumbering: changes in diet, 18; changes in technique, 18; economic pressures, 173; profitable areas, 129; transition period (1860-1890), 11
Lumbering business, pre-Civil War, 6
Lumbering crews, changes in provenance, 19-20
Lumbering economy, comments on, 226
Lumbering methods, pre-Civil War, 6
Lumberman's organizations: 109-110; convention 108-109; convention of 1872, 38; railroad shippers, 110-1; trade association, 114
Lumberman's food, 15-16
Lumberman's M u s e u m, (Patten, Maine), VIII
Lumberjack, 27n
Lure of the West, 185-6

Machias River: 2; Cut of 1883-1892, 46; fire, 126; operations in 1901, 417n; sawmill production, 125
McKee, Ralph, develops University of Maine Pulp and Paper course, 258
McNulty, James, testifies in 1903, 286
Manufacturer's Investment Co., 251-2
Maine: defined, 1-2, 7n; lumbering depends on good weather, 37-8; early visitors, 1; end of World War I, 370; forest area, 3-4; lumber, the central fact of history, 1; occupations 1820-1860, 6; population growth, 1820-1840, 5; terrain, 2; vegetation, 2-3; wild lands, 4
Maine Central Railroad: shipments, 224-5; tonnage, 400-1
Maine State Board of Agriculture: offers legislation, 339-340
Marsh, George Perkins, 333-4
Marsh V. Great Northern Paper Company: described, 289-292; Supreme Court hears case, 292; rules, 292; decision quoted, 263
Mary Ann McCann, 164n
Mattawamkeag River: driving methods, 76; cost 1878-1905; 77; operations, 48-9; continues business, 266
Mechanization: since World War II, 427
Millinocket: building, 253-5; description of plants, 255; impact on Bangor, 254; labor force, 254
Molasses cake, 28n
Moosehead Lake: ice free dates, 438-9
Moosehead Pulp and Paper Company: 1893 drive and jam, 74-5
Mount Katahdin State Park: defeated again, 323; killed, 316; killed again, 324
Muley saw, defined, 133n
Mutual Companies, formed, 263

Narraguagus River, 2
Nationality of Logs and Lumber sawed in border areas, 126; 155-6
New Brunswick, fire law, 338
New Brunswick Legislature, on St. John River, 303
New England Hurricane, 1938, 380n
New Hampshire, forest area, 3
New Sweden, 188-9
Newspapers, comment on forest depletion, 334
North Twin Dam: excursions to 84, 401
Northern Development Company, 252-253

Ohio Fever, 7

Palmer, Joab: interviewed, 51-2; operations described, 50-2; stumpage prices, 213-4
Palmers: as employees, 60-1n
Paper industry and conservation, 336-7
Passadumkeag River: driving, 76-7; cost 1885-1904, 78; operations, statistics, 1877-1887, 49; mentioned, 386
Passadumkeag Boom, statistics, 1887, 190n
Patten, 47
Payne, Oliver, role in Great Northern founding, 253
Pea Cove Boom, described, 85
Peavey Stick (cant dog); invention and use, 65
Peeling and fly time, 23
Penobscot River: 2; area described, 46; driving costs, 1878-1904, 80; navigation dates, 432-4; driving, 76-7; drive at boom, 81-2; 1879 drive lost, 82; 1901 drive, 271-5; ar-

Index

rives, 275; 1902 drive, 278; 1903 drive, 288-9; 1904 drive, 289; 1901 fall freshet, 275-6; 1902 freshet and aftermath, 278; East Branch, described 80; statistics, 1884-1889, 48, 49; operators, 47-8; operations, 46-53; rafting, 86-92, sawmill strike of 1889, 112-114; West Branch, Chesuncook Dam, 79; West Branch driving dates and totals, 1873-1900, 81; driving desribed, 79; last days, 293
Penobscot Chemical Fibre Corporation, 235
Penobsot Log Driving Company: annual meeting of 1901, 269-271; court decisions, 266; 1900 drive, 267; 1902 annual meeting, 277; proxy voting, 267
Pittston Farm, 33n
Pollution, impact on salmon, 339
Polly, 162
Poplar: driving, 69-70; as source for pulp, 233
Portable sawmills: 116; 128-9; 133n; on the Kennebec, 118
Portland and St. John competition, 154
Portland: as port, 144-5; shipping, 145-7
Portland *Eastern Argus*, editorial on depletion, 335
Presumscott River, 2, 70
Prices, 230-1n; log prices, 225, board prices, 225
Public Domain: Maine's, disposal of, 173-6; State's Purpose, 175
Public Lands, prices, 174
Public power, bills, 310
Pulp and Paper Industry: attacked, 366; Eve of World War II, 257-8; labor unrest, 256-7; need for consolidation, 247; need for water, 266-7; statistics in 1890, 247; strikes, 255-6; union growth, 256, working conditions, 255
Pulp and paper consolidation, early meetings, 248
Pulpwood: "craze", 242, 247; domestic shipments to Bangor, 161; Russian shipments to Bangor, 161; shipping 159-161; shipments to California, 159
Pulpwood drives, S. D. Warren, 241-2

Rafts: location for courtship, 91-2; seagoing, 156-159; difficulties with seagoing rafts, 156, 157; Maine comments on, 158
Rafting: areas and difficulties, 86-7;

immense amounts of work, 87-8, 89; logs replace boards, 90
Railroads: 220-5; in the woods, 396-400; lumber shipments, 162; shipments, 1876-7, 170n; shipments 1879-1887, 171n; ties, 384-5; transport of logs, 84; water transport beaten, 138; costs of driving compared, 399
Reforestation, 4
Religion, in the camps, 25
Ripogenus Gorge, 79
Ripogenus Dam, history and construction, 404-5
River Drivers: 64-5; in town, 67; source of supply, 82
River pollution, sawdust, 91
Robinson Manufacturing Company, 264
Road building, 14
Rollins, E. H. & H., description and production, 121
Ross, John, on the Connecticut, 78-9; sells logs, 271; at 1901 PLD meeting, 270
Ross, John, (Steamer), 83
Rumford, development, 250-1
Rumford Falls Boom Company, 265-6
Rumford Falls Sluice and Improvement Company, 265
Rumford Falls and Rangley Lake Railroad, 396-7

Saccarappa (Westbrook), 139
Saco River, 2; 40-1, 69
Sailing ships, demise, 144
St. Croix River, 2
St. John River; flowage described, 2; 1880's cut, 55; cut in 1890's, 55; late 1890's, 384; dam proposals, 302-3; drift drive, 307; driving difficulties, 93; described, 93-4; impact of Telos Canal, 92-3; mills described, 303-4; navigation dates, 436-7; operations, 53-5; river described, 53; organized for river driving, 94; port of St. John, 144; shipping, 153-6; 168n; to transatlantic ports, 155
St. John Log Driving Company: statistics, 304
St. John Lumber Company: booms and driving difficulties, 306-7; described, 305; 1905 drive, 305-6
Sandy River Lumber Company, 118
Saw, replaces ax, 389
Sawdust, 159
Sawmills: downeast, 124; economic development, 128; fires, 119; effect of depression, 108; economic infra-

structure, 107-8; sales, 124; small mills in northwestern Maine, 116; Calais, 1890, 114; working conditions, 112, 113
Saxby Gale, 24
Scaling, 212
Schenck, Garrett W.: biography, 252; comments on Baxter, 324; on 1926 primary 325
Shaw F. Brothers: described and history, 218-220; failure, 219-220
Sherman, 219
Shingles, northern Maine product, 127
Shipbuilding, 214-7; cooperatives, 215; statistics, 1875-1895, 215-6
Ships knees, 124
Shipping, destinations, 137-8
Shooks, defined, 163n
Silver Lake Farm, 33n
Size of logs, 88
Smallpox, 24
Smith, A. B., 406
Smith, A. Ledyard, biography, 279
Southern rivers, 40
Spanish American War, 387
Spruce budworm, 428
Spoolwood, 142-4
State Grange: Bangor meeting, 342-4
State of industry, eve of depression of 1873, 39
"Stay at Home" campaign, 186-7
Steamboats: Chesuncook Lake, 83-4; in spoolwood trade, 143-4; used as sawmill, 122
Stevedoring, 151-2; costs, 151; difficulty, 152
Stewart, T. J.: biography, 141; Box shook trade, 141-2; spoolwood trade, 142-4
Stilwell, E. M.: Commissioner of Fish and Game, campaign against pollution, 339, further comments, 352n
Stumpage: defined, 211; prices, 212, 213, 214
Survey designations: explained, 8n

Talbot, George F.: biography, 341; role in conservation, 341-2; Bangor meeting, 342; comments on Ring, 367, on depletion, 367; lobbyist, 343, woods' needs, 343
Tanneries: small, 220
Tanning: 217-220
Tariff and Forest depletion, 334-5
Telos: 217
Telos Canal, impact on St. John, 301-2
Thomas, W. W.: biography, 188
Thoreau, Henry David, quoted on Maine, 1

Tidewater mills, 116n
Topsfield, 219
Toting, 16
Toters: 22, 32n, 58n42
Town Credit: 128
Tracey, Murchie and Love, operations, 52-3
Treat's Falls Dam: interferes with rafts, 89; 92
Trees: large size, 26

Underselling, 109
Union River, 2
Union Water Power Company, described, 264-5; lawsuits, 265
University of Maine: forestry, 365; forestry education, 369-370; pulp and paper courses, 258; role in industry since World War II, 429

Van Buren, sorting gap, 302
Vessels, described, 140

Wages: in the woods, 21; 225-6
Wangan: 66-7
Warren, Samuel Dennis: biography, 238
Warren, S. D. Company: described, 238-9; in 1880's, 239; changes to electricity, 239; growth, 240; history, 238-241; in the 1870's, 239; Mutual Relief Society, 240, profit participation, 241; pulpwood procurement, 242; success reasons, 240; relations with employees, 240-1, 245n; relations, with Westbrook, 240; tree planting, 427; wages, 240, 241; World War I, 408-9; statistics, 409
Washington County men on river drives, 97-8n
Wassataquoik Stream operations, 52-3
Water pitch and driving, 68
Water power bills, 312-3
Waterways, storage, 264
Weather, success factor, 38-9
Websters, E. and J. F.: operations described, 128-9; holdings further described, 136n; pulpwood shippers, 159-160, 243n
West Branch Bill, 1903: compromise, 287-8; Hearings, postponed, 281-2; described, 283-6; lobbying, 280, 287; newspapers comment, 281, 283, 88
West Branch Driving and Reservoir

Index

Dam Company: directors, 267-8; remonstrances, 268; 1901 legislative campaign, 269
West Branch No. 2, 406
Western Maine operations, 1870-1890, 41
Whitneyville, 125-6
Wildlands, location and value in 1893, 194
Winslow, H. and Co., 1861 operations, 41
Winter port for Canada, 154
Winters, 23-4
Wiscasset, 117
Woodpulp expansion, 237, 238
Woodpulp mills, Norway, 234, 235; Topsham, 234-5
Woodpulp shipments: Europe, 247-8; London, 161; from sawdust, 130; rags, 233
Woods Crews: 411-412n; Twentieth Century, 388, 422
Woods Cut, 1969, 429
Woods Farms, 22-3
Woods fires: 365, 368-9
Woods Life: health, 391; holidays, 390; moral conditions of, 390
Woods supplies, 21-2
World War I, 387-8
World War II: French labor, 425-6; POW labor, 426-7
Working Conditions: improvement in Twentieth Century, 389-390
Work routine, 17-18
Wyman, J. F., 319

Spring Drive

634.9 S645   Me.Coll.
Smith, David Clayton,
1914-
A history of lumbering in
Maine, 1861-1960

In Memory Of
Harlan Sweetser
Staff of Bookmobile

Bangor Lumber Mill, Possibly Hathorn Mill, Near High Head

Bangor Public Library